ERGEBNISSE DER PHYSIOLOGIE

B OLOGISCHEN CHEMIE UND
EXPERIMENTELLEN PHARMAKOLOGIE

REVIEWS OF PHYSIOLOGY

BIOCHEMISTRY AND
EXPERIMENTAL PHARMACOLOGY

BAND 59

SPRINGER-VERLAG
BERLIN HEIDELBERG GMBH

Ergebnisse der Physiologie, biologischen Chemie und experimentellen Pharmakologie

Reviews of Physiology, Biochemistry and Experimental Pharmacology

Herausgegeben von K Kramer, O Krayer, E. Lehnartz, F. Lynen, A v Muralt, U G Trendelenburg, H. H. Weber und O. Westphal

56. Band

Mit 43 Abbildungen. IV, 380 Seiten
(davon 261 Seiten in englischer Sprache)
Gr.-8⁰. 1965
Gebunden DM 98,—; US $ 24 50

57. Band

Mit 41 Abbildungen IV, 248 Seiten
(davon 189 Seiten in englischer Sprache)
Gr.-8⁰ 1966
Gebunden DM 78,—, US $ 19 50

58. Band

Mit 67 Abbildungen XII, 580 Seiten Gr.-8⁰
1966 Gebunden DM 98,—; US $ 24 50

Inhaltsübersicht: Brenzkatechinamine und andere sympathicomimetische Amine Biosynthese und Inaktivierung, Freisetzung und Wirkung: Orientierende Übersicht. Vorkommen der korpereigenen Brenzkatechinamine. Biogenese der korpereigenen sympathicomimetischen Amine. Hemmstoffe der enzymatischen Biosynthese. Enzymatische Inaktivierung. Hemmstoffe der enzymatischen Inaktivierung Ausscheidung in den Harn Phaochromocytom und Neuroblastom. Subcellulare Lokalisation: Aufnahme, Speicherung und Freisetzung. Nichtenzymatische Inaktivierung und Noradrenalinspeicher des Gewebes Sekretion und Freisetzung. Zur Frage der Compartmentalisation des nervalen Noradrenalinspeichers Direkt und indirekt wirkende sympathicomimetische Amine Konstitution und Wirkung Über den Mechanismus noradrenergischer Erregungsubertragung in sympathischen Nerven Pharmakologische Wirkungen der Brenzkatechinamine Schlußbemerkungen Literatur.

SPRINGER-VERLAG

BERLIN HEIDELBERG GMBH

ERGEBNISSE DER PHYSIOLOGIE
BIOLOGISCHEN CHEMIE UND
EXPERIMENTELLEN PHARMAKOLOGIE

REVIEWS OF PHYSIOLOGY
BIOCHEMISTRY AND
EXPERIMENTAL PHARMACOLOGY

HERAUSGEGEBEN VON

R. JUNG   K. KRAMER   O. KRAYER   E. LEHNARTZ
FREIBURG/BR.   MÜNCHEN   BOSTON   MÜNSTER/WESTF.

F. LYNEN   A. v. MURALT
MÜNCHEN   BERN

U. TRENDELENBURG   H. H. WEBER   O. WESTPHAL
BOSTON   HEIDELBERG   FREIBURG/BR.

BAND 59

MIT BEITRÄGEN VON

T. J. MARCZYNSKI · U. TRENDELENBURG · W. VOGT
M. WALSER

MIT 16 ABBILDUNGEN

SPRINGER-VERLAG
BERLIN HEIDELBERG GMBH 1967

ISBN 978-3-662-31132-5   ISBN 978-3-540-34927-3 (eBook)
DOI 10.1007/978-3-540-34927-3

Library of Congress Catalog Card Number 62-37142.

Titel-Nr. 4779

Inhaltsverzeichnis

Seite

Some Aspects of the Pharmacology of Autonomic Ganglion Cells. By
U. TRENDELENBURG, Boston, Mass. With 9 Figures 1

Topical Application of Drugs to Subcortical Brain Structures and Selected
Aspects of Electrical Stimulation. By T. J. MARCZYNSKI, Chicago, Ill. 86

The Anaphylatoxin-Forming System. By W. VOGT, Gottingen. With
1 Figure 160

Magnesium Metabolism. By M. WALSER, Baltimore, Md. With 6 Figures 185

Namenverzeichnis 297

Sachverzeichnis 332

Berichtigung zu „Ergebnisse der Physiologie", Band 58

Seite 308, 1. Zeile, muß lauten:

„ .. 1955), könnte die nach α-Methyldopa über die Norm hinaus ge-steigerte Tyraminwirkung .. "

Auf Seite 416 sind unten die zwei letzten Zeilen versehentlich heraus-geblieben. Es muß heißen:

...; sie wäre aber nur schwer vereinbar mit der Beobachtung, daß die Implan-tation von Adrenalin- oder Noradrenalinkristallen in das hypothalamische Freßzentrum bei Ratten (GROSSMAN, 1960) ebenso wie die Verabfolgung von Phenobarbital (Luminal) in Dosierungen, die zu ausgeprägten motorischen Koordinationsstorungen fuhrten, die Nahrungsaufnahme *steigerten* (KUHN, 1959; SPENGLER u. WASER, 1959).

Seite 417, Zeile 15/16, muß heißen:

nicht „mobilitatssteigernd", sondern „motilitätssteigernd".

Some Aspects of the Pharmacology of Autonomic Ganglion Cells

U. TRENDELENBURG*

With 9 Figures

Table of Contents

I Introduction 2
II History of the "non-nicotinic ganglion-stimulating agents" 4
  A. Muscarinic agents 4
  B Histamine 6
  C. 5-Hydroxytryptamine 7
  D Polypeptides 8
III. Methods available for the study of ganglion-stimulating agents 9
IV. Nicotinic agents 11
  A. Nicotine 11
  B Tetramethylammonium (TMA) 17
  C. 1,1-Dimethyl-4-phenylpiperazinium (DMPP) 19
  D. Carbachol 19
V. Non-nicotinic ganglion-stimulating agents 20
  A. Action on the non-perfused superior cervical ganglion of the cat . . . 21
    1. Stimulation of the ganglion 21
    2 Facilitation of ganglionic effects by preganglionic stimulation 22
    3. The effect of chronic preganglionic denervation of the ganglion . . 23
    4 Delay of onset of response 23
    5. Slopes of dose-response curves 23
    6 Facilitatory effects of subthreshold doses of non-nicotinic agents . . . 24
    7. Tachyphylaxis 24
    8 Non-depolarizing ganglion-blocking agents 25
    9 Depolarizing ganglion-blocking agents 25
    10 Effects of non-nicotinic agents during the late, non-depolarizing phase of the block by nicotine 25
    11. Facilitation by nicotine of the ganglionic actions of some non-nicotinic agents 26
    12. Cocaine 27
    13. Morphine and methadone 29
    14. Calcium chloride 29
    15 Atropine 29
    16 Antihistaminic agents 29
    17. Antagonists to 5-hydroxytryptamine 30
    18. The location of non-nicotinic receptors 30
  B. Actions on the perfused superior cervical ganglion 30
  C. Actions on other ganglia 31

* Department of Pharmacology, Harvard Medical School, Boston, Massachusetts.

D. Actions on the adrenal medulla of the cat 32
   1. Sensitivity of the adrenal medulla to various agents 32
   2. Preferential release of epinephrine 32
   3 The effect of chronic denervation 33
   4 The effect of non-depolarizing ganglion-blocking agents   . . 33
   5. The effect of depolarizing ganglion-blocking agents 33
   6 The effects of cocaine, morphine and methadone   33
   7. The effects of specific blocking agents 34
   8. The adrenal medulla of other species 34
E. Pressor effects of non-nicotinic agents 34
   1. Adrenal medullary versus general ganglionic stimulation   35
   2. Non-depolarizing ganglion-blocking agents 37
   3 Depolarizing ganglion-blocking agents 37
   4 Effects of non-nicotinic agents during the late, non-depolarizing phase
      of the block by nicotine   37
   5. Adrenalectomy   38
   6. Xylocholine and bretylium 38
   7. Pretreatment with reserpine 38
   8. Cocaine, morphine and methadone 38
   9. Sympatholytic agents   38
   10. Pressor responses to angiotensin 38
F. Other sites of action of non-nicotinic agents   39
G. Agents pharmacologically or chemically related to non-nicotinic substances 40
   1. Muscarinic agents 40
   2. Histamine 41
   3 5-Hydroxytryptamine   41
   4 Polypeptides 42

VI. Acetylcholine 43

VII Transmission of preganglionic impulses through ganglia 47

VIII. Anticholinesterase agents 52

IX. The ganglionic effects of sympathomimetic amines   56

X. The effect of denervation   60
A. Presynaptic changes after preganglionic denervation 60
B Postsynaptic changes after preganglionic denervation 63
C. The effect of axotomy 66
D Homologous and heterogenous reinnervation of the superior cervical
   ganglion 67

XI. Synapses with ganglion cells of nerve fibers that do not belong to the classical
   preganglionic fibers   69
A Sensory innervation of ganglia   70
B. Adrenergic synapses in ganglia 71

XII. Conclusions 72

References 74

I. Introduction

A comprehensive review of the pharmacology of autonomic ganglia would require a monograph of considerable size. Hence, a subjective selection has been made of some topics that are close to the field of interest of the author. However, there are also objective reasons which indicate that it might be both useful and timely to review some aspects of the pharmacology of autonomic ganglia. In most fields of science there are periods during which certain facts accumulate which neither fully agree with established theories nor carry enough weight to require a new formulation of accepted theories. When enough facts of this kind have accumulated, they may well suddenly present a new picture which extends rather than contradicts accepted views. This has happened in recent years with regard to the ganglionic effects of numerous compounds which previously were believed to be devoid of ganglionic actions. It appears to be timely and useful to review the pharmacology of all those agents whose pharmacology differs from that of nicotine. Many of them are naturally occurring agents which might play a modulatory role in ganglionic transmission under physiological or pathological conditions. While it is now possible to review the pharmacology of their ganglionic actions, it should be realized that very little or nothing is known about their physiological or pathological significance. However, the analysis of these possible roles may well require the use of pharmacological tools; hence, this review may serve as a source of information for those who want to determine the possible modulating role of naturally occurring agents on normal or pathological ganglionic transmission.

In this review, the term "receptor" will be used, and it should be understood that the term is defined operationally as that site with which the agonist interacts to cause a response of the ganglion cell. Receptors will be described as being different from each other, when they can be differentiated pharmacologically. For instance, certain ganglionic effects of acetylcholine are specifically antagonized by small doses of hexamethonium, while others are specifically antagonized by small doses of atropine. The first type of action of acetylcholine will be assumed to be exerted on "nicotinic" receptors, while the second is exerted on "muscarinic" receptors. This empirical and pharmacological approach to the problem of receptors cannot contribute much to the elucidation of the biochemical identity of the receptor molecule, of its configuration and of its exact location on the cell membrane or some other cell constituent. However, the use of the pharmacologically defined term "receptor" has in

Table 1 *List of abbreviations used in this review*

AHR-602	N-benzyl-3-pyrrolidyl acetate methobromide
217 AO	2-diethoxyphosphenyl-thioethyldimethylamine acid oxalate
DCI	dichloroisoproterenol
DFP	diisopropyl phosphorofluoridate
DMPP	1,1-dimethyl-4-phenylpiperazinium iodide
5-HT	5-hydroxytryptamine
LSD	lysergic acid diethylamide
McN-A-343	4-(m-chlorophenylcarbamoyloxy)-2-butynyl-trimethylammonium chloride
TEA	tetraethylammonium ion
TEPP	tetraethyl pyrophosphate
TMA	tetramethylammonium ion

1*

the past greatly helped to clarify mechanisms of action of drugs, and it permits us to classify agonists into different groups.

A few abbreviations will be used in this review. For quick reference, they are presented in Table 1.

II. History of the "non-nicotinic ganglion-stimulating agents"

As mentioned in the Introduction, during several decades certain observations were made which failed to fit into the traditional concept that only nicotine and related compounds are able to stimulate autonomic ganglia. This chapter presents a brief survey of observations which were not consistent with the accepted theory and which failed to present a comprehensive picture. In sections V to VIII present views on the pharmacology of these agents will be discussed, and it will be shown that new concepts permit an explanation of the older observations.

A. Muscarinic agents

DALE and LAIDLAW (1912) were the first to describe the stimulation of a sympathetic ganglion by pilocarpine. When painted on the exposed superior cervical ganglion of the cat, pilocarpine appeared to cause ganglionic stimulation, since the resulting contraction of the nictitating membrane and the dilatation of the pupil were abolished by acute ganglionectomy. Similar responses appeared after the intravenous injection of pilocarpine and were reduced by both ganglionectomy and adrenalectomy. While the authors concluded that pilocarpine, like nicotine, stimulates sympathetic ganglia and the adrenals, they did not attempt a pharmacological differentiation of the effects of pilocarpine from those of nicotine.

These findings were overshadowed by DALE's subsequent classification of cholinomimetic agents into muscarine-like and nicotine-like compounds (DALE, 1914). Since DALE stated that "these 'muscarine' effects are purely peripheral in origin, unaffected by nicotine in large doses, but readily abolished by small doses of atropine", it was generally accepted that muscarine-like agents are devoid of ganglion-stimulating properties. In the following four decades there were only occasional reports on ganglionic effects of pilocarpine. DIXON and RANSOM (1924) and HOET (1928) observed stimulation of the coeliac ganglion of the cat when it was painted with a solution of pilocarpine as well as after the intravenous injection of this agent. The resulting contraction of the spleen was abolished by nicotine and by ergotamine, a sympatholytic agent. An antagonism of atropine to ganglionic actions of pilocarpine was first described by KOPPÁNYI (1932). With simultaneous recording of pre- and postganglionic electrical activity, MARRAZZI (1939a) demonstrated two effects of pilocarpine on the superior cervical ganglion of the rabbit: a) stimulation of the ganglion, and b) facilitation of the transmission of preganglionic impulses. Both effects of pilocarpine were abolished by small doses of atropine.

While STEWARD and ROGOFF (1921) failed to obtain evidence for a stimulant effect of pilocarpine on the adrenal medulla, FELDBERG et al. (1934) provided convincing evidence for such an action, which was found to be blocked specifically by atropine.

Another manifestation of the ganglionic effects of pilocarpine is its pressor effect in cats and dogs, especially in spinal preparations (HUNT, 1918; HEATON and MACKEITH, 1927). The secondary rise of blood pressure appears after an initial fall; it is reduced by adrenalectomy and abolished by atropine, and it is subject to tachyphylaxis (HEATON and MACKEITH, 1927). There was again the paradox that pilocarpine appeared to be nicotine-like because the secondary rise of blood pressure was found to be antagonized by large doses of nicotine (HOET, 1928; BACQ and SIMONART, 1938) and by ergotamine (HOET, 1928). It was unlike nicotine, on the other hand, because atropine abolished this pressor effect (KOPPÁNYI, 1939; VON EULER and DOMEIJ, 1945). In addition, some observations of KOPPÁNYI (1939) defied classical definitions: small doses of nicotine (that presumably failed to block ganglia) potentiated the secondary rise of blood pressure after pilocarpine, while cocaine antagonized it, although this agent potentiates the pressor response to nicotine. Pronounced potentiation of the pressor effects of pilocarpine by TEA was reported by ROOT (1951). Since TEA blocks ganglia to all nicotine-like agents but not to KCl (ACHESON and PEREIRA, 1946), and since TEA potentiates pressor responses in general, ROOT suggested that pilocarpine might have an action on sympathetic ganglion cells similar to that of KCl.

It was 37 years after DALE and LAIDLAW's original observations that an attempt was made to determine whether muscarine-like agents have ganglion-stimulating properties that are qualitatively different from those of nicotine; or in other words, to determine whether sympathetic ganglia have muscarinic receptors in addition to the well-known nicotinic receptors. AMBACHE (1949) found that injection of small amounts of pilocarpine into the fluid that perfused the isolated superior cervical ganglion of the cat caused stimulation. Atropine abolished the stimulant effect, but so did hexamethonium in higher doses. As a consequence, even then there remained some doubt whether there was a clear distinction between the nicotinic and the muscarinic receptors of ganglion cells. This doubt was not removed by a subsequent report in which the effects of a highly purified preparation of natural muscarine on the isolated and perfused superior cervical ganglion of the cat were described (AMBACHE et al., 1956). Muscarine was found to stimulate the ganglion; this action was antagonized by atropine (1 µg i.a.) and by larger doses of hexamethonium (0.1—1 mg i.a.) as well as by nicotine. Since subsequent reports, that describe results obtained with synthetic muscarine, did not confirm any antagonism by hexamethonium, the observations of AMBACHE et al may have been due to some impurities of the "purified muscarine" or to differences in methods. It is possible that weak atropine-like side-effects of hexamethonium are more pronounced in the isolated and perfused superior cervical ganglion than in ganglia which are left in their normal circulation. It is the latter type of preparation with which most of the more recent observations have been made.

In spite of the fact that pilocarpine and muscarine were found to stimulate the superior cervical ganglion, these agents were not yet accepted as a separate group of ganglion-stimulating compounds. In the intact cat, Waser (1955) failed to observe any signs of ganglionic stimulation by muscarine, because the pronounced vasodilatation overshadowed all other effects. And attempts to prevent these vascular effects of muscarine by administering atropine resulted in the absence of ganglionic effects of as much as 1 mg/kg of muscarine injected intraarterially to the inferior mesenteric ganglion of the cat (Herr and Gyermek, 1960).

B. Histamine

Feldberg and Vartiainen (1935) observed that the intraarterial injection of up to 100 µg of histamine into the isolated and perfused superior cervical ganglion failed to cause stimulation. Considering that the reactivity of sympathetic ganglia to various agents is rather similar to that of the adrenal medulla, this observation was puzzling, since stimulation of the latter by small doses of histamine had been demonstrated by Szczygielski (1932). In addition, Siehe (1934) found that chronic denervation of the adrenal medulla did not affect the stimulant action of histamine. It thus appeared that histamine had a direct stimulant effect on the adrenal medulla while failing to stimulate the ganglion. Weak "nicotine-like" effects were found to be a property of dimethyl-histamine (methylation of the nitrogen of the side chain), while the corresponding quaternary trimethylhistamine stimulated the perfused superior cervical ganglion of the cat (Vartiainen, 1935). In addition, this compound caused a pressor response in the spinal cat which was not entirely due to adrenal medullary stimulation. These observations appeared to indicate that even histamine might have some stimulant effects on sympathetic ganglia.

Konzett (1952) reinvestigated the action of histamine on the perfused superior cervical ganglion of the cat. In 18 of 22 experiments, there was no stimulant effect with doses of up to 100 µg of histamine. However, in 4 experiments, stimulation was observed. Moreover, in nearly all experiments histamine readily potentiated responses of the ganglion to acetylcholine, choline, nicotine and potassium.

It should be noted that the two reports dealing with the superior cervical ganglion are based on experiments with a perfused preparation, while reports on the effect of histamine on the adrenal medulla concern experiments with close-arterial injections of this agent. When Feldberg (1940) studied the effect of histamine in the *perfused* adrenal medulla, he found that "the perfused adrenal medulla was rather insensitive to histamine", an observation that differed very much from those of SZCZYGIELSKI and SIEHE. Consequently, it appears possible that the differences between the perfused superior cervical ganglion and the non-perfused adrenal medulla were mainly due to differences between perfused and non-perfused sympathetic structures. This view has recently been confirmed by VOGT (1965) who observed a great loss in sensitivity to histamine, angiotensin and bradykinin in the perfused adrenal medulla as compared to the non-perfused preparation, while the sensitivity to acetylcholine and potassium was found to be about the same in both types of preparations.

Like pilocarpine, histamine causes a secondary rise of blood pressure when injected intravenously into spinal cats (HOGBEN et al., 1924). Adrenalectomy abolishes this pressor response, while ergotamine causes its reversal (BURN and DALE, 1926; BEIN and MEIER, 1953).

The first evidence that suggested a "non-nicotine-like" mode of action of histamine came from GYERMEK and SZTANYIK (1952) who found that TEA failed to prevent the stimulation by histamine of the adrenal medulla. Simultaneously, SLATER and DRESEL (1952) observed that both TEA and hexamethonium greatly potentiated the secondary rise of blood pressure of histamine; their observations were very similar to those made by ROOT (1951) with pilocarpine. However, atropine failed to block the effects of histamine. SLATER and DRESEL's experiments also provided evidence for general stimulation of sympathetic ganglia after the intravenous injection of histamine into cats that were under the influence of either TEA or hexamethonium: under these experimental conditions, adrenalectomy reduced but did not abolish the secondary rise of blood pressure.

This short survey of the history of the ganglionic effects of histamine shows that the original negative observations of FELDBERG and VARTIAINEN (1935) were confirmed; however, they were relevant only to perfused preparations and not to sympathetic ganglia left in their normal circulation.

C. 5-Hydroxytryptamine

As soon as synthetic 5-hydroxytryptamine became available, there were numerous reports on various pharmacological actions of this agent. Some of these contained evidence in favor of ganglionic actions of 5-hydroxytryptamine, but the ganglion-stimulating properties of this compound were not immediately recognized. Stimulation of the cat's adrenal medulla by close-arterial injections of 5-HT was reported by LECOMTE (1953), but only WEIDMANN and CERLETTI (1957) found that the pressor effects of 5-HT in the spinal cat (which were predominantly due to adrenal medullary stimulation) were not blocked by hexamethonium.

HEYMANS and VAN DEN HEUVEL-HEYMANS (1953) failed to observe any stimulant effects of 5-HT on the chemoreceptors of the dog's carotid sinus, while McCubbin *et al.* (1956) found 5-HT to be more potent than lobelin in the same preparation and the stimulant effects of 5-HT were not antagonized by TEA. Since it was later found that the action of 5-HT on the superior cervical ganglion is blocked by morphine (see below), it may be relevant that Heymans and van den Heuvel-Heymans gave 100 μg/kg of morphine as premedication before anesthesia.

A third line of evidence came from experiments with isolated guinea-pig ileum. Robertson (1953, 1954) found the stimulant effect of 5-HT to be potentiated by a cholinesterase inhibitor (283 C 51) and antagonized by atropine. However, a ganglionic site of action was considered unlikely, since hexamethonium failed to antagonize the effects of 5-HT. Gaddum (1953) first provided evidence for the presence of specific tryptamine receptors in the isolated guinea-pig ileum which subsequently were found to represent two pharmacologically different types (Gaddum and Picarelli, 1957). The "D-receptors" appeared to belong to the smooth muscle and were blocked by phenoxybenzamine, LSD-25, dihydroergotamine and brom-LSD. The "M-receptors", on the other hand, appeared to be associated with the para-sympathetic innervation of the gut; actions of 5-HT on M-receptors were found to be antagonized by atropine, morphine, methadone, and cocaine. An action of 5-HT on the cholinergic innervation of the gut had already been postulated by Rocha e Silva *et al.* (1953), since they found the effects of 5-HT to be antagonized by certain concentrations of nicotine, by atropine and by cocaine. A ganglionic site of action was denied because hexamethonium failed to block the effects of 5-HT, and it was assumed that 5-HT stimulated the post-ganglionic axon. Since, in the isolated gut, it is impossible to distinguish between stimulation of the ganglion and stimulation of the postganglionic axon, decisive evidence for a ganglionic site of action had to come from experiments on the superior cervical ganglion.

The first report on a ganglionic action of 5-HT was misleading in so far as the effect of a large intraarterial injection of 5-HT was pure inhibition of the transmission of preganglionic impulses through the ciliary ganglion of the dog (Page and McCubbin, 1953); stimulation of this ganglion was not observed. However, soon thereafter stimulation by 5-HT of the perfused superior cervical ganglion of the cat was demonstrated by Robertson (1954).

D. Polypeptides

Peptone has long been known to stimulate the cat's adrenal medulla (Siehe, 1934) as well as sympathetic ganglia (Beraldo, 1958; Beraldo and Zanotto, 1960). However, evidence for ganglionic actions of well defined synthetic polypeptides has been obtained only very recently. Lecomte *et al.* (1961) were the first to draw attention to stimulation of the rat's and rabbit's adrenal medulla by bradykinin. Kaneko *et al.* (1961), on the other hand, observed adrenal medullary stimulation by angiotensin only after prior 'sensitization' by ganglion-stimulating agents such as nicotine or DMPP. This effect was not specific for angiotensin, since other pressor agents (norepinephrine, vasopressin) were also able to stimulate adrenal medullary secretion under these experimental conditions. Hexamethonium abolished the stimulation of the adrenal medulla by all pressor agents; consequently, the authors suggested that, after sensitization by nicotine and DMPP, vasoconstrictor agents cause anoxia of the adrenal medulla which results in an increased discharge of the presynaptic cholinergic fibers. The recent studies of FELDBERG and LEWIS (1964, 1965), STASZEWSKA-BARCZAK and VANE (1965a) and LEWIS and REIT (1965) provided much more direct evidence for the view that bradykinin, angiotensin and various related polypeptides stimulate both the adrenal medulla and the superior cervical ganglion of the cat, and that their mode of action differs from that of nicotine and DMPP.

III. Methods available for the study of ganglion-stimulating agents

For a pharmacological analysis of the mode of action of ganglion-stimulating agents, the superior cervical ganglion of the cat has long been found useful. Stimulation of the ganglion results in a contraction of the nictitating membrane which is easily recorded with accuracy. In this system, any drug must be suspected to have side effects on a) the smooth muscle of the nictitating membrane, b) the adrenal medulla, c) the central nervous system. In order to avoid these side effects the ganglion can be isolated and perfused according to the method first described by KIBJAKOW (1933). This method has certain disadvantages: a) it is laborious and requires experience, b) full isolation of the perfusion from the systemic circulation is difficult to accomplish (for discussion see AMBACHE, 1954; PATON, 1954), and most important c) the perfused preparation is much less sensitive to some agents than to others (TRENDELENBURG, 1956b; VOGT, 1965).

A much simpler method is available which avoids the disadvantages of the perfused preparation. Cannulation of the central stump of the lingual artery permits the intraarterial injection of drugs into the origin of the external carotid artery (TRENDELENBURG, 1959). The injection of drugs by this route delivers the material into the blood supply of the nictitating membrane, and the action of drugs on the smooth muscle of the membrane can be studied. If drugs are injected in the same way but during occlusion of the external carotid artery just distal to the lingual artery, the injected material reaches the ganglion. Injections "to the nictitating membrane" and "to the ganglion" have been found to give a reliable estimate of their actions on smooth muscle and on ganglia, respectively. Possible side effects on the adrenal medulla can easily be detected by recording the nictitating membrane of the other side (PATON, 1954) which responds to circulating catecholamines while not being affected by stimulation of the contralateral superior cervical ganglion. Side effects of the injected material on the central nervous system are easily avoided by section of the preganglionic fibers

With this method, no attempt is made to isolate the superior cervical ganglion from the cat, and drugs are deliberately injected "to the nictitating membrane" instead of being prevented from reaching the smooth muscle. With 5-hydroxytryptamine, for instance, injections "to the ganglion" and "to the nictitating membrane" result in contractions of the nictitating membrane which look very similar on the kymograph trace. However, simple pharmacological procedures reliably differentiate the ganglionic effects of 5-hydroxytryptamine from its smooth muscle effects: section of the postganglionic fibers, injections of depolarizing blocking agents (nicotine, DMPP, TMA) or of morphine, methadone or cocaine abolish the former but not the latter.

In addition, at the beginning of the experiment it can easily be verified that drugs injected "to the nictitating membrane" do indeed reach this organ and not the ganglion, and that the opposite is true for injections "to the ganglion"; intraarterial injections of small doses of norepinephrine and of nicotine establish this within a few minutes.

Although this preparation provides a simple, sensitive and reliable method suited to the pharmacological analysis of the mode of action of ganglion-stimulating drugs, it has its disadvantages. Drugs or procedures which change the sensitivity of the nictitating membrane to norepinephrine, which affect the amount of norepinephrine released from the nerve terminals in response to nervous impulses, and which affect conduction along the postganglionic fibers, must all result in a distortion of the response of the nictitating membrane which is then no longer a reliable indicator of stimulation of the ganglion. Furthermore, the time course of the contraction of the nictitating membrane cannot be assumed to reproduce the time course of ganglionic events, since the contractile tissue of the nictitating membrane responds much more slowly than the membrane potential of the ganglion cells does. Or in other words, responses of the nictitating membrane to stimulation of the ganglion represent a certain degree of extra-ganglionic "integration".

The recording of postganglionic electrical activity avoids many of these disadvantages. However, with this method it is nearly impossible to express responses quantitatively. Hence, this method is well suited to studies of qualitative changes in the response of the ganglion but rather unsuited to more quantitative studies.

Much could be learned from reliable recording of transmembrane potentials of ganglion cells. Unfortunately, the rigidity of the connective tissue of mammalian ganglia has foiled nearly all attempts to record with intracellular electrodes. While such recording is possible in ganglion cells of the frog, it is doubtful that the physiology and pharmacology of frog ganglia is identical with that of mammalian ganglion cells. At the present time, the best information about changes in membrane potentials of mammalian ganglion cells is obtained with surface electrodes. It should be remembered that they record changes not only in ganglion cells but also in all other cells present in ganglia. Furthermore, because of technical problems, most records yield reliable information only for rather rapid changes in membrane potential; very slow or very long lasting changes in membrane potentials are not reliably detected with conventional methods.

From this short survey of those methods which are frequently used in the study of ganglion cells, it is clear that not one is ideal and that only the combination of several methods will finally give a complete picture of the mode of action of a drug.

IV. Nicotinic agents

A. Nicotine

HIRSCHMANN (1863) was the first to study the influence of nicotine on sympathetic stimulation of the eye. Dilatation of the pupil in response to stimulation of the cervical sympathetic chain was abolished after nicotine, and miosis was observed on the unstimulated side. A mydriatic response to nicotine was not observed presumably because the drug was never administered intravenously. Since the nicotine-induced miosis was observed on the non-stimulated side even after section of the cervical sympathetic chain, HIRSCHMANN concluded that nicotine paralyzed the sympathetic nerve endings in the iris.

In 1870, SCHMIEDEBERG observed that concentrations of nicotine that blocked the response of the frog heart to vagal stimulation failed to block the effects of muscarine. As nicotine, applied to the trunk of the vagus, did not block the response of the heart to vagal stimulation, SCHMIEDEBERG concluded that nicotine paralyzed an intermediary mechanism (Zwischenapparat) which connects vagal fibers with the peripheral inhibitory mechanism of the heart. Since the vagal ganglion cells are dispersed in the wall of the heart, he was not able to identify this intermediate mechanism.

This was achieved by LANGLEY and DICKINSON (1889) in their classical study of the action of nicotine on the superior cervical ganglion of the rabbit. They showed that the intravenous injection of nicotine caused first mydriasis and vasoconstriction and then a block of the response to electrical stimulation of the preganglionic fibers, while responses to postganglionic stimulation remained unchanged. Painting the ganglion with a 1 % solution of nicotine produced similar results, but painting of the nerve trunks with the same solution of nicotine had no effect. The authors were thus able to localize the site of action of nicotine and to show that this agent first stimulates and then blocks the superior cervical ganglion.

In 1914, DALE provided conclusive evidence for the view that acetylcholine resembles muscarine in its peripheral effects on smooth muscle, heart muscle and gland cells, while it resembles nicotine in its ability to stimulate ganglia.

That nicotine and related agents depolarize ganglion cells has been demonstrated for the cat's superior cervical ganglion *in situ* with surface electrodes (PATON and PERRY, 1953), for the isolated superior cervical ganglion of the rat and the rabbit *in vitro* with surface electrodes (PASCOE, 1956; ECCLES,

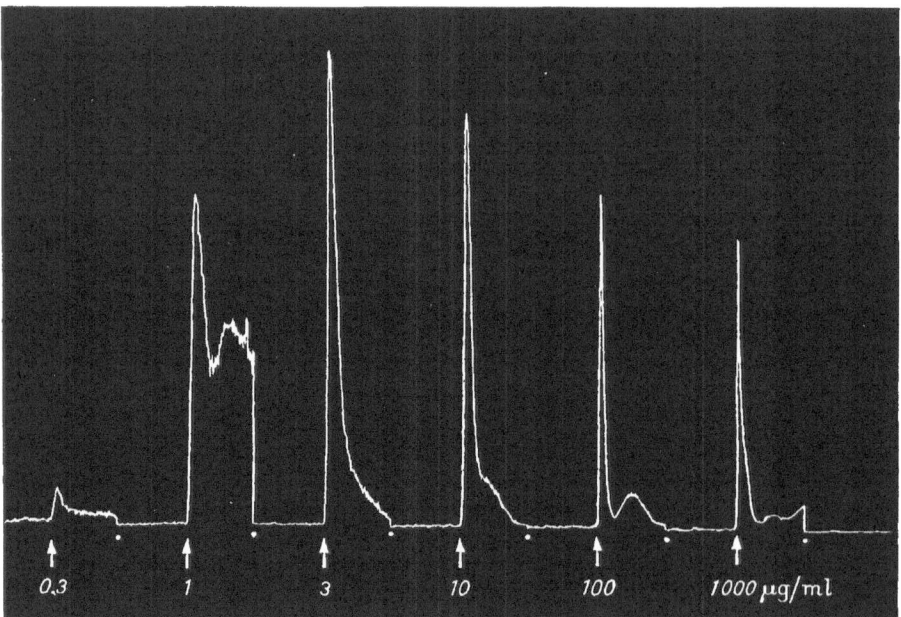

Fig 1 Response of the isolated guinea-pig ileum to increasing concentrations of DMPP. Tyrode's solution, 37° C Isotonic contractions of longitudinal muscle DMPP was added to the bath at arrows and washed out at points The drum was stopped between drug applications Intervals were 10 min between lower and 20 min between higher doses Note that dose-response curve is bell-shaped and that the duration of the response decreases with higher doses of DMPP

1956), and for frog sympathetic ganglia with intracellular recording (GINSBORG and GUERRERO, 1964).

Stimulant effects. When responses of an organ to increasing doses of nicotine (or of a related agent) are determined, the resulting dose-response curve is bell-shaped, *i.e.*, it has an ascending as well as a descending limb (TRENDELENBURG, 1961a). The shape of the dose-response curve is most probably due to a dissociation between the local response to nicotine (*i.e.*, the depolarization of the ganglion) and the "visible" response of the effector organ (which must be assumed to be related to the amount of peripheral transmitter released as a result of stimulation of the ganglion). Since increasing doses of nicotine cause increasing depolarization, not only the stimulant but also the blocking effects of nicotine increase. As a consequence, high doses of nicotine shorten the duration of the initial stimulation. Fig. 1 shows that responses of the effector organ to increasing doses of the nicotine-like agent DMPP yield a bell-shaped

dose-response curve. However, the figure also illustrates that responses obtained in the range of the ascending limb of the curve have a much longer
duration than responses of the same magnitude obtained in the descending
limb. Such observations are consistent with the view that with high doses of
nicotine the blocking effects of this drug cut short the stimulant effects.
VAN ROSSUM (1962) coined the term "autoinhibition" for this phenomenon.

The stimulant effects of nicotine and related agents on ganglia are not
blocked by atropine. However, in cholinergically innervated organs, atropine

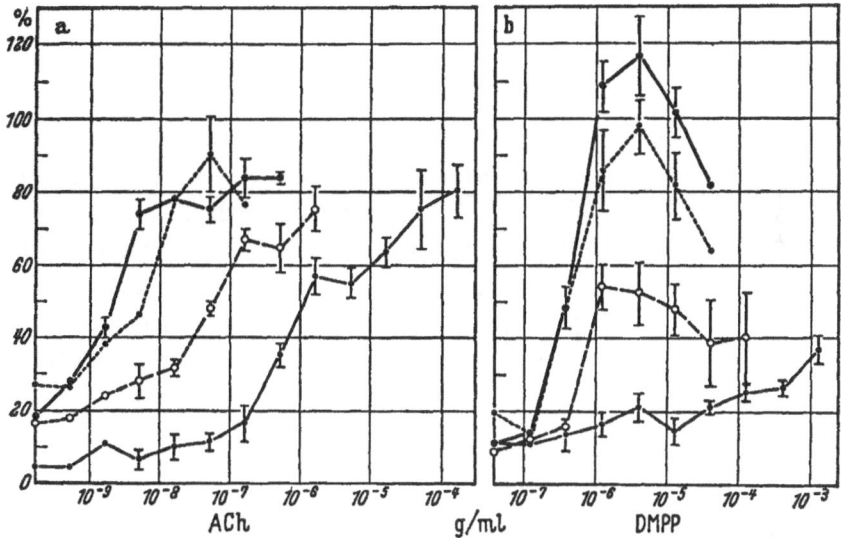

Fig. 2a and b Modification by atropine of dose-response curve for acetylcholine (a) and DMPP (b) Isolated
guinea-pig ileum. Plotted are the mean responses (± standard error as vertical bar, standard error is
omitted where it would obscure the graph) expressed as percent or initial control (for details see Methods)
Solid heavy line = control experiments, atropine 5 6 × 10⁻¹⁰ g/ml ●---●, 5.6 × 10⁻⁹ g/ml ○---○,
5.6 × 10⁻⁸ g/ml ●——● (Trendelenburg, 1961 a)

abolishes the effects of ganglionic stimulation by nicotine by antagonizing the
acetylcholine released from postganglionic nerve terminals. Although the antagonism of atropine against acetylcholine is of the surmountable type, that of
atropine against nicotine is not (TRENDELENBURG, 1961 a; VAN ROSSUM, 1962).
This phenomenon finds its explanation in Fig. 1, which demonstrates that the
amount of released acetylcholine reaches a maximum at the maximum of the
dose-response curve and appears to decline with increasing doses of the ganglion-
stimulating drug. A surmountable type of antagonism can only be expected
when the concentration of the agonist (in this case of acetylcholine) is increased
more and more. If increasing depolarization of ganglia by nicotine not only
results in no further increase in the release of acetylcholine but causes a reduced
output of transmitter from postganglionic nerve terminals, an antagonism of
atropine to nicotine should result in dose-response curves as those shown in
Fig. 2. In this experimental situation, nicotine must be regarded as an agent

with an indirect action, as far as the effector organ (the longitudinal muscle of the guinea-pig ileum) is concerned. It is noteworthy that the same type of insurmountable antagonism is observed when the effects on the heart of indirectly acting sympathomimetic amines (which release norepinephrine) are antagonized by β-receptor-blocking agents which cause a surmountable type of antagonism to norepinephrine (BENFEY and VARMA, 1964).

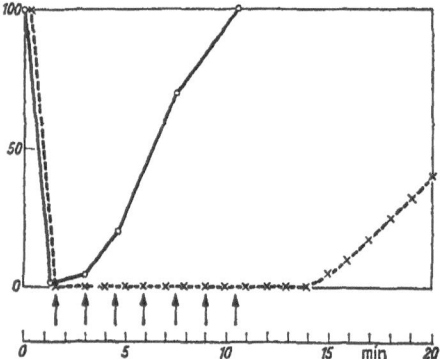

Fig 3 Biphasic nature of the block by nicotine of the cat's superior cervical ganglion Cat under chloralose Ordinates Response of nictitating membrane in percent of control Abscissae· Time (in minutes) Arrows indicate time of intraarterial injections of 100 μg of nicotine each to the superior cervical ganglion Solid line Response of the nictitating membrane to the intraarterial injection to the ganglion of 4 mg of potassium chloride, note that ganglionic response to potassium was abolished after *one* injection of nicotine (first arrow) but returned towards normal when potassium was injected after 2, 3 or 5 injections of nicotine and was normal after 7 injections of nicotine. Each point of this curve was obtained during a different series of injections of nicotine in order to keep 30-minute intervals between injections of potassium chloride. Broken line· Response of the nictitating membrane to preganglionic stimulation applied intermittently after 7 intraarterial injections of nicotine, note that ganglionic transmission began to recover about 5 min after the last injection of nicotine These results demonstrate that the ganglion regains its sensitivity to potassium *during* a series of intraarterial injections of nicotine, while ganglionic transmission is restored much later.
(From TRENDELENBURG, 1957c)

Blocking effects of nicotine. According to PATON and PERRY (1953) the onset of the inhibitory effects of a single injection of nicotine, acetylcholine or TMA to the superior cervical ganglion of the cat coincides with the depolarization of the ganglion (as determined with surface electrodes). However, with nicotine (but not with acetylcholine or TMA) the ganglionic block outlasts the depolarization of the ganglion. Moreover, when injections of nicotine are repeated at short intervals, they cause less and less depolarization of the ganglion while continuing to cause complete block of transmission. In other words, with single or with repeated injections of nicotine, it is possible, at certain times, to detect ganglionic blockade without depolarization. PATON and PERRY (1953) concluded that the block by nicotine has a tendency to pass from a depolarizing block to a late, non-depolarizing block which resembles that produced by hexamethonium or TEA.

Another tool that can be used to test for the depolarizing or non-depolarizing nature of ganglionic block is potassium chloride which is well known to retain its ganglion-stimulating effects in the presence of non-depolarizing ganglion-blocking agents (BROWN and FELDBERG, 1936; ACHESON and PEREIRA, 1946). The failure of a ganglion to respond to potassium chloride may be taken as suggestive evidence for block by depolarization. When repeated intraarterial injections of 100 μg of nicotine are given to the superior cervical ganglion of the cat at intervals of 90 seconds, the ganglion first fails to respond to injections of potassium chloride (*i.e.*, it is blocked by depolarization) but regains its responsiveness to this agent within 5 to 10 min, although there are no responses to preganglionic stimulation or to injections of nicotine (Fig. 3) (TRENDELEN-BURG, 1957c). During this late, non-depolarizing phase of the block by nicotine, ganglia respond not only to potassium chloride but to all those agents that are known to act on receptors which differ from the nicotinic acetylcholine receptors (see section V.A. 10).

The late, non-depolarizing phase of the block by nicotine can be obtained with great regularity by administering rather large doses of nicotine intravenously. Immediately after the initial period of stimulation, there is a ganglion-block of the depolarizing type. In spite of the continued intravenous administration of nicotine (ca. 4 mg/kg every 5 min), the responsiveness of the ganglion to potassium returns to normal (TRENDELENBURG, 1957c).

PATON and PERRY (1953) used the term "competitive block" to characterize this type of late block by nicotine. However, since there is an insurmountable type of antagonism to nicotine, DMPP and TMA during this late phase of the block by nicotine (TRENDELENBURG, 1966b), it might be preferable to use the purely descriptive term "non-depolarizing block".

Agents which depolarize the endplate region of skeletal muscle are known to cause a block of transmission that undergoes a similar change from "block by depolarization" to "late, non-depolarizing block". DEL CASTILLO and KATZ (1957) proposed the following extension of the receptor theory to account for this phenomenon:

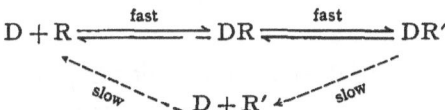

Combination of the drug with the receptor (D + R → DR) results in a response only after a further step has been taken, the activation of the receptor (DR → DR'). According to this view, the chain of events stops at the first step for non-depolarizing blocking agents, while it proceeds further for depolarizing agents. In order to account for the late phase of the block, a third step is postulated that leads to the dissociation of drug and receptor and leaves the receptor in a non-reactive form (R') which is then very slowly reversed to the reactive form R. The differences in rate lead to an accumulation of R' in the presence of depolarizing agents with the result that the depolarization wears off and that a non-depolarizing block develops. This concept was developed to account for phenomena observed at the endplate of skeletal muscle. It also fits the observations of GINSBORG and GUERRERO (1964) in frog ganglia which were initially depolarized by acetylcholine, carbachol, nicotine or TMA, and repolarized when these drugs were left in contact with the ganglion for 5 to 10 min.

While this concept, which should be regarded as an attempt to describe a phenomenon rather than to explain it, is based on the classical receptor theory as formulated by Clark, the more recent receptor theory of Paton (1961) provides another possible description of the phenomenon. According to Paton a response is the result of a high rate of drug-receptor association rather than of receptor occupancy. This means that agents with a high rate of association with the receptor are agonists (*i.e.*, depolarizing agents in the case of ganglia) which also must have a high rate of dissociation, if the response is sustained; according to this concept, free receptors must be available for interaction with the drug if a response is prolonged. Antagonists (*i.e.*, the non-depolarizing blocking agents), on the other hand, can be visualized as having a very low rate of association so as not to cause a visible response; in order to have a blocking action of any duration, they also must have a low rate of dissociation from the receptors. It is tempting to speculate that nicotine may have a high rate of association with the receptor (which would result in depolarization) combined with a low rate of dissociation from the receptor; the latter property would cause more and more receptors to be occupied by the drug with the consequence that depolarization (which is dependent on *new* drug-receptor interactions) would wear off and that a non-depolarizing block would ensue. This block should then be of the insurmountable or noncompetitive type encountered when the population of receptors is drastically reduced. Recent observations revealed that the effects of nicotine, DMPP and TMA are antagonized insurmountably during the late phase of the block by nicotine (Trendelenburg, 1966b).

Depending on the rate of dissociation of nicotine, it is possible that at any time of the late phase of the block a small number of receptors is free to react with free nicotine. This might cause a small and persistent depolarization of the ganglion cells throughout the later phase of the block. Some evidence (for instance, the facilitation by nicotine of the effects of various non-nicotinic agents, see V.A. 11) is compatible with such a proposal. In addition, the partial sustained contraction of the nictitating membrane observed during the late, non-depolarizing phase of the block by nicotine is also consistent with this proposal, especially since an intravenous injection of hexamethonium causes an immediate and complete relaxation of the nictitating membrane (Trendelenburg, 1966b). However, it should be noted that the proposals discussed in this section are pure speculation and are not yet supported by any direct experimental evidence.

Pelikan (1960) tried to explain the blocking effect of nicotine by a presynaptic action of the drug, namely by an impairment of the release of acetylcholine from preganglionic nerve terminals. The experimental evidence presented to support this hypothesis is not satisfactory; isolated and perfused superior cervical ganglia whose effluent has a hemoglobin content indistinguishable from that of venous blood cannot be regarded as satisfactory preparations. The criteria for the adequacy of a perfused preparation have been discussed in detail by PATON (1954) and AMBACHE (1954). In addition, a presynaptic type of block should not affect responses of ganglia to injected acetylcholine; however, nicotine undoubtedly antagonizes the effects of injected acetylcholine during the phase of depolarization of the ganglion by nicotine (TRENDELEN-BURG, 1966b). FELDBERG and VARTIAINEN (1935) found no effect of nicotine on the release of acetylcholine in response to preganglionic stimulation. Moreover, recent results indicate that during the late phase of the block by nicotine there is no impairment of the release of acetylcholine from preganglionic nerve terminals, since the ganglion continues to transmit preganglionic impulses although muscarinic mechanisms are involved rather than the nicotinic acetylcholine receptors (see below). In summary, there is no convincing evidence that nicotine impairs the release of acetylcholine from preganglionic nerve endings.

B. Tetramethylammonium (TMA)

As already mentioned, a transition from the depolarizing to the non-depolarizing type of block of the superior cervical ganglion of the cat has been described only for nicotine. According to PATON and PERRY (1953) TMA appears to block ganglia predominantly by depolarization, and attempts to detect a late, non-depolarizing phase of the block by TMA have failed (JONES et al., 1963).

Very recently GEBBER and VOLLE (1966) described a second type of ganglion block by TMA which is not accompanied by depolarization. Their evidence indicates that the intraarterial injection of 40 μg of TMA to the superior cervical ganglion of the cat causes an initial depolarization for about 30 sec which is accompanied by a block of transmission of slightly shorter duration (TMA-induced early block). During repolarization the block of transmission wears off. A period of hyperpolarization follows which lasts for a few minutes and which is accompanied by a second period of block of transmission (TMA-induced late block). Hyperpolarization accompanied by block of ganglionic transmission is also observed after intraarterial injections of methacholine, a muscarinic agent. However, the TMA-induced late block differs from that produced by methacholine in several respects: while the TMA-induced late block is prevented by a preceding period of preganglionic stimulation or by injections of pilocarpine, anticholinesterases or ouabain, these procedures either increase or do not affect the blocking effect of methacholine. Ouabain not only antagonizes the TMA-induced late block, it also prevents the appearance of the positive afterpotential that normally follows every ganglionic spike (elicited by preganglionic stimulation). The similarity of the effect of ouabain on the TMA-induced late block on the one hand, and on positive afterpotentials on the other, may indicate that both phenomena are due to the same mechanism. GEBBER and VOLLE (1966) suggest that both potential changes reflect an increase in cellular metabolism that is required for the restoration of the potassium content of the ganglion cells following their stimulation. It is noteworthy that no late block and no hyperpolarization was detected after injections of potassium chloride; it is possible that the injection of exogenous potassium chloride does not cause a loss of endogenous potassium from the excited ganglion cells.

More recently, VOLLE and his coworkers observed this type of ouabain-sensitive late block after various nicotine-like agents including nicotine, the block is accompanied by hyperpolarization of the ganglion (personal communication).

These observations as well as those of PATON and PERRY (1953) indicate that TMA has no pronounced muscarinic effects on the superior cervical ganglion of the cat. It is somewhat surprising that TMA appears to be a purely nicotinic agent in this preparation, because it has been found to have pronounced muscarinic effects in another preparation frequently used in the analysis of ganglion-stimulating agents, the isolated guinea-pig ileum. On the ileum, dose-response curves for TMA resemble those for acetylcholine and pilocarpine rather than those for nicotine and DMPP (TRENDELENBURG, 1961a). In addition, the antagonism of atropine to TMA and acetylcholine is of the surmountable type, while that of atropine to nicotine and DMPP is insurmountable (see above). Other muscarinic effects of TMA have been described for the cardiovascular system of the dog (WINBURY, 1958), for papillary muscle of the cat and for the heart-lung preparation of the dog (LEE and SHIDEMAN, 1959).

Very recently, a weak muscarinic effect of TMA has been observed even on the superior cervical ganglion of the cat. During the late phase of the block by nicotine (i e., when there is an insurmountable antagonism to the nicotinic effects of nicotine and TMA), small responses to TMA are observed that appear to be mediated through the muscarinic receptors of the ganglion (TRENDELENBURG, 1966b).

It is likely that certain actions of ganglion-stimulating agents can be observed only under certain experimental conditions. For instance, the weak muscarinic effects of TMA on the superior cervical ganglion are only detectable when a) the nicotinic acetylcholine receptors are blocked by nicotine, and b) the effects of muscarinic agents are potentiated by nicotine (see section V.A. 11). The TMA-induced late block is another example, since the failure of several investigators to detect this type of block may be due to the fact that, during the experiment, they applied preganglionic stimulation at high frequencies which, according to GEBBER and VOLLE (1966), antagonizes the TMA-induced late block. Apparent contradiction between observations of different groups of workers are probably due to differences in experimental procedures the importance of which has not yet been recognized.

C. 1,1-Dimethyl-4-phenylpiperazinium (DMPP)

Since the first description of the ganglion-stimulating actions of DMPP (CHEN et al, 1951) this agent has been widely used as a "typical" nicotinic drug. In a variety of preparations, the actions of DMPP have been found to be very similar to those of nicotine (LING, 1959; HERR and GYERMEK, 1960; TRENDELENBURG, 1961a; VAN ROSSUM, 1962); however, DMPP is unlike nicotine in that it fails to produce a late, non-depolarizing phase of ganglionic block.

Under conditions in which TMA exerts muscarinic effects, DMPP was found to have only nicotinic ones (WINBURY, 1958, TRENDELENBURG, 1961a).

On adrenergic nerve terminals, DMPP appears to exert some effects that are not observed with nicotine. For instance, LINDMAR (1962) found hexamethonium to be ineffective against the positive chronotropic action of DMPP on isolated guinea-pig and rat atria, while this agent abolished the effects of nicotine in the former species, and while nicotine failed to exert an effect in the latter species. Although this effect of DMPP is not "nicotine-like", it is mediated by the release of endogenous norepinephrine, since pretreatment with reserpine (to deplete the stores) prevents the positive chronotropic action. LINDMAR (1962) suggested that DMPP should be classified as a "tyramine-like" drug in its action on isolated atria of guinea-pigs and rats. On the isolated rat intestine, on the other hand, DMPP was found to antagonize the inhibitory effect of sympathetic nerve stimulation; this effect was not prevented by hexamethonium and was reversed by an exposure of the preparation to dopamine or amphetamine (BIRMINGHAM and WILSON, 1965). Since such a procedure does reverse the effects of bretylium and guanethidine, BIRMINGHAM and WILSON suggested that DMPP is "bretylium-like". We know too little about the effects of DMPP on adrenergic neurons to give a full description of this unexpected action. However, although the labels "tyramine-like" and "bretylium-like" may have to be revised, there is no doubt that DMPP can exert certain effects which are a) unlike those of nicotine, and b) resistant to hexamethonium.

D. Carbachol

VOLLE and KOELLE (1961) determined threshold doses for the stimulation of the superior cervical ganglion of the cat by intraarterial injections of acetylcholine and carbachol. On normal ganglia, acetylcholine had only 1/10 the potency of carbachol. Chronic denervation of the ganglion failed to affect the threshold dose of acetylcholine but reduced the potency of carbachol by a factor of 26. These observations are very similar to those obtained with

2*

"indirectly acting" sympathomimetic amines like tyramine which are known to exert their effects by releasing norepinephrine from adrenergic nerve terminals and which are known to lose their effects after degeneration of the nerve (FLECKENSTEIN and BURN, 1953; FLECKENSTEIN and STOCKLE, 1955). In analogy to this accepted view, it is possible that carbachol exerts its ganglion-stimulating action predominantly through the release of acetylcholine from the preganglionic nerve terminals. Some more direct evidence for this view was obtained by McKINSTRY et al. (1963), but there is not yet any evidence that provides a convincing correlation between the effects of carbachol on the superior cervical ganglion and changes in the acetylcholine content of the ganglion. For instance, the disappearance of endogenous acetylcholine from the superior cervical ganglion is known to happen between 24 and 72 hours after denervation (MACINTOSH, 1938); consequently, the decline in potency of carbachol should happen at the same time if the proposal is correct. Furthermore, hemicholinium (which is known to prevent the synthesis of acetylcholine) should be an antagonist to the ganglionic actions of carbachol.

V. Non-nicotinic ganglion-stimulating agents

This section summarizes observations with the following agents or groups of agents whose mode of action differs from that of nicotine: muscarinic agents (synthetic dl-muscarine, pilocarpine, 4-(m-chlorophenylcarbamoyloxy)-2-butynyl-trimethyl-ammonium chloride — McN-A-343, N-benzyl-3-pyrrolidyl ace-

Table 2. *Schematic representation of the effects of various drugs on the stimulation of the superior cervical ganglion of the cat by various ganglion-stimulating agents*

	Nicotine DMPP TMA	Muscarinic agents	Histamine	5-HT	Angiotensin	Bradykinin	Potassium chloride	For references see section
Non-depolarizing ganglion-blocking agents	inh.	0	0	0	0	0	0	V.A 8
Nicotine								
depolarizing phase	inh.	inh.	inh	inh.	inh.	inh	inh	V.A. 9
non-depolarizing phase	inh.	pot.	pot.	pot.	pot.	inh *	0	V A. 10. 11
followed by hexamethonium	inh.	return to normal					0	V.A. 10. 11
Cocaine	0	inh.	inh	inh.	inh.	inh.	0	V A 12
Morphine, methadone	0	inh	inh	inh	inh.	inh.	0	V.A. 13
Atropine	0	inh.	0	0	0	0	0	V.A. 14
Antihistaminic agents	0	0	inh.	0	inh.*	0	0	V.A. 15
5-Hydroxy-3-indolacetamidine	0			inh.				V A 16
Facilitation after								
preganglionic stimulation	0	pot.	pot.	pot.	pot.	pot.	0	V.A. 2

inh. inhibition of the effects of the ganglion-stimulating agent; pot. potentiation of the effects of the ganglion-stimulating agent; 0. no change in the effects of the ganglion-stimulating agent.

* partial inhibition.

tate methobromide — AHR-602), histamine, 5-hydroxytryptamine — 5-HT, and polypeptides (angiotensin and bradykinin). For quick reference, the most important findings are summarized in Table 2.

A. Action on the non-perfused superior cervical ganglion of the cat

1. Stimulation of the ganglion. For all non-nicotinic agents it has been established that section of the postganglionic fibers abolishes the response of the nictitating membrane elicited by the intraarterial injection of these drugs to the ganglion. Table 3 shows the ranges of doses reported to cause stimulation of the ganglion. The table also shows that not all preparations responded to some of the drugs; this indicates that the variability in the sensitivity of individual preparations to the non-nicotinic agents is much greater than the variability in sensitivity to nicotine. It is likely that at least two factors contribute to the great variability in sensitivity to non-nicotinic agents: a) the effect of a preceding period of preganglionic stimulation (see V.A. 2) and b) the very flat dose-response curve for these agents (see V.A. 5). Because of the variability, it is impossible to give exact relative potencies. However, it is clear from Table 3 that many agents have potencies which are not drastically different from those of nicotine and DMPP, and that angiotensin (which has the highest molecular weight of the agents shown in Table 3) is the most potent of all known ganglion-stimulating agents.

Another muscarinic agent, methacholine, also stimulates the superior cervical ganglion of the cat, since TAKESHIGE et al. (1963) recorded an increased activity in the postganglionic nerve after injections of this agent.

Table 3 *Range of doses of various agents required for stimulation of the superior cervical ganglion. Drugs were injected intraarterially into the blood supply of the ganglion*

Agent	Range of doses µg	Percentage of preparations which responded with stimulation* %	Reference
Muscarine	1—10	50	JONES, 1963
McN-A-343	1—10	100	ROSZKOWSKI, 1961; MURAYAMA and UNNA, 1963; JONES, 1963
Pilocarpine	25—40	60	TRENDELENBURG, 1954
AHR-602	5—1000	100	FRANKO et al , 1963, JONES, 1963
Histamine	2—55	82	TRENDELENBURG, 1954
5-HT	1—100	70	TRENDELENBURG, 1956a
Angiotensin	0 1—0 3	100	LEWIS and REIT, 1965
Bradykinin	0.5—10	100	LEWIS and REIT, 1965
Nicotine	ca. 3	100	JONES, 1963
DMPP	ca 1	100	TRENDELENBURG, 1966b
Acetylcholine	20—50	100	TRENDELENBURG, 1966b

* Any percentage value below 100% indicates that some ganglia failed to respond to that dose which is given as the upper limit of the range of doses used

2. Facilitation of ganglionic effects by preganglionic stimulation. A 5-second period of preganglionic stimulation (25 shocks/sec of supramaximal strength) causes a long-lasting and pronounced potentiation of ganglionic responses to McN-A-343, histamine, 5-HT, angiotensin and bradykinin (JONES, 1963; TRENDELENBURG and JONES, 1965; SMITH, 1966b, TRENDELENBURG, 1966a). The sensitization of ganglia results in a parallel shift of dose-response curves for histamine and pilocarpine to the left by about one log unit (Fig. 4); under identical conditions, potentiation of ganglionic effects is much less pronounced

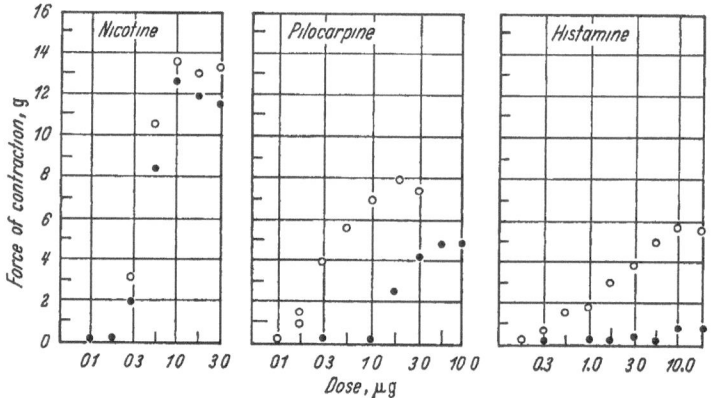

Fig. 4 The effect of preganglionic stimulation on the response of the nictitating membrane to injections of drugs to the superior cervical ganglion. Solid symbols represent responses of 3 preparations to intra-arterial injections of nicotine, pilocarpine and histamine, respectively, to the superior cervical ganglion of the spinal cat. Open symbols represent responses of the same preparations to injections administered after a prolonged period of preganglionic stimulation Note that this procedure potentiated the ganglionic responses to pilocarpine and histamine much more than those to nicotine (From IORIO and McIsAAC, 1966, with permission of the J Pharmac exp. Ther)

for nicotine or DMPP (TRENDELENBURG and JONES, 1965; IORIO and McIsAAC, 1966; SMITH, 1966b) and absent for potassium chloride (TRENDELENBURG and JONES, 1965).

Since the potentiating effects of one 5-second period of preganglionic stimulation last for more than one hour (VOLLE, 1962c; TRENDELENBURG and JONES, 1965; IORIO and McIsAAC, 1966; SMITH, 1966b), the simple procedure of ascertaining the normal function of the preparation before the actual beginning of the experiment must result in a pronounced change in the sensitivity of the preparation to non-nicotinic ganglion-stimulating agents. Since this procedure is by no means uncommon, it must be concluded that the values for minimal effective doses presented in Table 3 are too low for unstimulated preparations and too high for those which had been stimulated preganglionically. Or in other words, the heterogeneity of the material may well account for the variability of results.

Although the effect of a short train of preganglionic impulses is well defined, the mechanisms involved in this phenomenon are unknown. The potentiating effect of preganglionic stimulation is not abolished by doses of hexamethonium which block the response of the ganglion to the preganglionic impulses (TREN-DELENBURG and JONES, 1965; TRENDELENBURG, 1966a). Postganglionic stimu-lation or injections of acetylcholine, on the other hand, fail to potentiate the effects of non-nicotinic ganglion-stimulating agents. TAKESHIGE and VOLLE (1964a) concluded that a train of preganglionic impulses first causes hyper-polarization of the ganglion cells for 30 to 60 sec and that this is followed by a sustained low-amplitude depolarization, the duration of which cannot be measured because of the drift of the recording equipment. It is this postulated late phase of low-amplitude depolarization which would account for the poten-tiating effects of preganglionic stimulation; however, the origin of this depolari-zation remains obscure. According to TAKESHIGE and VOLLE (1964a) the low-amplitude depolarization is antagonized by atropine (1 to 5 μg i.a.) provided that this agent is injected soon after termination of the period of preganglionic stimulation (*i.e.*, during the phase of hyperpolarization), while no antagonism is observed when the atropine is injected prior to preganglionic stimulation. It remains obscure why the time of the injection of atropine is so important. It is unlikely that the effect of the intraarterially injected atropine wears off if it is injected prior to the period of preganglionic stimulation, since TRENDELEN-BURG and JONES (1965) found that pretreatment with as much as 200 μg/kg of atropine (i.v.) failed to antagonize the potentiation of the ganglionic effects of histamine by preganglionic stimulation.

3. *The effect of chronic preganglionic denervation of the ganglion.* There is no evidence that any of the non-nicotinic agents lose their ability to stimulate ganglion cells when tested after chronic denervation. On the contrary, super-sensitivity appears to develop to all those agents that have been tested (muscarine: perfused ganglion — AMBACHE et al., 1956; KONZETT and WASER, 1956; non-perfused ganglion — JONES, 1963; McN-A-343 and AHR-602: non-perfused ganglion — JONES, 1963; histamine: perfused ganglion — KONZETT, 1952; angiotensin and bradykinin: non-perfused ganglion — LEWIS and REIT, 1965). It should be noted that nearly all reports on denervation supersensitivity of ganglia are based on experiments in which the nictitating membrane was used as the indicator organ. The sources of error which arise from decentrali-zation of the nictitating membrane are discussed in section X.B.

4. *Delay of onset of response.* While intraarterial injections to the ganglion of nicotinic agents cause a nearly immediate response, there is a latency period of 6 to 10 sec for all non-nicotinic agents. For angiotensin, the latency period appears to be even longer (10—20 sec; LEWIS and REIT, 1965).

5. *Slopes of dose-response curves.* Maximal effects of nicotine are observed with doses about 30 times higher than the threshold dose, *i.e.*, the dose-response curve is very steep (JONES, 1963; TRENDELENBURG, 1966b). Dose-response curves determined with non-nicotinic agents, on the other hand, are rather flat, and their maxima do not approach the maximum response elicited

by nicotine (McN-A-343: Jones, 1963; pilocarpine and histamine: Iorio and McIsaac, 1966). In addition, while a descending limb of the dose-response curve for nicotine and related agents is a common observation (see IV.A.), a similar phenomenon has not been reported for non-nicotinic agents. The absence of this phenomenon may well be related to the fact that non-nicotinic agents even in very high doses usually fail to exert inhibitory effects on ganglia.

6. *Facilitatory effects of subthreshold doses of non-nicotinic agents.* In doses too small to cause stimulation of the ganglion, the non-nicotinic agents facilitate transmission of submaximal preganglionic impulses through the ganglion. This has been shown for histamine, pilocarpine and 5-HT (Trendelenburg, 1955, 1956a), for McN-A-343 (Murayama and Unna, 1963; Jones, 1963; Smith, 1966b), for AHR-602 and muscarine (Jones, 1963) and for angiotensin and bradykinin (Lewis and Reit, 1965). McN-A-343 also facilitates the response of the ganglion to potassium chloride (Murayama and Unna, 1963). The site of the facilitatory action must lie in the ganglion because histamine, pilocarpine and 5-HT fail to potentiate responses of the nictitating membrane to submaximal postganglionic stimulation, and because the height of postganglionic action potentials elicited by submaximal preganglionic stimulation is increased after injections of histamine, pilocarpine and 5-HT (Trendelenburg, 1957a).

When supramaximal stimulation is used, and when very high doses of non-nicotinic agents are injected, a depression of ganglionic transmission is observed. Gertner and Kohn (1959), for instance, found that doses of more than 150 μg of histamine depressed ganglionic transmission under these conditions, while Trendelenburg (1956b) observed facilitation of transmission of submaximal preganglionic impulses with 1—10 μg of histamine. Both observations were made in perfused ganglia.

7. *Tachyphylaxis.* Tachyphylaxis has been reported for all non-nicotinic agents, and time intervals of 15 to 40 min (depending on the dose) have to be kept between injections to prevent its appearance. With pilocarpine (Trendelenburg, 1954) and angiotensin (Lewis and Reit, 1965) tachyphylaxis is especially pronounced. Cross-tachyphylaxis is found for some of these agents when they are injected at a time at which they facilitate ganglionic transmission of submaximal preganglionic stimulation. Mutual cross-tachyphylaxis has been found for histamine, McN-A-343, pilocarpine and 5-hydroxytryptamine (Trendelenburg, 1956b; Smith, 1966b). Angiotensin becomes ineffective when injected soon after histamine, but the reverse is not true. Moreover, responses to bradykinin are not influenced by previous injections of either histamine or angiotensin (Lewis and Reit, 1965).

Whenever tachyphylaxis develops to one of the non-nicotinic agents, these drugs not only fail to stimulate the ganglion, they also fail to facilitate ganglionic transmission (Trendelenburg, 1956b; Iorio and McIsaac, 1966).

8. Non-depolarizing ganglion-blocking agents. The ganglionic effects of the non-nicotinic agents are not affected by injections of hexamethonium in doses that abolish responses of the ganglion to preganglionic stimulation and to nicotinic agents (for references see Table 2). TEA likewise fails to antagonize McN-A-343 (MURAYAMA and UNNA, 1963). While these observations were made with single doses of the non-nicotinic agents, IORIO and McISAAC (1966) determined dose-response curves for histamine, pilocarpine and nicotine; hexamethonium caused a parallel shift (to the right) of the dose-response curve of nicotine but did not alter the curves of histamine and pilocarpine.

Other ganglion-blocking agents of this group have not been investigated on the superior cervical ganglion, but TEA, chlorisondamine and mecamyl-amine failed to abolish pressor responses to some of the non-nicotinic agents (see section V.E.2).

Since none of the non-depolarizing ganglion-blocking agents antagonizes the ganglionic actions of the non-nicotinic agents, it is justifiable to postulate specific ganglionic receptors which differ from the nicotinic acetylcholine-receptors.

9. Depolarizing ganglion-blocking agents. The ganglion-stimulating effects of histamine, pilocarpine, 5-HT, angiotensin and bradykinin are prevented when these agents are injected about 60—90 sec after the intraarterial injection of nicotine which first stimulates and then blocks the ganglion (TRENDELEN-BURG, 1954, 1956a, 1966b). Ganglionic responses to muscarine, McN-A-343 and AHR-602 are similarly affected; these agents are also ineffective when injected immediately after TMA, another depolarizing blocking agent (JONES, 1963).

SMITH (1966b) found that an intraarterial injection of a rather large dose of DMPP (*i.e.*, 40 µg/kg i.a.) first antagonized the ganglionic effects of McN-A-343 and then caused a late potentiation of responses to McN-A-343 which in some animals lasted for hours. Since this type of late potentiation was observed only for the ganglionic effects of small doses of McN-A-343 and not for those of small doses of DMPP, it is possible that this phenomenon is related to effects of nicotine discussed in section V.A.11.

MURAYAMA and UNNA (1963) failed to observe full block of the effects of McN-A-343 when this compound was injected 5 min after nicotine. It is likely that the time interval between the injection of nicotine and that of McN-A-343 is crucial, since the depolarization produced by nicotine is of short duration; in fact, PATON and PERRY (1953) showed it to be shorter than the block of transmission (see IV.A.).

10. Effects of non-nicotinic agents during the late, non-depolarizing phase of the block by nicotine. When repeated intraarterial injections of nicotine are given at intervals of 90 sec, the depolarizing block by nicotine gradually changes into a block of the non-depolarizing type, since potassium chloride regains more and more of its normal stimulant effect when injected immediately after increasing numbers of injections of nicotine (TRENDELENBURG, 1957c). Under identical conditions, histamine (TRENDELENBURG, 1957c), McN-A-343, AHR-602, muscarine (JONES, 1963), angiotensin and bradykinin (TRENDELENBURG, 1966a) also regain their effectiveness at a time when the ganglion fails to respond to preganglionic stimulation and to nicotinic agents.

This late, non-depolarizing phase of the block by nicotine can also be demonstrated with intravenous injections of series of large doses of nicotine. Immediately after the intravenous injection of about 12 mg/kg of nicotine (administered in divided doses within 10 min) the ganglion fails to respond to any agent *i.e.*, block by depolarization appears to be present. However, immediately after a second series of injections of nicotine (administered 20 min later), injections of non-nicotinic agents are fully effective, while nicotine or DMPP remain ineffective. Reappearance of ganglion-stimulating effects during this late, non-depolarizing phase of the block by nicotine has been observed with histamine, pilocarpine, 5-HT, angiotensin and bradykinin (TRENDELENBURG, 1957c, 1966a) and with muscarine, McN-A-343 and AHR-602 (JONES, 1963).

While it is easy to demonstrate the late, non-depolarizing phase of the block by nicotine, attempts to produce a similar phenomenon with injections of TMA have been unsuccessful (TRENDELENBURG, 1957c). This may be attributed to the fact that TMA has little tendency to produce a non-depolarizing type of block (PATON and PERRY, 1953).

The discussion in this section is based on the classification of ganglion-blocking drugs proposed by PATON and PERRY (1953). It should be borne in mind that VOLLE and his coworkers have recently obtained evidence that some type of ganglion block may be due to a hyperpolarization of the ganglion cells. It is not yet possible to fit this new concept into a discussion of the effects of ganglion-blocking agents on the response of ganglia to non-nicotinic agents. For a discussion of the work of VOLLE, the reader is referred to section VI.

11. Facilitation by nicotine of the ganglionic actions of some non-nicotinic agents. During the late phase of the block by nicotine, the effects of histamine, pilocarpine, McN-A-343, 5-HT and angiotensin are not only restored but increased above the control level (TRENDELENBURG, 1957c, 1966a and b). Bradykinin appears to be the only non-nicotinic agent whose ganglionic actions are depressed rather than facilitated during the late phase of the block by nicotine (TRENDELENBURG, 1966a and b). These facilitatory and depressant effects observed during the late phase of the block by nicotine must be ascribed to some persisting action of nicotine, since they are readily abolished by an intravenous injection of 2 to 20 mg/kg of hexamethonium. Such an injection restores the effects of non-nicotinic agents to normal irrespective of whether they had been facilitated (histamine, 5-HT, McN-A-343, angiotensin) or depressed (bradykinin) by the prolonged administration of nicotine (TREN-DELENBURG, 1966a and b). Hence, it is clear that hexamethonium does not exert a direct influence on the action of non-nicotinic agents, but antagonizes the effects of previous injections of nicotine. For potassium chloride, there is neither a potentiation by nicotine nor a reduction of its stimulant effects by a subsequent injection of hexamethonium (TRENDELENBURG, 1957c, 1966b).

At the present time there is no conclusive explanation for the phenomenon of facilitation by nicotine of the effects of various non-nicotinic agents. It is tempting to speculate that during the late phase of the block by nicotine there is a small, persistent depolarization of the ganglion cells in response to nicotine; and that hexamethonium abolishes this effect. A small and persistent depolarization could account for the facilitation of the ganglionic effects of histamine, pilocarpine, McN-A-343 and angiotensin. However, it is difficult to understand why there is no potentiation of the effects of potassium chloride and an antagonism to the effects of bradykinin.

While the mechanism involved in the phenomenon of facilitation by nicotine remains obscure, it is clear that this type of facilitation differs from that produced by a period of preganglionic stimulation (see V.A.2). Three observations support this view: 1. While a preceding period of preganglionic stimulation potentiated the effects of both angiotensin and bradykinin, the effect of the former was increased during the late phase of the block by nicotine, while the effect of the latter was reduced (TRENDELENBURG, 1966a). 2. The facilitatory effect of preganglionic stimulation was not antagonized by hexamethonium (TRENDELENBURG and JONES, 1965), while that of nicotine was abolished (TRENDELENBURG, 1957c, 1966b). 3. Experiments with angiotensin showed that one can separate the two types of facilitation (Fig. 5). Angiotensin was injected in equal doses to both ganglia of the same preparation; while no preganglionic stimulation was applied to one side, all injections of angiotensin to the other side were preceded by a period of preganglionic stimulation. Before any nicotine was injected and during the late phase of the block by nicotine, the response of the stimulated ganglion to angiotensin was always greater than the response of the non-stimulated side; the responses became equal as soon as the first period of preganglionic stimulation was applied to the previously unstimulated ganglion (TRENDELENBURG, 1966a). Hence, the two types of facilitation are independent of each other.

12. Cocaine. Small doses of cocaine (0.2—1.0 mg/kg intravenously) do not affect responses of the ganglion to nicotinic agents, to potassium chloride or to preganglionic stimulation. However, such small doses of cocaine antagonize ganglionic responses to histamine, pilocarpine, 5-HT, angiotensin and bradykinin (TRENDELENBURG, 1954, 1956a, 1966a) as well as to muscarine, McN-A-343 and AHR-602 (JONES, 1963). In addition, MURAYAMA and UNNA (1963) found the intraarterial injection of as little as 1 µg of cocaine to antagonize the effect of McN-A-343 for one hour. Cocaine also abolishes the facilitatory effects of histamine, pilocarpine and 5-HT, although larger doses have to be injected (Trendelenburg, 1955, 1965a). Procaine has been found to antagonize the ganglionic effects of histamine, 5-HT and McN-A-343 (Smith, 1966b).

The mode of action of cocaine is obscure. It does not appear to interact with specific receptors, since it antagonizes not only the ganglionic effects of

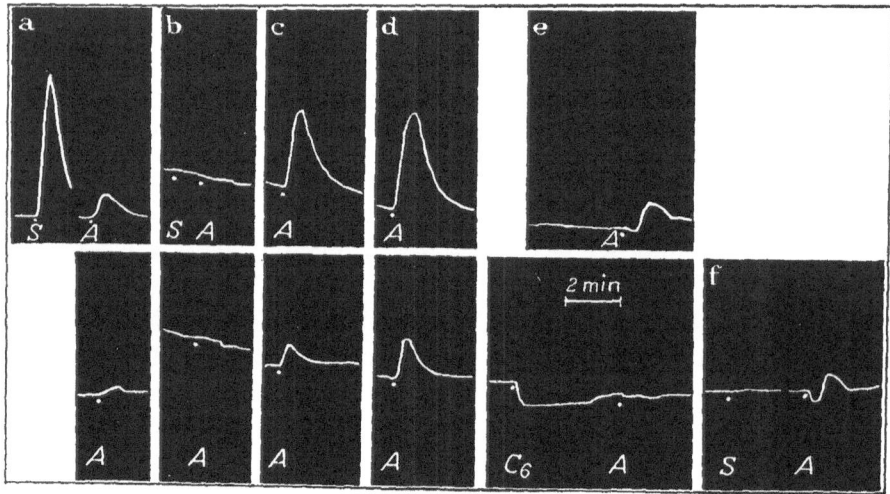

Fig. 5a—f. Modification of the response of the superior cervical ganglion to angiotensin by preganglionic stimulation, by nicotine and by hexamethonium. Spinal cat, left (upper trace) and right (lower trace) nictitating membrane. A intraarterial injection to the ganglion of 2 µg of angiotensin. S 10 sec of preganglionic stimulation (25 shocks/sec, supramaximal strength). A period of preganglionic stimulation was applied to the left side (upper trace) 5 min before *every* injection of angiotensin (even when not shown in figure), but stimulation was applied to right side only where shown (f). a Control period; the drum was stopped between S and A, upper trace. b Immediately after first series of intravenous injections of nicotine (total dose: 11 mg/kg). c After second series of intravenous injections of nicotine (total dose: 15.4 mg/kg). d After third series of intravenous injections of nicotine (total dose: 11.5 mg/kg). e After the intravenous injection of 20 mg/kg of hexamethonium (given at C6, lower trace). f Response to angiotensin after stimulation of hitherto unstimulated side. Intervals between injections to one side: 20 min. Injections to right ganglion were given 4 min before the injection to the left ganglion except in (b) where interval was only one minute. (From Trendelenburg, 1966a, with permission of the J. Pharmac. exp. Ther.)

muscarinic agents but also those of histamine, 5-HT and the polypeptides. It is likely that cocaine blocks a pathway common to the chain of events elicited by activation of the specific receptors with which the non-nicotinic agents react. Since cocaine fails to antagonize the effects of acetylcholine, nicotine, DMPP, TMA and potassium chloride, the pathway blocked by cocaine is apparently not involved in the chain of events elicited by activation of the nicotinic acetylcholine-receptors or by potassium chloride. This is shown schematically in Fig. 6. Since cocaine is known to affect the permeability of membranes to certain ions, it is possible that stimulation of ganglia by nicotinic agents on the one hand, and by non-nicotinic agents on the other, involves different species of ions. In this regard, it is of interest that the effects of cocaine are similar to those of calcium ions (see section V.A.14).

13. Morphine and methadone. The intravenous injection of small doses of morphine (10—30 µg/kg) or of methadone (10—100 µg/kg) has little effect on the response of the ganglion to nicotine, DMPP, acetylcholine, potassium chloride or preganglionic stimulation, but it strongly antagonizes the ganglionic effects of histamine, pilocarpine, 5-HT, angiotensin and bradykinin (TRENDELENBURG, 1957b, 1966a), and of muscarine, McN-A-343 and AHR-602 (JONES, 1963). While the intraarterial injection of up to 2 mg of morphine failed to affect responses of the perfused superior cervical ganglion to preganglionic stimulation, to acetylcholine or to potassium chloride (HEBB and KONZETT, 1949), as little as 1 µg of morphine (i.a. to the non-perfused ganglion) blocked the effects of McN-A-343 for one hour (MURAYAMA and UNNA, 1963). The mode of action of morphine and methadone appears to resemble that of cocaine (see above).

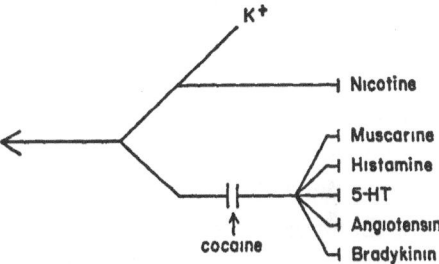

Fig 6 Schematic representation of possible site of action of cocaine in the superior cervical ganglion of the cat The lines indicate the postulated chains of events initiated by the action of various drugs on specific receptors of the ganglion cell, the final common pathway being the initiation of the propagated action potential conducted in the postganglionic axon. For details see text

14. Calcium chloride. The intraarterial injection of small doses of calcium chloride which do not affect the nicotinic response of the ganglion to acetylcholine, abolished the postganglionic discharge elicited by methacholine (TAKESHIGE and VOLLE, 1964c). The authors suggested that the ganglionic atropine-sensitive sites (*i.e.*, the muscarinic receptors) are very sensitive to block by nonspecific neuronal stabilizing agents like calcium chloride. Apparently, calcium ions equally affect the actions mediated through 5-HT-receptors of the ganglion (SMITH, 1966b). Hence, calcium seems to have an action very similar to that of cocaine (see above).

15. Atropine. Very small intravenous doses of atropine (1—10 µg/kg) abolish the ganglionic effects of the muscarinic agents without affecting those of histamine, 5-HT (TRENDELENBURG, 1954, 1956a), angiotensin or bradykinin (LEWIS and REIT, 1965), or of the nicotinic agents and potassium chloride. The antagonism of atropine to the ganglionic actions of pilocarpine is of the surmountable type (IORIO and McISAAC, 1966).

16. Antihistaminic agents. Pyrilamine (30 µg/kg i.v.) antagonizes the ganglionic effects of histamine without affecting those of pilocarpine and 5-HT (TRENDELENBURG, 1954, 1956a) or of bradykinin, while some antagonism of pyrilamine to angiotensin was observed (LEWIS and REIT, 1965). The antagonism of pyrilamine to histamine is specific and surmountable (IORIO and McISAAC, 1966). Diphenhydramine also antagonizes histamine but not McN-A-343 (MURAYAMA and UNNA, 1963).

17. Antagonists to 5-hydroxytryptamine. According to Gaddum and Pica-relli (1957) 5-HT causes a contraction of the isolated guinea-pig ileum by reacting with two types of receptors: the D-receptors of the longitudinal smooth muscle which are sensitive to one group of blocking agents (lysergic acid diethylamide-LSD; brom-LSD, phenoxybenzamine), and the M-receptors of the parasympathetic nervous elements which are sensitive to a different group of blocking agents (morphine, cocaine).

The antagonistic effects of morphine and cocaine to 5-HT indicate that the 5-HT-receptors of the superior cervical ganglion are of the M-type. The smooth muscle of the nictitating membrane, on the other hand, appears to have D-receptors (Thompson, 1958). None of the agents that block D-receptors has been found to antagonize ganglionic actions of 5-HT. 5-Hydroxy-3-indol-acetamidine blocks the effects of 5-HT on the superior cervical and on the inferior mesenteric ganglion of the cat without antagonizing DMPP and without blocking D-receptors (Gyermek, 1961). The evidence supports the view that ganglionic 5-HT-receptors are of the M-type.

18. The location of non-nicotinic receptors. Many of the observations reviewed in this section are compatible with the view that there might be two groups of ganglion cells, one that responds to nicotinic agents, and another responding to the non-nicotinic ganglion-stimulating substances. However, some observations are clearly contradictory to this concept. The ability of the non-nicotinic agents to enhance the ganglionic effects of nicotinic compounds, and the ability of nicotine, TMA, and DMPP to abolish the ganglionic effects of the non-nicotinic agents, seem to indicate that at least some of the ganglion cells have several types of receptors. This may not be true for all cells. Douglas and Poisner (1965) and Staszewska-Barczak and Vane (1965b) found that muscarinic agents and histamine cause a preferential release of epinephrine from the adrenal medulla of the cat, while splanchnic stimulation, nicotine and acetylcholine release both epinephrine and norepinephrine. Since there is some evidence for the presence of "epinephrine" and "norepinephrine" cells in the adrenal medulla, these observations may indicate that norepinephrine-containing cells have only nicotinic receptors, while the epinephrine-containing cells have both nicotinic and non-nicotinic receptors. At present, it is impossible to decide whether the cells of autonomic ganglia are similarly organized; the available evidence is not against such a view.

B. Actions on the perfused superior cervical ganglion

The perfused ganglion has only 1/10 to 1/100 the sensitivity of the non-perfused ganglion to histamine, pilocarpine and 5-HT (Trendelenburg, 1956b). Consequently, stimulation is rarely observed (Konzett, 1952). However, the facilitatory effects of the three agents are observed with regula   (Konzett, 1952; Trendelenburg, 1956b); there is not only facilitation of submaximal preganglionic stimulation but also enhancement of ganglionic responses to injected acetylcholine, nicotine, carbachol, choline and potassium chloride. A similar potentiation of the ganglionic effects of various drugs was observed after injections of muscarine (KONZETT and WASER, 1956).

These observations indicate that the perfused ganglion responds to non-nicotinic agents in the same way as the non-perfused ganglion does; however, the sensitivity to non-nicotinic agents is much lower in the former preparation than in the latter, although there are no pronounced differences in sensitivity to nicotine-like agents. It is of interest that similar findings were recently reported for the perfused and non-perfused adrenal medulla of the dog (VOGT, 1965).

C. Actions on other ganglia

Intraarterial injections of muscarine (GYERMEK et al., 1963), of pilocarpine and McN-A-343 (MURAYAMA and UNNA, 1963) stimulate the inferior mesenteric ganglion of the cat. The stimulant effect of 5-HT on this ganglion is specifically antagonized by 5-hydroxy-3-indolacetamidine, morphine and cocaine, while hexamethonium is a specific antagonist to DMPP (GYERMEK, 1961; GYERMEK and BINDLER, 1962a).

Intraarterial injections of 5-HT into the urinary bladder preparation of the dog cause a contraction with two components. The first is due to stimulation of ganglia; it is prevented by morphine, while hexamethonium, TEA, atropine and brom-LSD fail to affect it. The second component is not affected by morphine; it is prevented by brom-LSD (GYERMEK, 1960).

5-HT facilitates ganglionic transmission through the rat's stellate ganglion both *in situ* and in the organ bath (HERTZLER, 1961).

Drugs that stimulate the isolated guinea-pig ileum by an action on the intramural ganglion cells lose their effects after pretreatment of the intestine with botulinum toxin (AMBACHE and LESSIN, 1955). The effects of histamine and muscarine on this preparation cannot be of ganglionic origin, since pretreatment with botulinum toxin does not affect them. Pilocarpine likewise has a direct action on smooth muscle, since the response of the intestine to this agent is not antagonized by either cocaine or morphine in concentrations which reduce or abolish responses to nicotine (JONES, unpublished observations). VAN ROSSUM (1962) classified McN-A-343 as a ganglion-stimulating agent when testing its effects on the isolated guinea-pig ileum. However, in concentrations that clearly antagonize the effects of nicotine, morphine and cocaine fail to antagonize McN-A-343 (JONES, unpublished observations). SMITH (1966a) likewise failed to obtain any evidence for stimulation by McN-A-343 of cardiac or intestinal parasympathetic ganglia. For 5-HT, on the other hand, an action on the nervous structure of the gut has been proven (ROCHA E SILVA et al., 1953; ROBERTSON, 1953, 1954; GADDUM and PICARELLI, 1957).

With the classical "isolated guinea-pig ileum" movements are recorded of the longitudinal muscle only. The pharmacology of the circular muscle of the guinea-pig ileum is quite different. Both histamine and 5-HT act through cholinergic mechanisms, since their effects are abolished by atropine, morphine, procaine, botulinum toxin and hemicholinium. Specific histamine- and 5-HT-receptors appear to be involved, because pyrilamine antagonizes histamine but not 5-HT (Harry, 1963; Brownlee and Harry, 1963).

Intracellular recordings from sympathetic ganglion cells of the frog reveal that pilocarpine and McN-A-343 fail to cause either depolarization or stimulation of this preparation (Ginsborg, 1965). The only effect observed was a depression of ganglionic transmission with high concentrations of McN-A-343 (10^{-5} M) and pilocarpine (2×10^{-4} M).

D. Actions on the adrenal medulla of the cat

1. Sensitivity of the adrenal medulla to various agents. When intraarterial injections are made into the aorta (at the level of the superior mesenteric or coeliac artery), very small amounts of drugs cause stimulation of the adrenal medulla which is evident from the ensuing pressor response that is abolished by adrenalectomy. The following non-nicotinic agents have been found to be effective stimulants (with range of minimal effective doses in parentheses): pilocarpine (10—40 µg; Trendelenburg, 1954), McN-A-343 (10—30 µg; Lee and Trendelenburg, 1967), histamine (0.1—3.5 µg; Szczygielski, 1932; Trendelenburg, 1954), 5-hydroxytryptamine (1—50 µg; Reid, 1952; Lecomte, 1953), angiotensin (0.001—0.002 µg), bradykinin (0.05—1 µg; Feldberg and Lewis, 1964) and kallidin (Staszewska-Barczak and Vane, 1965a). Because of the pronounced vasodilatation produced by muscarine, the stimulant effect of this agent on the adrenal medulla cannot be demonstrated by recording changes in blood pressure; however, the supersensitive denervated nictitating membrane contracts when muscarine is injected to the adrenal medulla (Jones, unpublished observations). No stimulant effects were observed with up to 20 µg of elodoisin (Staszewska-Barczak and Vane, 1965a) and with up to 1 mg of AHR-602 (Jones, unpublished observations).

2. Preferential release of epinephrine. Fluorimetric determination of the catecholamines released into the perfusate of the cat adrenal gland showed that splanchnic stimulation, acetylcholine and nicotine released both epinephrine and norepinephrine; the latter amine contributed 55 to 61% of the total amount of catecholamines released. Muscarine and pilocarpine, on the other hand, caused a preferential release of epinephrine with only 4 to 16% of norepinephrine (Douglas and Poisner, 1965). By superfusing isolated organs with the venous effluent from the adrenal medulla, Staszewska-Barczak and Vane (1965b) found histamine to cause a preferential release of epinephrine. Results with other non-nicotinic agents have not been obtained. Nevertheless, it must be regarded as possible that only the epinephrine-containing cells of the adrenal medulla have receptors for the non-nicotinic agents.

3. The effect of chronic denervation. Chronically denervated preparations continue to respond to histamine (SIEHE, 1934) and to angiotensin and bradykinin (FELDBERG and LEWIS, 1965). Consequently, these agents must have a direct effect on the medullary cells. The effects of other agents on the denervated adrenal gland have not been studied.

4. The effect of non-depolarizing ganglion-blocking agents. Hexamethonium fails to block the stimulant effect of histamine and increases the pressor response to histamine as it increases pressor responses to epinephrine and norepinephrine (TRENDELENBURG, 1954). Hexamethonium also fails to block the effects of angiotensin and bradykinin (FELDBERG and LEWIS, 1965). The effects of hexamethonium (or of related drugs) on injections of other non-nicotinic drugs to the adrenal medulla have not been studied. However, results obtained with intravenous injections of these agents clearly indicate that non-depolarizing ganglion-blocking agents fail to antagonize the adrenal medullary effects of any of the non-nicotinic agents (see section V.E.2).

5. The effect of depolarizing ganglion-blocking agents. Intraarterial injections of histamine, pilocarpine or McN-A-343 become ineffective, when they are given immediately after the administration of ganglion-blocking doses of nicotine (SZCZYGIELSKI, 1932; TRENDELENBURG, 1954; LEE and TRENDELENBURG, 1967). While block by depolarization abolishes the effects of McN-A-343 on the adrenal medulla, this agent again becomes effective during the late, non-depolarizing phase of the block by nicotine at a time when the gland fails to respond to nicotine or DMPP (LEE and TRENDELENBURG, 1967). These observations are very similar to those made in experiments on the superior cervical ganglion (see section V.A.10 and 11), and this parallelism extends to the finding that responses to McN-A-343 are potentiated during the late phase of the block by nicotine, and that hexamethonium abolishes this potentiation without abolishing the stimulant effect of McN-A-343.

6. The effects of cocaine, morphine and methadone. In concentrations 20 times those required to antagonize the ganglionic effects of non-nicotinic ganglion-stimulating agents, cocaine and morphine fail to block the effects of histamine and pilocarpine on the adrenal medulla (TRENDELENBURG, 1954, 1961b). An analysis of the pressor response to the intravenous injection of the non-nicotinic agents (see section V.E.8) indicates a) that methadone likewise fails to antagonize adrenal medullary stimulation by histamine and pilocarpine, and b) that cocaine, morphine and methadone are unable to antagonize the adrenal medullary effects of other non-nicotinic agents as well.

It is of interest that cocaine, morphine and methadone fail to antagonize the effects of non-nicotinic agents on the adrenal medulla of the cat, while they are potent antagonists to the effects of the non-nicotinic agents on the superior cervical ganglion of the same species. This is the only qualitative difference between adrenal glands and sympathetic ganglia encountered so far in various studies with non-nicotinic agents.

7. *The effect of specific blocking agents.* Atropine antagonizes the effects of pilocarpine without interfering with those of histamine, while the reverse is true for the antihistaminic agent, pyrilamine (Trendelenburg, 1954).

8. *The adrenal medulla of other species.* While hexamethonium or section of the spinal cord has no effect on the stimulant effect of histamine on the adrenal medulla of the cat, the two procedures were found to antagonize the effects of histamine on the adrenal gland of the dog (Staszewska-Barczak and Vane, 1965 b). It is evident that one and the same agent may have a direct action on the adrenal medulla of one species (cat), while acting through reflexes in another (dog).

Stimulation of the adrenal gland by intraarterial injections of kallidin, bradykinin, angiotensin and elodoisin was observed in dogs; the response of the adrenal medulla to continuous infusions of angiotensin ceased within 5 to 10 min in spite of the continuing infusion (Staszewska-Barczak and Vane, 1965 a). Hence, it is unlikely that any adrenal medullary effects of angiotensin are involved in elevations of the blood pressure that are of long duration. While these observations were all made with non-perfused adrenal glands of dogs, Vogt (1965) found only a very poor stimulant effect of angiotensin and bradykinin in perfused preparations

The bronchoconstrictor effect of bradykinin in the guinea pig is increased by either adrenalectomy or by the administration of propranolol, a β-receptor blocking agent. Apparently, bradykinin stimulates the adrenal medulla of this species (Collier et al., 1965).

E. Pressor effects of non-nicotinic agents

In the spinal cat, most non-nicotinic ganglion-stimulating agents cause a rise in blood pressure which may or may not be preceded by an initial fall. For an analysis of drug-induced changes in this pressor response, the following three considerations are of importance:

1. Pressor responses mediated through the peripheral sympathetic system of the spinal cat can be due to stimulation of the adrenal medulla, to general stimulation of sympathetic ganglia or to both. The adrenal medullary component can be excluded by acute adrenalectomy. The ganglionic component can be excluded by two groups of drugs: a) those which prevent the liberation of norepinephrine from adrenergic nerve terminals without interfering with the liberation of catecholamines from the adrenal medulla (*e.g.*, xylocholine and bretylium); and b) those which prevent stimulation of ganglia by the non-nicotinic agents without antagonizing the adrenal medullary stimulation by these agents (*e g.*, cocaine, morphine and methadone).

2. Most of the non-nicotinic ganglion-stimulating agents have depressor effects unrelated to their ganglion-stimulating action. Hence, at any given time after the intravenous injection of one of these agents, the blood pressure is under the influence of two opposite effects: vasodilatation due to a direct effect of these agents on vascular smooth muscle and vasoconstriction due to the release of catecholamines as a consequence of stimulation of ganglia or adrenal medulla or both. Moreover, the time course of the two opposite effects may well differ. It is clearly impossible to present a rigorously quantitative analysis of the pressor effects of such an agent. For instance, it is inevitable that some of the pressor effects of such an agent are masked by the fall in blood pressure induced by its direct action on vascular smooth muscle; indeed, in the case of a very strong vasodilator (*e.g*, muscarine) it is impossible to obtain a rise of blood pressure above the pre-injection level in spite of stimulation of the adrenal medulla by this agent (see below). Nevertheless, it is quite possible to obtain semi-quantitative information on the action of drugs on pressor responses to non-nicotinic drugs.

3. Since the pressor responses discussed in this section are due to the release of norepinephrine, epinephrine or both, drugs which increase or decrease the sensitivity to these amines must increase or decrease pressor responses to ganglion-stimulating agents without necessarily influencing the amounts of catecholamines released. For instance, cocaine and the non-depolarizing ganglion-blocking drugs increase the sensitivity of the cardiovascular system to catecholamines, while all α-receptor-blocking agents reduce it. Hence, changes in pressor responses to non-nicotinic agents are only meaningful when they are compared with changes in pressor responses to about equieffective doses of catecholamines.

If these experimental limitations are kept in mind, the analysis of the pressor action of non-nicotinic agents provides a clear picture of their action in the spinal cat. For quick reference, the most important results are summarized in Table 4.

1. Adrenal medullary versus general ganglionic stimulation. Table 5 presents the intraarterial doses of four agents which are usually required to produce stimulation of the superior cervical ganglion and the adrenal medulla, respectively. If it is assumed that the sensitivity of the superior cervical ganglion to these agents is representative for the sensitivity of most, or all, sympathetic ganglia, agents with a high ratio of effective doses (last column, Table 5) should cause pressor responses mainly through adrenal medullary stimulation, when injected intravenously. A low ratio, on the other hand, should be indicative of an agent which causes pressor responses mainly by general stimulation of sympathetic ganglia. Experiments with procedures and agents that eliminate one or the other component of the pressor response, confirm that intravenous

Table 4. *Schematic representation of the effects of various drugs on the pressor response of the spinal cat to various ganglion-stimulating agents*

		Hist-amine	McN-A-343	AHR-602	Pilo-carpine	Nicotine DMPP	Sensiti-vity to norepi-nephrine
Adrenalectomy		↓↓↓	↓↓	0	↓	↓	0
Non-depolarizing							
ganglion-blocking agents	intact	↑	↑	↑	↑	abol.	↑
	adr	↑	↑	↑	↑	abol.	↑
Nicotine	intact						
depolarizing block		abol.	abol.	abol.	abol	abol.	0
non-depolarizing block		↑	↑	↑	↑	abol.	0
followed by hexamethonium		return to normal				abol.	0
Cocaine	intact	↑	↓↓	↓↓↓	↓↓	↑	↑
	adr.	abol.	abol.	abol.	abol.	↑	↑
Morphine, methadone	intact	0	↓	↓↓↓	↓↓	0	0
	adr	abol	abol	abol.	abol.	0	0
Xylocholine	intact	↓	↓↓↓	↓↓↓	↓↓↓	0	0
	adr.	abol	abol	abol.	abol.	abol.	0
α-blocking agents	intact} adr.	abol	abol.	abol.	abol	abol.	↓↓↓

Arrows indicate drug-induced increases or decreases of pressor responses to ganglion-stimulating agents. The number of arrows roughly indicates magnitude of change.

abol full abolition of pressor response, 0 no change in effect of ganglion-stimulating agent; intact· spinal cat; adr.: adrenalectomized spinal cat.

injections of histamine act mainly on the adrenal medulla, while intravenous injections of pilocarpine or AHR-602 act mainly on ganglia.

In order to study that component of the pressor response which is due to general stimulation of sympathetic ganglia, many of the experiments to be discussed below were performed both in normal and in adrenalectomized spinal cats. In addition, many of the experiments were conducted in preparations which had received large doses of non-depolarizing ganglion-blocking agents, because these agents fail to abolish the ganglionic effects of non-nicotinic drugs, and, more importantly, because they increase pressor responses to catecholamines. In other words, they magnify pressor responses which other-

Table 5 *Comparison of the sensitivity of the superior cervical ganglion (SCG) and of the sensitivity of the adrenal medulla (ADR) of the cat to some non-nicotinic agents*

Agent	Intraarterial dose required for stimulation of SCG	Intraarterial dose required for stimulation of ADR	Ratio of doses (SCG/ADR)
Histamine	2—55	0.1—3.5	about 20
McN-A-343	1—10	10—30	3—10
Pilocarpine	25—40	10—40	about 1
AHR-602	5—1000	>1000	<1

wise might be totally masked by the vasodilator effects of histamine and of the muscarinic agents.

2. *Non-depolarizing ganglion-blocking agents.* Hexamethonium and TEA increase pressor responses to histamine before and after adrenalectomy (SLATER and DRESEL, 1952), and this increase is fully accounted for by the increased pressor responses to epinephrine and norepinephrine (TRENDELEN-BURG, 1955). This also applies to the three muscarinic agents, pilocarpine, McN-A-343, AHR-602 (ROOT, 1951; TRENDELENBURG, 1955, 1961 b; JONES et al., 1963; ROSZKOWSKI, 1961; FRANKO et al., 1963). Chlorisondamine was found to potentiate pressor responses to pilocarpine and McN-A-343 (LEVY and AHLQUIST, 1962), while mecamylamine potentiated pressor responses to AHR-602 (FRANKO et al., 1963). Apparently, all non-depolarizing ganglion-blocking agents have qualitatively similar actions.

3. *Depolarizing ganglion-blocking agents.* The intravenous injection of nicotine in amounts which are too small to cause block of ganglia (0.6 mg/kg in 3 doses within 5 min) increases the pressor response to pilocarpine presumably by producing an amount of ganglionic depolarization which increases rather than decreases the ganglion-stimulating effect of this drug (TRENDELEN-BURG, 1961 b; confirming KOPPANYI, 1939).

The administration of full ganglion-blocking doses of nicotine (12 mg/kg, i.v., in divided doses within 10 min) abolishes pressor responses to subsequent injections of histamine, and pilocarpine (TRENDELENBURG, 1955, 1961 b) and of McN-A-343 and AHR-602 (JONES et al., 1963; FRANKO et al., 1963). TMA, administered in the same way, also abolishes pressor responses to these four compounds as well as to a nicotinic agent, DMPP.

Ganglion-blocking doses apparently block the effects of all non-nicotinic agents on sympathetic ganglia and on the adrenal medulla.

4. *Effects of non-nicotinic agents during the late, non-depolarizing phase of the block by nicotine.* If, after full ganglion-block by nicotine is achieved, further injections of nicotine are given intermittently for 20 to 30 min, injections of histamine, pilocarpine, McN-A-343 and AHR-602 cause large pressor responses at a time when the effects of nicotine and DMPP are abolished (TRENDELEN-BURG, 1955, 1961 b; JONES et al., 1963). These responses are larger than responses observed at that time without any intervening injections of nicotine. A subsequent injection of 5 mg/kg of hexamethonium depresses the increased responses to the level to be expected at that time without any injections of nicotine and hexamethonium. It is not possible to produce this phenomenon of reappearance of pressor responses to non-nicotinic agents with TMA. A second series of injections of TMA has the same blocking action as the first series (TRENDELENBURG, 1961 b).

These observations are very similar to those made on the superior cervical ganglion (see V.A.10 and 11). They emphasize that the findings obtained with one ganglion are indeed representative for the pharmacological behavior of the majority of sympathetic ganglia.

5. Adrenalectomy. According to the differences in sensitivity of ganglia and of the adrenal medulla to non-nicotinic agents (Table 3), exclusion of the adrenals from the circulation diminishes pressor responses in the following descending order: histamine > McN-A-343 > pilocarpine > AHR-602 (Tren-delenburg, 1955; Jones *et al.*, 1963). Intravenous injections of nicotine and DMPP act predominantly on the adrenal medulla (Jones *et al.*, 1963).

6. Xylocholine and bretylium. These agents prevent the release of norepi-nephrine from postganglionic fibers without affecting the release of catechol-amines from the adrenal medulla (Exley, 1957; Boura and Green, 1959). Xylocholine (10 mg/kg i.v.) only diminished the pressor response to histamine (Trendelenburg, 1961 b) but abolished those to pilocarpine, McN-A-343 and AHR-602 (Jones *et al.*, 1963). After adrenalectomy, however, xylocholine abolished pressor responses to *all* agents. Bretylium is effective against the pressor responses to McN-A-343 and AHR-602 (Roszkowski, 1961; Franko *et al.*, 1963). Because of its bretylium-like effects, guanethidine is also able to abolish pressor responses to pilocarpine, McN-A-343 (Levy and Ahlquist, 1962) and AHR-602 (Franko *et al.*, 1963).

7. Pretreatment with reserpine. Pretreatment with large doses that affect not only the adrenergic nerve endings but also the adrenal medulla, prevents the appearance of pressor responses to histamine and pilocarpine (Trendelen-burg, 1961b), McN-A-343 (Roszkowski, 1961) and AHR-602 (Franko *et al.*, 1963).

8. Cocaine, morphine and methadone. Since these agents antagonize the effects of non-nicotinic agents on sympathetic ganglia without interfering with their action on the adrenal medulla, their effects against pressor responses are virtually identical with those observed with xylocholine (Trendelenburg, 1961 b; Jones *et al.*, 1963; Franko *et al.*, 1963). The parallelism extends to observations after adrenalectomy (see above).

9. Sympatholytic agents. All of the following alpha-receptor-blocking, sympatholytic agents have been found to antagonize pressor responses to one or the other of the non-nicotinic agents: ergotamine (Burn and Dale, 1926; Hoet, 1928), ergotoxin (Root, 1951), Dibenamine (Root, 1951; Slater and Dresel, 1952; Franko *et al.*, 1963), dibozane (Roszkowski, 1961).

According to Smith (1966a), the β-receptor-blocking agent pronethalol antagonizes pressor responses to McN-A-343. It is not clear whether this effect is related to the cocaine-like local anesthetic properties of this compound or to block of the β-receptors.

10. Pressor responses to angiotensin. Since this polypeptide is well known to have a direct vasoconstrictor effect on vascular smooth muscle, the possible participation of adrenergic mechanisms in the pressor response to anigotensin has not been studied systematically. It is likely that stimulation of the adrenal medulla, of sympathetic ganglion cells, or both, contributes to the pressor response to angiotensin observed in the spinal cat, since BENELLI et al. (1964) reported that the pressor response is reduced immediately after a ganglion-blocking dose of nicotine. After subsequent injections of nicotine, however, the pressor response to angiotensin is increased. Thereafter, hexamethonium brings the pressor response to angiotensin back to the control level. These observations are not unlike those obtained with the other non-nicotinic ganglion-stimulating agents (JONES et al., 1963).

F. Other sites of action of non-nicotinic agents

KONZETT and ROTHLIN (1953) drew attention to the fact that agents which act on ganglia frequently also act on the chemoreceptors of the carotid body, on various nerve endings and possibly also on the neuromuscular junction. There is no systematic study which covers this topic, but it might be of interest to summarize some of the isolated and sometimes incomplete reports relevant to KONZETT and ROTHLIN's postulate.

Muscarinic agents. In the cat, pilocarpine has a weak stimulant action on the chemoreceptors of the carotid sinus (VON EULER and DOMEIJ, 1945), and in the "atropinized" dog, methacholine was found to have a similar action (DE WISPELAERE, 1937). However, in the latter report the dose of atropine is not specified, and records of the blood pressure show that methacholine was used in doses that caused a fall in blood pressure in spite of the pretreatment with atropine. Consequently, it is possible though by no means established, that muscarinic agents stimulate the chemoreceptors. On mammalian skeletal muscle, both AHR-602 (FRANKO et al., 1963) and McN-A-343 (ROSZKOWSKI, 1961) appear to have little or no effect. On the frog rectus, on the other hand, McN-A-343 was equipotent with acetylcholine (ROSZKOWSKI, 1961), while pilocarpine (VON EULER and DOMEIJ, 1945) and AHR-602 (FRANKO et al., 1963) had little effect.

Histamine. Histamine potentiates the response of the tibialis muscle of the cat to acetylcholine, and has some stimulant effects on the denervated muscle (MACMILLAN, 1956). Curarelike effects of histamine have been reported (MACMILLAN, 1956; SCHENK and ANDERSON, 1958; BOVET-NITTI et al., 1964). Some of these actions have been ascribed to the release of potassium by histamine (MACMILLAN, 1956; HANNA et al., 1959) Histamine appears to be able to stimulate sympathetic centers in the CNS (TRENDELENBURG, 1957d).

5-Hydroxytryptamine. On the chemoreceptors of the carotid sinus, 5-HT is more potent than lobeline and DMPP, and its effects are not abolished by TEA or hexamethonium (GINZEL and KOTTEGODA, 1954; McCUBBIN et al., 1956). Sensory receptors that respond to phenyldiguanide also respond to

5-HT (DOUGLAS and RITCHIE, 1957; FASTIER et al., 1959). 5-HT has facilitatory effects (of presumably presynaptic origin) on the neuromuscular junction of the crayfish (DUDEL, 1965).

G. Agents pharmacologically or chemically related to non-nicotinic substances

1. Muscarinic agents. Of the eight stereoisomers of muscarine, only l-muscarine is active on smooth muscle (GYERMEK and UNNA, 1958) and on the inferior mesenteric ganglion of the cat (GYERMEK et al., 1963). The isomers of muscarone, on the other hand, are about equipotent, and their ganglionic actions are not abolished by atropine (HERR and GYERMEK, 1960); they are abolished by TEA and have to be classified as "nicotine-like". These observations indicate that muscarinic receptors of ganglion cells are very similar to those of smooth muscle, heart muscle and gland cells. This view is also supported by the fact that the other muscarinic agents discussed in this section (pilocarpine, McN-A-343 and AHR-602) are known to exert muscarinic effects on smooth muscle, heart muscle and gland cells (ROSZKOWSKI, 1961; FRANKO et al., 1963). JONES (1963) drew attention to the fact that all four agents have a molecular structure that may be expected to fit the muscarinic receptor described by WASER (1961).

Although there are several arguments in favor of the view that all muscarinic receptors, whether ganglionic or located in other cells, are very similar, some observations indicate that there may be differences. For instance, there is no strict parallelism between the relative potencies of muscarinic agents in their effects on the superior cervical ganglion and in their depressor effects in the same species: while l-muscarine is a much more potent vasodilator than McN-A-343, the sensitivity of the ganglion is about equal to both agents. Such discrepancies in relative potencies may be due to small differences in the configuration of the muscarinic receptors in ganglia and in blood vessels, respectively. In this connection, it should be remembered that there are nicotinic receptors in ganglia and at the neuromuscular junction; there is good pharmacological evidence that they cannot be regarded as identical, since they are blocked by hexamethonium and d-tubocurarine, respectively.

Other agents which have ganglionic effects that are blocked by atropine (and therefore are presumably "muscarinic") are acetylcholine and methacholine (see section VI), various anticholinesterases (see section VIII) and oxotremorine (DE GROAT and VOLLE, 1963). As with the other muscarinic agents, the effects of the last compound were not abolished by previous denervation of the superior cervical ganglion; they were increased (in innervated ganglia) after repetitive preganglionic stimulation; and they were not antagonized by hexamethonium.

2. Histamine. Dimethylhistamine (substituted on the nitrogen of the side chain) causes contraction of the frog rectus muscle and of strips of the body wall of the leech; the corresponding quaternary trimethylhistamine also stimulates the perfused superior cervical ganglion of the cat as well as the denervated gastrocnemius of the cat (VARTIAINEN, 1935). Since at the time of these studies neither antihistaminic nor non-depolarizing ganglion-blocking agents were available, it is uncertain whether these two agents are nicotine- or histamine-like.

Two derivatives of histamine have been found to stimulate the adrenal medulla of the cat: 3-beta-aminoethyl-1,2,4-triazole has 1/40, and 2-beta-aminoethylpyridine has 1/100 the potency of histamine when injected intra-arterially (LINDELL *et al.*, 1960).

The histamine-releasing compound 48/80 blocks transmission through the perfused superior cervical ganglion of the cat (GERTNER, 1955). This effect is not due to a decrease in the release of acetylcholine in response to preganglionic nerve stimulation but to a decrease in sensitivity to acetylcholine. After an injection of compound 48/80, the effluent contained histamine, and repeated injections of the compound then released smaller amounts of histamine. GERT-NER pointed out that there was no relation of the block produced by compound 48/80 to the amounts of histamine released. However, since repeated injections of compound 48/80 caused increasing degrees of block, there might be a relation of degree of block and the degree of depletion of endogenous histamine. While it is tempting to speculate that endogenous histamine might have a facilitatory action in normal ganglia, there is not enough evidence available to consider such a possibility seriously.

In a recent study, GERTNER (1965) found a normal histamine content of the superior cervical ganglion of 4 μg/g, 25% of which was lost from the ganglion after injections of compound 48/80, acetylcholine, an unspecified reserpine analogue, or an inhibitor of amino acid decarboxylase (NSD 1055). Pre-ganglionic denervation, on the other hand, reduced the histamine content of the ganglion by 86%. There appear to be different pools of histamine in the ganglion the physiological or pharmacological importance of which remains to be determined.

3. 5-Hydroxytryptamine. GYERMEK (1964) studied four guanidine derivatives known to have stimulant effects on sensory nerve endings: phenyldiguanide, benzylguanidine, 2-naphthylguanidine and 2-anthrylguanidine. These guanidines were ineffective on the isolated rat uterus which is known to respond well to those actions of 5-HT that are mediated through the so-called D-receptors. On the pelvic nerve-bladder preparation of the dog, on the other hand, the guanidines caused effects that were similar to the ganglionic effects of 5-HT (mediated through M-receptors); brom-LSD (which blocks D-receptors) failed to antagonize their actions. Direct proof for ganglion-stimulating actions of the guanidines was obtained on the inferior mesenteric ganglion of the cat. The potency of the guanidines was between 5 times and 1/5 that of 5-HT. The guanidines appear to react with the ganglionic 5-HT receptors (M-receptors), because 1. they all specifically antagonize 5-HT, and because 2. cross tachyphylaxis was observed to develop between 5-HT and phenyldiguanide, and because 3. the guanidines (like 5-HT) exert their ganglion-stimulating effects even after hexamethonium or chlorisondamine (GYERMEK, 1964).

A series of 22 indol alkylamines and amidines was tested on the inferior mesenteric ganglion of the cat and 8 of these also on the isolated rat uterus (GYERMEK and BINDLER, 1962b). There was no parallelism between the ganglionic potency of these agents and their potency on the isolated rat uterus (which is known to have only D-receptors). Like 5-HT, N-methyl-5-hydroxy-tryptamine was blocked on the ganglion specifically by 5-hydroxy-3-indole-acetamine; however this is not true for other agents that are chemically related to 5-HT and that stimulate the inferior mesenteric ganglion: bufotenine, bufotenidine, N,N-dimethyl-tryptamine and N,N,N-trimethyl-tryptamine were all more easily blocked by hexamethonium than by 5-hydroxy-3-indoleacet-amidine. These observations suggest that some agents appear to combine "nicotinic" with "5-HT-like" ganglion-stimulating properties. This conclusion is supported by the observation that bufotenine is a much less specific blocking agent (when tested against 5-HT and DMPP) than 5-hydroxy-3-indoleacet-amidine. It is of interest that tryptamine, 6-hydroxytryptamine and 4-hydroxy-tryptamine were devoid of ganglion-stimulating properties.

Sympathetic ganglia do not appear to have measurable amounts of 5-HT (GADDUM and PAASONEN, 1955) while they are able to synthesize the amine from its precursor, 5-hydroxytryptophan (GADDUM and GIARMAN, 1956). However, ganglia are now known to be able to synthesize norepinephrine, and 5-hydroxytryptophan-decarboxylase and DOPA-decarboxylase are now regarded as one and the same enzyme. Consequently, the reported conversion of 5-hydroxytryptophan to 5-HT cannot be regarded as any evidence in favor of a physiological role of 5-hydroxytryptamine in sympathetic ganglia. This also applies to observations on perfused ganglia which were found to have no 5-HT in the effluent, neither under resting conditions nor during preganglionic stimulation (GERTNER et al., 1959). The amine was detected in the effluent only when the precursor (5-hydroxytryptophan) was perfused through the ganglion together with an inhibitor of monoamine oxidase, the enzyme responsible for the metabolism of the amine. It is concluded that, as with histamine, there is no reliable evidence in favor of a physiological role of 5-HT in sympathetic ganglia.

4. Polypeptides. For 13 angiotensin analogues, FELDBERG and LEWIS (1965) found a striking similarity between their potency in stimulating the adrenal medulla of the cat and their pressor potency in nephrectomized rats. One might speculate that angiotensin receptors of neural tissues do not differ from those of smooth muscle. Stimulation of the adrenal medulla was also observed after injections of renin, the enzyme which forms angiotensin.

Two bradykinin derivatives had some stimulant action on the cat's adrenal medulla: lysyl-bradykinin had 1/2 the potency of bradykinin, while the octapeptide lacking the single proline had only 1/400 the potency of bradykinin (FELDBERG and LEWIS, 1964).

Of other polypeptides, vasopressin, oxytocin and substance P were found to have no action on the cat's adrenal medulla (FELDBERG and LEWIS, 1964). Substance P also failed to stimulate the superior cervical ganglion of the cat (BELESLIN et al., 1960); however, substance P appears to have some ganglionic actions, since it potentiates responses of the ganglion to submaximal preganglionic stimulation and to acetylcholine (BELESLIN et al., 1960), while it was found to potentiate the effect of nicotine on the isolated guinea-pig ileum (BELESLIN and VARAGIĆ, 1960).

VI. Acetylcholine

The pharmacology of acetylcholine is discussed in a separate section, because there is now good evidence that it can exert effects on both the nicotinic and the muscarinic receptors of ganglia depending on the conditions of the experiment. Several experimental conditions have been described under which muscarinic effects of acetylcholine on ganglia can be observed; however, at present it is not clear whether these experimental conditions enable acetylcholine to exert muscarinic ganglionic effects which it normally lacks, or whether we simply unmask the normal muscarinic effects of acetylcholine by separating them from the nicotinic ones. Or in other words, while our knowledge of the pharmacology of the muscarinic effects of acetylcholine on ganglia is rapidly increasing, we know very little about the possible physiological importance of these actions.

There is no doubt that injected acetylcholine normally exerts nicotinic effects on various ganglia, i.e., its ganglionic effects are normally blocked by hexamethonium (and related agents) and resistant to small doses of atropine.

Some antagonistic effects of small doses of atropine have been reported by KONZETT and ROTHLIN (1949), but since in this case the injections of atropine (0.1 to 5 μg) were made into the perfused superior cervical ganglion of the cat, it is conceivable that in spite of the small dose the actual concentration of atropine in the ganglion was quite high at least for a certain period of time. That high concentrations of atropine can cause some block of ganglia is well known (BAINBRIDGE and BROWN, 1960). In non-perfused superior cervical ganglia, intravenous injections of small doses of atropine (100 μg/kg) fail to reduce the effects of intraarterial injections of acetylcholine to the ganglion, although they reduce or even abolish the effects of injections of acetylcholine to the nictitating membrane (Trendelenburg, 1962).

While there is good evidence for a predominantly nicotinic effect of acetylcholine on normal ganglia, muscarinic effects of acetylcholine on ganglia have been observed under certain experimental conditions. Probably the first report of such an action was that of Feldberg *et al.* (1934) who found after repeated

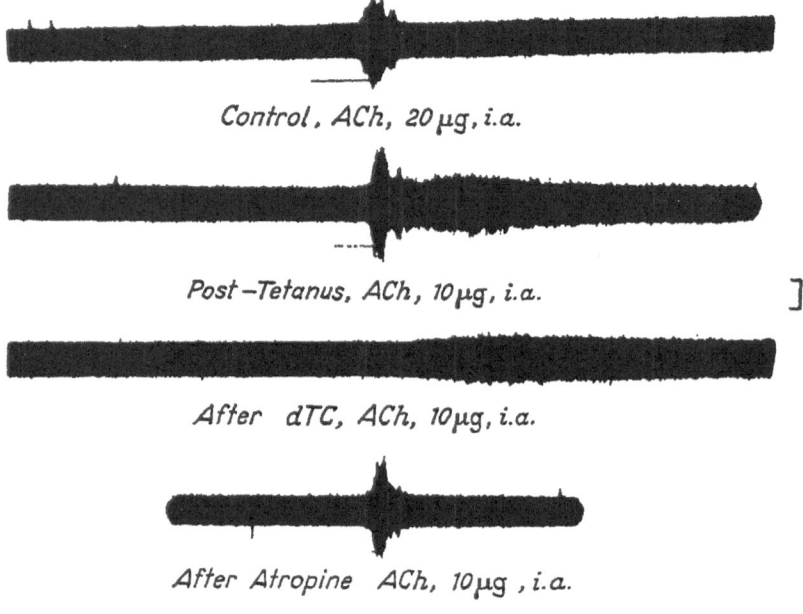

Control, ACh, 20 μg, i.a.

Post-Tetanus, ACh, 10 μg, i.a.

After dTC, ACh, 10 μg, i.a.

After Atropine ACh, 10 μg, i.a.

Fig. 7. Bimodal response to acetylcholine of the cat's superior cervical ganglion after repetitive preganglionic stimulation. Acetylcholine (ACh) was injected intraarterially to the ganglion where indicated by horizontal bar. Shown is the activity in the postganglionic fibers. Control: response to ACh prior to preganglionic stimulation. Post-tetanus: Response to a smaller dose of acetylcholine injected after 10 sec of preganglionic stimulation (60 shocks/sec, supramaximal strength). After dTC: Response to ACh after the intraarterial injection of 0.4 mg of *d*-tubocurarine. After atropine: response to ACh after the blocking action of *d*-tubocurarine had worn off and after an intraarterial injection of 1 μg of atropine. Vertical and horizontal calibrations are 10 μV and 1 sec, respectively. Note bimodal response after preganglionic stimulation, the first phase of which is sensitive to *d*-tubocurarine, while the second is sensitive to atropine. (From Takeshige and Volle, 1962; with permission of the J. Pharmac. exp. Ther.)

injections of nicotine (which abolished responses to nicotine) that acetylcholine was still able to stimulate the adrenal medulla; moreover, this remaining effect of acetylcholine was then abolished by the intravenous injection of small doses of atropine.

Volle and his coworkers studied the action of acetylcholine on the superior cervical ganglion of the cat by recording the activity of the postganglionic nerve. Without any pretreatment, the intraarterial injection of acetylcholine caused a short burst of postganglionic activity which was suppressed by hexamethonium and similar agents (Fig. 7). In other words, this short and early postganglionic discharge represented a nicotinic effect of acetylcholine. However, an additional late, low-amplitude discharge ("late firing") was observed in response to injections of acetylcholine under the following conditions: after a preceding period of preganglionic stimulation (Fig. 7) (TAKESHIGE and VOLLE, 1962), after premedication with physostigmine (TAKESHIGE and VOLLE, 1962) or neostigmine (TAKESHIGE and VOLLE, 1963a), and after chronic denervation of the ganglion (TAKESHIGE and VOLLE, 1963b). In all cases the "late firing" was not affected by injections of hexamethonium which abolished the "early firing", while small doses of atropine abolished the "late firing" without affecting the nicotinic response of the ganglion. It is of interest that at least two of these procedures (i.e., preganglionic stimulation and denervation) are known to enhance the effects of muscarinic agents on the superior cervical ganglion (JONES, 1963; TRENDELENBURG and JONES, 1965). Hence, it is likely that the "late firing" is observed in response to injections of acetylcholine whenever the sensitivity of the ganglion to muscarinic agents in general and to the muscarinic effects of acetylcholine in particular is increased.

Already in 1959 FEHER and BOKRI (1959a) provided evidence for the presence of two types of acetylcholine receptors in the superior cervical ganglion which they termed "innervated" (or nicotinic in the terminology used in this review) and "free" (or muscarinic) receptors, respectively. Subsequent observations were interpreted as indicating that the number of "free receptors" might be increased after denervation of the ganglion (BOKRI et al., 1963).

Further evidence for muscarinic effects of acetylcholine on the superior cervical ganglion of the cat stems from experiments with nicotine. During the depolarizing phase of the block by nicotine, the ganglion fails to respond to any drugs, and the effects of intraarterial injections of acetylcholine are abolished. In atropine-pretreated preparations (1 mg/kg i.v.), there is no response to acetylcholine during the late, non-depolarizing phase of the block by nicotine (TRENDELENBURG, 1957c). At that time the crucial importance of the preceding injection of atropine was not recognized. However, in recent experiments it has become evident that there are ganglionic responses to injections of acetylcholine during the late, non-depolarizing phase of the block by nicotine provided no atropine is injected (TRENDELENBURG, 1966b). The muscarinic nature of these responses is also clear from the following additional observations. 1. The latency period between the injection and the onset of the contraction of the nictitating membrane is as long as that observed with muscarinic agents and unlike the nearly immediate response to nicotinic agents. 2. The development of the contraction of the nictitating membrane is as slow and prolonged as that elicited by muscarinic agents and unlike the brisk response to nicotinic agents. 3. The slope of the dose-response curve of the muscarinic effects of acetylcholine is as flat as that for other muscarinic agents, while prior to any injections of nicotine, the dose-response curve for acetylcholine is as steep as that for nicotinic agents. 4. Morphine reduces the effects of acetylcholine during the late phase of the block by nicotine although it has no antagonistic effects against nicotinic agents. 5. The muscarinic effects of acetylcholine (*i.e.*, those observed during the late phase of the block by nicotine) are facilitated by a preceding period of preganglionic stimulation while the nicotinic effects of acetylcholine (*i.e.*, those observed prior to any injections of nicotine) are not. 6. During the late phase of the block by nicotine, the muscarinic responses to acetylcholine are potentiated as are the responses of the ganglion to other non-nicotinic agents, the intravenous injection of hexamethonium abolishes the facilitatory effects of nicotine. 7. All procedures which affect responses to a typical muscarinic agent (McN-A-343) affect the ganglionic responses to acetyl-choline qualitatively and quantitatively in the same manner (Trendelen-burg, 1966b). In summary, pure muscarinic effects of acetylcholine are observed in the superior cervical ganglion during the late, non-depolarizing phase of the block by nicotine, *i.e.*, when a) the nicotinic receptors are blocked insur mountably, and b) the effects of muscarinic agents are potentiated by nicotine.

The studies discussed in this section dealt mainly with a pharmacological analysis of the stimulant effects of acetylcholine injected to the ganglion. Recording of the surface potential of the ganglion revealed that acetylcholine (in doses of 2.5 to 20 µg injected intraarterially) caused a triphasic response: an initial depolarization was followed by a period of hyperpolarization with a subsequent late phase of depolarization. Smaller doses of acetylcholine tended to produce only the delayed depolarization, larger doses caused only an immediate (early) depolarization. The postganglionic discharge ("early firing") was observed with doses of 20 µg or more of acetylcholine, and it coincided with the early depolarization. Hexamethonium antagonized both the early depolarization and the early firing. Atropine, on the other hand, antagonized the hyperpolarization as well as the delayed phase of depolarization. Atropine also antagonized a period of block of transmission which was observed during hyper-polarization of the ganglion (Takeshige and Volle, 1964b). These observations indicate that there are three types of receptors with which acetylcholine can combine: one inhibitory receptor which is responsible for hyperpolarization accompanied by a block of transmission and which is sensitive to atropine (*i.e.*, is muscarinic), and two excitatory receptors which are responsible for depolarization; of the latter, one is sensitive to hexamethonium, the other to atropine. When the effects of acetylcholine were compared with those of the muscarinic agent methacholine, it was found that the latter compound caused first a period of hyperpolarization and then one of delayed depolarization; block of transmission coincided with the hyperpolarization of the ganglion, and atropine was found to antagonize *all* effects of methacholine. In other words, the effects of methacholine were similar to those of acetylcholine except that there was no initial period of depolarization, *i.e.*, there was no nicotinic effect of this drug.

The observations of TAKESHIGE and VOLLE (1964b) with acetylcholine do not necessarily contradict the concept of PATON and PERRY (1953) that acetylcholine can cause block of transmission by depolarizing ganglion cells. The recording of the surface potential of the ganglion does not permit any conclusions about the condition of single ganglion cells or parts thereof. Furthermore, it is evident from this short account that the response of the ganglion is dependent on the dose of acetylcholine. However, it is permissible to conclude from these observations that both acetylcholine and the muscarinic agent methacholine are able to produce a block of transmission which coincides with hyperpolarization of the ganglion cells and which appears to be sensitive to atropine. Hence, block of transmission and depolarization are not necessarily linked.

In isolated superior cervical ganglia of the rabbit exposed to d-tubocurarine, ECCLES and LIBET (1961) recorded the synaptic potential elicited by single preganglionic volleys. The compound potential of the ganglion consisted of an initial negative potential (N-wave), a subsequent slow positive potential (P-wave) and a late negative potential (LN-wave). Treatment of the preparation with botulinum toxin depressed all three waves. Hexamethonium antagonized only the N-wave, while atropine depressed both the P- and the LN-wave. These results are identical with those reported later by TAKESHIGE and VOLLE (1964b) with injections of acetylcholine. ECCLES and LIBET also observed that dibenamine antagonized the P-wave; and they suggested that the release of catecholamines might be involved in this hyperpolarization. However, pretreatment of cats with 1 mg/kg of reserpine (to deplete the stores of endogenous norepinephrine) failed to affect the hyperpolarization induced by either acetylcholine or methacholine (TAKESHIGE and VOLLE, 1964b); consequently, it is far from certain that catecholamines are involved in this type of hyperpolarization. It should not be forgotten that dibenamine is a notoriously non-specific blocking agent with high affinity not only to the adrenergic α-receptors but also to histamine-, 5-HT- and acetylcholine-receptors.

While it is clearly impossible to fit all recent observations into one comprehensive picture, it is equally clear that the superior cervical ganglion has atropine-sensitive receptor sites which can be activated by acetylcholine; or in other words: that the ganglion contains muscarinic receptors through which acetylcholine exerts certain effects.

VII. Transmission of preganglionic impulses through ganglia

From the fact that hexamethonium causes complete block of transmission through the superior cervical ganglion of the cat, it may be deduced that under normal conditions ganglionic transmission is mediated through the nicotinic receptors of this ganglion. However, since the discovery of the hexamethonium-like ganglion-blocking agents, there have been several reports that certain pathways appear to be resistant to the blocking action of these agents. For instance, Freyburger et al. (1950) found the pressor response of dogs and rabbits to asphyxia to be partly resistant to the administration of TEA, although the pressor response of cats and monkeys was abolished. Maxwell et al. (1956), on the other hand, observed that large doses of ganglion-blocking agents (hexamethonium, TEA and chlorisondamine) failed to suppress the pressor response of the cat to electrical stimulation of the splanchnic nerve.

Moe and Freyburger (1950) suggested "that there might exist two types of ganglia, one vulnerable to TEA and mediating the responses to moderate stress and the other being invulnerable to TEA and mediating the responses to severe stress". It is also possible that two types of receptors rather than of ganglion cells are involved, namely the nicotinic receptors (mediating low-frequency impulses) and muscarinic receptors (activated only when impulses reach the ganglion at high frequency). Eccles and Libet (1961) provided a basis for this speculation, since they found that one clearly defined part of the synaptic potential of the isolated superior cervical ganglion of the rabbit was antagonized by atropine and insensitive to hexamethonium and d-tubocurarine (the LN-wave, see section VI). Furthermore, Libet (1964) obtained evidence in favor of the view that the LN-wave (or slow excitatory postsynaptic potential in his terminology) is indeed involved in ganglionic transmission, since it is necessary for the maintenance of the postganglionic discharge in response to low frequencies of preganglionic stimulation. Libet distinguished between two types of posttetanic potentiation of ganglionic transmission: one is of presynaptic origin and unaffected by the administration of atropine, the other is of postsynaptic origin and abolished by atropine. These experiments demonstrate that endogenous acetylcholine released by preganglionic stimulation does react with atropine-sensitive or muscarinic receptors. Hence, it is not unreasonable to expect that the muscarinic receptors might be able to mediate preganglionic impulses when the nicotinic receptors of the ganglion are blocked by hexamethonium or related agents.

Recent experiments of Hilton and Steinberg (1966) demonstrated for the first time that the combination of chlorisondamine with atropine abolished pressor responses that were only partly antagonized by chlorisondamine alone. The authors elicited pressor responses in dogs by increasing the intracranial pressure. After partial block of the response by the intravenous injection of chlorisondamine, the additional administration of atropine abolished the rise in blood pressure. The central nervous system cannot have been the site of action of atropine, because identical results were obtained with valethamate, a quaternary compound which does not easily penetrate the blood-brain barrier. Hence, it is very likely that both chlorisondamine and the atropine-like agents acted on ganglia, the former to block the nicotinic receptors, the latter to block the muscarinic receptors.

More direct evidence was obtained by FLACKE and GILLIS (1966) who stimulated the preganglionic fibers of the right stellate ganglion of the dog. The resulting increase in heart rate was diminished but not abolished by hexamethonium. An increase in the dose of hexamethonium failed to cause further diminution of the response of the pacemaker. After the additional injection of a small dose of atropine (30 µg/kg i.v.), however, the response of the pacemaker was abolished.

All these observations illustrate that muscarinic mechanisms may well be involved in ganglionic transmission. It is of interest that most of these reports (FREYBURGER et al., 1950; ECCLES and LIBET, 1961; HILTON and STEINBERG, 1966; FLACKE and GILLIS, 1966) were based on experiments with dogs and rabbits. For the superior cervical ganglion of the cat, for instance, such a failure of conventional ganglion-blocking agents to abolish responses to preganglionic stimulation has not been reported. It is possible that there are species differences in the contribution which muscarinic receptors make to ganglionic transmission.

Nevertheless, it is possible to demonstrate a "muscarinic transmission" through the superior cervical ganglion of the cat. As mentioned above, the late, non-depolarizing phase of the block of this ganglion by nicotine is characterized by a) an insurmountable block of the nicotine receptors, and b) a facilitation of the effects of all muscarinic agents. In the past it had been reported that there is no response to preganglionic stimulation during the late, non-depolarizing phase of the block by nicotine (TRENDELENBURG, 1957c). However, all these experiments were carried out after pretreatment with 1 mg/kg of atropine, and the importance of this pretreatment was not realized at that time. More recently, similar experiments were conducted without any previous injections of atropine, and responses of the nictitating membrane to preganglionic stimulation were obtained regularly during the late phase of the block by nicotine (Fig. 8) (TRENDELENBURG, 1966b) Muscarinic rather than nicotinic mechanisms must be involved in this ganglionic response because of the following observations. 1. The response was abolished by atropine (0.1 mg/kg i.v.). 2. The latency between the beginning of stimulation and the onset of the response was increased from 1 to 2 sec (as observed in untreated preparations) to 4 to 8 sec (during the late phase of the block by nicotine); this long latency is typical for muscarinic agents (see above). 3. Instead of a brisk and short response (as before any injections of nicotine), there was a slowly developing response of long duration; this is typical for responses to muscarinic agents, as discussed above. 4. Morphine which normally has very little effect on responses to preganglionic stimulation, reduced responses elicited during the late phase of the block by nicotine as much as it reduced responses to muscarinic agents. 5. There was a highly significant correlation between the magnitude of responses to preganglionic stimulation and the magnitude of responses to a muscarinic agent (McN-A-343) injected to the ganglion It is concluded that ganglionic transmission of impulses from the preganglionic fibers can be observed in the superior cervical ganglion of the cat under experimental conditions in which the nicotinic receptors are blocked (by nicotine) and in which the effects of muscarinic agents are facilitated (by nicotine).

FLACKE and GILLIS (1966) obtained some evidence that under certain conditions, muscarinic receptors may be involved in transmission through vagal ganglia. In the heart-lung preparation of the dog, they obtained full block of the response of the cardiac pacemaker to vagal stimulation after the administration of 3 to 10 mg of hexamethonium Partial restoration of the response to

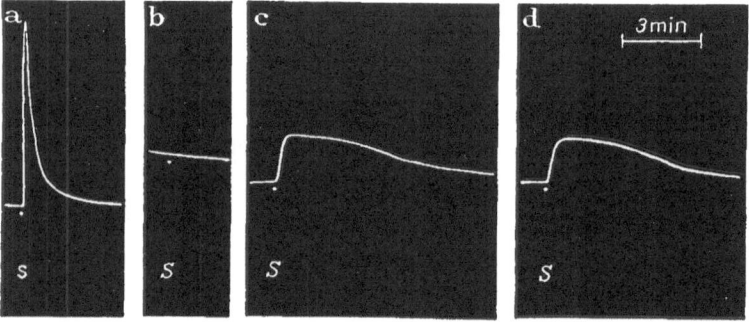

Fig 8a—d Response of the nictitating membrane of the spinal cat to preganglionic stimulation before and after intravenous injections of nicotine S preganglionic stimulation for 10 sec (25 shocks/sec, supramaximal voltage) Records were obtained before nicotine (a), immediately after the first series of intravenous injections of nicotine (b) (total dose 12.7 mg/kg), after the second series (c) (total dose 12 7 mg/kg), and after the third series of intravenous injections of nicotine (d) (total dose 13 3 mg/kg) (From TRENDELEN- BURG, 1966b, with permission of the J Pharmac exp Ther)

vagal stimulation was observed after the injection of 0.2 to 0.5 mg of physostigmine. This restored response to nerve stimulation was then completely blocked by a dose of atropine (5 µg) which caused only partial antagonism to the effects of injected acetylcholine. If physostigmine reversed the effect of hexamethonium by permitting the accumulation of endogenously released acetylcholine both in the ganglion and at the pacemaker, one would have expected atropine to be more effective against injected acetylcholine than against vagal stimulation. Since the opposite result was obtained, it is likely that the administration of physostigmine resulted in an unmasking of muscarinic mechanisms capable of transmitting impulses through vagal ganglion cells, although they may not be involved under physiological conditions.

Muscarinic transmission of preganglionic impulses has also been demonstrated for the adrenal medulla of the cat (DOUGLAS and POISNER, 1965). While normally splanchnic stimulation causes the release of both epinephrine and norepinephrine, after hexamethonium only epinephrine is released, and this response is abolished by atropine. Since nicotine causes the release of both amines, the initial response of the adrenal medulla to nerve stimulation represents a nicotinic effect. The post-hexamethonium response represents a muscarinic effect, because both muscarine and pilocarpine were found to cause a preferential release of epinephrine.

While stimulation of the splanchnic nerve of a spinal cat causes a brisk rise in blood pressure, this effect is not observed when splanchnic stimulation is applied immediately after the intravenous injection of ganglion-blocking doses of nicotine. However, when the administration of nicotine is continued so as

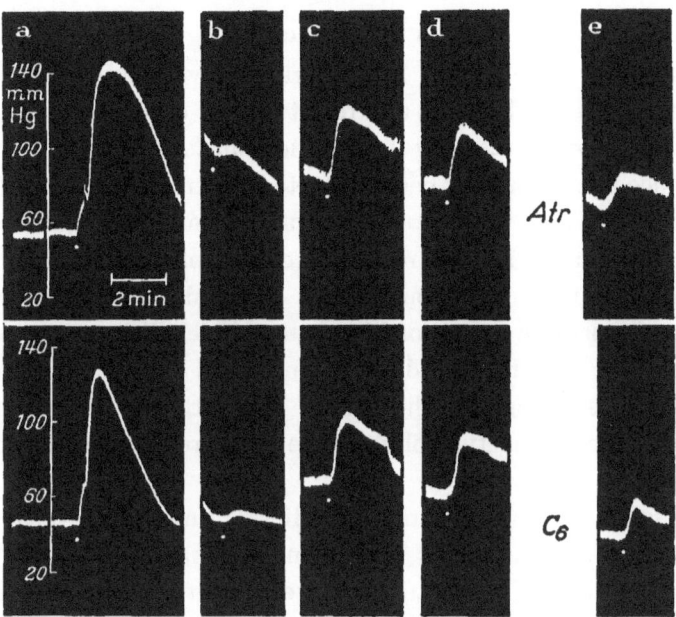

Fig 9a—e The influence of nicotine on the response of the blood pressure to splanchnic stimulation. Spinal cats, arterial blood pressure (in mm Hg) The left splanchnic nerve was stimulated for 30 sec (30 shocks/sec, supramaximal strength) at dots Responses in (a) were obtained before any nicotine was injected. Immediately after a series of intravenous injections of nicotine (14 mg/kg injected within 10 min) the response to splanchnic stimulation was abolished (b) After further injections of nicotine (ca 15 mg/kg each during the 15 min preceding traces c and d) splanchnic stimulation became again effective (c and d). This nicotine-resistant response is sensitive to atropine (100 μg/kg i v 5 min prior to upper) trace e) and to hexamethonium (3 mg/kg i v. 5 min prior to lower trace e) The results indicate that the response in c and d is mediated by muscarinic receptors whose sensitivity is increased by nicotine, for details see text.
(From LEE and TRENDELENBURG, 1967)

to produce the late, non-depolarizing phase of the block by nicotine (see V.A. 10), pressor responses to splanchnic stimulation reappear (Fig. 9), and these responses are abolished by the administration of small doses of atropine (LEE and TRENDELENBURG, 1967). This confirms and extends earlier observations by FELDBERG et al. (1934).

At the present time, it is impossible to assess the importance of the muscarinic mechanisms for normal (physiological) ganglionic transmission As already mentioned, it is possible that there are species differences. Nevertheless, it is intriguing that these recent observations reveal new possibilities by which ganglionic transmission can be influenced pharmacologically. The old concept of a "one type of receptor" mechanism involved in ganglionic transmission

4*

must be abandoned, and the role of acetylcholine in ganglionic transmission will have to be reinterpreted in accordance with a concept of "multiple receptors". It is likely that such a reinterpretation will strengthen the acetylcholine theory of ganglionic transmission, because it may provide explanations for certain observations which were thought to be inconsistent with the postulate that acetylcholine is the transmitter agent in ganglia.

VIII. Anticholinesterase agents

In recent years Koelle and his coworkers have combined histochemical and pharmacological studies, and as a result of their extensive investigations, a comprehensive theory concerning the function of acetylcholinesterase has been proposed. Since this theory has been presented in detail by Koelle (1963), this review will not deal with the various observations on which it is based. According to Koelle, the acetylcholine released from the preganglionic nerve endings serves not only as transmitter but also as a stimulant of the nerve endings, because "the immediate action of liberated acetylcholine on the presynaptic terminal causes it to release additional quanta of acetylcholine in order to facilitate transmission across the cleft". And "the major function of the presynaptically located acetylcholinesterase is to prevent perpetuation of this process of self-re-excitation".

Cohen and Hagen (1964) proposed a different function for the presynaptically located acetylcholinesterase. They point out that, because this enzyme is inhibited by an excess of substrate, the rate of hydrolysis of the acetylcholine released from presynaptic nerve terminals may well be quite low at the height of the release, i.e., when the local concentration of transmitter is very high. Hence, transmission across the cleft would be assured. On the other hand, the rate of hydrolysis of the released acetylcholine would increase, as soon as the transmitter is no longer released and begins to diffuse away. An increase in the rate of hydrolysis at that time would help to shorten the action of the transmitter.

These two proposals are contrasted simply to illustrate how difficult it is to decide on the validity of one versus the other theory. Moreover, many of the observations, on which any theory is based, are contradicted by other observations. For instance, Kamijo and Koelle (1952) concluded on the basis of experiments with DFP that acetylcholinesterase is essential for transmission through the superior cervical ganglion of the cat, while Fehér and Bokri (1959a) studied the effect of TEPP on this preparation and concluded that full block of cholinesterases did not interfere with ganglionic transmission. The problem arises whether DFP has actions other than inhibition of cholinesterases.

Whenever effects of anticholinesterase agents are analyzed, it should be borne in mind that for various members of this group of agents the following additional effects have been described: 1. the facilitation of the ganglionic effects of muscarinic agents; 2. direct inhibitory effects; 3. direct stimulant effects.

1. TAKESHIGE and VOLLE (1962) found that pretreatment of the ganglion with physostigmine leads to a qualitative rather than to a quantitative change in the action of acetylcholine injected intraarterially to the ganglion. While normally there is just a brief, hexamethonium-sensitive postganglionic discharge in response to acetylcholine, a bimodal response appears after premedication with physostigmine: the early discharge is followed by an atropine-sensitive discharge of longer duration. Similar observations were made after neostigmine (TAKESHIGE and VOLLE, 1963a) and 217 AO (TAKESHIGE and VOLLE, 1964a).

If it is true that anticholinesterase agents facilitate the muscarinic effects of acetylcholine on the ganglion, various old observations may have to be reevaluated. For instance, FELDBERG and VARTIAINEN (1935) found that physostigmine potentiated the transmission of submaximal preganglionic stimuli through the ganglion as indicated by an increase in the response of the nictitating membrane. Does this facilitation of transmission indicate an increase in the local concentration of the released acetylcholine with no qualitative change in its action (as the authors assumed at that time), or does this facilitation represent the participation of muscarinic in addition to the nicotinic mechanisms? There is a report that the facilitatory effect of DFP on ganglionic transmission is not abolished by pretreatment with atropine, but the number of observations (3 after atropine) was too small to permit any conclusions (HOLADAY et al., 1954).

There are various reports that anticholinesterase agents are able to antagonize the inhibition of ganglionic transmission by the non-depolarizing ganglion-blocking drugs. CHOU and ELIO (1948) observed antagonism of physostigmine to the ganglion-blocking effect of d-tubocurarine, KAMIJO and KOELLE (1952) of DFP to TEA, GRIMSON et al. (1955) of neostigmine to chlorisondamine. Here again, the question arises whether the classical interpretation (namely that anticholinesterase agents permit a more successful competition of the released acetylcholine with the non-depolarizing blockers) is adequate. Recent experiments by FLACKE and GILLIS (1966) indicate that in at least one experimental condition physostigmine antagonizes the ganglion-blocking effect of hexamethonium by opening muscarinic pathways. In the heart-lung preparation of the dog, these authors obtained full block of the effect of vagal stimulation on the cardiac pacemaker after an injection of 3 to 10 mg of hexamethonium (per 1 liter of blood). Partial restoration of vagal transmission was obtained after the injection of 0.2 to 0.5 mg of physostigmine. A subsequent injection of a small dose of atropine (5 μg) again abolished the response to vagal stimulation, although the effects of submaximal doses of acetylcholine were only reduced. It is likely that atropine abolished the response to preganglionic vagal stimulation by a block of the ganglionic muscarinic receptors rather than by an effect on the muscarinic receptors of the pacemaker.

Obviously, further experimentation must provide more evidence for the view that anticholinesterase agents may exert some of their effects by facilitating the muscarinic effects of acetylcholine on ganglia. However, at the present time, this concept must be regarded as a possibility (see also below: action of physostigmine on the blood pressure).

2. Paton and Perry (1953) described a depressant effect of high concentrations of physostigmine on the superior cervical ganglion which was not accompanied by depolarization. Because of the absence of any depolarization, it is difficult to visualize this ganglion-block as being due to an accumulation of acetylcholine. A direct postsynaptic action of physostigmine must be suspected. Depressant effects of high concentrations of DFP have also been described (Holaday et al., 1954).

3. For neostigmine a stimulant effect on the superior cervical ganglion accompanied by depolarization has been described by Mason (1962a and b). Hexamethonium antagonized this effect which was observed in normal and in chronically denervated ganglia. Takeshige and Volle (1963a), on the other hand, observed an asynchronous discharge in the postganglionic nerve of the superior cervical ganglion of the cat; this asynchronous firing was enhanced by d-tubocurarine and antagonized by atropine. In this regard, the effects of neostigmine resembled those of DFP (see below), but they differed in being observed both in normal and in chronically denervated ganglia. If it is accepted that denervated ganglia have lost their presynaptic nerve terminals, it must be concluded that neostigmine has direct stimulant effects on the superior cervical ganglion, possibly on both nicotinic and muscarinic receptors.

Such a proposal is supported by a wealth of very conflicting evidence concerning the pressor effect of neostigmine observed under certain conditions. There is agreement that the pressor response to neostigmine is not blocked by hexamethonium or chlorisondamine (Long and Eckstein, 1961; Hilton, 1961; Levy and Ahlquist, 1962). It has been reported not to be affected by nicotine (with no specification of the time interval between the injection of nicotine and that of neostigmine; Mendez and Ravin, 1941; Salerno and Coon, 1949) but TMA abolished it (Long and Eckstein, 1961). While some found it to be sensitive to atropine (Hilton, 1961; Long and Eckstein, 1961; Levy and Ahlquist, 1962), others found it resistant to this agent (Mendez and Ravin, 1941; Salerno and Coon, 1949). It is likely that the experimental conditions determined whether the nicotinic or the muscarinic effects of neostigmine were studied.

DFP also elicits a persistent firing of the ganglion cells which is detected by recording the electrical activity of the postganglionic fibers. However, this effect of DFP is absent in chronically denervated ganglia (Volle and Koelle,

1961). Hence, it was proposed at that time that DFP causes the persistent firing in normal ganglia by potentiating the effect of a continuous spontaneous release of acetylcholine. However, if this were the case, one would expect the DFP-induced postganglionic discharge to be pharmacologically similar to the discharge induced by acetylcholine or preganglionic stimulation. VOLLE (1962) found them strikingly different: the DFP-discharge was sensitive to atropine and insensitive to hexamethonium, while discharges elicited by preganglionic stimulation or by exogenous acetylcholine were sensitive to hexamethonium and resistant to atropine. Here again, a simple block of cholinesterases by DFP does not appear to be the adequate explanation, and it may be necessary to invoke the facilitation by anticholinesterase agents of the muscarinic effects of acetylcholine on the ganglion (see above).

KOMALAHIRANYA and VOLLE (1962) observed a further difference between the DFP-induced postganglionic discharge and that induced by either acetylcholine or preganglionic stimulation; intraarterially injected calcium antagonized the former without affecting the latter response. The selectivity of this effect of calcium and the fact that calcium is atropine-like in blocking the effects of muscarinic agents (TAKESHIGE and VOLLE, 1964c) suggested to KOMALAHIRANYA and VOLLE (1962) that DFP may elicit a postganglionic discharge through a muscarinic mechanism of the ganglion.

While it is difficult to account fully for the effects of anticholinesterase agents on the superior cervical ganglion, their effects on the blood pressure present additional problems. There is little doubt that the pressor response of the rat to physostigmine is of predominantly central origin, since it is not observed in the spinal rat (VARAGIĆ, 1955), and since it can be elicited by the injection of the agent into the fourth ventricle (DIRNHUBER and COLLUMBINE, 1955). The pressor response is mediated through the sympathetic system, since it is potentiated by cocaine and antagonized by dibenamine, tolazoline or ergotamine (VARAGIĆ, 1955; DIRNHUBER and COLLUMBINE, 1955). Recent experiments by GOKHALE et al. (1964) indicate that the mediation of the pressor response to physostigmine involves muscarinic ganglionic pathways, because a) hexamethonium caused only partial block, b) the combination of hexamethonium and atropine was fully effective, c) the combination of hexamethonium with cocaine (which blocks the ganglionic effects of muscarinic agents; see V.A. 12) was also fully effective in blocking the pressor response. It is quite possible that physostigmine has two different sites of action: one in the centers of the pons and medulla oblongata (KRISTIC and VARAGIĆ, 1965) to originate efferent impulses, and another at sympathetic ganglia to facilitate ganglionic transmission by simply increasing the effective amount of acetylcholine and/or to open up muscarinic pathways which are resistant to hexamethonium. It seems to be necessary to invoke multiple sites of action, since the experimental observations do not agree with the view that a single site of action in either the central nervous system or at autonomic ganglia is involved.

The author wishes to emphasize that the discussion presented in this section was not intended to provide answers to questions; on the contrary, the main aim was to pose questions that appear to be justified on the basis of recent observations. Even if all proposals made here find negative answers, clarification will have been provided in a field which presents very intricate problems.

IX. The ganglionic effects of sympathomimetic amines

While it is established that norepinephrine and epinephrine do not stimulate ganglion cells, the effects of these drugs on the transmission of impulses through autonomic ganglia have been a matter of considerable controversy. Marrazzi (1939) was the first to observe an inhibitory effect of epinephrine on transmission of impulses through sympathetic or parasympathetic ganglia (Marrazzi, 1939; Marrazzi and Marrazzi, 1947; tum Suden and Marrazzi, 1951). A period of post-inhibitory facilitation was ascribed to the vasoconstrictor effects of epinephrine rather than to an effect of epinephrine on ganglion cells.

Bülbring and Burn (1942) and Bülbring (1944), on the other hand, observed facilitatory effects with small doses of epinephrine and inhibition of ganglionic transmission with larger doses. Similar qualitative differences in effects that depended on the dose of the amine were described by Malméjac (1955), Trendelenburg (1956b) and Pardo et al. (1963). Since the facilitation of ganglionic transmission was observed with small doses of epinephrine, it was unlikely that the observed effect was entirely due to the vascular effects of this amine.

Konzett (1950) demonstrated that sympathomimetic amines potentiate the effects of acetylcholine injected into the perfusion of the superior cervical ganglion of the cat. For this facilitatory effect the relative potencies of the amines was: isoproterenol > epinephrine > norepinephrine. Since various α-receptor-blocking agents (ergotamine, dihydroergotamine, dibenamine) failed to antagonize the facilitatory effects of the catecholamines, Konzett suggested that this facilitation was mediated by ganglionic β-receptors. Isoproterenol also facilitates the transmission of impulses through ganglia (Matthews, 1956, Pardo et al., 1963).

The inhibitory effects of sympathomimetic amines, on the other hand, appear to be mediated by α-receptors because they are abolished by dihydroergotamine (Lundberg, 1952), phenoxybenzamine or dibenamine (Weir and McLennan, 1963), and because the relative potencies of the amines are: epinephrine > norepinephrine > isoproterenol (Weir and McLennan, 1963; Pardo et al., 1963). Lundberg (1952) observed occasional hyperpolarization of the ganglion cells in response to epinephrine but was unable to detect a correlation between this change in ganglionic potential and the inhibitory effect of the drug on ganglionic transmission. PATON and THOMPSON (1953) found that epinephrine had two effects on the superior cervical ganglion of the cat: to decrease the release of acetylcholine in response to preganglionic stimulation, and to decrease the sensitivity of the ganglion cells to injected acetylcholine.

The effects of epinephrine on ganglia are further complicated by the fact that any given dose of this amine may cause inhibition of ganglionic transmission when low frequencies of preganglionic stimulation are used, while facilitation is observed with high rates of stimulation (ELLIOTT, 1965).

In a recent systematic study of the ganglionic effects of sympathomimetic amines, DE GROAT and VOLLE (1966a) obtained good evidence for the presence of two different types of receptors which are both activated by norepinephrine and epinephrine. Interaction of the amines with α-receptors causes hyperpolarization of the ganglion cells accompanied by inhibition of ganglionic transmission; these effects are blocked by dihydroergotamine and not impaired by β-receptor blocking agents such as DCI and pronethalol. These α-effects are prominent after norepinephrine, detectable after epinephrine and absent after isoproterenol. Activation of the β-receptors, on the other hand, was most pronounced after isoproterenol, and was observed with norepinephrine only after block of the α-receptors. Interaction of the amines with β-receptors caused depolarization of the ganglion cells which was accompanied by facilitation of ganglionic transmission. While dihydroergotamine failed to impair these β-effects, they were blocked by DCI and pronethalol. The effect of the catecholamines on the response of the ganglion to nicotinic and muscarinic drugs was the subject of a subsequent study (DE GROAT and VOLLE, 1966b). The nicotinic postganglionic discharge observed after nicotine or acetylcholine was not affected by prior injections of either of the amines. However, administration of isoproterenol caused the appearance of a late, atropine-sensitive postganglionic discharge in response to injections of acetylcholine; or in other words, the catecholamine unmasked muscarinic effects of acetylcholine. This facilitatory effect of the catecholamine (like the others discussed above) was blocked by DCI and pronethalol and not affected by dihydroergotamine, i.e., it was a β-effect. Parallel to the changes in postganglionic discharge, there were changes in the ganglionic potentials set up by acetylcholine, isoproterenol tended to decrease the depolarizing and to increase the hyperpolarizing effects of acetylcholine. With epinephrine similar effects were observed, but they were obtained with regularity only after the administration of an α-receptor blocking agent. Norepinephrine caused the appearance of the late, atropine-sensitive postganglionic discharge in response to acetylcholine only in some of the experiments in which a α-receptor blocking agent had been given.

The β-effects of isoproterenol were not restricted to the unmasking of the muscarinic action of acetylcholine, since this agent caused consistent potentiation of the effects of other muscarinic agents (oxotremorine, anticholinesterases, methacholine). The effects of epinephrine and norepinephrine were either facilitatory or inhibitory depending on their effect on the ganglionic potential (depolarization or hyperpolarization), on the administration of blocking agents (block of α-receptors resulting in facilitatory effects, block of β-receptors in depressant effects of the catecholamines), and also on the intensity of the firing produced by the muscarinic stimulant. With small doses of the anticholinesterase 217 AO, a small postganglionic discharge was set up which tended to be increased by injections of the catecholamines. Large postganglionic discharges elicited by high doses of 217 AO, on the other hand, tended to be inhibited by injections of epinephrine or norepinephrine.

It is of interest that injections of various vasoconstrictor (angiotensin and vasopressin) and vasodilator agents (papaverine, nitroglycerin) failed to modify the discharge elicited by 217 AO. Hence, the vasoconstrictor or vasodilator effects of the catecholamines are probably not the cause of their ganglionic effects.

These observations by De Groat and Volle (1966a and b) resolved most of the contradictory reports of the past. Moreover, they revealed one further mechanism by which the latent muscarinic ganglionic effects of acetylcholine can be unmasked and lead directly to the question of a possible physiological role of the catecholamines in modifying ganglionic transmission. There is no definitive answer to this question, and the reader should be aware that to proceed from pharmacology to physiology may well mean to turn from facts to speculation.

While histochemical methods have revealed the presence of large numbers of adrenergic nerve terminals in the neighborhood of the ganglion cells of prevertebral ganglia of the cat (coeliac and inferior mesenteric ganglion), most workers agree that the superior cervical ganglion of the cat (like most of the paravertebral ganglia) has very few adrenergic nerve terminals (Hamberger et al., 1964; 1965). In the rabbit, on the other hand, the superior cervical ganglion is rich in adrenergic nerve endings associated with ganglion cells, and their number is not decreased after preganglionic denervation (Hamberger et al., 1965). This histological observation agrees with the electrophysiology of the isolated superior cervical ganglion of the rabbit, since Eccles and Libet (1961) detected a hyperpolarizing phase of the synaptic potential which was abolished by dibenamine.

The absence of histologically demonstrable adrenergic nerve terminals in the superior cervical ganglion of the cat is somewhat surprising in view of several reports that pre- or postganglionic stimulation of this ganglion causes the appearance of an epinephrine-like agent in the effluent of the perfused ganglion (Bülbring, 1944; Reinert, 1963; Weir and McLennan, 1963). Bülbring (1944) suggested that the ganglion may contain chromaffin cells, and

ECCLES and LIBET (1961) reached the same conclusion for rabbit ganglia because the hyperpolarizing phase of the synaptic potential was abolished not only by dibenamine but also by atropine. Very recently, ERANKÖ and HÁR-KÖNEN (1965) discovered near the superior cervical ganglion a well vascularized organ containing numerous cells that exhibited an intense yellow fluorescence when stained according to the method of FALCK. The authors suggested 5-hydroxytryptamine as a possible source of the fluorescence but did not exclude catecholamines.

While there is some evidence for a possible physiological role of catecholamines in transmission, it is completely undecided whether this role might be facilitatory or inhibitory. Experiments with norepinephrine-depleting agents have not clarified the possible physiological contribution of catecholamines to ganglionic transmission. COSTA et al. (1961) reported a very pronounced potentiation of the transmission of submaximal preganglionic stimuli after injections of reserpine, and LEI and MAENGWYN-DAVIES (1965) reported similar results after an injection of syrosingopine. However, these experiments are complicated by the fact that they have a very long duration (22 hours in the case of COSTA et al., 1961), and it is very questionable whether one can keep the experimental condition constant for such a long time. WEIR and MCLENNAN (1963) failed to observe any effects of reserpine which were not also observed in equally long experiments without reserpine. Similarly, REINERT (1963) obtained an increase in the height of postganglionic action potentials in experiments of long duration whether or not reserpine had been injected. The action potentials failed to increase when an infusion of saline solution prevented dehydration. In summary, it must be concluded that depletion of the ganglionic norepinephrine has so far failed to provide any evidence for a modulatory role of endogenous norepinephrine on ganglionic transmission.

Bretylium and guanethidine are known to prevent the release of norepinephrine from adrenergic nerve terminals in response to nerve stimulation. Both agents have been found to block ganglionic transmission (GERTNER and ROMANO, 1961). Recent experiments by SHAND (1965) demonstrate that bretylium probably acts on the postganglionic neuron, an observation that may be related to the finding by BOURA et al. (1960) that bretylium a) is taken up by adrenergic nerves, and b) has local anesthetic properties. Here again, the experimental findings do not support the view that endogenous norepinephrine is involved in modulating physiological transmission through the superior cervical ganglion.

It is likely that the prevertebral ganglia present better conditions for a study of this kind. It should be much easier to demonstrate a physiological role of endogenous norepinephrine in ganglia rich in adrenergic nerve terminals (HAMBERGER et al , 1965). Another promising object of study is the intestinal plexus of ganglion cells, since this also contains large numbers of adrenergic nerve terminals which appear to synapse with the ganglion cells of the Auerbach plexus (Norberg, 1964; Jacobowitz, 1965).

The effects of some sympathomimetic amines do not appear to fit into the picture drawn by De Groat and Volle (1966a). Kewitz and Reinert (1954), for instance, reported a stimulant effect of amphetamine and methamphetamine on the superior cervical ganglion of the cat. These agents, as well as hordenine and mescaline, appear to be nicotine-like, since they depolarize the ganglion and since their effects are blocked by hexamethonium (Reinert, 1959, 1960; Gold and Reinert, 1960). The stimulant effects of tyramine, amphetamine, phenylpropanolamine and phenylethylamine on the perfused adrenal medulla of the cat were also found to be sensitive to hexamethonium (Jaanus and Rubin, 1965). Furthermore, a study of the quaternary methyl derivatives of norepinephrine, dopamine and tyramine showed that the potency of their nicotinic effects increased in that order. These effects are nicotinic because they are antagonized by hexamethonium and because the quaternary derivatives are able to block ganglionic and neuromuscular transmission (Cuthbert, 1964; Atanackovic et al., 1965).

These last observations emphasize that all our attempts to classify drugs tend to oversimplify matters. It is quite possible that various sympathomimetic amines can exert effects which do not fit into the "α-receptor versus β-receptor" concept.

Various inhibitors of monoaminoxidase cause an irreversible block of ganglionic transmission (Gertner, 1961). The mechanism of this block is unknown; ganglionic transmission is depressed at a time when the sensitivity of the ganglion to acetylcholine is normal, when the amount of acetylcholine released by preganglionic stimulation is normal, and when the response of the nictitating membrane to postganglionic stimulation is unimpaired. It has been speculated that block of this enzyme has an effect on the norepinephrine stores of the ganglion, and that the pharmacological observations of Gertner are related to a physiological role of norepinephrine in ganglionic transmission (Costa et al., 1961). However, it would be premature to draw any conclusion from these experiments before we know much more about the adrenergic innervation of the superior cervical ganglion, about the ganglionic actions of norepinephrine and of inhibitors of monoaminoxidase.

X. The effect of denervation
A. Presynaptic changes after preganglionic denervation

The discussion of the pre- and postsynaptic changes produced by denervation of an organ is complicated by the fact that the time course of the events is greatly influenced by the length of the degenerating nerve. In a variety of adrenergically and cholinergically innervated organs, Cragg (1965) found very variable time intervals between the section of the postganglionic nerves and subsequent failure of the transmission of nerve impulses to the effector organ; for various species and various organs the intervals were as short as two days or as long as six to seven days with very little variability for any given organ. For the cholinergically innervated sweat glands of the cat, it has recently been found that the failure of peripheral transmission is delayed by one day when the section of the nerve is made 12 cm farther away from the effector organ (REAS and TRENDELENBURG, 1967). While these observations dealt with postganglionic nerves, DAVIDOVICH and LUCO (1956) compared the effect of a proximal section (7 cm from the ganglion) with that of a distal section (2 cm from the ganglion) of the preganglionic fibers of the superior cervical ganglion of the cat. The increase in the length of the degenerating preganglionic nerve delayed the failure of synaptic transmission by nearly one day. Hence, when results from different laboratories are compared, it should be borne in mind that considerable differences in the time course of the events after denervation may be due to the choice of the species, the ganglion and the length of the degenerating preganglionic nerve.

There is general agreement that synaptic transmission fails before the degenerating preganglionic nerve loses the ability to conduct nerve impulses (COPPÉE and BACQ, 1938; GIBSON, 1940; FELDBERG, 1943; DAVIDOVICH and LUCO, 1956). The failure of transmission appears to coincide with a loss of endogenous acetylcholine below 60 % of normal (MACINTOSH, 1938; FELDBERG, 1943). One day after denervation the acetylcholine in the ganglion declined to 60—90 % of normal, but transmission was still intact; two days after the operation transmission began to fail, and the endogenous acetylcholine fell below 60 % of normal; no transmission was observed on the third day, and the acetylcholine content of the ganglion was 10 to 15 % of normal.

Histological observations correlate well with the function of the ganglion. Nerve terminals (boutons) were detected during the second postoperative day when transmission was still intact; however, on the sixth postoperative day they had disappeared, and there was no transmission of impulses through the superior cervical ganglion of the cat (GIBSON, 1940). It is of interest that nerve terminals were detected again when transmission was restored after reinnervation (44 days after denervation). An electronmicroscopic study of a denervated sympathetic ganglion of the frog (HUNT and NELSON, 1965) showed swelling of mitochondria and clumping of the vesicles of preganglionic nerve terminals during the second postoperative day. Subsequent changes consisted of a growth of Schwann cells between the pre- and postsynaptic cells; some presynaptic terminals lost their cell membranes, while others remained relatively unchanged although they lost their synaptic vesicles and mitochondria. Intracellular recordings from single ganglion cells showed that functional changes coincided with the first morphological changes. During the second postoperative day the spontaneous miniature synaptic potentials (m. s. p.s) first increased in frequency and then disappeared. These changes were accelerated by repetitive preganglionic stimulation. Subsequent to a preganglionic tetanus there was for more than 90 minutes an increase in the rate of the spontaneous m.s.p.s, whereas, in normal ganglia, a similar phenomenon lasts only for a few seconds. The first sign of failure of transmission consisted in the appearance of fatigue, *i.e.*, while the ganglion responded normally to a single preganglionic shock, transmission began to fail when nerve stimulation was prolonged. The increased susceptibility of the denervated ganglion to fatigue was already reported by COPPÉE and BACQ (1938).

The increase in the rate of m.s.p s observed during the second postoperative day is probably related to a similar phenomenon described for degenerating postganglionic nerve endings. For the salivary gland of the cat (EMMELIN and STRÒMBLAD, 1958), for the eye of the rabbit (BÁRÁNY, 1962) and for the nictitating membrane of the cat (LANGER, 1966a), it has been shown that there is a leakage of transmitter from the degenerating nerve terminals at a time when the endogenous transmitter begins to decline (SMITH *et al.*, 1966). Since the increased rate of spontaneous m s.p s observed by HUNT and NELSON (1965) preceded failure of transmission by less than a day, it is likely that it coincides with that period during which MacINTOSH (1938) and FELDBERG (1943) observed a 10 to 40 % decline in the acetylcholine content of the ganglion. Or in other words, it is possible that the first decline in the acetylcholine content of the ganglion is due to a leakage of the transmitter from the degenerating fibers rather than to a failure of synthesis. HEBB and WAITES (1956) reported a gradual and nearly complete loss of the choline acetylase of the ganglion during the first 5 to 6 days after denervation, but their earliest observations were made two days after the operation. FELDBERG (1943), on the other hand, found in two experiments a nearly normal choline acetylase activity 24 hours after the operation, *i.e.*, at a time when the endogenous acetylcholine had started to decline. It is of interest that denervation of the cholinergically innervated sweat glands of the cat leads to a fall in acetylcholine which precedes the fall in choline acetylase activity (REAS and TRENDELENBURG, 1967).

The situation emerging from these various studies may be summarized as follows. Between the operation and the onset of morphological or functional changes there is a time interval of about 24 to 48 hours the length of which appears to depend on the length of the degenerating nerve. The first functional changes appear to involve the ability of the nerve terminals to store the transmitter, since there are various indications for a leakage of the transmitter from the terminals. The disappearance of the transmitter is subsequently accelerated by a failure of synthesis. Transmission fails when the transmitter falls below about 50 % of normal, and the tendency of the denervated ganglion to show fatigue as the first sign of failure of transmission may indicate that the store of "available" acetylcholine (PERRY, 1953) declines. Within three days, degeneration is complete and there is a total loss of function. While the spontaneous miniature endplate potentials have been found to reappear after chronic denervation of skeletal muscle (BIRKS *et al.*, 1960), a complete absence of spontaneous m.s.p.s has been reported for up to 30 days after denervation of a sympathetic frog ganglion (HUNT and NELSON, 1965).

Denervation causes a considerable loss of the acetylcholinesterase of the superior cervical ganglion of the cat (VON BRÜCKE, 1937) which amounts to 75 % of the activity of a normal ganglion (GROMADZKI and KOELLE, 1965). The loss involves only the cholinesterase associated with preganglionic nerve terminals (FREDRICSSON and SJOQVIST, 1962; KOELLE, 1963).

B. Postsynaptic changes after preganglionic denervation

Electronmicroscopy of a sympathetic ganglion of the frog showed no pronounced changes in the morphology of the denervated ganglion cells, excepting a disappearance of the special structures normally found in regions of close approximation of pre- and post-synaptic cells (HUNT and NELSON, 1965).

In the experiments in which the nictitating membrane is used as an indicator for the effects of drugs on the superior cervical ganglion, three serious complications arise in experiments with denervated ganglia. 1. The nictitating membrane acquires supersensitivity to the peripheral transmitter, norepinephrine; the increase is a linear function of time for up to 28 days, the longest time interval studied (LANGER and TRENDELENBURG, 1967). Hence, it is impossible to compare responses of the nictitating membrane to stimulation of the normal and of the denervated ganglion without making allowance for the change in sensitivity of the "indicator". 2. When this type of supersensitivity of the nictitating membrane develops, the membrane is in a prolonged state of partial contraction when studied in spinal or in anesthetized cats. According to LANGER (1966b) this partial contraction of the membrane is due to the fact that it has acquired enough additional sensitivity to catecholamines to respond to the catecholamines circulating in the blood. Since such a partial contraction is not observed with the normal nictitating membrane, responses to stimulation of normal and denervated ganglia must be expected to be influenced not only by different sensitivities of the indicator but also by differences in the state of contraction. For a more detailed discussion of this problem, see LANGER (1966b). 3. Recent experiments also indicate that, on the second day after denervation of the superior cervical ganglion, the response of the nictitating membrane to postganglionic stimulation is increased at all rates of stimulation (LANGER and TRENDELENBURG, 1967). So soon after the operation, there is not yet any significant increase in the sensitivity to norepinephrine; the increase in response must be attributed to an increased release of transmitter from the postganglionic nerve terminals per postganglionic impulse.

In the past, only the first of these three complications was recognized. In order to eliminate the influence of changes in sensitivity of the indicator organ (the nictitating membrane), it has been customary to relate responses of the nictitating membrane to stimulation of the ganglion to about equal responses to exogenous norepinephrine. Because of the other two complications, this procedure cannot any longer be regarded as satisfactory. Experiments of this kind should be performed in pithed cats to abolish tone (LANGER and TRENDELENBURG, 1957), and responses of the nictitating membrane induced by stimulation of the ganglion should be compared with responses to postganglionic stimulation.

It is likely that these hitherto unrecognized experimental difficulties account for some of the conflicts of reported evidence.

Both CHIEN (1958) and BOKRI et al. (1963) concluded that 12 to 38 days after the operation the denervated superior cervical ganglion of the cat is 3 to 40 times more sensitive to acetylcholine than the normal ganglion. In these experiments, the nictitating membrane was used as "indicator", and responses of ganglionic origin were compared with responses to injected norepinephrine. VOLLE and KOELLE (1961) on the other hand, recorded the postganglionic discharge elicited by injections of acetylcholine to the ganglion, and they were unable to detect any change in the sensitivity of the ganglion to threshold doses of acetylcholine after 6 to 29 days of denervation.

Because of the experimental difficulties enumerated in the first paragraph of this section, the conclusions of CHIEN (1958) and of BOKRI et al. (1963) must be questioned; it is likely that the superior cervical ganglion is an exception to CANNON's law of denervation in failing to become supersensitive to acetylcholine. However, while there were no quantitative changes in the sensitivity of the ganglion after denervation, various qualitative changes were reported.

PERRY and REINERT (1954, 1955) observed that pentamethonium, hexamethonium, decamethonium and pendiomid lose their ability to antagonize the ganglionic effects of acetylcholine when injected to ganglia denervated 20 to 40 days prior to the experiment; shorter time intervals were not studied. In about one third of the preparations, these agents not only failed to antagonize acetylcholine, they caused stimulation of the ganglion. Similar changes in the action of these drugs were observed in normal ganglia which were perfused with low-potassium salt solutions, while the perfusion of denervated ganglia with high-potassium salt solutions restored the effects of these drugs to normal. It remains a very puzzling phenomenon that these dramatic changes in the effects of the methonium compounds were not accompanied by any changes in the effects of other ganglion-blocking agents such as nicotine, TMA, TEA, and d-tubocurarine.

Qualitative changes of a different type were observed by TAKESHIGE and VOLLE (1963 b). A normal ganglion responds to acetylcholine with a brief postganglionic discharge ("early firing") which is blocked by hexamethonium. After physostigmine, the "early firing" is enhanced, and it is followed by a second, atropine-sensitive discharge ("late firing"). Six to 13 days after denervation of the ganglion, the response to injected acetylcholine is normal, but injections of hexamethonium or d-tubocurarine not only block the "early firing", they also cause the appearance of a "late firing". In addition, physostigmine now fails to enhance the "early firing" while still causing the appearance of the late, atropine-sensitive discharge. In other words, after the disappearance of the preganglionic acetylcholinesterase (see section X.A.), physostigmine appears to lose its ability to enhance the nicotinic response of the ganglion to acetylcholine, while there is no change in the ability of physostigmine to unmask the latent muscarinic actions of acetylcholine.

After 14 to 28 days of denervation the response of the ganglion to acetylcholine becomes bimodal without any pretreatment. In fact, the sensitivity of the muscarinic receptors to exogenous acetylcholine then surpasses that of the nicotinic receptors, since the threshold doses of intraarterial acetylcholine for the early and the late firing were 6.5 and 1.1 µg, respectively. Ganglia that had been denervated for 30 to 70 days showed responses to acetylcholine similar to those of normal ganglia, and it may be assumed that partial reinnervation had occurred.

These experiments demonstrate that denervation causes selective changes in the sensitivity of the two types of acetylcholine receptors: while the sensitivity of the nicotinic receptors remains unchanged, that of the muscarinic receptors increases greatly. Hence, the normal balance between nicotinic and muscarinic mechanisms is reversed. Similar conclusions were reached by BOKRI et al. (1963) who also reported that atropine abolished the "supersensitivity" of the ganglion to acetylcholine (see above for critical evaluation of the claim for supersensitivity). It is very tempting to speculate that at least some of the observations of PERRY and REINERT (1954, 1955) (namely the inability of hexamethonium to block the effect of acetylcholine after denervation) may be attributable to this change in balance between nicotinic and muscarinic mechanisms. However, before such a view can be accepted, an explanation has to be found for their observation that TEA and d-tubocurarine continue to block the effects of acetylcholine on the denervated ganglion, and also for their finding that perfusion of the ganglion with low-potassium salt solutions mimics the effects of denervation.

In spite of these reservations, two interesting facts emerge from these studies: 1. that the reported changes in the relative sensitivities of nicotinic and muscarinic mechanisms appear several weeks after denervation, i.e., long after the presynaptic changes discussed in section X.A. This observation indicates that there must be gradual changes in the function of the ganglion cells which continue to develop for several weeks after the preganglionic nerve fibers have degenerated. It is of interest to note that similar phenomena have been described for skeletal and smooth muscle (Trendelenburg, 1963). 2. The denervated superior cervical ganglion continues to defy our attempts to analyze its functions. A complete analysis of the changes caused by denervation should be rewarding, since recent observations on the effect of denervation on the sensitivity of smooth muscle (Trendelenburg, 1966c) has greatly contributed to our knowledge of the functions of normal smooth muscle.

C. The effect of axotomy

A few days after section of the postganglionic axon of the stellate or the inferior mesenteric ganglion of the cat or rabbit, the ganglion failed to respond to preganglionic nerve stimulation. Synaptic transmission was not restored by acetylcholine, physostigmine, neostigmine, epinephrine or potassium chloride, and the axotomized ganglia also became unresponsive to acetylcholine. Moreover, preganglionic stimulation caused the release of roughly normal amounts of acetylcholine (Brown and Pascoe, 1954)

While denervation causes a failure of transmission because of a loss of function of the presynaptic nerve terminals (see preceding section), axotomy appears to lead to a failure of synaptic transmission because of a change in the postsynaptic cells. Direct evidence for this view was obtained when the functional changes (Acheson and Remolina, 1955) were compared with the histological changes produced by axotomy of the inferior mesenteric ganglion of the cat (Acheson and Schwarzacher, 1956). In this ganglion, axotomy (section of the hypogastric nerve at least 3 cm from the ganglion) causes a fatigue of transmission of the second postoperative day which gradually develops to a failure of synaptic transmission which is complete after 2 weeks. Signs of spontaneous recovery are first observed during the third postoperative week, and recovery is complete after 72 to 100 days; however, only about 50% of the ganglion cells survive axotomy. Histological observations showed a progressive fragmentation and disappearance of the Nissl substance of axotomized ganglion cells which was maximal at the end of the second week. During the recovery period (40 to 100 days after axotomy) rather severe chromatolysis persisted. Consequently, a normal function of the cell is possible with histological changes of this kind. There was good correlation between the function of the cell and the position of the cell nucleus: loss of function was observed when the nucleus was displaced, and normal function returned when the nuclei began to recover normal positions. In about half of the ganglion cells the nucleus protruded through the cell wall, and these cells appeared to have been permanently damaged by axotomy.

While Brown and Pascoe (1954) reported a total loss of sensitivity of the axotomized ganglion cells to applied acetylcholine, Acheson and Remolina (1955) found the axotomized ganglion sensitive to acetylcholine, although it was necessary to increase the dose of this agent to elicit stimulation. It is of interest that there was a change in the sensitivity of the ganglion cells to the blocking action of TEA in the opposite direction: during the first week after axotomy the effective dose of TEA decreased to 1/10 of normal.

Axotomized ganglion cells lose their acetylcholinesterase activity. This change is especially pronounced for those ganglion cells which normally have a high cholinesterase content (FREDRICSSON and SJÖQVIST, 1962; GROMADZKI and KOELLE, 1965). The extent of total loss of acetylcholinesterase activity from the axotomized ganglion probably depends on the proportion of ganglion cells with high cholinesterase activity [which, for instance, is high in the superior cervical ganglion and low in the coeliac and inferior mesenteric ganglion (HOLM-STEDT and SJÖQVIST, 1959)] and on the number of axons in the axotomized postganglionic nerve (as compared to the total number of ganglion cells). After section of one of the postganglionic nerves of the superior cervical ganglion (the internal carotid nerve), the total acetylcholinesterase activity of the ganglion fell by only 6% as compared to a 75% decline in activity after preganglionic denervation (GROMADZKI and KOELLE, 1965).

D. Homologous and heterogenous reinnervation of the superior cervical ganglion

As pointed out in the preceding sections, reinnervation of the ganglion by the preganglionic fibers of the cervical sympathetic chain is observed a few weeks after denervation. GIBSON (1940), for instance, found in histological sections that reinnervating preganglionic fibers entered the ganglion 21 days after the operation; there were not yet any signs of synaptic contact, and these ganglia failed to respond to preganglionic stimulation. However, 44 days after the operation, nerve endings were histologically demonstrable, and transmission was reestablished. According to DE CASTRO (1951), reinnervation of the ganglion can also be accomplished when any other cholinergic nerve is joined to the lower pole of the ganglion. Eighty to 100% of the operations were successful when the central end of the cut glossopharyngeal, phrenic or vagal nerve was joined to the peripheral end of the cut cervical sympathetic chain. In all cases, physostigmine was found to enhance transmission after reinnervation, and this was taken as evidence in favor of a cholinergic mechanism of transmission.

DE CASTRO (1951) also reported successful reinnervation of the denervated ganglion in 30% of the operations in which the peripheral end of the vagus was joined to the lower pole of the ganglion in order to permit the ascending, afferent fibers of the vagus to promote "heterogenous" reinnervation. The relatively low proportion of successful operations (30%) may well be due to the fact that the procedure involves mobilization of the nodose ganglion, since the cell bodies of the reinnervating fibers must be preserved. According to DE Castro's observations, reinnervation by sensory vagal fibers is possible, he concluded that a non-cholinergic mechanism of transmission was involved after reinnervation, because physostigmine then failed to enhance transmission. Since the effects of specific ganglion-blocking agents (like hexamethonium) were not tested, the conclusions of DE Castro must be regarded as very interesting but cannot be accepted as definitive.

Functional evidence for the success of this type of reinnervation of the ganglion was provided by Guth (1956). Matsumura and Koelle (1961) also obtained reinnervation as shown by the fact that the nictitating membrane of the reinnervated side responded to dilatation of the oesophagus and to stimulation of the cervical vagus. Since physostigmine enhanced transmission of impulses through the ganglion, while TEA blocked it, the authors concluded that transmission through the reinnervated ganglion involved cholinergic mechanisms. At the present time, it is impossible to resolve the conflict between their and DE Castro's results. Since the chemical nature of the transmitter released by sensory fibers is unknown, and since there is no evidence that acetylcholine serves as a transmitter in this type of nerve fiber, the results of Matsumura and Koelle (1961) seem to indicate that it is the postsynaptic cell which determines the nature of the transmitter released by reinnervating nerve terminals. Such a view would be quite contrary to the concept that the enzyme involved in the synthesis of the transmitter liberated from the nerve endings is synthesized in the cell body and transported down the axon by axonal flow. According to the latter concept one would have expected sensory nerve terminals to continue to release the sensory transmitter irrespective of what kind of ganglion cell they contact with.

A possible explanation for the reported discrepancies may lie in observations by Murray and Thompson (1957a). After denervation of the superior cervical ganglion of the cat, these authors observed a contraction of the nictitating membrane during feeding of the animal. An analysis of the phenomenon showed that denervation of the superior cervical ganglion caused the efferent cholinergic fibers of the vagus to sprout into the neighboring superior cervical ganglion. The afferent impulses from the stomach induced by feeding traveled to the central nervous system and set up a descending discharge in the motor fibers of the vagus; these, in turn, activated the superior cervical ganglion through the sprouts.

Additional evidence for sprouting was obtained by partial preganglionic denervation of the ganglion (Murray and Thompson, 1957b). Histological observations showed that section of the preganglionic rami D1 to D3 caused the disappearance of 90 % of the preganglionic fibers innervating the ganglion, and, 4 to 5 days after the operation, the response of the nictitating membrane of the operated side to stimulation of the cervical sympathetic chain was much smaller than that of the control side. However, 10 to 20 days after the operation, the response on the operated side to nerve stimulation was bigger than that on the control side. Since by that time the "decentralization supersensitivity" of the nictitating membrane had developed, part of this reversal of the effect of partial denervation of the ganglion was due to changes in the sensitivity of the nictitating membrane, but some sprouting must have occurred. With intervals of more than 28 days after the operation, the responses of the operated side to nerve stimulation equaled the response of the control side, and there was no longer any supersensitivity of the nictitating membrane to norepinephrine. Since this time interval is too short for reinnervation by the sectioned nerve fibers, the 10% intact preganglionic fibers must have sent sprouts to the denervated ganglion cells. Experiments in which the release of acetylcholine by partially denervated ganglia was determined, support the view that sprouting of intact nerve fibers to denervated ganglion cells does occur.

XI. Synapses with ganglion cells of nerve fibers that do not belong to the classical preganglionic fibers

Classically, ganglion cells have been regarded as relay stations (with more or less pronounced properties of integration) between the cholinergic preganglionic fibers originating in the central nervous system and the effector organ. Within each preganglionic and postganglionic nerve trunk several groups of nerve fibers of different diameters and of different conductivities have been recognized, and it is known that the ultimate destination of these fibers may differ. For instance, the S1-fibers of the cervical sympathetic chain innervate the structures of the orbit (pupil, nictitating membrane, MÜLLER's muscle), while the more slowly conducting S2-fibers innervate blood vessels (BISHOP and HEINBECKER, 1932; ECCLES, 1935). The function of the S3- and S4-fibers remains unknown. FOLKOW et al. (1958) found that the cholinergic vasodilator fibers of the sympathetic system have a lower conductivity than the vasoconstrictor fibers, i.e., they may be equivalent to the S3- or S4-fibers of the cervical sympathetic chain. There are some indications that the pharmacology of the ganglion cells that give rise to these different groups of axons may differ (VOLLE, 1962b).

Evidence for two types of synapses that are quite different from the normal contacts between classical preganglionic fibers and ganglion cells comes from two types of observations: 1. the sensory innervation of the inferior mesenteric and coeliac ganglion, and 2. the histochemical evidence for adrenergic synapses in some types of ganglion cells. Although the evidence is very fragmentary, it will be discussed here in order to emphasize the desirability of further work on those aspects of the function of ganglia which go beyond the classical concept.

A. Sensory innervation of ganglia

After denervation of the inferior mesenteric ganglion of the cat, KUNTZ and
SACCOMANNO (1944) found histologically demonstrable nerve terminals in the
ganglion; they postulated that the cell bodies of these terminals were located
in the wall of the intestine and that the nerve fibers entered the ganglion with
the postganglionic trunks. This view was supported by their demonstration of
a reflex relaxation of the proximal colon in response to distension of the distal
colon, since the reflex was not abolished by preganglionic denervation of the
ganglion. Similarly, the relaxation of the proximal ileum in response to distension of the distal ileum was not abolished by preganglionic denervation of the
coeliac ganglion. SAKUSSOW (1959) described a similar viscero-visceral reflex
which caused relaxation of the intestine in response to distension of the bladder.
Chronic preganglionic denervation failed to abolish it, but removal of the
ganglion or an injection of a ganglion-blocking agent (TEA, hexamethonium or
pendiomid) abolished the response.

Electrophysiological evidence for this pathway was obtained simultaneously
by BROWN and PASCOE (1952) and by JOB and LUNDBERG (1952). The postganglionic ascending mesenteric nerve of the inferior mesenteric ganglion was
found to have both efferent and afferent fibers. Stimulation of the afferent
fibers caused a discharge in the efferent fibers of the same nerve branch. This
pathway remained intact after chronic preganglionic denervation of the ganglion, and it was blocked by injections of d-tubocurarine or nicotine. Hence, it
must be concluded that the afferent fibers synapse with the ganglion cells of the
inferior mesenteric ganglion, and that this synapse (or chain of synapses) is
sensitive to the classical ganglion-blocking agents of the hexamethonium-type.
After chronic denervation of this postganglionic branch, the outgoing signal
from the ganglion remained normal (in response to preganglionic stimulation),
while the ingoing signal was abolished. On the distal side of the cut, the
afferent fibers continued to conduct impulses. These and additional observations by MCLENNAN and PASCOE (1954) indicate that the cell bodies of the
afferent fibers are not located in the ganglion; the fibers arise somewhere in the
periphery, possibly in the intestinal wall. Recent histological observations
support these conclusions (KUZ'MINA, 1963).

Apparently the prevertebral ganglia have a sensory input that reaches the
ganglion via the postganglionic trunk. The pathway must have at least one
cholinergic synapse, since it is blocked by hexamethonium and related compounds. However, this does not necessarily indicate that the synapse between
the sensory fiber and the ganglion cell is cholinergic.

A sensory innervation of ganglion cells must also be present in the myenteric
plexus of the intestine. The adequate stimulus for induction of peristalsis is
distension of the intestinal wall, and this reflex is readily obtained in isolated
preparations suspended in an organ bath (P. TRENDELENBURG, 1917). Non-

depolarizing ganglion-blocking agents abolish the reflex (KOSTERLITZ and RO-BINSON, 1957). Hence, peristalsis appears to involve mechanisms very similar to those discussed for the prevertebral ganglia (see above).

B. Adrenergic synapses in ganglia

Various ganglia have been studied by fluorescence microscopy which permits the demonstration of nerve fibers containing monoamines. While very few fluorescent fibers were found in the paravertebral ganglia of the cat (superior cervical, stellate and various lumbar and sacral ganglia), a network of intensely fluorescent, varicose nerve fibers was found in the inferior mesenteric and coeliac ganglion of the cat. The fluorescent fibers were not related to the blood vessels of the ganglion. Some of the ganglion cells were surrounded by baskets of fluorescent fibers suggesting synapses between adrenergic fibers and ganglion cells (HAMBERGER et al., 1964). The adrenergic fibers in the inferior mesenteric ganglion of the cat are mainly of intraganglionic origin, because they do not disappear after chronic section of the preganglionic nerve (HAMBERGER and NORBERG, 1965). There appear to be species differences, since the superior cervical ganglion of the rabbit was rich in fluorescent fibers, while that of the cat was not (HAMBERGER et al., 1965).

A second type of ganglion cells with a possible adrenergic innervation is found in the Auerbach plexus of the intestine. Both NORBERG (1964) and JACO-BOWITZ (1965) found that the majority of fluorescent fibers that go to the intestine connect with the enteric plexus and form structures that may well be synapses. Very few fluorescent fibers penetrate into the smooth muscle layers, and it is likely that they innervate blood vessels rather than intestinal smooth muscle. This distribution of fluorescent fibers is strikingly different from that of cholinesterase-rich (and presumably cholinergic) nerves which can be found abundantly in the layers of intestinal smooth muscle. The adrenergic nature of the fluorescent fibers of the myenteric plexus has been demonstrated by the fact that removal of the coeliac ganglion causes their disappearance (JA-COBOWITZ, 1965).

Sympathomimetic amines cause relaxation of the intestine by acting on both α- and β-receptors (AHLQUIST and LEVY, 1959). It remains to be determined whether intestinal smooth muscle has both types of receptors or whether the α-receptors are associated with the myenteric plexus, while smooth muscle cells contain only β-receptors.

At the present time there is only one pharmacological observation which supports the view that adrenergic mechanisms may be involved in the regulation of ganglionic transmission, i.e., ECCLES and LIBET's (1961) evidence for the participation of an adrenergic mechanism in transmission through the superior cervical ganglion of the rabbit (see section VI). Because of the reported difference in density of adrenergic innervation (see above), it is of interest that this observation was made in the rabbit's rather than the cat's ganglion.

XII. Conclusions

During the third and fourth decades of this century, the acetylcholine theory of ganglionic transmission was established. During the fifth and sixth there has been a widespread interest in the ganglionic actions of agents unrelated to nicotine. These recent developments extend the classical acetylcholine theory, they do not contradict it. It must now be regarded as a possibility that various non-nicotinic ganglion-stimulating agents have modulatory effects on ganglionic transmission under physiological or pathological conditions. At the present time, it is impossible to ascribe with certainty a physiological or pathological role to any of these agents, perhaps with the exception of norepinephrine. In this case, the histochemical (and some electrophysiological) evidence is in favor of a physiological role of this amine in at least some types of ganglia (*i.e.*, the prevertebral ganglia) or in some species.

As a result of recent studies, a clear picture emerges of the pharmacology of various non-nicotinic ganglion-stimulating agents. This knowledge should enable us to use the appropriate pharmacological tools in a search for, for instance, the physiological importance of the muscarinic receptors of ganglia. It is likely that some of the recent observations will explain certain findings that were believed to be inconsistent with the acetylcholine theory of ganglionic transmission. KEWITZ (1954) found that atropine depressed ganglionic responses to acetylcholine, while not affecting responses to preganglionic stimulation. It is quite possible that the participation of muscarinic receptors in some of these effects is responsible for a discrepancy that appeared to be contradictory to the generally accepted theory.

The pharmacological approach to the problem of the physiological or pathological importance of histamine-, 5-HT-, polypeptide- or adrenergic receptors presents more difficulties than the pharmacology of the nicotinic receptors. This is due to the relative non-specificity of many of the agents which are available as pharmacological tools. For instance, not only cocaine, but many antihistaminic and β-receptor-blocking agents may well exert local anesthetic actions in addition to the ganglionic actions discussed in this review. Another example is morphine, which not only antagonizes the ganglionic action of non-nicotinic agents but also prevents the liberation of acetylcholine from the postganglionic fibers of the isolated guinea-pig ileum (PATON, 1957) and of norepinephrine from the postganglionic fibers of the nictitating membrane (TRENDELENBURG, 1957b). Hence, the pharmacological tools that we have to use in studies of non-nicotinic ganglion-stimulating agents lack the specificity so typical for the group of non-depolarizing ganglion-blocking drugs (*e.g.*, hexamethonium etc.). This complicates the task but also presents a challenge to the experimenter.

Histochemistry promises to be of increasing value in our search for the possible ganglionic functions of non-nicotinic agents, since the histochemical identification of cholinesterase-rich, of norepinephrine- and of 5-HT-containing nerve fibers should shed new light on problems of innervation by cholinergic and non-cholinergic fibers. In the past, the identification of transmitters was the exclusive domain of physiology and pharmacology. Recent developments in histochemistry promise immensely important contributions to this problem. Moreover, the combination of histochemistry and of electronmicroscopy may resolve many unsolved problems of innervation, since the former should be able to identify the nature of certain fibers, while the latter should provide a clear picture of the details of synaptic contacts.

For a pharmacologist, perhaps the most thrilling recent development is the realization that acetylcholine exerts both nicotinic and muscarinic effects on ganglion cells, and that both effects may well be of physiological importance. It is of great interest that certain nerve cells of the central nervous system have been found to have both nicotinic and muscarinic receptors. The two types of receptors can be differentiated pharmacologically in the Renshaw cells of the cat's spinal cord (CURTIS and RYALL, 1964), apparently the nicotinic receptors are responsible for quick synaptic responses, while the muscarinic receptors mediate delayed synaptic responses to acetylcholine. Similarly, both nicotinic and muscarinic receptors have been described for acetylcholine-sensitive cells in the pons and medulla oblongata of the cat (BRADLEY et al., 1966).

It has always been tempting to regard the cells of autonomic ganglia as "simple" models of the nerve cells of the central nervous system. It is satisfying to see that the "simple" model has ceased to be very simple, and that, because of the newly discovered sites of action (muscarinic receptors, α- and β-receptors, 5-HT-receptors), the pharmacology of peripheral ganglion cells presents problems which are very similar to those discussed for the central nervous system.

While any review must necessarily look back into the past in order to survey older observations and to fit them into a more or less coherent picture, it should be emphasized that the principal aim of such stock-taking of the past is to open up new approaches to unsolved problems. Histochemistry and electronmicroscopy have barely started to contribute to our field. It can be expected confidently that the old picture of ganglion cells as a "one receptor-type" relay station will soon be superseded by the concept of ganglion cells which are subject to various excitatory and inhibitory influences (of nervous or humoral origin) and which resemble the more intricate mechanisms of the central nervous system.

References

Acheson, G H , and S. A. Pereira The blocking effect of tetraethylammonium ion on the superior cervical ganglion of the cat. J. Pharmacol. exp. Ther. **87**, 273—280 (1946).

—, and J Remolina The temporal course of the effects of postganglionic axotomy on the inferior mesenteric ganglion of the cat J Physiol. (Lond.) **127**, 603—616 (1955).

—, and H. G. Schwarzacher Correlation between the physiological changes and the morphological changes resulting from axotomy in the inferior mesenteric ganglion of the cat. J. comp Neurol. **106**, 247—265 (1956)

Ahlquist, R P., and B. Levy Adrenergic receptive mechanism of canine ileum J Pharmacol exp Ther **127**, 146—149 (1959)

Ambache, N. The nicotinic action of substances supposed to be purely smooth muscle stimulating. B. Effect of $BaCl_2$ and pilocarpine on the superior cervical ganglion. J Physiol (Lond.) **110**, 164—172 (1949)

— Autonomic ganglionic stimulants. Arch int Pharmacodyn **97**, 427—446 (1954)

—, and A W Lessin Classification of intestinomotor drugs by means of type D botulinum toxin. J Physiol (Lond) **127**, 449—478 (1955).

— W. L. M. Perry, and P. A. Robertson. The effect of muscarine on perfused superior cervical ganglia of cats. Brit J. Pharmacol **11**, 442—448 (1956).

Atanackovic, D , A. F. de Schaepdryver, G. R de Vleeschhouwer, and C Heymans Effects of quaternary derivatives of tyramine and dopamine on sino-aortic chemoreceptors. Arch. int. Pharmacodyn. **157**, 202—206 (1965).

Bacq, Z M., et A Simonart L'action nicotinique de la pilocarpine. Arch. int. Pharmacodyn. **60**, 218—221 (1938)

Bainbridge, J. G., and D M Brown· Ganglion-blocking properties of atropine-like drugs. Brit. J Pharmacol **15**, 147—151 (1960)

Bárány, E. H. Transient increase in outflow facility after superior cervical ganglionectomy in rabbits. Arch. Ophthal. **67**, 303—311 (1962).

Bein, J. H., u R Meier Zur pharmakologischen Analyse der Blutdrucksteigerung durch Coramin, Histamin und Nicotin bei der narkotisierten Katze Naunyn-Schmiedebergs Arch exp. Path Pharmak **219**, 273—283 (1953).

Beleslin, D , B. Radmanović, and V. Varagić The effect of substance P on the superior cervical ganglion of the cat Brit. J Pharmacol **15**, 10—13 (1960).

—, and V. Varagić. The effect of substance P on the responses of the isolated guinea-pig ileum to acetylcholine, nicotine, histamine and 5-hydroxytryptamine Arch int. Pharmacodyn. **126**, 321—327 (1960).

Benelli, G , D Della Bella, and A Gandini Angiotensin and peripheral sympathetic activity. Brit. J. Pharmacol. **22**, 211—219 (1964)

Benfey, B. G., and D. R. Varma Inhibition of sympathomimetic effects on the heart Brit J Pharmacol **23**, 399—404 (1964)

Beraldo, W. T.· The pressor effect of peptone in the rat. Brit J. Pharmacol. **13**, 178—183 (1958)

—, and A Zanotto A ganglion-stimulating principle present in peptone Brit. J. Pharmacol. **15**, 224—229 (1960).

Birks, R , B Katz, and R. Miledi Physiological and structural changes at the amphibian myoneural junction in the course of nerve degeneration J Physiol (Lond) **150**, 145—168 (1960).

Birmingham, A. T., and A. B Wilson An analysis of the blocking action of dimethylphenylpiperazinium iodide on the inhibition of isolated small intestine produced by stimulation of the sympathetic nerves Brit J. Pharmacol. **24**, 375—386 (1965).

Bishop, G. H., and P. Heinbecker. A functional analysis of the cervical sympathetic nerve supply to the eye Amer. J. Physiol. **100**, 519—532 (1932).

Bokri, E , O. Fehér, and G Mózsik Investigation of denervation supersensitivity in a sympathetic ganglion Pflugers Arch ges Physiol **277**, 347—356 (1963)

BOURA, A. L. A , and A. F. GREEN The action of bretylium adrenergic neurone blocking and other effects Brit J. Pharmacol. **14**, 536—548 (1959).

— F. C. COPP, W. G. DUNCOMBE, A. F. GREEN, and A. MCCOUBREY. The selective accumulation of bretylium in sympathetic ganglia and their postganglionic nerves Brit. J. Pharmacol **15**, 265—270 (1960).

BOVET-NITTI, F., R. KOHN, M MAROTTA, W. P. SCOGNAMIGLIO, and B. SILVESTRINI Histamine induced modification of neuromuscular blockade by lepto- and pachy-curares Arch. int. Pharmacodyn. **149**, 308—317 (1964).

BRADLEY, P. B , B. N DHAWAN, and J H WOLSTENCROFT Pharmacological properties of cholinoceptive neurones in the medulla and pons of the cat. J. Physiol (Lond) **183**, 658—674 (1966).

BROWN, G L , and W. FELDBERG Differential paralysis of the superior cervical ganglion. J. Physiol. (Lond) **86**, 10P—11P (1936)

—, and J. E PASCOE Conduction through the inferior mesenteric ganglion of the rabbit J Physiol. (Lond) **118**, 113—123 (1952).

— — The effect of degenerative section of ganglionic axons on transmission through the ganglion. J. Physiol. (Lond.) **123**, 565—573 (1954).

BROWNLEE, G., and J. HARRY Some pharmacological properties of the circular and longitudinal muscle strips from the guinea-pig isolated ileum. Brit. J. Pharmacol. **21**, 544—554 (1963)

BRUCKE, F T. VON The cholinesterase in sympathetic ganglia J. Physiol (Lond) **89**, 429—437 (1937).

BULBRING, E The action of adrenaline on transmission in the superior cervical ganglion J. Physiol (Lond.) **103**, 55—67 (1944).

—, and J. H. BURN. An action of adrenaline on transmission in sympathetic ganglia, which may play a part in shock. J Physiol (Lond) **101**, 289—303 (1942).

BURN, J. H., and H. H. DALE. The vaso-dilator action of histamine, and its physiological significance. J. Physiol (Lond.) **61**, 185—214 (1926).

CASTRO, F. DE: Aspects anatomiques de la transmission synaptique ganglionnaire chez les mammifères. Arch. int. Physiol. **59**, 479—513 (1951).

CHEN, G , R. PORTMAN, and A WICKEL Pharmacology of 1,1-dimethyl-4-phenyl-piperazinium iodide, a ganglion-stimulating agent. J. Pharmacol. exp. Ther. **103**, 330—336 (1951).

CHIEN, S. Sensitization of denervated superior cervical ganglia to ACh. Fed Proc. **17**, 25 (1958).

CHOU, T. C., and F J. DE ELIO The anticurare activity of eserine on the superior cervical ganglion of the cat. Brit J. Pharmacol. **3**, 113—115 (1948).

COHEN, L. H , and P. B HAGEN A physiological role for the presynaptic localization of acetylcholinesterase and for its inhibition by excess substrate. Canad. J. Physiol **42**, 593—594 (1964).

COLLIER, H. O J., G. W. L. JAMES, and P J. PIPER Intensification by adrenalectomy or by β-adrenergic blockade of the bronchoconstrictor action of bradykinin in the guinea-pig. J Physiol (Lond.) **180**, 13P—14P (1965).

COPPÉE, G., et Z. M. BACQ Dégénérescence, conduction et transmission synaptique dans le sympathique cervical Arch. int. Physiol. **47**, 312—320 (1938)

COSTA, E., A. M REVZIN, R KUNTZMAN, S SPECTOR, and B. B. BRODIE Role for ganglionic norepinephrine in sympathetic synaptic transmission Science **133**, 1822—1823 (1961)

CRAGG, B G.. Failure of conduction and of synaptic transmission in degenerating mammalian C fibers. J. Physiol. (Lond.) **179**, 95—112 (1965)

CURTIS, D. R., and R. W. RYALL Nicotinic and muscarinic receptors of Renshaw cells Nature (Lond.) **203**, 652—653 (1964).

CUTHBERT, M. F. Relative actions of quaternary methyl derivatives of tyramine, dopamine and noradrenaline Brit. J. Pharmacol. **23**, 55—65 (1964).

Dale, H. H. The action of certain esters and ethers of choline, and their relation to muscarine. J. Pharmacol. exp. Ther. 6, 147—190 (1914).

—, and P. P Laidlaw The significance of the suprarenal capsules in the action of certain alkaloids. J. Physiol. (Lond.) 45, 1—26 (1912)

Davidovich, A., and J. V. Luco. The synaptic transmission of sympathetic ganglia during Wallerian degeneration. Effect of length of degenerating nerve fibers. Acta physiol. lat.-amer. 6, 49—59 (1956).

de Groat, W. C, and R L. Volle Ganglionic actions of oxotremorine. Life Sci. 8, 618—623 (1963).

—, and R. L. Volle The actions of the catecholamines on transmission in the superior cervical ganglion of the cat J Pharmacol. exp. Ther. 154, 1—13 (1966a).

— — Interactions between the catecholamines and ganglionic stimulating agents in sympathetic ganglia. J Pharmacol exp Ther. 154, 200—215 (1966b).

Del Castillo, J., and B. Katz Interaction at end-plate receptors between different choline derivatives. Proc. roy. Soc. B 146, 369—381 (1957).

Dirnhuber, P., and H. Collumbine. The effect of anticholinesterase agents on the rat's blood pressure. Brit. J. Pharmacol 10, 12—15 (1955)

Dixon, W. E, u. F. Ransom Pilocarpin, Physostigmin, Arecolin. Gifte, welche bestimmte Nervenendigungen erregen Handb. exp. Pharmak. 2, 746—816 (1924)

Douglas, W. W., and A. M. Poisner Preferential release of adrenaline from the adrenal medulla by muscarine and pilocarpine. Nature (Lond) 208, 1102—1103 (1965).

—, and J. M Ritchie· On excitation of non-medullated afferent fibres in the vagus and aortic nerves by pharmacological means. J. Physiol. (Lond.) 138, 31—43 (1957)

Dudel, J. Facilitatory effects of 5-hydroxytryptamine on the crayfish neuromuscular junction. Naunyn-Schmiedebergs Arch. exp. Path. Pharmak. 249, 515—528 (1965).

Eccles, J. C. The action potential of the superior cervical ganglion J. Physiol (Lond) 85, 179—206 (1935).

Eccles, R. M. The effect of nicotine on synaptic transmission in the sympathetic ganglion J. Pharmacol exp Ther 118, 26—38 (1956).

—, and B. Libet· Origin and blockade of the synaptic responses of curarized sympathetic ganglia. J. Physiol (Lond) 157, 484—503 (1961)

Elliott, R C Centrally active drugs and transmission through the isolated superior cervical ganglion of the rabbit when stimulated repetitively Brit J. Pharmacol 24, 76—88 (1965)

Emmelin, N., and B. C. R Stromblad A "paroxysmal" secretion of saliva following parasympathetic denervation of the parotid gland. J. Physiol. (Lond.) 143, 506—514 (1958)

Eranko, O, and M. Harkonen· Monoamine-containing small cells in the superior cervical ganglion of the rat and an organ composed of them Acta physiol. scand 63, 511—512 (1965).

Euler, U. S. von, and B. Domeij Nicotine-like actions of arecoline Acta pharmacol. (Kbh) 1, 263—269 (1945).

Exley, K. A. The blocking action of choline 2 6-xylyl ether bromide on adrenergic nerves Brit. J. Pharmacol. 12, 297—305 (1957).

Fastier, F N, M A. McDowall, and H Waal Pharmacological properties of phenyldiguanide and other amidine derivatives in relation to those of 5-hydroxytryptamine Brit. J. Pharmacol. 14, 527—535 (1959)

Fehér, O., u. E. Bokri· Über die Rolle der Cholinesterase im Prozeß der ganglionaren Erregungsubertragung Pflugers Arch. ges Physiol. 269, 55—67 (1959a)

— — Über die Acetylcholinreceptoren der sympathischen Ganglien. Pflugers Arch. ges Physiol 269, 68—76 (1959b).

Feldberg, W. The action of bee venom, cobra venom and lysolecithin on the adrenal medulla. J. Physiol (Lond.) 99, 104—118 (1940).

— Synthesis of acetylcholine in sympathetic ganglia and cholinergic nerves J Physiol (Lond.) 101, 432—445 (1943).

FELDBERG, W., and G. P. LEWIS The action of peptides on the adrenal medulla. Release of adrenaline by bradykinin and angiotensin. J. Physiol. (Lond) **171**, 98—108 (1964).

— — Further studies on the effects of peptides on the suprarenal medulla of cats. J. Physiol. (Lond.) **178**, 239—251 (1965).

— B. MINZ, and H. TSUDZIMURA The mechanism of the nervous discharge of adrenaline J. Physiol. (Lond.) **81**, 286—304 (1934)

—, and A. VARTIAINEN: Further observations on the physiology and pharmacology of a sympathetic ganglion J. Physiol. (Lond.) **83**, 103—128 (1935).

FLACKE, W., and R. A. GILLIS Muscarinic ganglionic transmission. Pharmacologist **8**, 193 (1966)

FLECKENSTEIN, A., and J. H. BURN The effect of denervation on the action of sympathomimetic amines on the nictitating membrane. Brit. J. Pharmacol. **8**, 69—78 (1953).

—, u. D. STOCKLE Zum Mechanismus der Wirkungsverstarkung und Wirkungsabschwachung sympathomimetischer Amine durch Cocain und andere Pharmaka. II. Die Hemmung der Neuro-Sympathomimetica durch Cocain Naunyn-Schmiedebergs Arch exp Path Pharmak. **224**, 401—415 (1955).

FOLKOW, B., B. JOHANSSON, and B. ÖBERG The stimulation threshold of different sympathetic fiber groups as correlated to their functional differentiation. Acta physiol scand. **44**, 146—156 (1958).

FRANKO, B V., J W. WARD, and R. S. ALPHIN Pharmacologic studies of N-benzyl-3-pyrrolidyl acetate methobromide (AHR-602), a ganglion stimulating agent. J. Pharmacol. exp. Ther **139**, 25—30 (1963).

FREDRICSSON, B, and F. SJOQVIST A cytomorphological study of cholinesterase in sympathetic ganglia of the cat Acta morph neerl -scand. **5**, 140—166 (1962).

FREYBURGER, W. A, C. C. GRUHZIT, B R. RENNICK, and G. K MOE· Action of tetraethylammonium on pressure response to asphyxia Amer. J. Physiol. **163**, 554—560 (1950).

GADDUM, J H Tryptamine receptors J Physiol (Lond) **119**, 363—368 (1953)

—, and N. J. GIARMAN Preliminary studies on the biosynthesis of 5-HT. Brit J. Pharmacol **11**, 88—92 (1956).

—, and M. K PAASONEN The use of some molluscan hearts for the estimation of 5-HT. Brit. J Pharmacol. **10**, 474—483 (1955)

—, and Z P. PICARELLI· Two kinds of tryptamine receptors. Brit. J. Pharmacol. **12**, 323—328 (1957).

GEBBER, G. L, and R. L. VOLLE Mechanisms involved in ganglionic blockade induced by tetramethylammonium. J. Pharmacol. exp. Ther. **152**, 18—28 (1966).

GERTNER, S B. The effect of compound 48/80 on ganglionic transmission. Brit J. Pharmacol. **10**, 103—109 (1955).

— The effects of monoamine oxidase inhibitors on ganglionic transmission. J. Pharmacol exp Ther **131**, 223—230 (1961).

— Histamine pools in sympathetic ganglia. Pharmacologist **7**, 153 (1965).

—, and R KOHN Effect of histamine on ganglionic transmission. Brit. J. Pharmacol. **14**, 179—182 (1959)

— M K PAASONEN, and N. J GIARMAN Studies concerning the presence of 5-hydroxytryptamine (serotonin) in the perfusate from the superior cervical ganglion J. Pharmacol exp. Ther **127**, 268—275 (1959).

—, and A. ROMANO. Action of guanethidine and bretylium on ganglionic transmission Fed Proc. **20**, 319 (1961)

GIBSON, W. C.· Degeneration and regeneration of sympathetic synapses. J. Neurophysiol. **3**, 237—247 (1940).

GINSBORG, B L. The actions of McN-A-343, pilocarpine and acetyl-β-methylcholine on sympathetic ganglion cells of the frog J. Pharmacol. exp Ther **150**, 216—219 (1965).

—, and S GUERRERO On the action of depolarizing drugs on sympathetic ganglion cells of the frog J Physiol. (Lond.) **172**, 189—206 (1964).

GINZEL, K. H., and S R KOTTEGODA· The action of 5-hydroxytryptamine and tryptamine on aortic and carotid sinus receptors in the cat J. Physiol (Lond) 123, 277—
288 (1954).

GOKHALE, S. D , O D GULATI, and N. Y. JOSHI Participation of an unusual ganglionic
pathway in the mediation of the pressor effect of physostigmine in the rat Brit. J.
Pharmacol. 23, 34—42 (1964)

GOLD, D., and H REINERT: The depolarizing and blocking action of some sympathomimetic amines in the cat's superior cervical ganglion. J. Physiol (Lond) 151, 3 P—4 P
(1960)

GRIMSON, K. S , A. K TARAZI, and J. W FRAZER· A new orally active quaternary
ammonium, ganglion blocking drug capable of reducing blood pressure, SU-3088.
Circulation 11, 733—741 (1955).

GROMADZKI, C. G., and G B. KOELLE The effect of axotomy on the acetylcholinesterase
of the superior cervical ganglion of the cat Biochem Pharmacol. 14, 1745—1754
(1965).

GUTH, L.· Functional recovery following vagosympathetic anastomosis in the cat. Amer.
J. Physiol. 185, 205—208 (1956).

GYERMEK, L. The action of 5-hydroxytryptamine on the urinary bladder of the dog
Pharmacologist 2, 89 (1960).

— 5-Hydroxy-3-indoleacetamidine. a new type of blocking agent against 5-HT. Nature
(Lond) 192, 465 (1961)

— Action of guanidine derivatives on autonomic ganglia. Arch. int. Pharmacodyn. 150,
570—581 (1964).

—, and E. BINDLER Blockade of the ganglionic stimulant action of 5-hydroxytryptamine
J. Pharmacol. exp Ther. 135, 344—348 (1962a).

— — Action of indole alkylamines and amidines on the inferior mesenteric ganglion
of the cat. J. Pharmacol exp. Ther. 138, 159—164 (1962b).

— E B SIGG, and E. BINDLER Ganglionic stimulant action of muscarine. Amer. J
Physiol. 204, 68—70 (1963).

—, u. L. SZTANYIK Die Wirkung des Histamins auf die Nebenniere Kisérl Orvostud.
4, 1—8 (1952).

—, and K R. UNNA Relation of structure of synthetic muscarines and muscarones to
their pharmacological action Proc. Soc. exp. Biol. (N.Y.) 98, 882—885 (1958)

HAMBERGER, B , and K A. NORBERG Studies on some systems of adrenergic synaptic
terminals in the abdominal ganglia of the cat. Acta physiol. scand 65, 235—242
(1965)

— —, and F. SJOQVIST Evidence for adrenergic nerve terminals and synapses in
sympathetic ganglia Int J. Neuropharmacol 2, 279—282 (1964)

— —, and U. UNGERSTED Adrenergic synaptic terminals in autonomic ganglia Acta
physiol. scand. 64, 285—286 (1965).

HANNA, C., P B. McHUGO, and W H. MACMILLAN Cardiovascular actions of histamine
and potassium Amer. J Physiol 197, 1005—1007 (1959).

HARRY, J. The action of drugs on the circular muscle strip from the guinea pig isolated
ileum. Brit. J. Pharmacol. 20, 399—417 (1963).

HEATON, T B., and M H MACKEITH On the action of pilocarpine J Physiol (Lond.)
63, 42—50 (1927).

HEBB, C. O., and H KONZETT The effect of certain analgesic drugs on synaptic transmission as observed in the perfused superior cervical ganglion of the cat. Quart. J
exp. Physiol. 35, 213—217 (1949).

—, and G. M. H. WAITES Choline acetylase in antero- and retrograde degeneration of a
cholinergic nerve. J. Physiol. (Lond) 132, 667—671 (1956).

HERR, F., and L GYERMEK Action of cholinergic stimulants on the inferior mesenteric
ganglion of the cat J. Pharmacol exp. Ther. 129, 338—342 (1960)

HERTZLER, E. C. 5-Hydroxytryptamine and transmission in sympathetic ganglia Brit.
J. Pharmacol. 17, 406—413 (1961).

HEYMANS, C, et G. VAN DEN HEUVEL-HEYMANS Sur la pharmacologie de l'hydroxy-tryptamine et d'une substance analogue. Arch int Pharmacodyn. 93, 95—104 (1953).

HILTON, J G The pressor response to neostigmine after ganglionic blockade. J. Pharmacol. exp. Ther. 132, 23—28 (1961).

—, and M. STEINBERG Effects of ganglion- and parasympathetic-blocking drugs upon the pressor response elicited by elevation of the intracranial fluid pressure. J Pharmacol. exp. Ther. 153, 285—291 (1966).

HIRSCHMANN, L. Zur Lehre von der durch Arzneimittel hervorgerufenen Miosis und Mydriasis Arch. Anat. Physiol. 1863, 309—318.

HOET, J · Sur les effets de la pilocarpine Ann Physiol. 4, 714—715 (1928).

HOGBEN, L T , W SCHLAPP, and A D MacDONALD Studies on the pituitary. IV Quantitative comparison of pressor activity. Quart J exp Physiol 14, 301—318 (1924)

HOLADAY, D. A., K. KAMIJO, and G. B. KOELLE Facilitation of ganglionic transmission following inhibition of cholinesterase by DFP. J. Pharmacol exp. Ther. 111, 241—254 (1954)

HOLMSTEDT, B , and F SJOQVIST Distribution of acetocholinesterase in the ganglion cells of various sympathetic ganglia. Acta physiol scand. 47, 284—296 (1959).

HUNT, C. C., and P. G NELSON Structural and functional changes in the frog sympathetic ganglion following cutting of the presynaptic nerve fibers. J. Physiol. (Lond.) 177, 1—20 (1965)

HUNT, R Vasodilator reactions II Amer J Physiol. 45, 231—267 (1918).

IORIO, L C , and R J McISAAC Comparison of the stimulating effects of nicotine, pilocarpine, and histamine on the superior cervical ganglion of the cat J Pharmacol. exp. Ther. 151, 430—437 (1966)

JAANUS, S D , and R. P. RUBIN The action of sympathomimetic amines on the adrenal medulla. Pharmacologist 7, 168 (1965).

JACOBOWITZ, D Histochemical studies of the autonomic innervation of the gut. J. Pharmacol. exp. Ther. 149, 358—364 (1965).

JOB, C , and A LUNDBERG Reflex excitation of cells in the inferior mesenteric ganglion on stimulation of the hypogastric nerve. Acta physiol. scand. 26, 366—382 (1952)

JONES, A.· Ganglionic actions of muscarinic substances J Pharmacol exp. Ther. 141, 195—205 (1963).

— B. GOMEZ ALONSO DE LA SIERRA, and U. TRENDELENBURG The pressor response of the spinal cat to different groups of ganglion-stimulating agents. J. Pharmacol. exp Ther 139, 312—320 (1963)

KAMIJO, K , and G B. KOELLE· The relationship between cholinesterase inhibition and ganglionic transmission. J Pharmacol exp Ther 105, 349—357 (1952).

KANEKO, Y., J. W McCUBBIN, and I. H PAGE Ability of vasoconstrictor drugs to cause adrenal medullary discharge after 'sensitization' by ganglion stimulating agents. Circulat Res 9, 1247—1254 (1961)

KEWITZ, H Zur Bedeutung des Acetylcholins als Übertragersubstanz sympathischer Gangien. Naunyn-Schmiedebergs Arch exp. Path. Pharmak 222, 323—329 (1954).

—, u. H. REINERT Wirkung verschiedener Sympathicomimetica auf die chemisch und elektrisch ausgeloste Erregung des oberen Halsganglions Naunyn-Schmiedebergs Arch. exp. Path. Pharmak 222, 311—314 (1954)

KIBJAKOW, A W.. Über humorale Übertragung der Erregung von einem Neuron auf das andere. Pflugers Arch ges Physiol. 232, 432—443 (1933).

KOELLE, G. B Cytological distribution and physiological functions of cholinesterases. Cholinesterases and Anticholinesterase Agents. (Ed G B. KOELLE) Handbuch der experimentellen Pharmakologie, Erg.-Bd. 15, p. 1220. Berlin-Gottingen-Heidelberg. Springer 1963.

KOMALAHIRANYA, A , and R. L. VOLLE Actions of inorganic ions and veratrine on asynchronous postganglionic discharge in sympathetic ganglia treated with diisopropyl phosphofluoridate (DFP). J. Pharmacol. exp. Ther 138, 57—65 (1962).

Konzett, H   Sympathomimetica und Sympathicolytica am isoliert durchstromten Ganglion cervicale superius der Katze. Helv. physiol. pharmacol. Acta **8**, 245—258 (1950).
— The effect of histamine on an isolated sympathetic ganglion. J. Mt Sinai Hosp. **19**, 149—153 (1952).
—, u. E. Rothlin   Beeinflussung der nikotinartigen Wirkung von ACh durch Atropin. Helv physiol pharmacol. Acta **7**, C46 (1949)
— — Die Wirkung synaptotroper Substanzen auf gewisse efferente und afferente Strukturen des autonomen Nervensystems. Experientia (Basel) **9**, 405 (1953).
—, u. P. G. Waser   Zur ganglionaren Wirkung von Muskarin. Helv. physiol pharmacol Acta **14**, 202—206 (1956).
Koppanyi, Th. · Studies on the synergism and antagonism of drugs I. The non-parasympathomimetic antagonism between atropine and the miotic alkaloids J. Pharmacol exp Ther. **46**, 395—405 (1932).
— The hemodynamic effect of pilocarpine J Pharmacol. exp. Ther **65**, 19—20 (1939)
Kosterlitz, H W., and J. A Robinson Inhibition of the peristaltic reflex of the isolated guinea-pig ileum. J Physiol (Lond.) **136**, 249—262 (1957)
Kristic, M., and V Varagić Difference between the blood pressure responses to intrajugular and intracarotid injections of physostigmine, and the syndrome of contralateral torsion in the rat. Brit J Pharmacol **24**, 132—137 (1965).
Kuntz, A , and G Saccomanno. Reflex inhibition of intestinal motility mediated through decentralized prevertebral ganglia. J Neurophysiol. **7**, 163—170 (1944).
Kuz'mina, S. V · Structural organization of inferior mesenteric ganglion. Arkh. Anat Gistol. Embriol. **45**, 51 (1963) Fed Proc. **23**, T706—T710 (1964).
Langer, S Z The degeneration contraction of the nictitating membrane in the unanesthetized cat. J. Pharmacol exp Ther. **151**, 66—72 (1966a).
— Presence of tone in the denervated and the decentralized nictitating membrane of the spinal cat and its influence on determinations of supersensitivity J Pharmacol exp. Ther. **154**, 14—34 (1966b)
—, P R Draskóczy, and U Trendelenburg Time course of development of supersensitivity of the denervated or decentralized nictitating membrane of the cat. J Pharmacol. exp Ther. in press (1967)
Langley, J N., and W L Dickinson On the local paralysis of peripheral ganglia, and on the connexion of different classes of nerve fibres with them. Proc. roy. Soc Series B **46**, 423—431 (1889).
Lecomte, J. · Sensibilisation à l'adrénaline par la 5-HT Arch int Physiol. **61**, 84—85 (1953).
— J Troquet et A. Dresse Stimulation médullo-surrénalienne par la bradykinine Arch. int Physiol. **69**, 89—91 (1961)
Lee, F. L., and U. Trendelenburg· Transmission of preganglionic impulses through the muscarinic receptors of the adrenal medulla of the cat In preparation, 1967.
Lee, W. C, and F. E. Shideman The inotropic actions of quaternary ammonium compounds J. Pharmacol exp Ther. **127**, 219—228 (1959).
Lei, B W , and G D. Maengwyn-Davies Adrenergic receptors in the superior cervical ganglion (SCG) of the cat Pharmacologist **7**, 185 (1965).
Levy, B., and R P. Ahlquist A study of sympathetic ganglionic stimulants J Pharmacol exp. Ther **137**, 219—228 (1962).
Lewis, G P , and E. Reit: The action of angiotensin and bradykinin on the superior cervical ganglion of the cat. J. Physiol. (Lond) **179**, 538—553 (1965).
Libet, B. Slow synaptic responses and excitatory changes in sympathetic ganglia. J. Physiol. (Lond.) **174**, 1—25 (1964).
Lindell, S E , H. Westling, and T White Release of pressor amines from the adrenal glands of the cat by two histamine analogues. Arch. int. Pharmacodyn. **127**, 410—412 (1960).

LINDMAR, R.: Die Wirkung von 1,1-Dimethyl-4-Phenyl-Piperazinium-Jodid am isolierten Vorhof im Vergleich zur Tyramin- und Nicotinwirkung. Naunyn-Schmiedebergs Arch. exp. Path. Pharmak. **242**, 458—466 (1962).

LING, H W.. Actions of dimethylphenylpiperazinium. Brit J. Pharmacol. **14**, 505—511
(1959).

LONG, J. P., and J. W. ECKSTEIN. Ganglionic actions of neostigmine methylsulfate.
J. Pharmacol. exp. Ther. **133**, 216—222 (1961).

LUNDBERG, A.: Adrenaline and transmission in the sympathetic ganglion of the cat.
Acta physiol scand **26**, 252—263 (1952).

MACINTOSH, F. C.: L'effet de la section des fibres préganglionnaires sur la teneur en
acétylcholine du ganglion sympathique. Arch. int. Physiol. **47**, 321—324 (1938).

MACMILLAN, W. H.· The effect of histamine on skeletal muscle contraction. Arch. int.
Pharmacodyn. **108**, 19—26 (1956).

MALMÉJAC, J.· Action of adrenaline on synaptic transmission and on adrenal medullary
secretion J. Physiol. (Lond.) **130**, 497—512 (1955).

MARRUZZI, A. S. Electrical studies on the pharmacology of autonomic synapses. I. The
action of parasympathomimetic drugs on sympathetic ganglia J. Pharmacol. exp.
Ther **65**, 18—35 (1939a)

— Electrical studies on the pharmacology of autonomic synapses. II. The action of a
sympathomimetic drug (epinephrine) on sympathetic ganglia. J. Pharmacol. exp.
Ther. **65**, 395—404 (1939b).

—, and R. N. MARRAZZI. Further localization and analysis of adrenergic synaptic inhibition J Neurophysiol. **10**, 165—178 (1947).

MASON, D. F. J.: A ganglion stimulating action of neostigmine. Brit. J. Pharmacol. **18**,
76—86 (1962a).

— Depolarizing action of neostigmine at an autonomic ganglion. Brit. J. Pharmacol.
18, 572—587 (1962b).

MATSUMURA, M., and G. B. KOELLE. The nature of synaptic transmission in the superior
cervical ganglion following reinnervation by the afferent vagus. J Pharmacol. exp.
Ther. **134**, 28—46 (1961).

MATTHEWS jr., R. J. The effect of epinephrine, levarterenol and dl-isoproterenol on
transmission in the superior cervical ganglion of the cat. J. Pharmacol. exp. Ther.
116, 433—443 (1956).

MAXWELL, R. A, A. J. PLUMMER, S. D. ROSS, and M. W. OSBORNE. Effect of ganglionic
blocking agents on pressor responses induced by splanchnic faradization. Proc. Soc.
exp. Biol. (N.Y) **92**, 225—227 (1956).

McCUBBIN, J. W., J. H. GREEN, G. C. SALMOIRAGHI, and I H PAGE The chemoreceptor
stimulant action of serotonin in dogs. J Pharmacol exp. Ther. **116**, 191—197 (1956).

McKINSTRY, D. N., E. KOENIG, W. A. KOELLE, and G. B KOELLE· The release of
acetylcholine from a sympathetic ganglion by carbachol. Canad. J. Biochem. **41**,
2599—2609 (1963).

McLENNAN, H., and J. E PASCOE· The origin of certain non-medullated nerve fibres
which form synapses in the inferior mesenteric ganglion of the rabbit. J. Physiol.
(Lond.) **124**, 145—156 (1954).

MENDEZ, R., and A. RAVIN On the action of prostigmine on the circulatory system.
J. Pharmacol. exp. Ther. **72**, 80—89 (1941).

MOE, G. K, and W. A FREYBURGER: Ganglionic blocking agents. Pharmacol. Rev. **2**,
61—95 (1950).

MURAYAMA, S., and K R.UNNA: Stimulant action of 4-(m-chlorophenyl-carbamoyloxy)-
2-butynyltrimethylammonium chloride (McN-A-343) on sympathetic ganglia. J.
Pharmacol. exp. Ther. **140**, 183—192 (1963).

MURRAY, J. G., and J. W. THOMPSON. Collateral sprouting in response to injury of the
autonomic nervous system and its consequences. Brit. med. Bull. **13**, 213—219 (1957a).

— — The occurrence and function of collateral sprouting in the sympathetic nervous
system of the cat. J Physiol. (Lond) **135**, 133—162 (1957b).

Norberg, K. A   Adrenergic innervation of the intestinal wall studied by fluorescence microscopy. Int J. Neuropharmacol. 3, 379—382 (1964)

Page, I H , and J W. McCubbin   The variable arterial pressure response to serotonin in laboratory animals and man   Circulat Res. 1, 354—362 (1953).

Pardo, E G , J Cato, E Gijón, and F Alonso de Florida   Influence of several adrenergic drugs on synaptic transmission through the superior cervical ganglion and the ciliary ganglion of the cat   J Pharmacol. exp Ther 139, 296—303 (1963).

Pascoe, J. E.   The effects of acetylcholine and other drugs on the isolated superior cervical ganglion. J. Physiol. (Lond.) 132, 242—255 (1956).

Paton, W D M   Types of pharmacological action at autonomic ganglia   Arch. int. Pharmacodyn 97, 267—281 (1954).

— The action of morphine and related substances on contraction and on acetylcholine-output of coaxially stimulated guinea-pig ileum. Brit. J Pharmacol 12, 119—127 (1957).

— A theory of drug action based on the rate of drug-receptor combination   Proc. roy Soc B 154, 21—69 (1961).

—, and W. L M Perry   The relationship between depolarization and block in the cat's superior cervical ganglion. J. Physiol. (Lond) 119, 43—57 (1953).

—, and J. W. Thompson   The mechanism of action of adrenaline on the superior cervical ganglion of the cat. XIX. Int. Congress, Montreal, p 664, 1953

Pelikan, E W.   The mechanism of ganglionic blockade produced by nicotine   Ann. N Y Acad Sci 90, 52—69 (1960).

Perry, W L M   Acetylcholine release in the cat's superior cervical ganglion   J Physiol (Lond) 119, 439—454 (1953)

—, and H Reinert   The effects of preganglionic denervation on the reactions of ganglion cells. J Physiol (Lond.) 126, 101—115 (1954)

— — On the metabolism of normal and denervated sympathetic ganglion cells   J Physiol. (Lond) 130, 156—166 (1955).

Reas, H W , and U Trendelenburg   Denervation supersensitivity of the sweat glands of the cat. J. Pharmacol exp Ther in press (1967).

Reid, G   Circulatory effects of 5-hydroxytryptamine   J Physiol (Lond) 118, 435—453 (1952).

Reinert, H.   The effect of amphetamines on peripheral synaptic structures   In Neuropsychopharmacology, p 399—404. Amsterdam   Elsevier Publ Co. 1959

— The depolarizing and blocking action of amphetamine in the cat's superior cervical ganglion   In Adrenergic Mechanisms, p 373—379. Boston   Little, Brown & Co. 1960

— Role and origin of noradrenaline in the superior cervical ganglion   J Physiol. (Lond.) 167, 18—29 (1963).

Robertson, P. A..   An antagonism of 5-hydroxytryptamine by atropine   J. Physiol (Lond) 121, 54P—55P (1953).

— Potentiation of HT by the true cholinesterase inhibitor 284C 51   J. Physiol (Lond) 125, 37P—38P (1954)

Rocha e Silva, M , J R Valle, and Z P. Picarelli   A pharmacological analysis of the mode of action of serotonin (5-hydroxytryptamine) upon the guinea-pig ileum. Brit J Pharmacol. 8, 378—388 (1953)

Root, M A   Certain aspects of the vasopressor action of pilocarpine. J Pharmacol. exp Ther 101, 125—131 (1951).

Rossum, J M van   Classification and molecular pharmacology of ganglionic blocking agents   Part I. Mechanism of ganglionic synaptic transmission and mode of action of ganglionic stimulants. Int J. Neuropharmacol. 1, 97—110 (1962).

Roszkowski, A P   An unusual type of ganglionic stimulant   J Pharmacol. exp Ther 132, 156—170 (1961)

Sakussow, W. W   Die Wirkung von ganglienblockierenden Substanzen auf visceroviscerale Reflexe   Naunyn-Schmiedebergs Arch. exp. Path Pharmak 236, 156—157 (1959)

SALERNO, P. R , and J. M. COON· A pharmacologic comparison of hexaethyl tetraphosphate (HETP) and tetraethyl pyrophosphate (TEPP) with physostigmine, neostigmine and DFP. J. Pharmacol. exp Ther. **95**, 240—255 (1949).

SCHENK, E. A , and E. G ANDERSON. The effect of histamine on neuromyal blocking agents. J Pharmacol. exp Ther **122**, 234—238 (1958).

SCHMIEDEBERG, O.. Ber. kgl. sachs Ges. Wiss. **22**, 130 (1870). Quoted by W HEUBNER· Historische Notiz uber die Entdeckung der Synapsen im autonomen Nervensystem mit Hilfe des Nikotins. Naunyn-Schmiedebergs Arch exp. Path. Pharmak. **204**, 33—35 (1947).

SHAND, D G The mode of action of drugs blocking ganglionic transmission in the rat. Brit J. Pharmacol **24**, 89—97 (1965).

SIEHE, H. J Die Reaktion des denervierten Nebennierenmarkes auf humorale Sekretionsreize Pflugers Arch ges Physiol. **234**, 204—209 (1934)

SLATER, I H , and P. E. DRESEL· The pressor effect of histamine after autonomic ganglionic blockade. J. Pharmacol exp Ther **105**, 101—107 (1952).

SMITH, C. B., U TRENDELENBURG, S Z LANGER, and T. H. TSAI The relation of retention of norepinephrine-H³ to the norepinephrine content of the nictitating membrane of the spinal cat during development of denervation supersensitivity J. Pharmacol exp Ther **151**, 87—94 (1966)

SMITH, J. C. Observations on the selectivity of stimulant action of 4-(m-chlorophenylcarbamoyloxy)-2-butynyltrimethylammonium chloride on sympathetic ganglia. J. Pharmacol exp Ther **153**, 266—275 (1966a).

— Pharmacologic interactions with 4-(m-chlorophenylcarbamoyloxy)-2-butynyltrimethylammonium chloride, a sympathetic ganglion stimulant. J Pharmacol. exp Ther. **153**, 276—284 (1966b)

STASZEWSKA-BARCZAK, J., and J. R VANE The release of catecholamines from the adrenal medulla by peptides. J. Physiol (Lond) **177**, 57P—58P (1965a).

— — The release of catechol amines from the adrenal medulla by histamine Brit. J. Pharmacol. **25**, 728—742 (1965b).

STEWARD, G N , and J. M ROGOFF The action of drugs upon the output of epinephrine from the adrenals. VII Physostigmine. J. Pharmacol exp. Ther. **17**, 227—248 (1921).

SUDEN, C. TUM, and A S. MARRAZZI Synaptic inhibitory action of adrenaline at parasympathetic synapses Fed Proc **10**, 138 (1951).

SZCZYGIELSKI, J. Die adrenalinabsondernde Wirkung des Histamins und ihre Beeinflussung durch Nikotin. Naunyn-Schmiedebergs Arch exp Path. Pharmak. **166**, 319—332 (1932).

TAKESHIGE, C., A. J. PAPPANO, W C DE GROAT and R L VOLLE Ganglionic blockade produced in sympathetic ganglia by cholinomimetic drugs. J Pharmacol. exp. Ther **141**, 333—342 (1963).

—, and R. L VOLLE· Bimodal response of sympathetic ganglia to acetylcholine following eserine or repetitive preganglionic stimulation J. Pharmacol. exp. Ther. **138**, 66—73 (1962)

— — Asynchronous postganglionic firing from the cat superior cervical sympathetic ganglion treated with neostigmine. Brit. J. Pharmacol. **20**, 214—220 (1963a).

— — Cholinoceptive sites in denervated sympathetic ganglia. J. Pharmacol. exp Ther **141**, 206—213 (1963b)

— — Modification of ganglionic responses to cholinomimetic drugs following preganglionic stimulation, anticholinesterase agents and pilocarpine J. Pharmacol. exp. Ther. **146**, 335—343 (1964a).

— — A comparison of the ganglion potentials and block produced by acetylcholine and tetramethylammonium. Brit J. Pharmacol. **23**, 80—89 (1964b).

— — Similarities in the ganglionic actions of calcium ions and atropine. J. Pharmacol. exp Ther. **145**, 173—180 (1964c).

Thompson, J. W.. Studies on the response of the isolated nictitating membrane of the cat J. Physiol. (Lond) 141, 46—72 (1958).

Trendelenburg, P Physiologische und pharmakologische Versuche uber die Dünndarmperistaltik Naunyn-Schmiedebergs Arch. exp Path Pharmak. 81, 55—129 (1917).

Trendelenburg, U.. The action of histamine and pilocarpine on the superior cervical ganglion and the adrenal glands of the cat. Brit J. Pharmacol. 9, 481—487 (1954).

— The potentiation of ganglionic transmission by histamine and pilocarpine J. Physiol (Lond.) 129, 337—351 (1955).

— The action of 5-hydroxytryptamine on the nictitating membrane and on the superior cervical ganglion of the cat. Brit. J. Pharmacol. 11, 74—80 (1956a).

— Modification of ganglionic transmission through the superior cervical ganglion of the cat. J. Physiol. (Lond) 132, 529—541 (1956b).

— The action of histamine, pilocarpine and 5-HT on transmission through the superior cervical ganglion. J. Physiol. (Lond) 135, 66—72 (1957a).

— The action of morphine on the superior cervical ganglion and on the nictitating membrane of the cat. Brit. J. Pharmacol. 12, 79—85 (1957b).

— Reaktion sympathischer Ganglien wahrend der Ganglienblockade durch Nikotin Naunyn-Schmiedebergs Arch exp. Path. Pharmak. 230, 448—456 (1957c).

— Stimulation of sympathetic centres by histamine. Circulat. Res. 5, 105—110 (1957d).

— Non-nicotinic ganglion-stimulating substances. Fed. Proc. 18, 1001—1005 (1959)

— Observations on the mode of action of some non-depolarizing ganglion-blocking substances. Naunyn-Schmiedebergs Arch. exp. Path. Pharmak 241, 452—466 (1961a).

— The pressor response of the cat to histamine and pilocarpine. J. Pharmacol. exp Ther. 131, 65—72 (1961b).

— The action of acetylcholine on the nictitating membrane of the spinal cat J. Pharmacol. exp. Ther. 135, 39—44 (1962).

— Supersensitivity and subsensitivity to sympathomimetic amines. Pharmacol. Rev. 15, 225—276 (1963).

— Observations on the ganglion-stimulating action of angiotensin and bradykinin. J Pharmacol. exp Ther. 154, 418—425 (1966a).

— Transmission of preganglionic impulses through the muscarinic receptors of the superior cervical ganglion of the cat. J Pharmacol. exp. Ther. 154, 426—440 (1966b).

— Mechanism of supersensitivity and subsensitivity to sympathomimetic amines. Pharmacol. Rev. 18, 629—640 (1966c).

—, and A. Jones. Facilitation of ganglionic responses after a period of preganglionic stimulation. J. Pharmacol. exp Ther. 147, 330—335 (1965).

Varagić, V · The action of eserine on the blood pressure of the rat Brit. J. Pharmacol. 10, 349—353 (1955).

Vartiainen, A.· The action of certain new histamine derivatives J. Pharmacol. exp. Ther. 54, 265—282 (1935).

Vogt, M. Release of medullary amines from the isolated perfused adrenal gland of the dog. Brit. J Pharmacol. 24, 561—565 (1965).

Volle, R. L.: The actions of several ganglion blocking agents on the postganglionic discharge induced by diisopropylphosphorofluoridate (DFP) in sympathetic ganglia J. Pharmacol. exp. Ther. 135, 45—53 (1962a).

— The responses to ganglionic stimulating and blocking drugs of cell groups within a sympathetic ganglion. J. Pharmacol exp. Ther. 135, 54—61 (1962b).

— Enhancement of postganglionic responses to stimulating agents following repetitive preganglionic stimulation. J. Pharmacol exp. Ther. 136, 68—74 (1962c).

—, and G. B. Koelle. The physiological role of acetylcholinesterase (AChE) in sympathetic ganglia. J. Pharmacol. exp. Ther. 133, 223—240 (1961).

WASER, P. G. Fehlende ganglionare (nikotinische) Wirkungen des Muscarins Experientia (Basel) **11**, 452—453 (1955).

— Chemistry and pharmacology of mucarine, muscarone, and some related compounds. Pharmacol. Rev. **13**, 465—516 (1961)

WEIDMANN, H., u A. CERLETTI · Untersuchungen uber die pressorische Wirkung des 5-Hydroxytryptamins (Serotonin). Arch. int. Pharmacodyn. **111**, 98—107 (1957).

WEIR, M C L , and H. McLENNAN: The action of catecholamines in sympathetic ganglia. Canad. J. Biochem. **41**, 2627—2636 (1963).

WINBURY, M M. Comparison of the vascular actions of 1,1-dimethyl-4-phenylpiperazinium and tetramethylammonium. J. Pharmacol. exp Ther **124**, 25—34 (1958).

WISPELAERE, H DE · Actions de l'acétyl-β-méthylcholine, de l'éthyl-β-méthylcholine et de l'éthylcholine sur la circulation et sur la respiration Arch. int. Pharmacodyn. **56**, 363—375 (1937)

Topical Application of Drugs to Subcortical Brain Structures and Selected Aspects of Electrical Stimulation

T. J. MARCZYNSKI*

Table of contents   Page

Introduction 87

I. Limitations in investigations on the central effects of neurohormonal substances administered into the systemic circulation 90
- A The blood brain barrier 90
- B Synaptic barrier 90
- C. Osmoreceptors 92
- D Intracranial pressoreceptors . . . 93
- E Intracranial chemoreceptors 93
- F. Transport and selected actions 94
  - 1 Catecholamines 94
  - 2 5-OH tryptamine 95
  - 3 Acetylcholine 96

II. Intraventricular and intracisternal administration 97

III. Direct application of drugs into the brain tissue . 99
- A. General considerations 99
  - 1. Methodological remarks 99
  - 2 Potential side-effects 100
  - 3 Main advantages 101
- B. Central effects of steroid hormones 104
  - 1 Hypothalamic control of the adenohypophysis . . . 105
    - a) Adrenal steroids 105
    - b) Gonadal steroids 106
  - 2 Behavioral effects of gonadal steroids . 108
- C Evidence of neurohormonal regulation of food and water intake . 109
  - 1 Introductory remarks 109
  - 2 Chemical stimulation of the hypothalamic "feeding area" 111
- D. Electrophysiological and neurohormonal evidence of the existence of two functionally opposed ascending reticular systems . 113
  - 1 EEG and behavioral effects of surgical ablation and electrical stimulation 114
  - 2 Alterations in incoming signals 116
  - 3. Cortical photic responses influenced by pontine transection . 117
  - 4 Action of drugs upon cortical optic responses 117
    - a) Drug-induced interruption of ascending reticular pathways 117
    - b) Cholinergic and adrenergic drugs 118

* Assistant Professor of Pharmacology, University of Illinois, College of Medicine.

Page

    5. EEG and behavioral effects of neurohormonal substances . . .   119
       a) Adrenergic stimulation   120
       b) Cholinergic stimulation .   .   .   120
       c) Tryptaminergic stimulation . .   .   124
    6 Concluding remarks   125
  E. Electrophysiological and neurohormonal evidence of the existence of the
    forebrain descending inhibitory system 127
    1 EEG and behavioral effects of electrical stimulation of the forebrain
      hypnogenic structures 127
    2. Neurohormonal stimulation of the forebrain limbic–midbrain limbic
      pathway of NAUTA   128
       a) Cholinergic and adrenergic stimulation   . .   128
       b) Tryptamine and related compounds   131

IV. Speculations on the relationship between central neurohormonal mechanisms
  and effects of electrical stimulation   132
    1. Extraneuronal influences   133
    2. Possible role of acetylcholine in the liberation of neurotransmitters . . . 134
    3. The relationship between impulse frequency and preferential liberation
      of acetylcholine or noradrenaline.   136
    4 Cholinesterase inhibitors and "cholinergic link" in central adrenergic
      mechanisms.   138
    5. Possible central neurohormonal effects of electrical stimulation at
      different frequencies   139
  V. Summary and conclusions 141

Bibliography 143

Introduction

Since the demonstration that catecholamines, acetylcholine, and serotonin (5-OH tryptamine) are naturally occurring constituents of nervous tissue, a tremendous amount of work has been done to shed light on their modes of action both at the peripheral synapses and in the central nervous system. However, in spite of the efforts, the picture emerging from results obtained in the peripheral system is becoming even more complicated than it was several years ago. Suffice it to mention the interesting hypothesis advanced by BURN and RAND (1959, 1960, 1965; see also BURN, 1961) and extended by KOELLE (1961, 1962), which assigns to acetylcholine a much wider role in synaptic events by postulating its presynaptic action in triggering the release of transmitter substances. It is not surprising, therefore, that experiments on the central nervous system which involve much greater technical difficulties and subtleties, have also yielded even more diverse and contradictory results.

The central actions of neurohormonal substances administered into the systemic circulation are extremely difficult to interpret because of the indiscriminate and presumably indirect effects on functionally different neuronal substrates. Conceivably, the overall behavioral and EEG effects are the outcome of various actions on intracranial and extracranial receptors associated mainly with the vascular bed, and only a relatively small part of the central action may be attributed to the more direct effect on synaptic mechanisms in the brain. However, most studies on the central action of neurohormonal substances administered into the systemic circulation, following studies by DELL, BONVALLET and HUGELIN (1954, 1956) and ROTHBALLER (1956) with catecholamines, and those by RINALDI and HIMWICH (1955a and b) with acetylcholine, respectively, tended to regard results as the consequence of direct and specific action on central synaptic junctions.

It is not the aim of this review to give an exhaustive account of work on central effects of drugs administered into the vascular bed for the author regards himself as neither competent nor prepared to do so. It is also felt that such an attempt would be superfluous, since many symposia and reviews have already dealt with the subject of central actions of drugs. For more comprehensive literature the reader is referred to the reviews by HIMWICH (1955), SEGUNDO (1956), UNNA (1957), FRENCH (1958), BRADLEY (1958, 1960), DELL (1958, 1960), PATON (1958), ROTHBALLER (1959), CROSSLAND (1960), DEWS and MORSE (1961), DEWS (1962), KILLAM (1962), DOMINO (1962), MACHNE and UNNA (1963), CURTIS (1963), TOMAN (1963), SALMOIRAGHI, COSTA and BLOOM (1965), GOLLUB and BRADY (1965), and KARCZMAR (1966) as well as to the monographs of WIKLER (1957), LONGO (1962), ROBSON and STACEY (1962), McLENNAN (1963), and ECCLES (1964).

The purpose of this review is to discuss the merits of topical administration of neurohormonal substances into the brain tissue in exploring the functional organization of highly heterogenous and intermingled populations of neurones. During the last few years evidence has accumulated that local application of drugs to subcortical areas may produce behavioral and EEG effects that are more specific to a given functional system than effects induced by electrical stimulation of the same area. Thus, it seems that the complicated integrative processes that determine the sleep-wakefulness cycle and patterns of goal-oriented behavior, are neurohormonally coded within the neuronal networks and can be selectively activated by local application of drugs (cf. MILLER, 1965). In contrast, electrical impulses may indiscriminately affect all kinds of synapses and also fibers that are merely passing through the stimulated area. The author does not intend to minimize either the classical studies of HESS (cf. 1954) and his followers or the results obtained by intravenous or intra-arterial injections of drugs. Much knowledge on central action of drugs has accumulated from their systemic administration both in experiments on animals and in therapy of diseases. For the sake of objectivity it must be said that in spite of all possible indirect and diffuse actions on brain structures produced by systemic administration of drugs, a careful analysis may reveal differential and selective action on certain neuronal systems. For instance, LONGO and SILVESTRINI (1957) have shown that physostigmine produced more widespread neocortical desynchronization than amphetamine, and, in contrast to the latter, it blocked the cortical recruiting response obtained by stimulation of the anteromedial thalamic nuclei. SAILER and STUMPF (1957) demonstrated that following the systemic administration of lysergic acid diethylamide (LSD-25) in rabbits a striking dissociation may be disclosed between the electrical activity of the hippocampus and neocortex. Normally, during arousal reaction caused by nociceptive stimuli or amphetamine the neocortical desynchronization is always associated with rhythmic theta waves (synchronized activity) in the hippocampus. However, following the administration of LSD both the neocortex and hippocampus exhibit desynchronized activity.

The close arterial injection of drugs, especially of neurohormonal substances, has advantages over the intravenous administration. Numerous interesting results have been obtained from the experiments in which drugs have been administered into the external carotid artery (EVARTS, LANDAU, FREYGANG and MARSHALL, 1955; BISHOP, FIELD, HENNESY and SMITH, 1958; DAVID, MURAYAMA, MACHNE and UNNA, 1963; MONNIER and ROMANOWSKI, 1963). Moreover, it seems that these results could not have been obtained by employing the local application of drugs. Both methods have their merits and limitations in exploring the synaptic events in the brain.

Microelectrophoretic application of drugs directly into the immediate external environment of nerve cells with simultaneous recording of the electrical activity constitutes an important methodological advance in the study of neurohormonal effects on single neurones (CURTIS and ECCLES, 1958a and b; BRADLEY and WOLSTENCROFT, 1962, 1963; BRADLEY, DHAWAN and WOLSTENCROFT, 1964, CURTIS and WATKINS, 1961, KRNJEVIĆ and PHILLIS, 1963a and b; SPEHLMANN, 1963; BLOOM, OLIVER and SALMOIRAGHI, 1963; SALMOIRAGHI, 1964; SALMOIRAGHI and STEINER, 1963; and others). The ejected ions may be restricted within a radius of a few microns. Quantitative data show that the amount released ranges roughly from 10 to 100 pmols following the application of a current of 100 µA (KRNJEVIĆ, LAVERTY and SHARMAN, 1963). Much more diffuse penetration of drugs can be obtained by their direct application in much greater amounts (5—30 µg) through macrocannulas either in concentrated solutions or microcrystals. A considerable body of data indicates that this method of local application of drugs as opposed to microelectrophoresis produces specific activation (or inhibition) of a whole population of neurons which may constitute a functional system or pathway, thus causing specific behavioral and EEG changes in experimental animals and man. It seems very likely that this method of drug application to discrete brain areas will widen our knowledge of mechanisms of brain function beyond the scope attained by the methods which have been dominant until recently: electrical stimulation and electrolytic as well as surgical lesions.

I. Limitations in investigations on the central effects of neurohormonal substances administered into the systemic circulation

A. The blood brain barrier

It is generally accepted that the main function of the blood brain barrier is to protect and buffer the central nervous system from noxious fluctuations and environmental changes, and thereby to achieve a maximum signal-to-noise ratio of the electrophysiological events. There is a great deal of information showing the existence of the blood brain barrier (see FRIEDMANN, 1942; KROGH, 1946; TSCHIRGI, 1952, 1958; DAVSON, 1956; DOBBING, 1961; JEPPSON, 1962; BERTLER, FALCK and ROSENGREN, 1963). Relatively little is known about its mode of functioning although it was shown recently that the endothelium of the capillaries of the brain, in contrast to other organs, contains dopa-decarboxylase and monoaminoxydase. The last enzyme probably constitutes the main obstacle to penetration of biogenic amines (BERTLER, FALCK and ROSENGREN, 1963).

It is not the aim of this section to discuss the problem of whether there exists a blood brain barrier for many substances in terms of an anatomic structure, or whether the phenomena which justify its assumption are to be considered only as consequences of the cerebral metabolism. These and related problems have been discussed and reviewed elsewhere (DAVSON and SPAZIANI, 1959, DOBBING, 1961). Many compounds administered into the systemic circulation are very quickly metabolized, which does not necessarily mean that they are impeded from penetrating the nervous tissue of the brain by the special functional barrier (cf. DOBBING, 1961). On the other hand, if they do not cross the blood brain barrier, they may produce striking but indirect effects on the brain function through intracranial and extracranial receptors in the vascular bed.

B. Synaptic barrier

There is also strong evidence from the experiments of ECCLES, FATT and KOKETSU (1954) and CURTIS and ECCLES (1958a and b) that two functionally independent barriers may be interposed between the systemic circulation and the postsynaptic regions of the Renshaw cell surface. One is the generally accepted blood brain barrier, and the second might be the "synaptic barrier" which has been postulated by CURTIS and ECCLES (1958b) to surround the axonal endings and synaptic areas of the postsynaptic neurone. According to these authors, a synaptic barrier which surrounds the Renshaw cell would prevent the diffusion of the acetylcholine released at the endings of the motor axon collaterals. The existence of this barrier would account for: a) a relatively long-lasting (at least 50 msec) repetitive discharge of the Renshaw cell; and b) the inability of drugs, which are active in depressing the response of Renshaw cell, to prevent the first two or three spikes in the ensuing burst of discharges (ECCLES, FATT and KOKETSU, 1954; CURTIS and ECCLES, 1958b; MACLENNAN, 1963).

As suggested by MACLENNAN (1963), synaptic barriers are likely to exist in many other structures of the central nervous system, and it seems probable that the negative results of electrophoretic application of acetylcholine, nor-adrenaline, and 5-OH tryptamine to the mesencephalic and medullary reticular formation by CURTIS and KOIZUMI (1961) might be due to this.

Other observations indicate the existence of synaptic barriers in the central nervous system which might account for the long-lasting changes in single unit activity following a single centripetal stimulus. It is known, for instance, from the experiments of MACHNE, CALMA and MAGOUN (1955) that a single reticulopetal shock applied to the sciatic nerve may produce either long-lasting (for 2—30 sec) repetitive discharges, subsequent increase of firing of a given unit, or reduction of spontaneous activity of another unit. Also ADEY, SEGUNDO and LIVINGSTON (1957) found that single electric stimuli applied to the gyrus hippocampi markedly depress the conduction of impulses from the midbrain reticular formation to the diencephalon for 1.5 to 2 sec. Further-more, SCHLAG (1959) has shown that a very stable pattern of a single unit activity picked up from the brain stem reticular formation may be markedly changed for several ensuing minutes in response to a single reticulopetal stimulus. Also SCHLAG and FAIDHERBE (1961) demonstrated that low frequency stimulation of the unspecific thalamic nuclei, although delayed in its effect, gradually changes single unit activity in the reticular forma-tion; the inhibitory effect was successively built up following each single stimulus.

Many experiments with electrophoretic administration of neurohormonal substances to the brain stem reticular formation which followed the negative report of CURTIS and KOIZUMI (1961), have shown that there are within this structure both acetylcholine and adrenaline-sensitive neurones which, however, respond sometimes after a very long delay, up to 20 seconds (BRADLEY and WOLSTENCROFT, 1962, 1963, SALMOIRAGHI and STEINER, 1963) instead of a few hundred milliseconds as in a typical Renshaw cell (CURTIS and ECCLES, 1958b).

Neurones which showed such markedly delayed response to electrophoreti-cally administered acetylcholine, have also been found in the cat's inferior colliculi (CURTIS and KOIZUMI, 1961), in the cerebellum (CRAWFORD, CURTIS, VOORHOEVE, and WILSON, 1963), ventrobasal complex of the thalamus (CURTIS and ANDERSEN, 1962), hypothalamus (BLOOM, OLIVER and SAL-MOIRAGHI, 1963) and in the cortex (KRNJEVIĆ and PHILLIS, 1961, 1963a—c; SPEHLMANN, 1963).

The concept of widespread existence of synaptic barriers in the central nervous system offers at least a partial explanation of the above-mentioned electrophysiological events. It should be noted, however, that there may be other possible mechanisms besides synaptic barriers, which account for the delayed responses, as for instance an action on the "electrically excitable" portion of the cell membrane or the action on neighboring glia cells which in turn might influence the behavior of the nerve cell (SALMOIRAGHI and STEINER, 1963). Nonetheless, it seems that the synaptic area of at least some populations of central neurones may be very well protected against the action of drugs, even when applied in their close vicinity.

C. Osmoreceptors

The following examples of effects produced by injection of hypertonic solutions show how much care must be exercised in interpreting the results obtained from injecting drugs into the systemic circulation.

Since experiments of VERNEY (1947) and JEWELL and VERNEY (1957) which led them to postulate the existence of osmoreceptors in the hypothalamus responsible for the release of antidiuretic hormone, it has been shown repeatedly that injections of hypertonic solutions into the carotid artery produce characteristic responses in the electrical activity of the brain. v. EULER (1953) demonstrated direct current shifts or "osmopotentials" in the supraoptic and preoptic region. SAWYER and GERNANDT (1956) observed EEG changes in various parts of the rhinencephalon following the injection of hypertonic solutions into the carotid artery or cerebral ventricles. Furthermore, CROSS and GREEN (1959) found striking changes in firing frequency of neurones in the supraoptic and paraventricular nuclei and in their vicinity resulting from very small changes in the osmotic pressure of the circulating blood.

Osmosensitive receptors have been detected not only in the hypothalamus CLEMENTE, SUTIN and SILVERSTONE (1957) demonstrated that intravenous administration of hypertonic solutions produced marked increase in the EEG activity in the floor of the fourth ventricle at the level of the obex. Another active area situated more rostrally in the medulla and close to the midline was suggested by CLEMENTE et al. to represent rostral projection to the higher centers of the above mentioned osmosensitive region. HOLLAND, CROSS and SAWYER (1959) were able to correlate the bursts of high amplitude fast activity in the preoptic region with the release of neurohypophyseal hormones, and HOLLAND, SUNDSTEN and SAWYER (1959) found that hypertonic solutions injected into the carotid artery may activate certain osmoreceptor areas in the hindbrain which are capable of producing dramatic changes in the blood pressure. This in turn may influence the general pattern of the EEG through an action on baroreceptors of the carotid sinuses (BONVALLET, DELL and HIEBEL, 1954) or may activate pressoreceptors elsewhere.

D. Intracranial pressoreceptors

Recently, careful studies by BAUST, NIEMCZYK and VIETH (1963) and BAUST and NIEMCZYK (1963) demonstrated that single units of the reticular formation which had been observed to be sensitive to blood pressure elevation produced by small doses of adrenaline or vasopressin, showed the same changes in their discharge rate following a mechanically produced rise in blood pressure caused by compression of the descending aorta. Moreover, when blood presure was kept constant artificially, no changes in electrical activity could be disclosed following the intravenous injection of adrenaline. During an elevation of blood pressure, 80 % of the reticular units showed a decrease and 20 % an increase in their firing rate.

E. Intracranial chemoreceptors

In addition to the blood brain barrier and the osmo- and pressoreceptors the existence of special chemoreceptors associated with the vascular bed should be considered in the evaluation of central drug effects. By the term "intracranial chemoreceptors", WINTERSTEIN (1961) means, "specific nervous formations which are especially adapted to receive chemical stimuli and to transmit these excitations to other centers, in the same way as the nerve endings of the sensory organs are adapted to their specific stimuli".

The vomiting center is a good example. It has been shown by WANG and BORISON (1952) that the medullary surface of the lateral reticular formation contains a selective chemoreceptor trigger zone, and its destruction causes loss of reactivity to intravenously or orally administered apomorphine. However, this trigger zone is merely a selective receptor which by itself cannot initiate integrated autonomic responses for it is unable to actuate emesis independently without the mediation of an emetic center in the reticular formation. These authors concluded that the concept of direct action of central emetics on the vomiting center is no longer tenable. The same conclusion has been reached recently by HAYAMA and OGURA (1963) following their study of the site of emetic action of a fish poison, tetrodotoxin, in dogs.

WINTERSTEIN and GOEKHAN (1953), while investigating the central effects of NH_4Cl-induced acidosis, found that H^+ ions of the blood also act as specific chemoreceptor-stimulating substances which lose their effect after deafferentation of appropriate areas. They also demonstrated that following their direct administration into the cerbrospinal fluid, H^+ ions regain their activity, presumably acting on special intracranial receptors.

The hypothalamus is the seat of selective receptors controlling food and water intake (ANAND and BROBECK, 1951; ANAND, 1961; ANDERSSON and LARSSON, 1961), constant body temperature (v. EULER, 1961; HARDY, 1961), hormonal "negative feedback" mechanism (SAWYER, 1962; MICHAEL, 1962; FLERKO, 1962; DAVIDSON and FELDMAN, 1963) as well as sexual behavior

(HARRIS, 1958, FISHER, 1960; LISK, 1962a and b). The effects of direct application of neurohormonal substances and gonadal steroids have revealed the
discrete localization of these receptor areas (see Section III). The hypothalamus
presumably exhibits a special kind of selective permeability to blood-borne
chemicals and with its multiple receptors it may be regarded as the primary
stage in triggering the integrated goal-oriented behavior. Besides the hypothalamus the following parts of the mammalian brain exhibit a marked permeability to blood-borne chemicals: the area postrema, the paraphysis, the
wall of the optic recesses, the eminentia saccularis of the hypophyseal stem,
the neurohypophysis, and the pineal body (see DAVSON, 1960). There is
evidence that at least some of these areas contain selective chemoreceptors or
trigger zones which transmit their impulses to appropriate autonomic centers
(see the exhaustive and excellent review by WINTERSTEIN, 1961).

F. Transport and selected actions

1. Catecholamines

Studies with C^{14}-labelled adrenaline by SCHAYER (1951) have shown that
after infusion of this amine into the systemic circulation only negligible radio-
activity could be detected in the brain of rats, although very marked radio-
activity was found in the kidney and liver. RAAB and GIGEE (1951) also were
unable to show any increase in brain noradrenaline following the intravenous
infusion of this amine in rats. Studies by WEIL-MALHERBE, AXELROD and
TOMCHICK (1959) and by WEIL-MALHERBE, WHITBY and AXELROD (1961;
see also WEIL-MALHERBE, 1960) with tritium labelled adrenaline and noradrenaline injected intravenously into rats, have shown that the hypothalamus
and pituitary are the only parts of the brain in which significant uptake of
circulating amines could be detected. The hypothalamus may, therefore, be
regarded as a possible trigger area for central effects of circulating catecholamines.

MANTEGAZZINI, PIECK and SANTIBAÑEZ (1959) performed a series of
elaborate experiments in which the mechanism of the EEG activation induced
by intravenous or intraarterial injections of adrenaline and noradrenaline
was investigated in the curarized *encéphale isolé* cat preparation, following
bilateral vagotomy, cervical sympathectomy, and carotid sinus denervation.
In these preparations the intravenous injection of adrenaline or noradrenaline
in amounts of 2 to 7 μg/kg regularly induced EEG activation, which was more
pronounced over, or sometimes limited to the hemisphere contralateral to the
mesencephalic lesion. It must be emphasized, however, that the activating
effect coincided with the rise in the systemic blood pressure. Furthermore, the
injections into the ascending aorta usually caused much smaller systemic
hypertension and no detectable EEG changes. Small doses (1 to 3 μg) of either

amine injected directly into the cerebral circulation produced a contraction of the nictitating membrane comparable in amplitude to that observed after intravenous injection, but no desynchronizing effect could be observed.

CAPON (1960), in an extensive study on unanesthetized curarized rabbits using a technique which allowed the simultaneous recording of EEG, carotid blood pressure, and ear transparency, was able to show that cutaneous vaso-constriction and carotid hypertension induced by the intravenous injection of adrenaline or noradrenaline always preceded the beginning of the EEG activation by 1 to 8 sec. Vasopressin likewise induced marked EEG activation similar to that following adrenaline. Furthermore, both vasomotor reactions and EEG activations were modified to the same extent by the adrenolytic drug phentolamine. The author concluded that adrenaline injected into the systemic circulation does not act directly upon the postulated adrenoceptive neurones of the reticular activating system.

GOLDSTEIN and MUNOZ (1961) also failed to obtain clear-cut results in unanesthetized rabbits. LONGO and SILVESTRINI (1957) even found a syn-chronizing action of these amines following intravenous injection in noncurar-ized and unanesthetized rabbits; in curarized cats after intracarotid administra-tion, the EEG activation was delayed and appeared together with the increase in the blood pressure. LONGO (1962) emphasized three points: 1. the action of adrenaline on the EEG of experimental animals can be either desynchronizing or synchronizing; 2. the action always appears with a 10 to 20 sec latency, which is greater than the time required for the amine injected i.v. to reach the brain, 3. intraarterial (centrifugal) injections are much less effective in pro-ducing the EEG changes.

In view of what is known about the blood brain barrier and the existence of intracranial chemo- and pressoreceptors it is difficult to accept the conclu-sions of LING and FOULKS (1959) drawn from results obtained with the injections into the innominate artery in the very elaborate, chronically deafferented cat preparation, or those of BRADLEY and MOLLICA (1958) from the changes of single unit activity of the brain stem reticular formation following the administration of neurohormonal drugs into the systemic circulation. Recently BRADLEY and WOLSTENCROFT (1964), in discussing BRADLEY and MOLLICA's results, pointed out that presumably only the immediate effects following the injection might be regarded as specific central actions, whereas all the markedly delayed effects were probably associated with blood pressure changes.

2. 5-OH tryptamine

Since there is evidence that 5-OH tryptamine circulating in the blood stream does not cross the blood brain barrier in contrast to its precursor, 5-hydroxy-tryptophan (UDENFRIEND, WEISSBACH and BOGDANSKI, 1957), GLAESSER and

MANTEGAZZINI (1960) have compared the central effects of intracarotid injections of both compounds on the EEG of the midpontine pretrigeminal cat preparation. 5-OH tryptamine in doses of 2 to 20 µg did not alter the activated EEG pattern in the midpontine pretrigeminal preparation, while it desynchronized the EEG pattern after it had been synchronized by additional mesencephalic hemisection or by visual and olfactory deafferentation.

It is noteworthy that opposite results have been obtained with 5-hydroxy-tryptophan (5-HTP) which synchronized the spontaneously activated EEG pattern of the midpontine pretrigeminal preparation. Moreover, the same doses of 5-HTP also increased the synchronization of the EEG pattern after it had been deactivated already by mesencephalic hemisection. Since the EEG changes were limited predominantly to the hemisphere corresponding to the side of the intracarotid injection, neither the extracerebral metabolites of 5-HTP nor substances liberated by it which might circulate in the blood were presumably responsible for the EEG changes.

These results seem to show that 5-OH tryptamine formed in the brain after decarboxylation of 5-HTP can produce effects diametrically opposite to those seen after 5-OH tryptamine injection into the systemic circulation. Since it is known that 5-OH tryptamine is a powerful stimulant of chemo-receptors associated with the extracranial vascular bed (see the review of PAGE, 1958), it presumably acts also on intracranial chemo- and presso-receptors, the existence of which is highly probable.

3. Acetylcholine

RINALDI and HIMWICH (1955a and b) have shown in cats, and LONGO (1955) in rabbits, that doses of acetylcholine as small as 0.5 to 1 µg injected into the carotid artery induce instantaneous activation of the EEG pattern. Atropine prevents and abolishes these effects. RINALDI and HIMWICH postu-lated that this activation of the EEG is induced by the direct excitatory action of acetylcholine on cholinoceptive mechanisms in the mesodiencephalic reticular activating system. It was shown by the same authors that doses of acetylcholine as small as 0.5 µg injected into the carotid artery evoke marked EEG activation also in the cat *cerveau isolé* preparation. This is in contrast to adrenaline-induced EEG activation, which is abolished in such a cat prepara-tion unless the remaining part of rostral reticular formation is sensitized by the administration of amphetamine (HIEBEL, BONVALLET, HUVÉ and DELL, 1954). However, no EEG changes can be induced by larger doses of acetyl-choline in the cortex of a hemisphere isolated by a cut running through the corpus callosum, anterior commissure, fornix and capsula interna, externally to the thalamus but through the medial part of the corpus striatum (RINALDI and HIMWICH, 1955b). This complete lack of EEG changes in the isolated hemisphere following relatively large doses of acetylcholine injected into the blood stream casts doubt on the specificity of the acetylcholine-induced EEG activation in both intact animal and *cerveau isolé* preparation, since it is known that several areas of the isolated hemisphere, especially the cortex, contain neurones which markedly respond to the electrophoretic administration of acetylcholine (KRNJEVIĆ and PHILLIS, 1961, 1963 a—c; SPEHLMANN, 1963).

LONGO and SILVESTRINI (1957), working with the rabbit *cerveau isolé* preparation, were not able to demonstrate any clearcut cortical reaction after small doses (1 to 2 µg) of acetylcholine injected into the carotid artery.

When comparison was made between the central actions of acetylcholine and noradrenaline in cat *encéphale isolé* preparations, it was found that the intracarotid administration of either acetylcholine or noradrenaline in doses of 5 µg produced essentially identical activation of the cortical EEG pattern. However, these compounds act with different latencies; acetylcholine causes an immediate effect whereas adrenaline-induced activation is delayed in onset for about 20 sec (BRADLEY, 1960). This might suggest that in spite of identical effects, the mechanism or the site of action of these two substances is different.

It seems very unlikely that the above-mentioned effects are due to the direct and specific action of acetylcholine on synaptic events. It is known that quaternary ammonium compounds such as acetylcholine, tubocurarine, neostigmine, succinylcholine and carbaminoylcholine, although active at peripheral sites, do not cross the blood brain barrier readily, in contrast to tertiary amines such as physostigmine and dihydro-beta-erythroidine (ECCLES, ECCLES and FATT, 1956; CURTIS, ECCLES and ECCLES, 1957). KOELLE and STEINER (1956) have also shown that, in contrast to a tertiary amine anticholinesterase compound which was capable of producing 90 % inactivation of the total brain acetylcholinesterase following intravenous injection of the LD 50, the corresponding LD 50 of the quaternary derivative caused no measurable inhibition. Furthermore, MAYER and BAIN (1956) demonstrated that the quaternary analogue of the convulsant acridones does not penetrate into the central nervous system, and that the chief barrier for the quaternary ammonium compounds exists between the outer wall of the capillary endothelium and the glial plasma membrane.

II. Intraventricular and intracisternal administration

It may be assumed that, following intraventricular administration of drugs, the affected neural tissue is that adjacent to the linings of the ventricles and aqueduct. As shown by DOMER and FELDBERG (1960), bromophenol blue, which does not pass the blood brain barrier, when perfused from the lateral ventricle to the aqueduct, penetrates to a certain extent into the hypothalamus and into the gray stratum around the aqueduct. Moreover, from the picture shown by Domer and Feldberg it may be assumed that superficial layers of the caudate nucleus and hippocampus may also be involved. Feldberg and Fleischhauer (1962, 1963) have shown on the basis of careful EEG analysis that tubocurarine, which also does not readily cross the blood brain barrier, acts directly on several structures including the hippocampus and amygdala when administered into the cerebral ventricles. Recent studies by Feldberg and Myers (1963) using either the same technique or injecting drugs directly into the hypothalamus, indicate that antagonistic effects of adrenaline to 5-OH tryptamine on shivering and the rise in body temperature have their seat presumably in the hypothalamus; nevertheless, other diencephalic structures may also be involved. If adrenaline, acetylcholine and 5-OH tryptamine injected intraventricularly penetrate the wall of the ventricle like bromophenol blue or tubocurarine, the overall effect on brain function might be the outcome of their effects on functionally different structures. In this situation no decisive conclusions can be drawn as to specific site and mode of action of these substances.

Apparently drugs applied to the inner surface of the brain affect functional systems other than those influenced by drugs administered into the vascular bed. It is well known, for instance, that both small and large doses of adrenaline or noradrenaline injected into the cerebral ventricles of the unanesthetized animals cause marked depressant and analgesic effects (Leimdorfer and Metzner, 1949) and that the general state of the animals resembles light pentobarbital anesthesia (Feldberg and Sherwood, 1954). In contrast, the same amines injected in small doses into the systemic circulation induce consistent behavioral and EEG arousal effects in sleeping cats (Rothballer and Sharpless, unpublished results cited by Rothballer, 1959). In this connection, it seems worthwhile to note that adrenaline administered intravenously in newborn kittens or chicks in which the blood brain barrier is immature, also produces effects opposite to those consistently observed in adult animals, i.e., marked sedation instead of excitation. Newborn guinea pigs, however, in which the blood brain barrier is mature from birth, always respond like adult animals to the intravenous administration of adrenaline (Key and Marley, 1962; Marley and Key, 1963). Thus it appears that adrenaline injected into the systemic circulation in the presence of a mature blood brain barrier is not able to reach in sufficient concentration those adrenoceptive brain areas which are responsible for the central depressant action. Another alternative is that the depressing central action is masked by excitatory effects on the brain stem reticular system via the multiple intracranial and extracranial receptors associated with the vascular bed.

For complete information on the effects elicited by drugs injected into the cerebral ventricles the reader is referred to the monograph by Feldberg (1963) and recent review by Feldberg and Fleischhauer (1965).

III. Direct application of drugs into the brain tissue

A. General considerations

1. Methodological remarks

In "chronic" animals, the best available technique for repeated direct deposition of crystalline substances into subcortical target areas seems to be a double-walled and stereotaxically implanted stainless steel cannula attached to the skull by means of dental cement (GROSSMAN, 1960, 1962a; FISHER, 1960; STEIN and SEIFTER, 1962). These cannulas usually consist of two modified concentric syringe needles, the hubs of which are exactly adjusted to each other and threaded so that after they have been screwed together, the tips of both needles are flush with each other. For the purpose of stimulation, the inner cannula is removed and cleaned from the crust of dried tissue fluid in order to allow minute amounts (1 to 5 μg) of crystalline drugs to be tapped into its tip. After returning the inner needle to its usual position, the drug is not expelled by the stylet as in the original "semi-acute" experiments of MacLEAN (1957), but is allowed to diffuse out of the cannula without any additional tissue damage. Following the insertion of metal cannulas and electrodes the neural tissue always shows some recession and slight proliferation of adjacent glia cells. This process represents an advantage of "chronic" implants over "acute" or "semi-acute" ones, in which crystalline substances or concentrated solutions are introduced into the freshly injured neural tissue. It seems that chemicals tamped into the inner cannula, which has exactly the same length as the outer guide cannula, penetrate far more slowly through the barrier of glia cells into the surrounding tissue resulting in more physiological concentrations of the applied drugs. HERNANDEZ-PEON, CHAVEZ-IBARRA, MORGANE and TIMO-IARIA (1963) modified the double-walled cannula in such a way as to permit the inner needle to descend by 1 mm steps into the brain, thus making possible investigations of several points with the same cannula. YAMAGUCHI, LING and MARCZYNSKI (1964) used cannulas provided with a stylet of exactly the same or even slightly shorter length so that the solid chemicals pushed to the tip do not mechanically injure the barrier of glia cells.

Another somewhat different technique, described by FOX and HILTON (1958), GADDUM (1961) and DELGADO (1962; for complete bibliography see DELGADO, 1963), is based on implantation of a double cannula (push-pull) which permits the injection and circulation of liquids as well as collecting them after they have been perfused through the chosen subcortical structure.

A special cannula "chemitrode" has been described by DELGADO (1962), consisting of two stainless steel tubes plus a Teflon-covered stainless steel wire with a 1 mm bare tip; it permits injections, collection of fluids, as well as EEG recording and electrical stimulation of the same subcortical area. Delgado also developed a radio-controlled chemitrode pump with two compartments

7*

separated by an elastic membrane. One compartment is filled with the solution to be injected, and the other with a solution of hydrazine. The current controlled by the relay of a micro-radio receiver may be passed through the hydrazine to release gas which in turn expels and injects the liquid of the other compartment through the chemitrode connected with the pump by means of polyethylene tubing. Thus, the device permits the application of drugs by remote control and avoids direct handling of animals.

For the purpose of administration of steroid hormones into a particular area of the brain, the crystalline substance may be tamped into the end of a thin stainless steel needle from which, after stereotaxical orientation of the implant, the substance is expelled by an appropriate stylet as in the original experiments described by MacLean (1957). Also, small quantities of the crystalline steroids, after heating to the melting point, may be drawn by capillary action into the end of a thin stainless steel tubing (Harris, 1958; Lisk, 1960; Davidson and Sawyer, 1961a and b). After cooling, the outside of the tubing must be carefully cleaned, and a minute amount of solid hormone remains at the tip. Such tubes are implanted into the brain and fixed to the skull with the aid of dental cement for the appropriate periods of time required to obtain specific central or centrally mediated peripheral effects. At the conclusion of experiments, the tubes are removed and the amount of absorbed hormone may be estimated. However, measurements of microamounts of hormone utilized during the experimental period were unsuccessful (Lisk, 1962b).

2. Potential side-effects

As has been pointed out by Steiner and Seifter (1962), there is an immediate tendency to reject as unpharmacological and injurious any method which involves application of crystals directly into the neural tissue. Admittedly, several non-specific effects are to be expected mainly resulting from mechanical and osmotic changes. Furthermore, relatively small changes in acid-base composition of the stimulated area, as well as unspecific chemical stimulation, may play an important role in producing uncontrollable side-effects.

Osmotic changes may be especially relevant in the study of the neuro-hormonal control of food and water intake (Grossman, 1962a and b), since both a polydipsia and hyperphagia may be produced in satiated animals by microinjections of hypertonic saline or sugar solutions into the area caudal to the optic chiasma (Larsson, 1954; Andersson and MacCann, 1955b; Epstein, 1960).

Furthermore, substances such as noradrenaline, adrenaline, acetylcholine, and 5-OH tryptamine are known to exert pronounced vascular effects which can be expected to contribute significantly to the observed functional changes in neuronal activity.

Consideration should also be given to the possibility that some of the active material deposited in highly vascularized neuronal tissue may penetrate to the systemic circulation and thus produce indiscriminate actions on remote structures, or evoke effects through action on extra- and intracranial chemoreceptors.

A large body of interesting data obtained during the last few years offers convincing evidence that practically all the above-mentioned potential nonspecific effects can be differentiated from specific ones, and, in consequence, ruled out by appropriate control experiments.

As pointed out by GROSSMAN (1962a, 1964), it seems highly unlikely that active neural elements were in direct contact with the deposited substance before a considerable dilution had taken place during the relatively slow diffusion through the intercellular space and "barrier" of injured cells. This can be inferred from latency periods of behavioral and EEG effects which usually ranged from 5 to 25 min, depending on the character of the substance used and minor modifications of the construction of the cannulas employed.

It may be argued, however, that the observed EEG and behavioral effects should be attributed rather to inactivation or abnormal functioning of adjacent neural tissue. It is conceivable that a small population of neurones in immediate contact with the tip of the cannula, being exposed to a redundant concentration of the applied substance, might be totally inactivated. Also, it may be anticipated that beyond this region another population of neurones will be exposed to unspecific excitatory or depressant effects. However, it seems highly unlikely that the overall qualitative behavioral and EEG changes can be ascribed solely or primarily to these actions, since: 1. in many experiments appropriate parameters of electrical stimulation are capable of producing identical or very similar effects from exactly the same anatomic area; 2. both the EEG pattern and behavioral changes usually do not exhibit any marked abnormalities; 3. EEG recordings picked up from the tip of the cannula after depositing biologically active substances usually do not show any gross disturbances of electrical activity in "chronic" experiments, provided that the substance is not pushed out far beyond the tip of the cannula through the protective wall of the glia cells (GROSSMAN, 1960, 1962a; STEIN and SEIFTER, 1962; YAMAGUCHI, LING and MARCZYNSKI, 1963, 1964).

3. Main advantages

MacLEAN (1957) was the first to emphasize the main advantage of using crystalline substances for studying the function of subcortical structures as compared to microinjection. He was able to show that the non-yielding character of the brain tissue precludes any strictly localized deposits of even such small amounts of fluid as 0.05 ml. Seemingly negligible quantities of fluid, e.g. 0.01 to 0.02 ml, tended to leak out by following the path of least resistance along the shank of the needle. In contrast, the main advantage of depositing crystalline substances is their strictly limited spread. MacLean found that powdered Nile blue, when deposited in amounts comparable to those of cholinergic drugs, spreads in living brain tissue over a radius of little more than 1 mm during 40 min. Also Lockhart (quoted by MacLean, 1957), after depositing a mixture of equal quantities of powdered Nile blue and radio-phosphorus in the cortex, was able to show, by radiographic technique, that radiophosphorus diffused only little more than the dye.

As pointed out by MacLean, local chemical stimulation of the neural tissue has several other important advantages over electrical stimulation, especially in the study of functional localization. It does not seem to affect fibers of passage. One consequence is that it does not produce antidromic effects, which in many instances might lead to functional suppression of structures which otherwise might participate in overall reactions. Chemical stimulation enables prolonged observation of the animal's behavior and EEG changes.

However, microinjections of very concentrated solutions of neurohormonal substances and other drugs, provided that they are used in amounts not larger than 0.02 ml, also yielded reliable and consistent results in the hands of experienced investigators (Fisher, 1956, 1960, Rech and Domino, 1959; Cordeau, Moreau, Beaulnes and Laurin, 1962, 1963; Courville, Walsh and Cordeau, 1962a and b; cf. Cordeau, 1962; Myers, 1963, 1964). It appears that in some experimental situations the microinjection technique has the advantage over administration of solid substances in that it avoids direct handling of the animal, since it can be easily carried out by remote control through fine flexible polyethylene tubing. The ingenious method of self-injection in rats deserves special attention (Olds and Olds, 1958). This method permits self-injections of a volume of about 3 µl of the test solution. Since approximately 10 such injections were required to yield a behavioral effect, the threshold quantities ranged from 0.36 to 4.5 times 10^{-9} moles of test neurohormonal substances; the threshold concentration of these substances and other chemicals ranged from 12 to 150 mM (Olds, Yuwiler, Olds and Chang Yun, 1964). Cordeau and his group (1963) obtained reproducible effect in the brain stem reticular formation after injecting 20 µg of acetyl-choline hydrobromide or l-adrenaline bitartrate in a volume of 0.02 ml

Although in many instances the exact extent of the effective diffusion of various drugs exhibiting various physico-chemical properties cannot be directly measured, nonetheless, it seems that in general their spread must be very limited even in the case of readily water soluble substances This can be inferred from analyzing the negative and positive placements of solid substances or microinjections based on histological data collected from a series of animals. It appears that some preliminary data concerning functional delineation of a neurone population can be obtained even from very few animals bearing several chronically implanted cannulas if the position of the specially devised ejecting needle is changed, as those, for example, which have been used by HERNANDEZ-PEON, CHAVEZ-IBARRA, MORGANE and TIMO-IARIA (1963). All ivestigators who employed direct chemical stimulation of subcortical structures of the brain have found that points which yield characteristic EEG and behavioral responses occur in very close proximity, since, by moving the cannula by 1 mm steps, one can change a positive effect into a negative one, and vice versa (see GROSS-MAN, 1962a and b, 1964; CORDEAU, MOREAU, BEAULNES and LAURIN, 1963; OLDS, YUWILER, OLDS and CHANG YUN, 1964; COURVILLE, WALSH and CORDEAU, 1962b).

Apparently, direct chemical stimulation has another very important advantage over the electrolytic ablation method. The latter permits observation only of negative results, this being its main drawback in the studies of motivated goal-oriented behavior. Furthermore, the ablation method does not prove conclusively that the observed effects are the result of destruction of particular cell populations rather than of fiber tracts which merely pass through the affected area.

In many subcortical structures and within exactly the same anatomical locus a consistent and apparent correlation can be observed between hypnogenic action of cholinergic drugs and low frequency electrical impulses on the one hand, and the arousing effects of adrenergic drugs and high frequency electrical stimulation on the other. This seems relevant in a consideration of the behavioral and EEG "tonic" effects, i.e. those which require some latency to appear and outlast for relatively long times the several trains of electrical impulses and the mechanism of which can best be explained in terms of biochemical changes or mobilization of neurohormonal substances. The differential "tonic" EEG and behavioral effects produced by changing the impulse frequency assume special theoretical significance in the light of BURN and RAND's (1959) and KOELLE's (1961) hypothese concerning the intermediate and triggering role of acetylcholine in the release of transmitting substances in cholinergic and non-cholinergic synapses. Differential EEG and behavioral effects of low and high frequency impulses, and cholinergic and adrenergic stimulation respectively, may be consistently obtained from those subcortical areas where at least two functionally different populations of neurones, with regard to the regulation of the sleep-wakefulness cycle, can be anticipated, as for instance the preoptic region and brain stem reticular formation in its whole length. The preoptic region has been shown to be a nodal areas of many converging influences (NAUTA, 1958); when stimulated by the topical administration of acetylcholine in the conscious cat (YAMAGUCHI, LING and MARCZYNSKI, 1963, 1964; HERNANDEZ-PEON, CHAVEZ-IBARRA, MORGANE and TIMO-IARIA, 1963) or by low frequency electrical impulses (STERMAN and CLEMENTE, 1962a and b; CLEMENTE and STERMAN, 1963) it yields behavioral and EEG sleep effects.

The same area, however, when stimulated with adrenergic drugs or by high frequency electrical impulses, evokes behavioral and EEG arousal (YAMA-GUCHI et al., 1964; HERNANDEZ-PEON *et al.*, 1963).

Another example is the brain stem reticular formation. Here also an overlap of at least two functionally opposed ascending systems can be anticipated on the basis of many experimental approaches which will be discussed later. It appears that these two ascending components may be selectively activated by adrenergic compounds or high frequency electrical stimulation on the one hand, and by cholinergic drugs or low frequency electrical stimulation on the other (see Section III D).

One of the most striking examples showing the high selectivity of direct chemical stimulation is the observation of FISHER and COURY (1962). They found in rats that from the same permanently implanted cannula three different effects could be obtained. Microamounts (1 to 3 μg) of carbachol applied into this region produced repeatable drinking responses in water-satiated animals, noradrenaline caused marked eating, and testosterone consistently evoked typical maternal behavior characterized by nest building. These effects were obtained from a locus at the junction of the area of the diagonal band of Broca and the medial preoptic region. It is noteworthy that these effects were lost when the position of the cannula was changed by 1 mm. As will be discussed later in more detail, the correlation between drinking and cholinergic stimulation on the one hand, and between eating and adrenergic stimulation in satiated animals on the other, is highly specific with regard to the lateral hypothalamus (GROSSMAN, 1960, 1962a, 1964). Only a few of about 100 rats were found by FISHER and COURY (1962) to respond by both drinking and eating following carbachol administration to the hypothalamus.

Summing up, the above-mentioned examples of the effects of direct application of drugs to the subcortical structures, as well as the general considerations of the possible side-effects and main advantages, point to the value of this method, and emphasize the chemical specificity of the receptor sites. It seems important to stress that this method apparently allows pharmacological analysis of functionally overlapping different systems of neurones within the same anatomical area. The limited spread of topically administered drugs, and high specificity of both electrophysiological and behavioral responses obtained from inhomogenous structures constitute major advantages over electrical stimulation, which usually produced an unknown mixture of ortho- and antidromic activation and inhibition, respectively, of different populations of neurones over a wide field.

B. Central effects of steroid hormones

The wealth of data obtained from lesion experiments, direct electrical stimulation, and electrical self-stimulation experiments, and from electroencephalographic recording from subcortical structures, indicates that the hypothalamus is the most important area in the regulation of secretory activity of the adenohypophysis, and in integrating reproductive behavior. The experimental evidence has been discussed in the authoritative monograph of HARRIS (1955), in the symposium on hypothalamicohypophyseal inter-relationship (FIELDS, GUILLEMIN and CARTON, 1956), and in the reviews of BENOIT and ASSENMACHER (1955), SAWYER and CRITCHLOW (1957), HARRIS (1960), SAWYER (1960, 1962, 1963), SAWYER and KAWAKAMI (1961), FLERKO (1962) and GUILLEMIN (1963).

Our discussion will be confined to the recent experiments employing direct administration of steroid hormones to subcortical structures, and to their contribution in studying the functional localization of the anatomical substrates governing the secretion of trophic hormones from the adenohypophysis as well as sexual and reproductive behavior.

1. Hypothalamic control of the adenohypophysis

It is well known that a direct anatomical link between the adenohypophysis and the hypothalamus is essential for normal functioning of the pituitary-gonadal axis. The view is also generally accepted that the adenohypophysis remains under control of several specific receptor structures, located in the hypothalamus, which are sensitive to variations in the level of circulating steroid hormones and respond accordingly by modifying the secretion of the adenohypophysis.

Discrete electrolytic lesions in various parts of the hypothalamus may produce ovarian atrophy or constant estrus, and this depends upon the location of the lesion. Small lesions in the arcuate nucleus induce atrophy of gonads in both males and females (DEY, 1943; McCANN, 1953; BOGDANOVE and SCHOEN 1959), as well as atrophy of the adenohypophysis (TALEISNIK and McCANN, 1961).

a) Adrenal steroids. Experiments with direct steroid implants in the brain contributed to delineate the seat of the "negative feedback" receptor area involved in the control of the pituitary secretion. Variations in the circulating level of adrenocortical hormones profoundly affect the secretion of adreno-corticotrophin (GANONG and FORSHAM, 1960; HARRIS, 1960). The exact location of the responsible "feedback receptor area" is still a matter of controversy. The lesion studies by HUME (1958) and his group implicated the anterior median eminence as the area responsible for the control of the pituitary adrenocortico-trophin secretion, but there are also suggestions that the mesencephalon and pituitary itself may contain receptor areas (cf. discussion of DAVIDSON and FELDMAN, 1963). SMELIK and SAWYER (1962) found in rabbits that the stress-induced adrenal activation, as measured by the level of plasma corticosterone, was inhibited following the implantation of crystalline cortisol into the anterior portion of the median eminence, post-optic region and, to a lesser extent, antero-medial part of the hypothalamus. No effects could be disclosed by similar implants into the posterior hypothalamus, mesencephalon, or adeno-hypophysis. From the very close proximity of the effective and non-effective sites it can be inferred that the diffusion of the implanted hormone through the brain tissue must be very slight. Since sham implants as well as even much larger electrolytic lesions than those caused by implants did not produce changes in normal responses of adrenal function, it does not seem likely that the local damage of neural tissue could account for the blockade of cortico-trophin secretion. In considering the specificity of the results and functional localization, the following points should be emphasized: 1. the most effective area in inhibiting adrenal activation was found to be situated anterior to the analogous "negative feedback" estrogen-sensitive region in the median eminence (DAVIDSON and SAWYER, 1961a); and 2. cortisol implants placed in the estrogen-sensitive area did not mimic the specific inhibitory effect of estrogen on gonadotrophic action of the pituitary.

Recent experiments of CHOWERS, FELDMAN and DAVIDSON (1963) have shown that cortisol acetate implanted in the median eminence region of the hypothalamus also markedly inhibits the depletion of adrenal ascorbic acid, which normally follows the acute surgical stress of unilateral adrenalectomy. Similar implants of cortisol in the pituitary itself were ineffective as were also testosterone implants in the median eminence area. Furthermore, DAVIDSON and FELDMAN (1963) have reported that single hydrocortisone implants in the median eminence and in the anteromedial hypothalamus in rats prevented the compensatory adrenal hypertrophy following unilateral adrenalectomy, and produced atrophy of the remaining adrenal gland, which points to a significant inhibition of adrenocorticotrophin secretion. Similar implants in the lateral basal hypothalamus induced only partial inhibitory effects, and cholesterol implants in the most effective median eminence area were without any noticeable effect. Furthermore, double mesencephalic and triple sub-cutaneous deposits were ineffective. Provided that the compensatory adrenal hypertrophy following unilateral adrenalectomy is a reliable criterion of the adrenocorticotrophin release by the pituitary, the experiments adduce evidence that not the pituitary itself but the median eminence region of the hypo-thalamus is the main "feedback receptor area" for circulating adrenal steroid hormones.

b) Gonadal steroids. In intact female rabbits and male dogs (DAVIDSON and SAWYER, 1961a and b) and female and male rats (LISK, 1960, 1962c) the same general region of the arcuate nucleus responds specifically to direct implants of crystalline estradiol and testosterone by inhibition of normal gonadotrophin synthesis and secretion from the adenohypophysis, respectively, resulting in gonadal atrophy. Similar implants in the thalamus, in other parts of the hypothalamus (ventromedial region) or empty control tubes in the arcuate nucleus had no such effects. Characteristically, in dogs, implants of testosterone placed in the pituitary itself induced no gonadal or prostatic atrophy. These experiments indicate that the arcuate nucleus in rats and the posterior median eminence of the basal tuberal region in rabbits and dogs, play a key role in the "negative feedback" mechanism by which a relatively small rise of steroid hormones in the blood inhibits the release of gonadotrophins. They also suggest that the same neuronal elements when "disinhibited" by castration are responsible for enhanced gonadotrophin release.

KANEMATSU (1963) investigated in rabbits the effects of brain estrogen implants on the pituitary content of prolactin and of luteotrophic hormone. Estrogen implantations into several parts of the brain other than the postero-median eminence did not change the content of the luteotrophic hormone. However, the same amount of estrogen implanted into the posteromedian eminence produced a sharp fall of this hormone, and this was in contrast to the simultaneous significant elevation of the prolactin content of the pituitary. Characteristically, estrogen implantation into the pituitary failed to change the content of luteotrophic hormone in this organ, although it appeared to cause discharge of prolactin.

Experiments on estrogen-sensitive areas in the rat brain by LITTLEJOHN and DE GROOT (1963) indicate that, besides the mammillary complex, where estradiol implants were most effective in inhibiting the compensatory ovarian hypertrophy following hemiovariectomy, there are also several extrahypothalamic receptor areas capable of influencing the gonadotrophin-releasing activity of the hypothalamo-hypophyseal complex No suppression of compensatory ovarian hypertrophy was observed after estradiol implanted in the preoptic region, lateral hypothalamus, basolateral amygdala, fornix-hippocampus, caudate-putamen or in the subarachnoidal space. However, they obtained greater than normal hypertrophy in animals bearing estradiol implants in the anterior and anteromedial amygdala. This implies that there are structures which contain specific receptors, the stimulation of which by circulating gonadal hormones, instead of inhibiting the release of gonadotrophins, may initiate their discharge from the pituitary gland, and subserve a "positive feedback" mechanism in the hypothalamo-hypophyseo-gonadal system.

LISK and NEWLON (1963) carried out a careful karyometric analysis of the effects produced by implants of estrogen in the arcuate nucleus of the rat, and showed that certain neurones in this region react very consistently with a decrease in nuclear size. Since these changes were always associated with atrophy of the ovaries and uterus, it seems probable that the affected neurones are the substrate of the "negative feedback" receptor area.

Results opposite to those mentioned above have been obtained in castrated animals following steroid implantation in the same general region of the arcuate nucleus. In ovariectomized rats, small amounts of estradiol placed in this area resulted in regression of the castration changes already present in the pituitary (LISK, 1962a, 1963). It is noteworthy that rats even 60 days after castration responded markedly in that 100 days after estradiol implants the cytologic characteristics of the pituitary returned to normal. Interestingly enough, in some instances the complete regression of the pituitary castration changes occurred first in that side of the pituitary which corresponded to the side of estradiol implants in the arcuate nucleus.

Subcutaneous administration of comparable amounts of estradiol was completely ineffective in alleviating the castration changes in the pituitary. Implants in the mammillary bodies or blank tubes placed in either the arcuate nucleus or mammillary bodies did not prevent the normal castration changes. Estradiol implants in the pituitary itself were completely ineffective in inhibiting the characteristic sequence of castration changes in this organ.

These findings in castrated animals indicate that the presence of estrogen in the region of the arcuate nucleus is necessary for the maintenance of normal cytology of the pituitary. It was shown that the stalk-median eminence region of the rat contains a factor(s) initiating the release of the luteinizing hormone from the pituitary (McCANN, 1962; NIKITOVITCH-WIENER, 1962; cf. GUILLE-MIN, 1963), and it seems that an optimal level of estrogen in this area might be necessary for the production and release of this hypophysiotropic substance(s). It seems also highly probable that the same region is essential in the "negative feedback" mechanism by which high levels of gonadal hormones suppress the normal gonadotrophic function of the hypophysis in non-castrated animals (LISK, 1960; DAVIDSON and SAWYER, 1961a and b).

2. Behavioral effects of gonadal steroids

It was known from the early experiments that stilbestrol implants in the posterior hypothalamus, but not elsewhere, markedly influence behavior in spayed female cats (HARRIS, 1958). These animals displayed full sexual receptivity in the absence or only minimal development, of the vaginal epithelium and uterus. More recent investigations on the estrogen-sensitive neuronal systems subserving sexual receptivity in cats revealed that sexual "system" of neurones extends from the posterior mamillary region to the anterior preoptic region (MICHAEL, 1962 and 1965).

Interesting effects of gonadal steroid hormones have been reported by FISHER (1956, 1960). In male rats, remote microinjections of water-soluble sodium testosterone sulfate through chronically implanted cannulas in the medial preoptic region produced an integrated and long-lasting goal-oriented behavior. The characteristics of this androgen-induced effect were rather unexpected. After a few minutes of latency, the male rats began to retrieve litters of newborn pups to the nests which they had just built. Curiously enough some of the test males even "retrieved" full-size adult animals. In contrast, microinjections of the same dose of testosterone through cannulas implanted in the lateral preoptic region markedly increased the sexual drive which led them to display copulatory behavior even to 20-gram rat puppies. FISHER found also that in some hypothalamic loci, strong male sex behavior could be elicited in female rats. Control experiments with physiological saline injected into the same brain areas never produced a significant effect. However, the effects of central microinjections of a chelating agent, ethylenediaminetetraacetic acid, were indistinguishable from the behavioral effects of testosterone. This observation casts some doubts on the specificity of the androgen-induced effect. On the other hand, in view of the well known action of steroid hormones on ionic transport mechanism across the cellular membrane, a common denominator in the action of the chelating agent and testosterone can be hypothesized.

LISK (1962b) and LISK and WEIN (1962) have shown that estradiol implants in the medial basal preoptic or anterior hypothalamic region in spayed female rats induced sexual behavioral receptivity, as indicated by the presence of the lordosis reflex. Estradiol implants in other hypothalamic sites or empty tubings were ineffective, as was subcutaneous steroid administration. Daily subcutaneous injections of estradiol (up to 5 µg), although much more effective in increasing the uterine weight than direct diencephalic administration of the drug, never produced behavioral receptivity. The shortest latency (3 days) of the first manifestation of the lordosis reflex was obtained after implantation in the basal medial preoptic region and in the area above the optic tract at the level of the filiform nucleus.

Summing up, the above-mentioned data indicate the existence of dual estrogen- and testosterone-sensitive centers in the diencephalon. One, located in the basal medial and lateral preoptic region, is responsible for initiation and integration of overt sexual and parental behavior, and the other, situated in the area of the basal tuberal median eminence, is concerned with the physiological mechanism of the estrus cycle, through the regulation and maintenance of the pituitary secretion. The latter region is also presumably the seat of the "negative feedback" receptor area that is selectively sensitive to adrenal steroids. Preliminary data indicate that some extrahypothalamic structures contain "positive feedback" receptors sensitive to estradiol.

C. Evidence of neurohormonal regulation of food and water intake

1. Introductory remarks

It is known from numerous studies that the lateral and medial hypothalamus at the level of the ventromedial nuclei is concerned with the regulation of food and water intake. Exhaustive reviews of the pertinent literature on this subject have been published by ANDERSSON and LARSSON (1961 b) and by ANAND (1961). The relationship between body temperature and food and water intake has been recently discussed by ANDERSSON, GALE and SUNDSTEN (1963).

A number of different factors and mechanisms have been implicated in the hypothalamic control of feeding. Among the most important are: a) the mechanism based on the cerebral arteriovenous glucose differences (MAYER, 1955); b) blood temperature (BROBECK, 1958; ANDERSSON and LARSSON, 1961a); c) lipids (KENNEDY, 1953, HERVEY, 1959); and d) afferent impulses from the viscera (MILLER, 1957; JANOWITZ and GROSSMAN, 1949).

Furthermore, "cellular dehydration" and the presence of "osmoreceptors" seem to constitute a crucial factor regulating thirst, since the urge to drink can be produced by microinjections of hypertonic saline solution directly into the medial hypothalamus of goats (ANDERSSON, 1953; ANDERSSON and McCANN, 1955a).

Bilateral ablation of the lateral hypothalamic areas produces complete, long-lasting aphagia and starvation, in a number of animal species (ANAND and BROBECK, 1951; TEITELBAUM and STELLAR, 1954; MORRISON and MAYER, 1957; TEITELBAUM, 1961; MORGANE, 1961 b and c). Conversely, activation of the same lateral area by electrical stimulation produced overeating (BRUEGGER, 1943; MILLER, 1957; ROBINSON and MISHKIN, 1962) and food intake in a stimulus-bound manner in satiated animals (DELGADO and ANAND, 1953; LARSSON, 1954; HOEBEL and TEITELBAUM, 1962).

It was shown also that the same general lateral hypothalamic area is involved in the regulation of water intake, since discrete bilateral lesions may lead to complete adipsy (WITT, KELLER, BATSEL and LYNCH, 1950; TEITEL-BAUM, 1961; MORGANE, 1961a), whereas its electrical stimulation as well as direct microinjections of hypertonic saline, as already mentioned, evoke copious polydipsia in water-satiated animals (ANDERSSON, 1953; ANDERSSON and McCANN, 1955a and b). Furthermore, it has been demonstrated that self-injections of cholinergic drugs into the hypothalamic "drinking" area in rats have reinforcing effects, and can serve to reinforce instrumental responses (MORGANE, 1962).

In contrast to the lateral area, the ventromedial hypothalamus is regarded as the "satiety center", which upon stimulation is capable of inhibiting the lateral "feeding area" (ANAND and BROBECK, 1951; OLDS, 1958; EPSTEIN, 1960; HOEBEL and TEITELBAUM, 1962). Bilateral destruction of the ventro-medial hypothalamus leads to obesity (HETHERINGTON and RANSON, 1942) as the consequence of hyperphagia (BROBECK, TEPPERMAN and LONG, 1943). The excessive food consumption, however, does not seem to be the effect of an intensified hunger "drive", since there is evidence that the "drive" is even decreased following this lesion, and the hyperphagia should be ascribed rather to the lack of ability to achieve satiety during feeding (TEITELBAUM, 1955, 1961). When the ventromedial hypothalamus is electrically stimulated, the hunger "drive" seems to disappear (OLDS, 1958; WYRWICKA and DOBRZECKA, 1960; HOEBEL and TEITELBAUM, 1962).

From the morphological point of view, the lateral hypothalamus by no means has the characteristics of a "center" since through this area numerous nerve tracts pass connecting the frontal brain, basal ganglia, and temporal lobes with the hypothalamus and lower brain stem. Between these fiber systems there are relatively few nerve cells. Therefore, evidence for the physiological role of this area based only on electrical stimulation and ablation methods does not seem to establish conclusively whether the observed effects are due to destruction or stimulation of the fiber tracts or of the cell concentrations. Hence, it appears that direct pharmacological analysis of this inhomogenous region by application of drugs is particularly suited for this area (see also the review by MILLER, 1965).

2. Chemical stimulation of the hypothalamic "feeding area"

GROSSMAN (1960, 1962a; see also the recent review of his own experiments, 1964) showed that the placement of small amounts (1 to 5 µg) of crystalline adrenaline or noradrenaline in the area located between the fornix and mammillo-thalamic tract, lateral and dorsal to the ventromedial nuclei, induces vigorous and prolonged eating in satiated rats. The same area when stimulated by comparable quantities of acetylcholine ("capped" by physostigmine) or carbachol, produced a marked polydipsic effect.

STEIN and SEIFTER (1962) confirmed the polydipsic action of cholinergic drugs, and extended these studies by showing that the cholinergic effects are based on muscarinic and not on nicotinic action. In their hands, a few µg of muscarine produced an effect equal to that of the same dose of carbachol. Nicotine administered through the same cannulas and in the same animals exhibited small and apparently nonspecific effects, which could be compared with the action of sodium chloride. 5-OH tryptamine, potassium chloride, and sucrose did not give any appreciable effect. The action of carbachol and particularly of muscarine was blocked by topical atropine administration 30 min prior to these drugs.

To rule out several possible nonspecific effects, GROSSMAN (1962a) carried out control experiments. Strychnine sulfate in the same hypothalamic loci previously shown to affect water and food intake failed to produce any significant effects. Crystalline NaCl (expected to cause osmotic stimulation), vasoconstrictors (barium chloride and posterior pituitary extract), a vasodilator, sodium nitrite, as well as substances with a wide range of pH values, were not able to duplicate the effects of neurohormonal stimulation.

Grossman (1962b) extended the series of control experiments to prove further the specificity of central actions of adrenergic and cholinergic drugs in rats on water and food intake evoked from exactly the same lateral hypothalamic area. Systemic administration of adrenergic or cholinergic blocking agents selectively reduced the central actions of both groups of neurohormonal drugs.

Intraperitoneal injection of 50 mg of atropine sulfate per kg one hour before direct central stimulation, completely abolished the polydipsic effect of carbachol in all the experimental animals. However, these large doses of atropine also slightly reduced the noradrenaline-induced eating effect. This was probably due to a strong parasympathetic blockade, and concomitant motor hyperactivity and restlessness. However, relatively lower doses of atropine (10 and 25 mg/kg, i.p.) selectively depressed only cholinergic-induced drinking effects (after both carbachol and acetylcholine) without significantly changing eating produced by adrenergic stimulation.

On the other hand, the systemic administration of 5 mg of ethoxybutamoxane per kg, a central adrenergic blocking agent (Slater and Jones, 1958; Henderson, Martz and Slater, 1958; Verhave, Owen, Fadely and Clark, 1958), one hour before direct central adrenergic stimulation, prevented the previously observed eating effect. Here again, it appeared that such large doses of the adrenergic blocking compound produced several side-effects of strong sympathetic blockade, such as general hypoactivity, and this in turn might have reduced also secondarily the cholinergically induced drinking. However, smaller doses of ethoxybutamoxane (2.5 and 1 mg/kg) produced very selective and statistically significant blockade of only adrenergic-induced eating, without markedly affecting the cholinergic-induced drinking. Another adrenergic blocking agent, phenoxybenzamine (Dibenzyline), also abolished completely and selectively the adrenergic eating effect.

Similar selective effects of both systemic and central administration of adrenergic and cholinergic blocking agents have been demonstrated with regard to spontaneous hunger and thirst.

In hypothalamic loci previously shown to yield positive responses direct central administration of microgram amounts of dopamine or dimethylaminoethanol, presumed precursors of noradrenaline and acetylcholine, respectively, also caused characteristic and selective effects on food or water intake. These results have not yet been confirmed by independent groups of investigators. It appears rather unlikely that precursors of either noradrenaline or acetylcholine could have produced effects comparable to those observed following administration of neurohormonal substances administered in the same dose range.

Fisher and Coury (1962), using a technique similar to that employed by Grossman (1962a) and Stein and Seifter (1962), obtained confirmatory cholinergic drinking effects in rats from the same general perifornical and lateral hypothalamic area. In the same preliminary report, they showed that some interrelated limbic and diencephalic structures, when stimulated with microgram amounts of cholinergic drugs caused marked increase in water consumption. Since many of the highly localized positive points have been found within the circuit postulated by PAPEZ (1937), they tentatively suggested that besides other functions this circuit may subserve the mediation and integration of the thirst drive.

More recently, a dose-response study employing liquid microinjections of carbachol and noradrenaline into the lateral hypothalamic "feeding" area in rats showed that the threshold dose of carbachol is about 2.7×10^{-10} mole (*i.e.* 0.047 µg), and maximum drinking was observed after 24×10^{-10} mole. In contrast, progressively higher doses elicited less drinking and, finally, they produced convulsions (MILLER, GOTTESMAN and EMERY, 1964). Noradrenaline produced a specific eating effect in doses 24 through 216×10^{-10} mole. However, the most effective dose (648×10^{-10} mole) was much higher than that of carbachol. Larger doses of noradrenaline elicited less eating effect, lethargy and eventually convulsions in some of the rats. It should be pointed out that the optimal dose of noradrenaline (22 µg) injected into the jugular vein via a permanently implanted catheter caused hungry rats to stop eating or to stop pressing a bar for food delivery. These results indicate that exactly the same dose of noradrenaline administered peripherally produces effects opposite to those elicited centrally (MILLER, 1965).

Highly interesting but undocumented, preliminary results have been reported by MILLER (1965) concerning the effects of carbachol injected into the lateral hypothalamus of rats on urine secretion, level of blood glucose, shivering and temperature regulation. He claims that carbachol greatly reduced the volume of urine and increased its concentration. It also raised the blood glucose and lowered the body temperature. Peripheral administration of similar doses of this compound had no such action.

D. Electrophysiological and neurohormonal evidence of the existence of two functionally opposed ascending reticular systems

It is too early for a coherent physiological and anatomical picture of the reticular formation, since so many contradictory observations have been published regarding its function, intrinsic organization, and pharmacology. In view of the extreme cytoarchitectural heterogeneity of this system, some investigators have even refrained from regarding it as an anatomical unit, and have suggested that, from the physiological standpoint, the term "central internuncial system" should be substituted for "reticular formation" (OLSZEWSKI and BAXTER, 1954; OLSZEWSKI, 1958). This point of view should not cause surprise, since, as was pointed out by ROSSI and ZANCHETTI (1957) in their excellent review, "there is no other part of the central nervous system which fulfills so many functions with so little evidence of segregation". Methods employing electrical stimulation, and surgical and electrolytic ablation, however, have given results which have permitted the gross differentiation of the ascending component of the reticular formation. More recently, some promising results have been obtained from direct chemical stimulation of this structure which have both confirmed implications of previous electrophysiological studies and shed new light on the possible neurohormonal mechanisms involved in the functioning of this system (see also the recent review by Cordeau, 1962).

1. EEG and behavioral effects of surgical ablation and electrical stimulation

In spite of the cytoarchitectural and anatomical evidence of complexity, the ascending reticular system was regarded until recently as functionally a rather homogeneous structure. Since the classic work of Bremer (1935) and pioneer investigations of Moruzzi and Magoun (1949), who described the ascending activating reticular system as extending from the medulla oblongata up to the thalamus, it was usually assumed that the whole structure is significant and more or less equipotential in the maintenance of wakefulness, because high frequency electrical stimulation of the whole system elicited electrocortical desynchronization and arousal. However, more recent experiments established that two diametrically opposite EEG patterns may be obtained, depending upon the level of the transection of the pons (Batini, Moruzzi, Palestini, Rossi and Zanchetti, 1959a and b; see also Morruzzi, 1960). The most rostral transection of the pons (rostropontine pretrigeminal cat preparation) or in the adjacent part of the mesencephalon, induces the well-known state of unconsciousness associated with EEG synchrony. On the other hand, transection a few millimeters caudad but still pretrigeminal (midpontine pretrigeminal preparation) leaves the brain almost permanently desynchronized, although in both cases the tonic sensory inflow mediated by the same cranial nerves did not change. Moreover, there is also more recent evidence that the midpontine transection leaves the brain capable even of perceiving and learning (Affani, Marchiafava and Zernicki, 1962a and b). It follows from these experiments that the rostral part of the pons is the seat of strong tonic activating influences which maintain the EEG patterns of wakefulness. This predominance of behavioral and EEG signs of wakefulness in the midpontine preparation obviously contrasts with the *encéphale isolé* preparations, which despite much greater afferent input often show long periods of sleep. Based on these observations, Batini *et al.* (1959a and b) suggested that structures lying between the midpontine section and the rostral end of the spinal cord should have an opposite, *i.e.* tonic synchronizing influence on the EEG patterns.

Simultaneous and subsequent work of CORDEAU and MANCIA (1958, 1959) confirmed these suggestions and demonstrated that hemisection of the brain stem at any level between the pretrigeminal area and the nucleus paraolivaris medialis induces marked asymmetry in electrocortical activity. Following sectioning on the same side, a persistent desynchronized EEG pattern was observed, which contrasted with the synchronized activity of the intact side.

Further investigations carried out by MAGNES, MORUZZI and POMPEIANO (1960, 1961) on the effects of low frequency electrical stimulation of the reticular formation in acute *encéphale isolé* cat preparations have shown that the synchronizing structures are mainly located in the caudal brain stem in the region of the solitary tract which receives afferent impulses from the baroceptive zones of the carotid sinus. Low frequency stimulation of this area produced marked electrocortical synchronization. It was also shown that these "synchronizing" structures exhibit some autonomic tonicity, and do not depend exclusively on the sensory input from the baroceptive zone of the carotid sinus. Consequently, MAGNES et al. (1961) postulated that there exist several groups of neurones within the brain stem reticular formation which exert an influence opposite to that attributed to the activating ascending reticular formation.

In confirmation and extension of the foregoing studies, BONVALLET and BLOCH (1960, 1961) and BLOCH and BONVALLET (1961), using either localized lesions or direct procaine injections, were also able to show that phasic inhibitory influences originate in areas of the medullary reticular formation. Hence, it appears probable that a constant interplay exists between functionally opposed components of the ascending reticular system, and that a dynamic balance determines the level of wakefulness.

More recently, BONVALLET and ALLEN (1963) have demonstrated that, although the afferent impulses from the vagus and glosso-pharyngeal nerves contribute to the tonicity of the "synchronizing" bulbar reticular formation, this input is not required for the maintenance of the ascending inhibitory influences which originate in this region

The hypothesis of MAGNES et al. (1961) has been corroborated and extended by findings of FAVALE, LOEB, ROSSI and SACCO (1961), ROSSI (1962), MONNIER, HOESLI and KRUPP (1963), and YAMAGUCHI, LING and MARCZYNSKI (1963, 1964) who obtained both EEG and behavioral drowsiness or sleep following low frequency electrical stimulation not only of the bulbopontine, but also of the mesencephalic reticular formation. This implies that the "synchronizing" neurones, although confined predominantly to the bulbopontine area, are presumably intermingled with the whole classic activating system of MORUZZI and MAGOUN. It must be pointed out, however, that these effects could be obtained only in a rather narrow range of frequencies of electrical stimulation (5 to 20 c/sec), whereas, in the same anatomical loci higher frequencies always induced opposite reactions, *i.e.* desynchronization of the electrocortical activity and behavioral arousal. It should be noted that as early as 1943 BUERGI and MONNIER reported that stimuli of low frequency, low voltage, and long pulse duration, when applied in the freely moving cat to the dorsolateral pontine reticular formation, induced a particular adynamic reaction and progressive drowsiness with drooping of the head due to hypotonia of the neck muscles.

2. Alterations in incoming signals

It is generally accepted that inhibition and facilitation of sensory transmission are the important functions of the reticular formation. The first evidence of central regulation of sensory receptors was adduced by GRANIT and KAADA (1952) in their studies of muscle spindles. The discharges of muscle spindles were inhibited by stimulation of medial bulbar areas and enhanced by electrical impulses delivered to the lateral and rostral parts of the reticular system.

More recently, advances in the neurophysiology of sensory transmission have demonstrated the existence of descending inhibitory influences which are capable of modifying the afferent inflow to the central nervous system at the first synapse of the specific sensory pathways (HAGBARTH and KERR, 1954; GRANIT, 1955; HERNANDEZ-PEON, 1955, 1959, 1960, 1961). During stimulation of the reticular formation, sensory input has been shown to be blocked at the nucleus gracilis (HERNANDEZ-PEON, SCHERRER and VELASCO, 1956) and in the sensory nucleus of the trigeminal nerve (HERNANDEZ-PEON and SCHERRER, 1955). Optical responses were shown to be reduced at the lateral geniculate nucleus (BREMER and STOUPEL, 1959), whereas retinal ganglion cell discharge was found to be either enhanced or inhibited by centrifugal influences (GRANIT, 1955, 1959). Auditory responses may be depressed at the geniculate (BREMER and STOUPEL, 1959) and at the cochlear nucleus (GALAMBOS, 1956; JOUVET and DESMEDT, 1956).

Psychophysiological evidence indicated that some perceptual processes are markedly facilitated in a state of reticulo-cortical arousal (see LINDSLEY, 1958; LIVINGSTON, 1959). It has also been postulated that reticulo-cortical discharges which arouse the brain not only facilitate, but also differentiate thalamocortical responses involved in perceptual integration (see BREMER, 1960). The differentiation of cortical neuronal activity which follows the stimulation of the reticular formation has been demonstrated by FUSTER (1961) and by CREUTZFELDT and JUNG (1960).

During intense attention focused upon a given external stimulus, many sensory impulses which are not closely related to the general direction of the attention reaction, may be partially or even completely blocked (HERNANDEZ-PEON, SCHERRER and JOUVET, 1956; HERNANDEZ-PEON, 1959, 1961). On the other hand, simultaneously with the inhibitory processes accompanying the arousal reaction, marked facilitation occurs of the other sensory inputs and motor systems which might be relevant to well adapted behavioral reaction in a given situation. Many experimental findings indicate that this highly selective "filtering" and probably "amplifying" mechanism has its main seat within the reticular formation (cf. HERNANDEZ-PEON, 1963a—c). To quote from HERNANDEZ-PEON (1963c): "when we conceive the excitatory nerve impulses as carrying minute spots of light which illuminate only those pathways with significant traffic, the awake brain would not be brightly illuminated as SHERRINGTON (1906) believed, but on the contrary, within the extensive and generalized darkness resulting from inhibition in the attentive brain only a stream of bright light would be constantly moving as the beacon ceaselessly explores ocean and land in a moonless night". From this point of view, it should not be surprising that some authors have found a decrease in the amplitude of peripherally evoked sensory cortical responses in "acute" cat preparations during reticulocortical arousal (HERNANDEZ-PEON, SCHERRER and VELASCO, 1956; BREMER and STOUPEL, 1959; BREMER, STOUPEL and VAN REETH, 1960; see also BREMER, 1960) and others have described no change in primary, but evident augmentation in secondary photic responses to flash in waking rabbits (FUSTER and DOCTER, 1962).

3. Cortical photic responses influenced by pontine transection

ARMENGOL, LIFSCHITZ and PALESTINI (1961) demonstrated that both primary and secondary evoked photic responses of the cortex to flash in non-anesthetized cats were markedly augmented after midpontine and rostro-pontine pretrigeminal transection. Especially the positive deflections 3, 4 and the negative 5 [according to SCHOOLMAN and EVART'S (1959) nomenclature] were enhanced, the amplitude being usually doubled in secondary photic responses. These findings led ARMENGOL et al. to suggest that the surgical procedure eliminates certain ascending tonic inhibition originating from below the lesion, i e. from the caudal pontine and bulbar reticular formation.

4. Action of drugs upon cortical optic responses

a) Drug-induced interruption of ascending reticular pathways. As mentioned previously, peripherally evoked cortical responses have been shown to be either enhanced or suppressed during reticulo-cortical arousal. In contrast to these often seemingly contradictory observations, there is agreement with regard to the effect of arousal on "centrally" evoked cortical sensory responses, *i.e.* those induced by stimuli which, bypassing the sensory receptors, are applied directly to the sensory pathways or relay nuclei. Since centrally evoked cortical optic potentials have been found to be regularly augmented during reticulo-cortical arousal (BREMER and STOUPEL, 1959; DUMONT and DELL, 1960;

STERIADE and DEMETRESCU, 1960; BREMER, STOUPEL and VAN REETH, 1960) COURVILLE, WALSH and CORDEAU (1962a and b; see also CORDEAU, 1962) have chosen cortical potentials produced by single shock stimulation of the optic chiasma in an attempt to test further the hypothesis concerning the existence of two functionally opposed ascending reticular systems. In curarized or *encéphale isolé* cat preparations the direct injection of procaine into the "synchronizing" caudal pontine or rostrobulbar reticular formation produced a constant and marked increase in the evoked potentials, associated with electrocortical desynchronization. In contrast, similar injections into the rostral part of the tegmental mesencephalic reticular formation consistently produced a sharp fall in the amplitude of evoked potentials accompanied by electrocortical synchronization. These findings were interpreted by COURVILLE *et al.* (1962a and b) as the results of removal of a tonic inhibitory influence of the bulbopontine part of the reticular formation and a consequent release of the more rostral activating part of this system, whereas opposite effects were interpreted as due to the blocking of the activating system itself.

b) Cholinergic and adrenergic drugs. CORDEAU, MOREAU and BEAULNES (1961) not only confirmed the findings of ROTHBALLER (1955, quoted by JASPER, 1958) concerning the arousing effects of adrenaline injection directly into the reticular formation, but they also extended these studies to direct cholinergic stimulation of this system. Since the injection of acetylcholine (20 µg) into the bulbar reticular formation was usually followed by marked electrocortical synchronization (in curarized and *encéphale isolé* preparations, as well as in freely moving cats), they advanced a working hypothesis according to which the ascending "synchronizing" and inhibitory reticular system is cholinergic in its nature, as opposed to the activating and facilitatory adrenergic component. They anticipated that by the direct injection of adrenergic or cholinergic substances these two antagonistic systems might be differentially activated and influence cortical responses produced by single shock stimulation of the optic chiasma. This assumption was fully substantiated since it has been shown in a more detailed study by COURVILLE, WALSH and CORDEAU (1962a and b) that microinjections of adrenaline and acetylcholine, even when performed in the same loci of the pontine or bulbar reticular formation, produced opposite effects on the amplitude of the evoked potentials. Adrenaline usually caused a significant increase with concomitant electrocortical desynchronization, whereas acetylcholine consistently produced a marked decrease in these potentials, associated with simultaneous synchrony of the EEG. Injections of the solvent alone had no effect.

Although in the foregoing experiments of COURVILLE *et al.* there was a typical and obvious parallelism between the acetylcholine-induced inhibition of the evoked cortical responses and electrocortical synchronization, nevertheless, in some instances a rather short lasting dissociation between these two linked phenomena was observed. In several cases, following the injection of acetylcholine the typical synchronized EEG pattern appeared after a period of latency as compared to the much more quickly established inhibition of the evoked responses. This dissociation — according to COURVILLE et al. — might be tentatively explained as a possible direct and quicker effect on thalamic relays of some inhibitory ascending groups of neurones which had been activated by acetylcholine, before the inhibition or moderation of the ascending activating and facilitatory system could be accomplished. It is well known from observations on the terminal distribution of the ascending reticular fibers that both thalamic nuclei and subthalamic region may be influenced by direct projections of some cell groups scattered throughout the whole brain stem. The bulk of these fibers, however, originate in the medulla oblongata and pons (BRODAL and ROSSI, 1955; ROSSI and ZANCHETTI, 1957; NAUTA and KUYPERS, 1958; BRODAL, 1958).

5. EEG and behavioral effect of neurohormonal substances

The working hypothesis of CORDEAU, MOREAU and BEAULNES (1961) found further support in a series of experiments carried out with unrestrained, freely moving cats with chronically implanted cannulas for chemical stimulation and recording electrodes (CORDEAU, MOREAU, BEAULNES and LAURIN, 1962, 1963; see also CORDEAU, 1962; YAMAGUCHI, LING and MARCZYNSKI, 1963, 1964).

The technique used by CORDEAU et al. (1962) consisted in employing a microinjector assembly placed outside the soundproof cage and connected with the permanently implanted cannula by means of polyethylene tubing. This system served for injecting microamounts of neurohormonal substances, dissolved in Tyrode solution of pH 7.4, into different parts of the reticular formation. YAMAGUCHI et al. (1963, 1964) applied crystalline substances directly.

CORDEAU et al. explored three main levels of the brain stem reticular system: the rostral pontine and caudal mesencephalic tegmentum (F 0 to 2), the caudal pontine reticular formation (P 5 to P 7) and the bulbar part of this system (P 12 to 15). YAMAGUCHI et al. investigated more rostrally located areas (F 2 to 3.5). In most cases control administration of crystalline glucose or sodium chloride did not evoke any noticeable EEG or behavioral changes. Also, injections of the solvent alone in a sleeping cat produced only negligible reactions in the form of short lasting desynchronization of the EEG without any overt behavioral manifestations. These changes occurred during the second half of the period of injection (the whole period lasting 100 sec), and the sleeping animal sometimes showed some slight movements of the head without even opening its eyes. After 10 to 20 sec the EEG usually returned to its previous fully synchronized pattern.

a) Adrenergic stimulation. In the hands of CORDEAU *et al.*, injection of 20 μg of l-adrenaline bitartrate into the mesencephalic or pontine, as well as into the rostral or caudal bulbar reticular formation, always produced very obvious behavioral and EEG arousal of the sleeping cat which, for 15 to 30 min following the injection, exhibited an "insomniac" type of behavior. The animals usually curled up after a few minutes in an attempt to resume sleep, but it was evident that they were unable to do so. They peered around the cage, sat up, and tried another position for sleep. This "tossing and turning" could usually be observed for 30 min, until the animals returned to their preinjection behavioral and EEG patterns. Similar injections into the bulbar area produced arousal which was sometimes accompanied by retching, vomiting and salivation, and therefore — as was pointed out by CORDEAU *et al.* —, this activation might have been partly due to these disturbing autonomic effects. Surprisingly, noradrenaline bitartrate administered into the rostral parts of the mesencephalic reticular formation did not produce consistent results, whereas similar doses of *d*-amphetamine sulfate elicited very strong and long lasting EEG activation and behavioral excitement (YAMAGUCHI *et al.*, 1964). Hence, these results obtained from "chronic" animals in conjunction with the previously described "acute" experiments seem to confirm the view already expressed by BONVALLET, DELL and HIEBEL (1954) and by ROTH-BALLER (1956), based on the effects of intrasystemic administration of drugs, that the ascending activating system of the reticular formation operates through an adrenergic or adrenoceptive mechanism.

b) Cholinergic stimulation. Microinjections of acetylcholine bromide (20 μg) as well as deposits of crystalline acetylcholine chloride (15 to 30 μg) into the same previously explored areas of the reticular formation of awake but quietly resting cats always induced effects diametrically opposite to those described after adrenaline and noradrenaline administration, *i.e.* behavioral and EEG sleep patterns, which in some instances progressed after several minutes to the so-called "paradoxical" stage characterized by desynchronized electro-cortical activity and rhythmic theta waves in the hippocampus. It should be noted that similar injections of acetylcholine in experiments on *encéphale isolé* cat preparations always produced EEG synchronization comparable to that observed in unanesthetized intact cats without causing any peripheral blood pressure changes; this seems to be a convincing argument in favor of the specificity of the observed central effects.

The typical hypnogenic effect of cholinergic stimulation of the brain stem reticular formation appeared usually about 60 to 120 sec after the end of the injection, or in some cases even earlier. It should be emphasized that injections into the bulbar reticular formation, although in some instances associated with disturbing autonomic reactions such as salivation and retching, also produced rather abrupt EEG and behavioral transition from wakefulness to sleep. Even from this area, which is the seat of many autonomic responses, "pure" hypnogenic and arousing effects can be observed following cholinergic and adrenergic stimulation, respectively; this seems to be another argument in favor of the selectivity of this procedure, which presumably affects only synaptic areas and not the *"fibres du passage"*.

Observations by CORDEAU and his group on acetylcholine-induced sleep support the suggestion previously made by HUBEL (1960) and JOUVET (1961), that "desynchronized" fast-wave sleep represents a deeper stage than sleep characterized by high voltage slow-wave electrocortical activity because: 1. mild external stimuli applied during "desynchronized" sleep usually shift the electrocortical activity for a moment to the "synchronized" pattern; 2. the "desynchronized" phase never occurs without passing through the "synchronized" one; and 3. animals on waking pass through the "synchronized" phase.

Concerning the correlation between the injection sites and the effectiveness in producing typical EEG and behavioral changes, CORDEAU and his group suggested that acetylcholine is capable of inducing more consistent and more obvious sleep effects when administered into the caudal part of the reticular formation and close to the midline, whereas adrenaline and noradrenaline produced more evident arousal effects when injected into the rostral parts of this system.

In the hands of YAMAGUCHI, LING and MARCZYNSKI (1963, 1964), crystalline acetylcholine when tamped in to areas of the reticular formation (F 2 to 3.5) more rostrally located than those investigated by CORDEAU *et al.*, also produced unmistakable sleep. Carbachol, even in much smaller doses (5 μg) did not mimic the acetylcholine effects. After carbachol the animals tended to retreat to a corner of the cage, exhibited a catatonic-like behavior, and, although they appeared to have lost interest in the immediate environment, their EEG displayed continuous activation patterns which lasted for about 1 hour. Also, CORDEAU, MOREAU and BEAULNES (1961) found that administration of acetylcholine with physostigmine produced strong electrocortical desynchronization and behavioral arousal *i.e.*, effects opposite to those previously observed by these authors after acetylcholine alone. In their hands, acetylcholine when applied to all areas of the reticular system, from bulbar to mesencephalic parts, always produced unmistakable sleep and electrocortical synchronization. The paradoxical effect of acetylcholine in the presence of a cholinesterase inhibitor was particular striking after administration into the mesencephalic reticular formation. In this connection it should be noted that carbachol administered into other structures involved in the regulation of the sleep-wakefulness cycle, such as the basal forebrain inhibitory system delineated by CLEMENTE and STERMAN (1963; see also STERMAN and CLEMENTE, 1962a and b), non-specific thalamic nuclei, and the preoptic region, not only failed to mimic the previously observed hypnogenic action of acetylcholine but consistently produced

EEG and behavioral arousal effects (CORDEAU, MOREAU, BEAULNES and LAURIN, 1963; YAMAGUCHI, LING and MARCZYNSKI, 1964; MARCZYNSKI, CLEMENTE and STERMAN, unpublished observations).

The arousal effect of carbachol applied to the lateral mesencephalic reticular formation was reported by HERNANDEZ-PEON, CHAVEZ-IBARRA, MORGANE and TIMO-IARIA (1963). This seems to indicate, as pointed out by these authors, that the mesencephalic arousing system is likely to operate through cholinergic synapses in addition to an adrenergic mechanism. Such an interpretation would lend support to the view already expressed by RINALDI and HIMWICH (1955 a and b), based on observations of electrocortical effects obtained after systemic administration of acetylcholine, physostigmine and atropine, which led them to postulate the existence of a "cholinergic mesodiencephalic arousing system". It seems, however, that in view of the conflicting results of direct application of different cholinergic drugs to the mesencephalic reticular formation, the evidence for the existence of the cholinergic or cholinoceptive arousal system is inconclusive.

It seems that there are several possible tentative explanations of the above mentioned discrepancies, if we accept the possibility that two or even three distinctive kinds of cholinoceptive sites can be present in subcortical brain structures in analogy to the autonomic sympathetic ganglia in which on the postsynaptic membrane the presence of two or even three distinctive cholinoceptive sites has been postulated (ECCLES and LIBET, 1961; cf. ECCLES, 1964; TAKESHIGE and VOLLE, 1964; see also recent review by KARCZMAR, 1966). A depolarizing block can be produced at cholinergic ganglionic junctions by an excess of acetylcholine injected into the ganglionic blood supply (PATON and PERRY, 1953). It is also known that intravenous infusion of physostigmine has clearly biphasic action on electrocortical activity: it first increases the tendency to synchronized activity, and with a greater dose, it produces desynchronized electrocortical patterns (TRACZYK and SADOWSKI, 1962). More recently, suggestive evidence has been presented that a cholinergic step may be involved both in cortical spindling and recruitment induced by electrical stimulation of the caudate nucleus (TRACZYK and SADOWSKI, 1964; McLENNAN 1964). Nonetheless, both electrocortical phenomena can be abolished by physostigmine introduced in microamounts directly into the caudate nucleus (RAKIĆ, BUCHWALD and WYERS, 1962). Furthermore, recruiting responses abolished by intravenous administration of physostigmine can be partially restored by subsequent administration of atropine (TRACZYK and SADOWSKI, 1962). In view of these results it can be tentatively suggested that conflicting data obtained in experiments on the brainstem reticular formation after direct administration of acetylcholine in association with physostigmine, and after carbachol, might be due to the functional blockade of the inhibitory and hypnogenic influences mediated through the postulated cholinergic mechanism rather than to the activation of a cholinergic arousal system of the lateral mesencephalic reticular formation as suggested by HERNANDEZ-PEON et al.

There is, however, another possible explanation of the above mentioned conflicting data. Acetylcholine and carbachol exhibit both "muscarinic" and "nicotinic" actions on the peripheral nervous system. It is, therefore, not unlikely that when they are administered directly into the brain stem reticular formation, the overall EEG and behavioral effects are the outcome of actions on both kinds of receptors, the existence of which can be anticipated on the basis of comparative studies of the central actions of muscarine, arecoline, and other drugs (RIEHL, PAUL-DAVID and UNNA, 1962) including nicotine (KNAPP and DOMINO, 1962).

The receptors responsible for the previously discussed inhibitory and hypnogenic action of the ascending reticular system after direct administration of acetylcholine may be essentially muscarinic. This may be inferred from the central action of oxotremorine [1(2-oxopyrrolidino)-4-pyrrolidino-2-butene], an active metabolite of tremorine (CHO, HASLETT and JENDEN, 1961). In a series of careful experiments on the peripheral nervous system, it has been demonstrated that oxotremorine is completely lacking in nicotinic activity, and its strong muscarinic action can be compared to that of acetylcholine and carbachol (CHO, HASLETT and JENDEN, 1962). Microinjections of this compound into the brain stem reticular formation in conscious cats and dogs produce flaccid paralysis and areflexia, associated with deep sleep. In these animals the eyes show oneiric movements while the EEG patterns are characterized by low voltage, fast electrocortical activity. This typical syndrome which is indistinguishable, as pointed out by HASLETT (1963), from that of "paradoxical sleep" was also obtained by CORDEAU et al. (1962) after administration of acetylcholine into the brain stem reticular formation in conscious cats, and by HERNANDEZ-PEON et al. (1963) following administration of cholinergic drugs into the forebrain limbic — midbrain limbic pathway of NAUTA, as well as by STERMAN and CLEMENTE (1962b) in response to low frequency electrical stimulation of the basal forebrain. It should be noted that all the above-mentioned central effects of oxotremorine could be prevented by prior administration of atropine (either peripherally i.v., i.m. or centrally into the brain), which seems to add another argument in favor of the view that the powerful central hypnogenic action of oxotremorine is mediated selectively by muscarinic receptors within the reticular system (HASLETT, 1963; HASLETT, GEORGE and JENDEN, 1963 a and b; GEORGE et al., 1964).

As to the nicotinic receptors, results obtained following the intrasystemic administration of drugs suggest that they also exist in the brain stem reticular formation (STUMPF, 1959, 1962; RIEHL, PAUL-DAVID and UNNA, 1962), and that they are confined predominantly to the ponto-mesencephalic site (KNAPP and DOMINO, 1962). It can be inferred from these studies that stimulation of nicotinic receptor sites in the brain stem raises the tonicity of the arousing component of the reticular system. The existence of nicotinic receptors intermingled with the muscarinic ones in this area of the brain may account at least in part for the discrepancies of results obtained with acetylcholine alone, on the one hand, and with carbachol or acetylcholine plus physostigmine, on the other.

Furthermore, if one accepts the proposal of CORDEAU *et al.* that the cholinergic ascending component of the brainstem reticular formation represents the inhibitory and "synchronizing" system as opposed to the adrenergic one, there seems to be another tentative explanation of the arousal effect produced by carbachol or by acetylcholine administered in association with physostigmine. Although any inferences with regard to the mechanism of central action of drugs on the basis of results obtained in the peripheral nervous system must be drawn with reservation, nonetheless, it is tempting to suggest that the general proposal of BURN and RAND (1959, 1960) concerning the intermediary action of acetylcholine in postganglionic adrenergic transmission may be applicable to the central action of physostigmine and carbachol on the postulated adrenergic mechanism. (See the last section of this review.)

c) Tryptaminergic stimulation. Marked hypnogenic effects with concomitant EEG changes similar to those observed following acetylcholine administration, were noted after the application of 5-hydroxytryptamine creatinine sulfate (approximately 20 μg) to the mesencephalic reticular formation (F 2 to 3.5), although these effects were less apparent than following implantation of the same quantity of this substance into the preoptic region or nucleus centralis medialis (YAMAGUCHI, LING and MARCZYNSKI, 1963, 1964). No clear-cut results were obtained following the administration of similar or even greater amounts of other endogenous tryptamine derivatives such as bufotenine (5-hydroxy-N-dimethyltryptamine oxalate) or melatonin (5-methoxy-N-acetyl-tryptamine) into the same mesencephalic part of the reticular formation although these compounds were very effective in producing EEG and behavioral changes when placed at the preoptic region (MARCZYNSKI *et al.* 1962, 1964). Also several other non-endogenous tryptamine derivatives such as 3-beta-piperidinoethylindole, 5-methoxy-3-(beta-piperidinoethyl)-indole, and related compounds, have proved rather ineffective in the brainstem reticular formation in contrast to their strong "trophotropic" and hypnogenic action following administration into the preoptic region (MARCZYNSKI and YAMAGUCHI, 1963; MARCZYNSKI, 1965). These findings suggest an uneven distribution of tryptamine receptor sites within subcortical structures involved in the regulation of the sleep-wakefulness cycle.

It is probable that the uneven distribution of monoamines reflects their functional significance in various parts of the central nervous system. It may also be suggested that the distribution reflects the density of specific receptor sites. The hypothalamus is known to contain much higher levels of 5-OH tryptamine than the brainstem reticular formation (for references see ROBSON and STACEY, 1962; MACLENNAN, 1963), and it should not be unexpected that the administration of tryptamine derivatives to the hypothalamus would be more effective in producing behavioral and EEG changes than their administration to the reticular system. Furthermore, it seems that the parallelism between the effectiveness of tryptamine derivatives and the distribution of 5-OH tryptamine suggests that the observed central effects are mediated through a specific system of neurones, the functional tonicity of which can be regulated through the level of active 5-OH tryptamine. Such a proposal, already expressed by BRODIE and SHORE (1957) with regard to the hypothalamus, has recently received strong support from histochemical findings which for the first time directly implicate 5-OH tryptamine in the processes of synaptic transmission in the brain and spinal cord. These studies of the distribution of 5-OH tryptamine (as well as of catecholamines) point directly to the existence of tryptaminergic central neurones, since it was shown that this amine is preferentially accumulated in the varicosities of certain nerve terminals (CARLSSON, FALCK, HILLARP and TORP, 1962; FALCK, 1962; CARLSSON, FALCK, FUXE and HILLARP, 1964; BERTLER, FALCK and OWMAN, 1963; DAHLSTROM and FUXE, 1964), which, on the basis of electronmicroscopic studies, can be regarded as synaptic junctions (ELFVIN, 1963). Moreover, a few days after transection of corresponding fiber systems, 5-OH tryptamine disappears from synaptic terminals (CARLSSON et al., 1964; ANDEN et al., 1964). The same findings relate to both noradrenaline and dopamine. Thus, for the first time direct evidence has been produced that monoaminergic neurones may exist in the central nervous system, although it remains to be shown whether these amines are released upon appropriate stimulation of central neurones.

6. Concluding remarks

The effects of microinjection of adrenergic drugs on the local gross EEG activity (TRZEBSKI, 1960), changes in single unit activity following the intracarotid injection of physostigmine (DESMEDT and SCHLAG, 1957), and particularly the effects of electrophoretic application of adrenergic and cholinergic drugs to single neurones of various parts of the reticular system, point to the existence of both adrenoceptive and cholinoceptive neurones (BRADLEY and WOLSTENCROFT, 1962, 1963, 1964; SALMOIRAGHI and STEINER, 1963; BRADLEY, DHAWAN and WOLSTENCROFT, 1964). These studies have shown that within the reticular formation there are neurones which can be both excited and inhibited by acetylcholine, and the same is true of adrenaline and noradrenaline. Furthermore, some units have been found the firing of which can be affected in the same direction by either cholinergic or adrenergic substances. Moreover, some of them can be excited by adrenergic drugs, inhibited by cholinergic, and vice versa. This implies that both kinds of receptors may be present on the same neuron (BRADLEY and WOLSTENCROFT, 1962, 1964). Such an interpretation cannot be accepted without reservation, since, as pointed out by BLOOM, OLIVER and SALMOIRAGHI (1963; see also SALMOIRAGHI, 1964), the possibility cannot be excluded that one type of neurohormonal stimulation directly affects the neuron the activity of which is being recorded, while the other type influences its firing rate indirectly through axonal collaterals or through the action on neighboring neurones synapsing with the former one. In connection with the possibility of the existence of two different receptor sites on one neuron, it may be recalled that RENSHAW cells appear to be sensitive to two different neurohormonal substances. The effect of ventral root stimulation can be blocked by the administration of cholinergic blocking agents, but, on the other hand, in spite of the cholinergic blockade, these cells can still be activated by stimulating the dorsal root afferents (CURTIS, PHILLIS and WATKINS, 1961).

Recent studies by BRADLEY, DHAWAN and WOLSTENCROFT (1964), in which a special electronic counter and print-out unit was used for recording the frequency of unit activity (BRADLEY and WOLSTENCROFT, 1963), clearly revealed a much higher proportion of cholinoceptive neurones present in the reticular formation of the medulla and pons when compared with previous findings (BRADLEY and WOLSTENCROFT, 1962, 1964). These results indicate that approximately 47% of reticular units of this area are cholinoceptive. Since previous studies of this area showed that about 46% of the units are adrenoceptive (BRADLEY and WOLSTENCROFT, 1964), it would appear that the number of both types of neurones remains in a relative balance.

The foregoing findings, which demonstrate the existence of two main populations of neurones within the brain stem reticular formation, appear to be consistent with the hypothesis of CORDEAU, MOREAU, BEAULNES and LAURIN (1963), according to which "both adrenaline and acetylcholine raise the level of activity of two functionally different but closely intermixed systems spatially; the first being responsible for behavioral arousal and EEG desynchronization (adrenergic system), and the second having an opposite effect, *i.e.* producing sleep and electrocortical synchronization (cholinergic system)".

The existence of one functional system, the ascending tonic activity of which may be changed in opposite directions by both types of neurohormonal stimulation, seems rather unlikely. As was pointed out by CORDEAU *et al.*, if acetylcholine injected into the bulbar reticular formation reduced the level of activity of this area, which was shown previously by surgical elimination to exert a tonic synchronizing effect (BATINI, MORUZZI, PALESTINI, ROSSI and ZANCHETTI, 1959; and others), one could expect desynchronized EEG patterns and behavioral arousal instead of a marked "synchronizing" action to be produced by cholinergic stimulation of this area.

E. Electrophysiological and neurohormonal evidence of the existence of the forebrain descending inhibitory system

The existence of hypnogenic subcortical structures was suggested long before the inhibitory and "synchronizing" component of the ascending reticular system was postulated. Suggestions of the existence of active "sleep centers" were made on the basis of clinical observations on insomnia accompanying pathologic lesions of subcortical structures (VON ECONOMO, 1918), and behavioral changes produced by experimental lesions in rats (NAUTA, 1946).

1. EEG and behavioral effects of electrical stimulation of the forebrain hypnogenic structures

HESS in his well known diencephalic studies (1928, 1929, 1944a and b, 1954) was the first who showed that the experimental state of adynamia, drowsiness and sleep may be elicited in a freely moving cat by stimulating subcortical areas, particularly the paramedian thalamic and anterior hypothalamic regions. He coined the term "trophotropic zone" for that area, which upon low frequency electrical stimulation gave behavioral inhibitory and parasympathetic visceral responses. The dorsocaudal region of the hypothalamus was named the "ergotropic zone" by HESS, because its electrical stimulation produced opposite reactions, i.e., alertness, defense responses, and sympathetic activation.

The original findings of HESS have been repeatedly confirmed and extended, and it has been shown that there is no difference between electroencephalographic patterns observed during artificially induced and spontaneous sleep (HUNTER and JASPER, 1949; MONNIER, 1950; AKERT, KOELLA and HESS, 1952; AKIMOTO, YAMAGUCHI, OKABE, NAKAGAWA, ABE, TORII and MASAHASHI, 1956; CLEMENTE and STERMAN, 1963; MONNIER, HOSLI and KRUPP, 1963; YAMAGUCHI, LING and MARCZYNSKI, 1963, 1964; and others).

Special attention, however, should be accorded to the very effective hypnogenic area of the medial forebrain bundle at the preoptic region (HERNANDEZ-PEON, 1962; YAMAGUCHI, MARCZYNSKI and LING, 1963), and particularly to the more rostrally located area described by STERMAN and CLEMENTE (1962a and b). The last named authors found that the greater part of this hypnogenic zone lies rostrally to the levels investigated by HESS (1944a and b) and NAUTA (1946), in close vicinity to the diagonal band of BROCA and even partially within its fibers. In studying the effectiveness of this area in producing sleep on electrical stimulation, one is impressed by the rapid transition (during 0.5 to 1 min) from an alert waking behavior to electroencephalographic and behavioral sleep (see CLEMENTE and STERMAN, 1963; STERMAN and CLEMENTE, 1962b). Moreover, when the intermittent electrical stimulation is continued, the animals usually progress to the deeper and "desynchronized" stage of sleep described by DEMENT (1958), HUBEL (1960) and JOUVET (1961). To emphasize the physiological nature of the artificially induced sleep, it should be noted that the deep sleep produced by stimulation of the basal forebrain synchronizing area of Clemente and Sterman can be easily interrupted at any time by nociceptive stimuli. On the basis of the foregoing observations it has been suggested that electrical stimulation of the abovementioned brain areas might result in active descending inhibition of the arousal system of the brain stem (Hernandez-Peon, 1962; Sterman and Clemente, 1962b). More recently, Clemente and Sterman (1963) obtained electrophysiological evidence that the basal forebrain-induced cortical synchronization may be mediated through the unspecific thalamic projection system. It is therefore not unlikely that the demonstrated interaction between basal forebrain structures and arousal system may occur at the thalamic level.

2. Neurohormonal stimulation of the forebrain limbic — midbrain limbic pathway of NAUTA

a) Cholinergic and adrenergic stimulation. Studies on freely moving cats to delineate more exactly the hypnogenic pathway and to test the nature of the possible neurohormonal mechanism involved, showed that parenteral administration of atropine in a dose of 1.5 mg/kg prevented sleep which was otherwise induced by electrical stimulation of the lateral preoptic region (Hernandez-Peon and Chavez-Ibarra, 1963). Further extensive studies by Hernandez-Peon et al. have shown that sleep may be induced by direct cholinergic stimulation of many points located along the discrete pathway within the limbic forebrain — limbic midbrain circuitry of Nauta (1958). Both behavioral and EEG sleep effects could be produced by insertion of micro-amounts of crystalline acetylcholine, eserine, or carbachol into the precommissural fornix system, septal fibers (which form the roots of the medial forebrain bundle), upper medial preoptic region, lateral hypothalamic area, supramammillary commissural fibers of the medial forebrain bundle, posterior dorsal hypothalamic area, the periventricular system, and into the posterior area "located around the origin of the mammillothalamic tract at the level where it passes through the posterior hypothalamic periventricular system of fibers". Furthermore, the very rapid onset of "cholinergic" sleep could be induced by tamping acetylcholine into different points located within the mesencephalon, e.g., in the interpenduncular nucleus, and Gudden's ventral and dorsal tegmental nuclei. The latencies of the hypnogenic effects of acetylcholine in the foregoing experiments showed marked and rather consistent differences, which characteristically depended on the area stimulated. These latencies were very short (20 to 30 sec) in cases in which the limbic midbrain was stimulated, and were much longer (1 to 4 min) following the stimulation of the limbic forebrain.

Unless otherwise stated the following data and arguments are based on the work of VELLUTI and HERNANDEZ-PEON (1963), HERNANDEZ-PEON, CHAVEZ-IBARRA, MORGANE and TIMO-IARIA (1963), HERNANDEZ-PEON and CHAVEZ-IBARRA (1963) and HERNANDEZ-PEON (1965). They support the view (expressed by the same authors) that the limbic hypnogenic pathway is a descending trajectory, cholinergic in nature, which conveys inhibitory, *i.e.* hypnogenic influences, from the limbic forebrain structures to the brainstem:

1. Small electrolytic bilateral lesions within the medial forebrain bundle, in areas slightly posterior to the points of cholinergic stimulation, abolished previously obtained hypnogenic effects of crystalline acetylcholine. This fact seems to rule out the possibility of a non-specific action of acetylcholine through widespread vascular changes, or through indiscriminate effects on other remote cerebral structures.

2. In the same animals, cholinergic stimulation of loci immediately caudad to the lesion still evoked typical sleep effects with concomitant EEG changes.

3. Evident and consistent differences in latencies of the sleep responses related to the stimulated area; short latencies in the limbic midbrain area as compared to much longer ones in the forebrain suggest that the target region of the inhibitory, *i.e.* hypnogenic, influences lies close to, or even within, the ascending reticular system.

4. Topical application of minute amounts of atropine at several points along the previously described pathway, for instance in the upper medial preoptic region or in the interpeduncular nucleus, usually evoked a state of alertness associated sometimes with motor hyperactivity in a relaxed but conscious animal. This seems to indicate the tonic character of the descending inhibitory influences.

5. Acetylcholine, previously shown to be active in producing sleep when tamped in the upper preoptic region or in the interpeduncular nucleus, was completely ineffective when applied to the same loci after atropine.

6 The crucial point is, however, that atropine is capable of blocking cholinergically induced sleep also from a distant locus, as long as it is placed within the postulated hypnogenic pathway. For instance, acetylcholine tamped into the cannula implanted in the rostral segment of the medial forebrain bundle is no longer effective after the atropinization of the pontine caudal end of the pathway, *i e.* of the interpeduncular nucleus, or BECHTEREW's or GUDDEN's nuclei. This confirms not only the caudad direction of the hypnogenic influences, but also strongly suggests that cholinergic transmission may be involved at the most caudal part of the hypnogenic trajectory in the brainstem in response to the cholinergic stimulation of the rostral part of the NAUTA pathway.

It should, however, be noted that arguments based on blocking effects of atropine are weakened by the possibility of a nonspecific action, especially when atropine is applied in relatively high concentration. As was shown by CURTIS and PHILLIS (1960), both atropine and procaine, when applied electrophoretically to the spinal cord, are capable of suppressing the responses of interneurones, motoneurones, and RENSHAW cells, evoked synaptically, antidromically, or by the administration of excitatory substances. Concentrations effective in blocking the responses of RENSHAW cells to cholinergic stimulation were lower than those which prevented the excitatory effect induced by glutamate by a factor of only two or three. In this particular case, susceptibility to locally administered atropine which has marked local anesthetic properties does not necessarily mean that the affected synaptic transmission is cholinergic in nature. On the other hand, the finding of HERNANDEZ-PEON and CHAVEZ-IBARRA (1963) that relatively low doses of atropine injected parenterally were capable of preventing sleep otherwise induced by low frequency electrical stimulation of the preoptic region, seems to lend support to the arguments for a cholinergic nature of the descending pathway.

HERNANDEZ-PEON et al. (1963) proposed that the descending hypnogenic influences travelling along NAUTA'S pathway impinge upon the mesodiencephalic arousal system. Although the anatomical substrate of this final link between the two systems remains to be settled, it has been suggested that "the massive projections of the radiatio grisea tegmenti of WEISSCHEDEL (1937), extending from the region of the periaqueductal gray substance to the entire cross section on the midbrain tegmentum, appears anatomically suited for that functional role" (HERNANDEZ-PEON, 1963 c). Another alternative, namely that the hypnogenic system is capable of exerting inhibitory effects on the neocortex without its intermediary action on the brain stem arousal system, seems rather unlikely. However, studies on latencies and distribution of evoked potentials produced by electrical stimulation of the basal forebrain suggest that some inhibitory influences originating in this area may affect the unspecific thalamic nuclei more directly (CLEMENTE and STERMAN, 1963).

Diametrically opposite reactions have been produced by the topical administration of noradrenaline into the medial and lateral preoptic region, in the same animals via the same cannulas from which cholinergic agents induced sleep. In this region noradrenaline always induced EEG and behavioral arousal, motor hyperactivity, and nondirected rage behavior (HERNANDEZ-PEON et al., 1963; YAMAGUCHI et al., 1964). The specificity of this reaction is striking in that the same dose of noradrenaline (15—30 μg) administered into the slightly rostrally located area, i e. close to the diagonal band of BROCA, loses its arousing capacity and even induces drowsiness (MARCZYNSKI, STERMAN and CLEMENTE — unpublished results). Parenthetically, it should be recalled that similar doses of noradrenaline administered into the nucleus centralis medialis of the thalamus produced unmistakable deep sleep characterized by long lasting phases of desynchronized electrocortical activity, hippocampal theta rhythm and disappearance of electromyographic potentials, *i.e.* a typical "paradoxical" or "rhombencephalic" sleep (YAMAGUCHI *et al.*, 1964). According to JOUVET (1962), the descending and ascending influences originating in the nucleus reticularis pontis caudalis are solely responsible for these phenomena. He also suggested that the triggering of the "rhombencephalic" sleep may depend upon a cholinergic mechanism, since the peripheral administration of atropine prevented these effects. It remains to be shown whether noradrenaline administered into the midline thalamic nuclei may influence or trigger the activity of pontine reticular neurones. Recent electrophysiological evidence strongly suggests that there exist both poly- and oligosynaptic descending pathways from the midline thalamic nuclei to the brain stem reticular formation (SCHLAG and FAIDHERBE, 1961; SCHLAG and CHAILLET, 1963; MARCZYNSKI *et al.*, 1966).

HERNANDEZ-PEON *et al.* (1963) also reported that the administration of carbachol into the dorsal anterior commissure produced a state of specific alertness "during which the animal's gaze was focused on and followed any object placed in its visual field". They called this phenomenon "magnetic attention". Alertness was also obtained after administration of cholinergic agents into such areas as dorsal anterior commissure, dorsal perifornical area, habenulointerpeduncular tract, lateral habenular nucleus, stria medullaris, red nucleus, medial and lateral mesencephalic reticular formation and ventral central gray substance.

It is difficult, however, to draw any conclusions from these results, since the authors did not state even the approximate dose of cholinergic agents employed. Moreover, they did not distinguish between the effects of acetylcholine (administered alone or mixed with physostigmine) and carbachol, which, as already mentioned, produces different and opposite results in the hands of other investigators.

b) Tryptamine and related compounds. As already pointed out, microgram amounts of crystalline 5-OH tryptamine and melatonin, as well as several non-endogenous tryptamine derivatives, administered directly into the mesencephalic reticular formation, nucleus centralis medialis, and preoptic region, produce behavioral and EEG effects comparable to those observed after acetylcholine, *i.e.*, electrocortical synchronization and sleep which last 1 to 2 hours (YAMAGUCHI, LING and MARCZYNSKI, 1963, 1964, MARCZYNSKI and YAMAGUCHI, 1963; MARCZYNSKI, 1965). MARCZYNSKI, CLEMENTE and STERMAN (unpublished data) obtained a similar hypnogenic effect of 5-OH tryptamine also from the rostral part of the "basal forebrain synchronizing area" described by CLEMENTE and STERMAN (1963).

It has been shown by WADA (1963) that the systemic administration of atropine prevents central "trophotropic" and peripheral parasympathetic manifestations produced normally by the systemic administration of large doses of 5-hydroxytryptophan (5-HTP). Furthermore, it has been demon-

9*

strated by the same author that atropine is also capable of reversing the 5-HTP-induced deterioration of both avoidance and approach behavior in monkeys whereas physostigmine further enhances the central depressant effects of 5-HTP. The above-mentioned data suggest that 5-OH tryptamine may owe its central hypnogenic or "trophotropic" action, at least partially, to the activation of the cholinergic system of neurones which, according to HERNANDEZ-PEON et al., mediates the descending forebrain limbic inhibitory influences on the one hand, and, according to CORDEAU et al. (1963), represents the neuronal substrate of the inhibitory component of the ascending brain stem reticular system, on the other.

The "trophotropic" and hypnogenic effects obtained after the administration of acetylcholine and 5-OH tryptamine into the preoptic region in unrestrained cats, and the typical "ergotropic" effects produced by the administration of noradrenaline into the same anatomical loci (YAMAGUCHI et al., 1963, 1964; HERNANDEZ-PEON et al., 1963), point to the overlap of at least two functionally opposite populations of neurones in this area involved in the regulation of the level of wakefulness. This seems to support both the classical idea of HESS (1954) concerning the existence of two opposing autonomic integrative hypothalamic systems, as well as the concept of BRODIE and SHORE (1957) which implicates 5-OH tryptamine in the maintenance of the tonicity of the "trophotropic" system, and noradrenaline in the same capacity within the "ergotropic" system.

The association of the central action of 5-OH tryptamine with the cholinergic system of neurones seems to be further supported by the findings of APRISON (1962), who showed that this amine has a strong and reversible anticholinesterase action in vitro. However, similar inhibition of cholinesterase in vivo could not be demonstrated, although following the administration of large doses of 5-OH tryptophan the experimental animals exhibited marked behavioral changes which could be related to the elevated level of 5-OH tryptamine in the brain. In this connection, it should be mentioned that differential and density gradient centrifugation revealed that the largest proportion of this amine is in the cholinergic fraction of nerve endings (ZIEHER and DE ROBERTIS, 1963).

IV. Speculations on the relationship between central neurohormonal mechanisms and effects of electrical stimulation

It has already been pointed out in the section "General considerations" that a parallelism was found between the hypnogenic action of acetylcholine and "tonic" EEG and behavioral effects of low frequency electrical stimulation, on the one hand and the action of adrenergic drugs and "tonic" effects of high frequency stimulation, on the other. The aim of this discussion is to consider additional experimental evidence for the assumption that differential activation of two different neurohormonal mechanisms may be anticipated when different frequencies are employed in stimulation of brain structures.

1. Extraneuronal influences

Many observations indicate that electrical stimulation of subcortical structures brings about the release of neurohormonal substances which may influence other remote neural structures through extraneuronal action. The experiments of INGVAR (1955; cf INGVAR, 1958) were the first to show that both normal and epileptic activity of a neuronally isolated slab of the cat's cortex, with its intact pial circulation, may be markedly changed by electrical stimulation of the brain stem reticular formation. Further experiments by PURPURA (1956), using a cross-circulation technique, led him to conclude that unidentified humoral factors are involved in the mechanism of the electro-cortical arousal observed after electrical stimulation of the reticular formation.

Not only arousal effects can be obtained by extraneuronal influences following high frequency electrical stimulation of subcortical structures, but also opposite effects, *i.e.*, electrocortical synchronization, may be produced by low frequency electrical stimulation of intralaminar thalamic nuclei. KORNMÜLLER, already in 1947, advanced a hypothesis to explain electroencephalographic phenomena in terms of neurosecretory activity and postulated the existence of a "sleep hormone". Some years later he showed in cross-circulation experiments that: 1. striking differences in the EEG patterns of two animals disappear shortly after the establishment of the cross-circulation (KORNMULLER, LUX and WINKEL, 1961) and 2. electrical stimulation at low frequency (4 c/sec) of intralaminar thalamic nuclei in the donor cat produces marked synchronization of the EEG pattern in the cat receiving the blood, 20 to 30 sec after the onset of stimulation (KORNMULLER, LUX, WINKEL and KLEE, 1961). Also, KOLLER and MONNIER (1962), and MONNIER and HOSLI (1965) showed that in rabbits the low frequency electrical stimulation (6 c/sec) of some hypnogenic intralaminar thalamic structures of the donor animal gives rise to significant electrocortical changes in the animal receiving the blood. These changes were manifested in the increased amplitude of cortical delta waves. Moreover, when the venous blood collected from the cranial sinus of an animal kept asleep by electrical stimulation of the thalamus is pumped through a dialysing system (artificial kidney of KUHN *et al.*, 1957), it still produces hypnogenic effect in the recipient animal. Conversely, injection of dialysate from a donor kept alert by electrical stimulation of the midbrain elicits a behavioral and electrographic arousal. These results indicate that relatively stable and dialysable hormonal factors may play a role in the regulation of sleep and wakefulness. Obviously, under these experimental conditions it is unlikely that the potential neurohormonal substances involved in hypnogenic action might be acetylcholine, 5-OH tryptamine, dopamine or noradrenaline since they undergo rapid deactivation in the blood, do not cross the blood brain barrier, and when injected into the systemic blood circulation all of them are known to produce opposite effects, *i.e.*, electrocortical desynchronization and behavioral arousal, as discussed previously.

Another compound that has been implicated in extraneuronal inhibitory effects is gamma-hydroxybutyric acid. This compound has been reported to produce surgical anesthesia in man (LABORIT, JOUANY, GERARD and FABIANI, 1960), and it has been recently shown to occur in relatively high concentrations in mammalian brain (BESSMAN and FISHBEIN, 1963; BESSMAN and SKOLNIK, 1964). It is interesting that a natural brain component can produce sleep following intravenous injection. Levels exceeding 1 millimolar concentration in the circulating blood are effective in man. No data as yet are available showing whether this compound or its close metabolite is released into the blood stream upon appropriate electrical stimulation of subcortical brain structures, or whether this compound gradually accumulates in some brain areas as a "sleep hormone" during the state of wakefulness, and is subsequently metabolized below the threshold level during sleep.

2. Possible role of acetylcholine in the liberation of neurotransmitters

BURN and RAND (1959) postulated an intermediary role of acetylcholine in adrenergic transmission in postganglionic sympathetic nerves. This hypothesis has been extended by KOELLE (1962) who proposed on the basis of several lines of indirect evidence that at a wide variety of synapses, both central and peripheral, there is cholinergic mediation in the process of liberation of neuro-transmitters. This implies that small amounts of acetylcholine or related cholinesters are released by the nerve action potential to act presynaptically so as to trigger the liberation of the "true" synaptic transmitter. In cholinergic synapses, the latter would be acetylcholine itself, and in this particular case such a mechanism would provide a self-amplifying system. The first electro-physiological pieces of evidence for presynaptic action of acetylcholine were reported by MASLAND and WIGTON (1940), and by RIKER et al. (1957, 1959). KOELLE and his associates provided ample evidence for prejunctional role of acetylcholine along several lines of investigations (KOELLE, 1961, 1962, 1963).

A comprehensive discussion of the controversial hypothesis of BURN and RAND is beyond the scope of this review and the reader is referred to their recent article (1965). However, for objective and critical appraisal of the validity of their theory the reader is referred to excellent reviews by SCHÜMAN (1963), FERRY (1966) and KARCZMAR (1966).

With regard to the possible prejunctional role of acetylcholine in the central nervous system, the following evidence can be cited:

Histochemical study of the mammalian brain showed that besides neurones which are known to be cholinergic and which exhibit high concentrations of acetylcholinesterase, the majority of the remaining neurones also show appreciable straining for this enzyme (KOELLE, 1954). Hence, the extrapolation of the BURN and RAND hypothesis to the central nervous system would provide a tentative explanation of the function of these lower concentrations of acetylcholinesterase.

Studies on the hypothalamic-hypophyseal tract gave indirect evidence that the neurosecretory cells are themselves cholinergic in nature. This means that acetylcholine is presumably released by these fibers, acting in turn on their membranes and causing liberation of the endocrine secretion. These neurones exhibit throughout their length moderate concentrations of acetylcholinesterase, whereas all neighboring neurones synapsing with them seem to be devoid of this enzyme (ABRAHAMS, KOELLE and SMART, 1957; KOELLE, GEESEY and SCHMIDT, 1961; for references concerning the effect of anticholinesterase compounds injected into the supraoptic nuclei, also indicative of the cholinergic mechanism involved in the liberation of the endocrine secretion, see KOELLE, 1962, 1963). In this connection it may be recalled that the postulated cholinergic nature of the neurosecretory fibers received independent support also from electronmicroscopic studies (PALAY, 1957; GERSCHENFELD, TRAMEZZANI and DE ROBERTIS, 1960), which showed that these neurones contain two distinct populations of vesicles: one group accumulated at the terminals and resembling the synaptic vesicles usually seen at presynaptic sites of neurones, and another, presumably representing the endocrine secretion, which can be traced back to the hypothalamic nuclei.

For the sake of objectivity it must be noted, however, that the widespread distribution of cholinesterase in certain brain areas which contain little acetylcholine or cholinacetylase may also be interpreted as coincidental property of cellular organels. It is known, for instance, that 50% of the cholinesterase activity in the muscle is due to myosin content and its ability to hydrolyze acetylcholine (VARGA, KOVÉR, KOVÁCS and HETÉNYI, 1957).

One of the main points supporting the general concept of BURN and RAND is the observation that after reserpine-induced depletion of adrenergic transmitter, stimulation of postganglionic sympathetic nerves produced cholinomimetic effects. This observation is consistent not only with the well-known central "trophotropic" action of reserpine (BRODIE and SHORE, 1957; ANAND, DUA and MALHOTRA, 1957) but also with the central hypnogenic effects of directly administered acetylcholine into many subcortical structures involved in the regulation of the sleep-wakefulness cycle, such as the basal forebrain, unspecific thalamic nuclei, brainstem reticular formation (CORDEAU et al., 1961, 1962, 1963; YAMAGUCHI et al., 1963, 1964) and the forebrain limbic — midbrain limbic pathway of NAUTA (HERNANDEZ-PEON et al., 1963).

3. The relationship between impulse frequency and preferential liberation of acetylcholine or noradrenaline

It was shown by MITCHEL (1963) that the amount of acetylcholine released from the surface of the cerebral cortex is roughly proportional to the electrical activity of the brain. The same author also demonstrated that direct electrical stimulation of the homolateral or contralateral cortex, or of the peripheral sensory nerves, markedly increases the release of acetylcholine per unit of time, the increase being strictly dependent on the frequency of impulses applied. Interestingly the most effective frequency in releasing the maximum of acetylcholine was very low, and, depending on the experimental variant, ranged between 0.25 and 30 c/sec. At higher frequencies of about 50 to 100c/sec no increase in acetylcholine output from motor nerves has been found by STRAUGHAN (1960) and KRNJEVIĆ and MITCHEL (1961). More recently, McLENNAN (1964) using a pushpull cannula implanted in the caudate nucleus of the cat, to irrigate this structure with minute amounts of LOCKE's solution, has shown that low frequency stimulation (3 to 5 c/sec) of the nucleus ventralis anterior or anterior sigmoid gyrus was more effective in releasing acetylcholine per unit of time from the irrigated caudate nucleus than higher frequencies (7 to 10 c/sec). On the other hand, 7 to 10 c/sec stimulation of thalamic nuclei appeared to be more effective in releasing 3-hydroxytyramine from the irrigated caudate nucleus.

Several experimental results reported by BURN and his associates have been interpreted by them as indirect evidence supporting the view that "sympathetic cholinergic fibres in many situations release acetylcholine to act directly only at low frequencies, and that at higher frequencies the acetylcholine is almost entirely used to release noradrenaline". It should be pointed out that along this highly speculative reasoning no reasonable tentative explanation has been offered of the proposed mechanism according to which, in response to low frequency stimulation, the liberated acetylcholine acts directly on muscarinic receptors, and, as the stimulus frequency rises, the same acetylcholine is supposedly entirely used up in the process of liberation of noradrenaline. If we anticipate with BURN et al. that, as the stimulus frequency rises, the level of acetylcholine surrounding the adrenergic nerve terminals also rises, then this in turn would imply that the process of liberation of noradrenaline might depend merely on the concentration of the active acetylcholine at the nerve terminals. Consequently, it might be speculated that after a threshold has been reached, acetylcholine might be actively used in the process of noradrenaline liberation, and therefore cannot be readily detected pharmacologically unless the studied organs were pretreated with guanethidine (DAY and RAND, 1961) or reserpine (BURN and RAND, 1960).

The result of experiments on the hypogastric nerve-vas deferens preparation of the guinea pig has been interpreted along this line of speculative reasoning (Burn and Weetman, 1963; Burn, Dromey and Large, 1963). In the presence of scopolamine to block muscarinic effects, the potentiating action of physostigmine has been tentatively explained as being due to its strengthening effect on the cholinergic link in adrenergic transmission and resulting increase of noradrenaline output by low frequency stimulation of the hypogastric nerve. At higher frequencies no effect of this cholinesterase inhibitor could be observed "because inhibition of cholinesterase was of less importance when, due to the increased frequency, there was less time for cholinesterase to act" (Burn, Dromey and Large, 1963).

Objections can be raised based on the observation that the hypogastric nerve is likely to contain a significant number of preganglionic fibers, the existence of which can be demonstrated both histologically (Ohlin and Stromblad, 1963) and pharmacologically (Sjostrand, 1962a; Schümann and Grobecker, 1963). Furthermore, evidence obtained by Sjostrand (1962b), and Schumann and Grobecker (1963) indicates that in the hypogastric nerve preparation, liberated noradrenaline originates predominantly from cholinergically innervated chromaffin cells, and the relative participation of adrenergic nerve terminals is presumably very small. Therefore, further experiments were carried out on the nictitating membrane of the cat (Burn, Dromey and Large, 1963) which contains very few chromaffin cells, and 95 % of noradrenaline may be regarded as present in nerve terminals (Kirpekar, Cervoni and Furchgott, 1962, cf. Schümann, 1963). After preganglionic fibers had been cut which might terminate beyond the superior cervical ganglion, it was demonstrated that low frequency stimulation of fibers leaving the ganglion produced contractions of the nictitating membrane which could be markedly reduced by scopolamine. At higher frequencies (as in the hypogastricnerve-vas deferens preparation) no blocking effect of scopolamine occurred at all. Here again, physostigmine in the presence of scopolamine potentiated the effect only of the low frequency stimulation of postganglionic nerves, this action being negligible as the stimulus frequency was raised. An increased output of noradrenaline has been postulated to occur in the presence of a cholinesterase inhibitor (Burn, Dromey and Large, 1963) but no attempts have been made to substantiate this view. It should be noted that the experiments with scopolamine were not confirmed by Reas and Tsai (1966).

In the above mentioned series of experiments by Burn et al. one may object to the absence of control observations with no administration of scopolamine. A possibility of a "spontaneous" loss of responsiveness of nictitating membrane to repetitive stimulation might be anticipated. Also a comparison of antagonistic effect of scopolamine against exogenous acetylcholine and noradrenaline should be made. The non-specific blocking action of this compound should be ruled out since atropine is known to antagonize both neurally and noradrenaline-induced responses of the nictitating membrane (MIRKIN and CERVONI, 1962). In general, it must be stated that most of the evidence supplied in support of the BURN and RAND hypothesis needs to be explored further. The effect of anticholinesterases on sympathetically innervated organs cannot be used unreservedly as evidence for the presence of a cholinergic link in adrenergic transmission. Several other experimental results militate against the BURN and RAND hypothesis (cf. KARCZMAR, 1966; FERRY, 1966). In conclusion it may be noted that following a careful histochemical study of the presence of acetylcholinesterase and catecholamines in postganglionic autonomic nerves of different animal species JACOBOWITZ and KOELLE (1965) believe that "the participation of acetylcholine in adrenergic transmission might be considered tentatively as facultative rather than obligative". Subsequent study of the inhibitory effect of hemicholinium (HC-3) on sympathetic transmission at the nictitating membrane of the cat and rabbit lends further support to this view (JACOBOWITZ, JOHNSON, KITCHNER and KOELLE, 1965). Their results are consistent with the BURN and RAND hypothesis solely in vas deferens of the guinea pig and nictitating membrane of the rabbit but not in the same organ of the cat.

4. Cholinesterase inhibitors and "cholinergic link" in central adrenergic mechanisms

Admittedly, it is difficult to explain the well-known dissociation between the normal behavior and significant electroencephalographic arousal patterns observed following the systemic administration of physostigmine (cf. BRADLEY, 1958; KILLAM, 1962). It was suggested by BRADLEY and KEY (1958) that lowering by physostigmine (injected peripherally) of the arousal threshold to electrical stimulation of the brainstem reticular formation previously elevated by atropine, might be due to an action on the diffuse thalamic projection system or on the not yet either anatomically or functionally defined cholinergic "mesodiencephalic activating system" postulated by RINALDI and HIMWICH (1955a and b). The interpretation is incompatible with previously discussed effects of direct administration of acetylcholine which, in contrast to that of either carbachol or acetylcholine plus physostigmine, never produced behavioral and EEG arousal effects (YAMAGUCHI et al., 1964; CORDEAU et al., 1963). Central electroencephalographic effects of physostigmine are, however, compatible with the effects induced by directly applied noradrenaline to the brain stem reticular formation and preoptic region, where an overlap of hypnogenic cholinoceptive systems with adrenoceptive arousal systems is not unexpected. It is tempting to suggest that the extrapolation of the general proposal of BURN and RAND to the central adrenergic mechanisms may offer a tentative explanation of seemingly contradictory central action of physostigmine. Hence, it is suggested that the EEG and behavioral dissociation induced by systemic administration of physostigmine may be due to the raised tonicity of the inhibitory, hypnogenic and cholinoceptive systems on the one hand, and to the specific action on the "cholinergic link" in adrenergically mediated activating influences on the other.

In addition to the previously discussed indirect evidence obtained from the peripheral adrenergic system reported by BURN et al. (1963; see also BURN and RAND, 1965) the following findings may be adduced in favor of the postulated action of physostigmine on central ergotropic and presumably adrenoceptive mechanisms:

a) The electrical activity of preganglionic sympathetic fibers is increased after physostigmine administration; bretylium and choline-2,6-xylyl ether bromide, which causes a failure of the release of adrenergic transmitter, depresses or even abolishes the presumably centrally mediated hypertensive effect of physostigmine (VARAGIĆ, LESIĆ, VUCO and STEMENOVIĆ, 1962);

b) Physostigmine and other cholinesterase inhibitors raise the blood pressure of anesthetized rats (VARAGIĆ, 1955), and augment the hypertensive reflexes elicited from the carotid sinus in anesthetized (HORNYKIEWICZ and KOBINGER, 1956) as well as in conscious animals (MEDUKOVIĆ and VARAGIĆ, 1957); and

c) After relatively small doses of physostigmine given intravenously, which are insufficient to produce electrocortical changes, a significant lengthening of the activation time is observed in response to both external stimulation and electrical impulses delivered to the brainstem reticular formation (LONGO,1962), the activating ascending component of which is thought to be adrenergic in nature.

However, in view of the fact that the desynchronizing electrocortical action of physostigmine can still be observed in *cerveau isolé* preparations (see BRADLEY, 1958), it seems reasonable to suggest that the postulated adrenergic neuronal substrates responsible for its action are not exclusively located below the transection of the brain stem. Such a possibility is indicated by the early findings of HIEBEL, BONVALLET, HUVÉ and DELL (1954) that amphetamine-induced electrocortical desynchronization may be still present in a classical *cerveau isolé* preparation of the cat. WHITE and DAIGNEAULT'S (1959) results in the rabbit point in the same direction.

5. Possible central neurohormonal effects of electrical stimulation at different frequencies

In conjunction with the previously discussed indirect evidence for the existence of a "cholinergic link" in central and peripheral neurohormonal mechanisms, the possibility of differential activation of different central neurohormonal mechanisms by different frequencies of electrical stimuli, can be summarized as follows:

1. It is probably not a mere coincidence that the most effective frequencies of electrical stimulation, applied either centrally or peripherally to the afferent

nerves, producing the liberation of acetylcholine from the cerebral cortex (MITCHEL, 1963) and subcortical areas (MITCHEL and SHERB, 1962; McLENNAN, 1964) do not exceed 20 to 30 c/sec. The most effective frequencies in producing "tonic", *i.e.*, outlasting the trains of stimuli, behavioral and EEG signs of sleep from widespread subcortical areas (including the brainstem reticular formation) also do not exceed this limit of frequency. The optimal frequency in producing cortical recruitment (5 to 7 c/sec) also has been shown to produce maximum acetylcholine release from the caudate nucleus Since the direct administration of acetylcholine into the structures involved in the regulation of the sleep-wakefulness cycle also induces marked sleep effects, it might be suggested that the central "tonic" behavioral and EEG effects produced by low frequency electrical impulses are mediated by the preponderance of the released acetylcholine over noradrenaline. Conceivably, the raised level of active acetylcholine increases the tonicity of cholinergic pathways discussed previously, which may be responsible for several inhibitory processes connected with the production of sleep.

2. It is not unlikely that acetylcholine released during low frequency electrical stimulation of brain structures involved in the regulation of wakefulness, may liberate in turn not yet identified and relatively more stable modulator substance(s) capable of influencing the synaptic events of remote brain structures. The anticipation of such substance(s) seems to be a necessity, since no other tentative explanation can be offered of the extraneural effects of low frequency electrical stimulation of brain structures in previously discussed cross-circulation experiments of KORNMULLER *et al.* (1961), KOLLER and MONNIER (1962) and MONNIER and HÓSLI (1965).

3. The following phenomena have been reported to occur in parallel as the stimulus frequency of electrical impulses rises when applied either to the brain or peripheral afferent nerves:

a) a decline of the acetylcholine liberation from the cerebral cortex (MITCHEL, 1963) and subcortical structures (McLENNAN, 1964);

b) indirect evidence which may be interpreted as pointing to a selective release of noradrenaline from peripheral postganglionic sympathetic nerves, in which low frequency stimulation is thought to liberate both acetylcholine and noradrenaline (BURN *et al.*, 1963);

c) in many nonhomogeneous subcortical structures a hypnogenic (*i.e.* trophotropic) effect of low frequency impulses is changed to "tonic" arousal (*i.e.* ergotropic) effect, which can be mimicked by direct application of noradrenaline into the same structures (preoptic region, brain stem reticular formation). It is, therefore, tempting to suggest that high frequency stimulation may produce a significant preponderance of the released noradrenaline, which exerts its action on the adrenoceptive structures responsible for EEG and behavioral arousal.

V. Summary and conclusions

Limitations in investigations on central effects of neurohormonal substances administered into the systemic circulation have been considered. More recent methods have been discussed which permit direct application of drugs into the brain tissue. Possible side-effects of topically administered substances, criteria for their elimination and the specificity of central action and main advantages of this relatively new approach have been emphasized. It was concluded that the administration of micro-amounts of neurohormonal substances directly into the brain tissue by means of specially devised cannulas in "chronic" and "acute" experiments, yields reproducible and specific central effects, through an action on a large but still restricted population of neurones which may constitute a functional system determining characteristic EEG and behavioral changes. It appears that such "chemical stimulation" with micro-amounts of neurohormonal substances in "chronic" animals does not cause any gross unphysiological disturbances in the electrical activity of the stimulated area and allows a chemical "dissection" of functionally overlapping different systems of neurones within a given anatomic region. The limited spread, and specificity of behavioral and EEG responses, which may usually be obtained from highly inhomogeneous structures, constitute a major advantage over indiscriminate effects of electrical stimulation, which usually affects also the fibers of passage. The observation of positive effects instead of negative ones, which result from electrolytic lesions, seems especially important in investigations of the existing functional links between peripheral steroid hormones and the central nervous system, on the one hand, and neurohormonal mechanisms by which functional neuronal systems are integrating both the goal-oriented behavior and various levels of wakefulness on the other.

Direct application of steroid hormones made it possible to delineate with a high degree of precision the location of specific neuronal substrates responsible for the "negative feedback" mechanism which is of considerable importance in the regulation of adenohypophyseal secretion. Some preliminary indications were obtained which point to the existence of extrahypothalamic "positive feedback" neuronal receptors. Substrates involved in the initiation and integration of the overt sexual and parental behavior have also been delineated with high degree of precision.

Convincing evidence has been obtained that hypothalamic "feeding centers" involved in the regulation of food and water intake are operated selectively by adrenergic and cholinergic mechanisms, respectively.

Pharmacological "dissection" of the ascending brainstem reticular system strongly indicates that within this highly inhomogeneous structure there exist functionally opposite cholinergic and adrenergic components, the tonicity of which may be selectively enhanced by direct application of acetylcholine and noradrenaline, respectively. This results in marked EEG and behavioral effects, as well as in changes in the evoked cortical potentials. Thus, it appears that the existence of two functionally opposite ascending reticular systems has been proven neurohormonally, in addition to the considerable electrophysiological evidence accumulated during the last few years and based mainly on ablation and electrical stimulation methods

Evidence has been also discussed in favor of the cholinergic nature of the basal forebrain inhibitory and hypnogenic system, which acts antagonistically to the ascending activating brainstem reticular formation Also, results obtained from direct application of cholinergic and anticholinergic drugs point to the possibility that the forebrain limbic — midbrain limbic pathways of NAUTA serve as the main descending multisynaptic inhibitory trajectory, cholinergic in nature, which presumably affects the overall tonicity of the ascending reticular system. There is, however, recent electrophysiological evidence, that the basal forebrain inhibitory system may influence more directly neocortical function through an action on the diffuse thalamic projection system.

The "synchronizing" cholinergic component may also be expected to exist in the diffuse thalamic projection system, since its "stimulation" by direct application of acetylcholine produces behavioral and EEG signs of sleep, *i.e.* effects similar to those obtained after direct application of acetylcholine into the mesencephalic, pontine, and bulbar reticular formation, as well as into the basal forebrain inhibitory system.

The preoptic region and the brainstem reticular system in its whole length have been shown to produce behavioral and EEG arousal in response to the administration of noradrenaline, which seems to indicate the existence of ascending activating neuronal substrates that are adrenergic or adrenoceptive in nature.

Characteristically, carbachol, which is known to be resistant to action of cholinesterase and acetylcholine administered in association with physostigmine, even in much smaller amounts, produced central effects opposite to those seen after acetylcholine alone with regard to the regulation of level of wakefulness.

There are also some indications that the anterior-commissural septal region contains neuronal elements, cholinergic or cholinoceptive in nature, which may be involved in psychophysiological processes of attention.

Since parallelism seems to exist between the central "tonic" behavioral and EEG effects of high frequency electrical stimulation of subcortical structures involved primarily in the regulation of the sleep-wakefulness continuum and the effects of adrenergic drugs, on the one hand, and between the "tonic" action of low frequency impulses and the effects of acetylcholine, on the other, it has been suggested by several lines of indirect evidence that electric stimulation, depending upon the frequency employed, may activate preferentially cholinergic and adrenergic mechanisms, respectively. Indirect and debatable evidence has been discussed in an attempt to justify the extrapolation to the central nervous system of the general proposal of BURN *et al.* concerning the intermediary role of acetylcholine in the peripheral adrenergic mechanism. This working hypothesis is in agreement with the more general concept advanced by KOELLE concerning the prejunctional role of acetylcholine in the process of liberation of both cholinergic and noncholinergic (central and peripheral) neurotransmitters

Acknowledgement The author of this review wishes to express his gratitude for constructive criticism to Dr. KLAUS R UNNA, Dr GEORGE B KOELLE and Dr CHRISTOPH STUMPF.

Bibliography

ABRAHAMS, V. C., G. B. KOELLE, and P. SMART· Histochemical demonstration of cholinesterase in the hypothalamus of the dog. J. Physiol. (Lond) **139**, 137—144 (1957)

ADEY, W R., J. P. SEGUNDO, and R. B. LIVINGSTON Corticifugal influences on intrinsic brain stem conduction in cat and monkey J Neurophysiol. **20**, 1—16 (1957).

AFFANI, J , P. L MARCHIAFAVA, and G ZERNICKI Orientation reaction in the midpontine pretrigeminal cat. Arch. ital. Biol **100**, 297—304 (1962a)

— — — Conditioning in the midpontine pretrigeminal cat Arch. ital. Biol. **100**, 305—310 (1962b).

AKERT, K , W KOELLA, and R HESS jr. Sleep produced by electrical stimulation of the thalamus Amer J. Physiol **168**, 260 (1952)

AKIMOTO, H , N YAMAGUCHI, K OKABE, T. NAKAGAWA, I. NAKAMURA, K ABE, H TORII, and K MASAHASI On the sleep induced through electrical stimulation on dog thalamus Folia psychiat. neurol. jap. **10**, 117—146 (1956).

ANAND, B. K. Nervous regulation of food intake Physiol Rev **41**, 677—708 (1961)

—, and J. R. BROBECK Hypothalamic control of food intake in rats and cats Yale J Biol. Med. **24**, 123—140 (1951).

— S DUA, and C L. MALHOTRA Effects of reserpine on blood pressure responses evoked from the hypothalamus Brit J. Pharmacol. **12**, 8—11 (1957).

ANDÉN, N E , J HAGGENDAL, T. MAGNUSSON, and E. ROSENGREN The time course of the disappearance of noradrenaline and 5-hydroxy-tryptamine in the spinal cord after transection Acta physiol. scand. **62**, 115—118 (1964)

ANDERSSON, B The effect of injections of hypertonic NaCl-solutions into different parts of the hypothalamus of goats Acta physiol scand. **28**, 188—201 (1953).

— C C GALE, and J. W. SUNDSTEN The relationship between body temperature and food and water intake. In Proceedings of the First Int Symposium on Olfaction and Taste, p 361—375 Oxford, London, New York, Paris Pergamon Press 1963.

—, and B LARSSON Influence of local temperature changes in the preoptic area and rostral hypothalamus on the regulation of food and water intake Acta physiol. scand **52**, 75—89 (1961a)

— — Physiological and pharmacological aspect of the control of hunger and thirst Pharmacol. Rev. **13**, 1—16 (1961b)

—, and S M. McCANN Hypothalamic control of water intake J. Physiol (Lond.) **129**, 44P (1955a).

— — A further study of polydipsia evoked by hypothalamic stimulation in the goat Acta physiol. scand **33**, 333—346 (1955b).

APRISON, M. H. On a proposed theory for the mechanism of action of serotonin in the brain. Recent Advanc Biol Psychiat **4**, 133—146 (1962).

ARMENGOL, V., W. LIFSCHITZ, and M. PALESTINI Inhibitory influences on primary and secondary cortical photic potentials originating in the lower brain stem J Physiol (Lond.) **159**, 451—460 (1961)

BATINI, C , G MORUZZI, M. PALESTINI, G. F. ROSSI, and A. ZANCHETTI Effects of complete pontine transection on the sleep-wakefulness rhythm in the midpontine pretrigeminal preparation. Arch. ital. Biol. **97**, 1—22 (1959).

— M PALESTINI, G. F ROSSI, and A ZANCHETTI Neural mechanisms underlying the enduring EEG and behavioral activation in the midpontine pretrigeminal cat. Arch. ital. Biol **97**, 13—25 (1959).

BAUST, W , and H NIEMCZYK Studies on the adrenaline-sensitive component of the mesencephalic reticular formation J Neurophysiol. **26**, 692—704 (1963).

— —, and J. VIETH The action of blood pressure on the ascending reticular activating system with special reference to adrenaline induced EEG arousal Electroenceph. clin. Neurophysiol **15**, 63—72 (1963).

BENOIT, J., et J ASSENMACHER Le controle hypothalamique de l'activité préhypophysaire gonadotrope J. Physiol (Paris) **47**, 427—567 (1955)

BERTLER, O , B FALCK, and CH. OWMAN Cellular localization of 5-hydroxytryptamine in the rat pineal gland Kgl. Fysiogr. Sallskapets Hdl **33**, 13—16 (1963).

— —, and E. ROSENGREN The direct demonstration of a barrier mechanism in the brain capillaries. Acta pharmacol (Kbh) **20**, 317—321 (1963).

BESSMAN, S P , and W N FISHBEIN Gamma-hydroxybutyrate, a normal brain metabolite Nature (Lond) **200**, 1207—1208 (1963)

—, and S. J. SKOLNIK Gamma hydroxybutyrate and gamma butyrolactone Concentration in rat tissues during anesthesia. Science **143**, 1045—1047 (1964)

BISHOP, P. O , G. FIELD, B. L. HENNESY, and J R. SMITH Action of D-lysergic acid diethylamide on lateral geniculate synapses. J Neurophysiol. **21**, 529—549 (1958).

BLOCH, V., et M BONVALLET · Interactions des formations réticulaires mésencéphalique et bulbaire. J. Physiol. (Paris) **53**, 280—281 (1961)

BLOOM, F E., A. P. OLIVER, and G C. SALMOIRAGHI The responsiveness of individual hypothalamic neurons to microelectrophoretically administered endogenous amines Int J. Neuropharmacol **2**, 181—193 (1963).

BOGDANOVE, E. M , and H C. SCHOEN. Precocious sexual development in female rats with hypothalamic lesions. Proc. Soc. exp Biol (N Y) **100**, 664—689 (1959)

BONVALLET, M., and M. B. ALLEN jr Prolonged spontaneous and evoked reticular activation following discrete bulbar lesions Electroenceph. clin Neurophysiol. **15**, 969—988 (1963).

—, and V. BLOCH Le controle bulbaire des activations corticales et sa mise en jeu C.R. Soc Biol (Paris) **154**, 1428—1431 (1960).

— — Bulbar control of cortical arousal. Science **133**, 1133—1134 (1961).

— — P DELL et G HIEBEL · Tonus sympathique et activité electrique corticale Electroenceph. clin Neurophysiol. **6**, 119—144 (1954).

BRADLEY, P B The central action of certain drugs in relation to the reticular formation of the brain In Reticular formation of the brain, ed by H H JASPER, L D PROCTOR, R. S. KNIGHTON, W. C. NOSHAY and R T. COSTELLO, p. 123—150 Boston Little, Brown & Co. 1958

— Electrophysiological evidence relating to the role of adrenaline in the central nervous system In Ciba Foundation Symposium on Adrenergic Mechanisms (ed. J R. VANE, G. E W WOLSTENHOLM and M O'CONNOR), p. 410—420 London Churchill 1960

— B N DHAWAN, and J H WOLSTENCROFT Some pharmacological properties of cholinoceptive neurones in the medulla and pons of the cat. J Physiol. (Lond) **170**, 59P (1964)

—, and B J. KEY The effects of some drugs on the electrical activity of the brain. Brain **80**, 77—117 (1957)

— — The effect of drugs on arousal responses produced by electrical stimulation of the reticular formation of the brain Electroenceph clin Neurophysiol **10**, 97—110 (1958).

BRADLEY, P B , and A MOLLICA The effect of adrenaline and acetylcholine on single unit activity in the reticular formation of the decerebrated cat Arch. ital Biol **96**, 168—186 (1958)

—, and J H WOLSTENCROFT Excitation and inhibition of brainstem neurones by noradrenaline and acetylcholine. Nature (Lond) **196**, 840—873 (1962)

— — A counter and print-out unit for recording the frequency of neuronal action potentials J Physiol (Lond) **170**, 2—3P (1963)

— — The action of drugs on single neurones in the brain stem In Neuropsychopharmacology, vol 3 (eds P B BRADLEY, F FLUEGEL and P HOCH) Amsterdam Elsevier Publ Co 1964

BREMER, F Cerveau isolé et physiologie du sommeil C R Soc Biol (Paris) **118**, 1235—1242 (1935)

— Neurophysiological mechanism in cerebral arousal In The nature of sleep (ed G E W WOLSTENHOLM and M O'CONNOR), p 30—50 Boston Little, Brown & Co 1960

—, et N STOUPEL Facilitation et inhibition des potentials évoqués corticaux dans l'éveil cérébral Arch int Physiol **67**, 240—275 (1959)

— — et P E VAN REETH Sur la facilitation et l'inhibition des potentials évoqués corticaux dans l'éveil réticulaire Arch ital Biol **98**, 229—247 (1960)

BROBECK, J R . Food intake as a mechanism of temperature regulation Yale J Biol Med **20**, 545—552 (1948)

— J TEPPERMAN, and C N H LONG Experimental hypothalamic hyperphagia in the albino rat Yale J Biol Med **15**, 831—853 (1943)

BRODAL, A The reticular formation of the brain stem Anatomical aspects and functional correlations, p 23—29 Edinburgh and London Oliver & Boyd 1958

—, and G F ROSSI· Ascending fibers in the brain stem reticular formation of cat Arch Neurol Psychiat (Chic) **74**, 68—87 (1955).

BRODIE, B B , and P A SHORE A concept for a role of serotonin and norepinephrine as chemical mediators in the brain Ann N Y Acad Sci **66**, 631—643 (1957)

BRUEGGER, M Freßtrieb als hypothalamisches Symptom Helv physiol pharmacol Acta **1**, 183—198 (1943)

BUERGI, S , u M MONNIER Motorische Erscheinungen bei Reizung und Ausschaltung der substantia reticularis pontis Helv physiol. pharmacol Acta **1**, 489—510 (1943)

BURN, J H A new view of adrenergic nerve fibers explaining the action of reserpine, bretylium, and guanethedine Brit. med J **1961**I, 1623—1627.

— J J DROMEY, and B J LARGE The release of acetylcholine by sympathetic nerve stimulation at different frequencies Brit. J Pharmacol **21**, 97—103 (1963)

—, and H FROEDE· The action of substances which block sympathetic postganglionic nervous transmission Brit. J Pharmacol **20**, 378—387 (1963)

—, and M J RAND· Action of nicotine on the heart Brit med J **1958**aI, 137—139.

— — Noradrenaline in artery walls and its dispersal by reserpine Brit med. J. **1958**bI, 903—908

— — Sympathetic postganglionic mechanisms Nature (Lond.) **184**, 163 (1959)

— — Sympathetic postganglionic fibers Brit J Pharmacol **15**, 56—66 (1960)

— — Acetylcholine in adrenergic transmission Ann Rev Pharmacol **5**, 163—182 (1965).

— —, and R WIEN The adrenergic mechanism in the nictitating membrane Brit J Pharmacol **20**, 83—94 (1963)

—, and D F WEETMAN The effect of eserine on the response of the vas deferens to hypogastric nerve stimulation Brit J Pharmacol **20**, 74—82 (1963)

CAPON, A Analyse de l'effet de éveil exercé par l'adrenaline et d'autres amines sympathicomimétiques sur l'electrocorticogramme du lapin non narcotise Arch int Pharmacodyn **127**, 141—162 (1960)

CARLSSON, A , B FALCK, K FUXE, and N Å HILLARP Cellular localization of monoamines in the spinal cord Acta physiol scand **60**, 112—119 (1964)

CARLSSON, A , B FALCK, N. Å HILLARP, and A. TORP Histochemical localization at the cellular level of hypothalamic noradrenaline Acta physiol. scand 54, 385—386 (1962)

CHO, A. K , W L HASLETT, and D. L. JENDEN The identification of an active metabolite of tremorine. Biochem. biophys. Res. Commun. 5, 276—279 (1961).

— — — The peripheral actions of oxotremorine, a metabolite of tremorine J Pharmacol exp. Ther 138, 249—257 (1962).

CHOWERS, J , S. FELDMAN, and J. M. DAVIDSON Effects of intrahypothalamic crystalline steroids on acute ACTH secretion. Amer J. Physiol. 205, 671—673 (1963)

CLEMENTE, C D., and M. B. STERMAN. Cortical synchronization and sleep patterns in acute restrained and chronic behaving cats induced by basal forebrain stimulation. Electroenceph. clin Neurophysiol., Suppl. 24, 172—187 (1963)

— J. SUTIN, and J T. SILVERTONE Changes in electrical activity of the medulla on intravenous injection of hypertonic solutions. Amer. J. Physiol. 188, 193—198 (1957)

CORDEAU, J. P Functional organization of the brain stem reticular formation in relation to sleep and wakefulness. Rev. canad. Biol 21, 113—125 (1962)

—, and M MANCIA Effect of unilateral chronic lesions of the midbrain on the electrocortical activity of the cat. Arch ital. Biol. 96, 374—399 (1958)

— — Evidence for the existence of an electroencephalographic synchronization mechanism originating in the lower brain stem. Electroenceph. clin Neurophysiol. 11, 551—564 (1959).

— A. MOREAU, and A BEAULNES EEG synchronization following microinjections of acetylcholine in the caudal brain stem In· Proceedings of the Fifth Internat Congr. of Electroencephalography and Clinical Neurophysiology Excerpta Med , Int Congress, Series 37 (1961)

— — — EEG and behavioral effect of microinjection of drugs in the brain stem of cats. In Proceedings of the Fifth Internat Congr. of Electroencephalography and Clinical Neurophysiology. Excerpta Med., Int. Congress, Series 37 (1961).

— — —, and C LAURIN EEG and behavioral effects of microinjection of drugs in the brain stem of cats. In Proceedings of the XXII Internat. Congr of Physiological Sciences. Excerpta Med , Int. Congress, Series 48 (1962)

— — — — EEG and behavioral changes following microinjections of acetylcholine and adrenaline in the brain stem of cats. Arch ital Biol. 101, 30—47 (1963)

COURVILLE, J , J. WALSH, and J P CORDEAU Functional organization of the brain stem reticular formation and specific sensory input. Proc. of the Canad. Fed. of Biol. Soc. 5, 22 (1962a).

— — — Functional organization of the brain stem reticular formation and sensory input. Science 138, 973—975 (1962b).

CRAWFORD, J. M , D. R. CURTIS, P E. VOORHOEVE, and V. J. WILSON· Excitation of cerebellar neurones by acetylcholine. Nature (Lond.) 200, 579—580 (1963).

CREUTZFELDT, O , and R JUNG Neuronal discharge in the cat's motor cortex during sleep and arousal. In The nature of sleep, ed. by G E. W. WOLSTENHOLME and M. O'CONNOR, p. 131—170. Boston Little, Brown & Co. 1960

CROSS, B. A , and J. D GREEN Activity of single neurones in the hypothalamus effect of osmotic and other stimuli. J. Physiol. (Lond) 148, 554—569 (1959).

CROSSLAND, J Chemical transmission in the central nervous system J. Pharm. Pharmacol. 12, 1—36 (1960).

CURTIS, D. R. The pharmacology of central and peripheral inhibition. Pharmacol. Rev 15, 333—364 (1963).

—, and P. ANDERSEN· Acetylcholine — a central transmitter. Nature (Lond) 195, 1105 (1962).

— J. C. ECCLES, and R. M ECCLES Pharmacological studies on spinal reflexes. J. Physiol. (Lond.) 136, 420—434 (1957).

—, and R. M ECCLES The excitation of Renshaw cells by pharmacological agents applied electrophoretically. J Physiol. (Lond.) 141, 435—445 (1958a).

CURTIS, D. R., and R. M. ECCLES· The effect of diffusional barriers upon the pharmacology of cells within the central nervous system. J. Physiol. (Lond.) 141, 446—463 (1958b).

—, and K. KOIZUMI· Chemical transmitter substances in the brain stem of cat. J. Neurophysiol. 24, 80—90 (1961).

—, and J. W. PHILLIS: The action of procain and atropine on spinal neurones J Physiol. (Lond) 153, 17—34 (1960).

— —, and J. C WATKINS· Cholinergic and non-cholinergic transmission in the mammalian spinal cord. J. Physiol. (Lond.) 158, 269—323 (1961).

DAHLSTROM, A., and K. FUXE. Evidence for the existence of monamine-containing neurons in the central nervous system. Acta physiol. scand. 62, Suppl. 232, 1—55 (1964)

DAVID, J. P., S. MURAYAMA, X. MACHNE, and K. R. UNNA· Evidence supporting cholinergic transmission at the lateral geniculate body of the cat. Int. J Neuropharmacol. 2, 113—125 (1963).

DAVIDSON, J. M., and S. FELDMAN· Cerebral involvement in the inhibition of ACTH secretion by hydrocortisone. Endocrinology 72, 936—946 (1963).

—, and CH. H. SAWYER Effects of localized intracranial implantation of estrogen on reproductive functions in the female rabbit Acta endocr. (Kbh.) 37, 385—393 (1961a).

— — Evidence for an hypothalamic focus of inhibition of gonadotropin by androgen in the male. Proc. Soc. exp. Biol. (N.Y.) 107, 4—7 (1961b).

DAVSON, H. Physiology of the ocular and cerebrospinal fluids. London Churchill 1956.

— Intracranial and intraocular fluids. In: Handbook of physiology, Sect. 1 Neurophysiology (ed. J. FIELD, H. W. MAGOUN and V. E. HALL), vol. 3, p. 1760-1788. Washington, D.C.. Amer. Physiol. Soc. 1960.

—, and E. SPAZIANI Blood-brain barrier and the extracellular space of brain. J. Physiol. (Lond.) 149, 135—143 (1959).

DAY, M. D., and M. J. RAND. Effect of guanethidine in revealing cholinergic sympathetic fibers. Brit. J. Pharmacol. 17, 245—260 (1961).

DELGADO, J. M R.· Pharmacological modifications of social behavior. In Proceedings of the First Int. Pharmacol. Meeting, vol. 8, p. 265—292. Oxford, London, New York, Paris Pergamon Press 1962.

— Telemetry and telestimulation of brain. In Biotelemetry, p. 231—249. Oxford, London, New York, Paris. Pergamon Press 1963.

—, and B. D. ANAND· Increase of food intake induced by electrical stimulation of the lateral hypothalamus Amer J. Physiol. 172, 162—168 (1953).

DELL, P. C. Humoral effects on the brain stem reticular formation. In Reticular formation of the brain, ed by H. H. JASPER, L. D. PROCTOR, R. S. KNIGHTON, W. C NOSHAY and R T. COSTELLO, p. 365—379 Boston Little, Brown & Co. 1958.

— Intervention of an adrenergic mechanism during brain stem reticular activation. In Ciba Foundation Symposium on Adrenergic Mechanisms (ed. J. R. VANE, G E W. WOLSTENHOLME and M. O'CONNOR), p. 393—409. London Churchill 1960.

— M. BONVALLET et A HUGELIN. Tonus sympathique, adrénaline, et controle réticulaire de la motricité spinale. Electroenceph. clin. Neurophysiol. 6, 599—618 (1954).

— — — Demonstration d'un mécanisme adrénergique au niveau des formations réticulaires du tronc cérébral. Conséquences pharmacologiques Encéphale 45, 1119—1123 (1956).

DEMENT, W.· The occurrence of low voltage, fast electroencephalogram patterns during behavioral sleep in the cat. Electroenceph. clin Neurophysiol. 10, 291—296 (1958).

DESMEDT, J. E., et J. SCHLAG Mise en évidence d'éléments cholinergiques dans la formation réticulée mésencéphalique. J. Physiol. (Paris) 49, 136—138 (1957).

DEWS, P B.: Psychopharmacology. In Experimental foundations of clinical psychology, ed. by A. J. BACHRACH, p 423—441. New York: Basic Books 1962.

—, and W. H. MORSE· Behavioral pharmacology. Ann. Rev. Pharmacol. 1, 145—174 (1961).

DEY, F. L.   Evidence of hypothalamic control of hypophyseal gonadotropic function in the female guinea pig Endocrinology 33, 75—82 (1943).

DOBBING, J.: The blood barrier. Physiol. Rev. 41, 130—188 (1961).

DOMER, F. R., and W. FELDBERG. Some central actions of adrenaline and noradrenaline when administered into the cerebral ventricles. In Adrenergic mechanisms (J R VANE, ed.), p 386—392. London Churchill 1960.

DOMINO, E. F.   Sites of action of some central nervous system depressants Ann. Rev Pharmacol. 2, 215 (1962).

DUMONT, S , et P. DELL: Facilitation réticulaire des mécanismes visuels corticaux Electroenceph. clin. Neurophysiol. 12, 769—796 (1960).

ECCLES, J. C · The physiology of synapses New York Academic Press, Inc , Berlin-Gottingen-Heidelberg· Springer 1964.

— R. M. ECCLES, and P. FATT· Pharmacological investigations on a central synapse operated by acetylcholine. J. Physiol (Lond.) 131, 154—169 (1956)

— P. FATT, and K KOKETSU Cholinergic and inhibitory synapses in a pathway from motor-axon collaterals to motoneurones J. Physiol (Lond) 126, 524—562 (1954)

—, and B. LIBET Origin and blockade of the synaptic responses of curarized sympathetic ganglia. J. Physiol. (Lond) 157, 484—503 (1961).

ECONOMO, C. v.: Die Encephalitis Lethargica. Wien Franz Deuticke 1918.

ELFVIN, L. G.   The ultrastructure of the superior cervical sympathetic ganglion of the cat II. The structure of the preganglionic and fibers and their synapses as studied by serial sections. J Ultrastruct. Res. 8, 441—476 (1963).

EPSTEIN, A N.· Reciprocal changes in feeding behavior produced by intra-hypothalamic chemical injections. Amer. J. Physiol. 199, 969—974 (1960).

EULER, C. v.: A preliminary note on slow hypothalamic "osmopotentials". Acta physiol scand 29, 133—136 (1953)

— Physiology and pharmacology of temperature regulation. Pharmacol. Rev. 13, 361—398 (1961).

EVARTS, E. V., W. LANDAU, W. FREYGANG jr., and W H MARSHALL Some effects of lysergic acid diethylamide and bufotenine on electrical activity in the cat's visual system. Amer. J. Physiol. 182, 594—598 (1955).

FALCK, B.   Observations on the possibilities of the cellular localization of monamines by flourescene method. Acta physiol. scand. 56, Suppl 197, 1—25 (1962).

FAVALE, E., C. LOEB, G. F. ROSSI, and G SACCO EEG synchronization and behavioral signs of sleep following low frequency stimulation of the brain stem reticular formation. Arch. ital. Biol. 99, 1—22 (1961).

FELDBERG, W.   A pharmacological approach to the brain from its inner and outer surface Evarts Graham Memorial Lectures. London Arnold 1963.

—, and K FLEISCHHAUER The site of origin of the seizure discharge produced by tubocurarine acting from the cerebral ventricles. J Physiol. (Lond.) 160, 258—283 (1962).

— — The hippocampus as the site of origin of the seizure discharge produced by tubocurarine acting from the cerebral ventricles J. Physiol. (Lond.) 168, 435—442 (1963).

— — A new experimental approach to the physiology and pharmacology of the brain Brit. med. Bull. 21, 36—43 (1965).

—, and R. D. MYERS: Changes in temperature produced by microinjections of amines into the anterior hypothalamus of cats. J. Physiol. (Lond.) 177, 239—245 (1963).

—, and S.L. SHERWOOD. Injections of drugs into the lateral ventricle of the cat. J.Physiol (Lond.) 123, 148—167 (1954).

FERRY, C. B.   Cholinergic link hypothesis in adrenergic neuroeffector transmission Physiol. Rev. 46, 420—456 (1966).

FIELDS, W. S , R. GUILLEMIN, and C. A. CARTON   Hypothalamic-hypophysial interrelationships, vol. 1. Springfield (Ill.): Ch. C. Thomas 1956.

FISHER, A. E. · Maternal and sexual behavior induced by intracranial chemical stimulation. Science **124**, 228—229 (1956)

— Behavior as function of certain neurobiochemical events. Current trends in psychological theory, p. 70—86 Pittsburgh Pittsburgh University Press 1960.

—, and J. N. COURY Cholinergic tracing of a central neural circuit underlying the thirst drive. Science **138**, 692—693 (1962)

FLERKO, B . Hormonal feedback and gonadotrophin secretion. In· Proceedings of the Int. Union of Physiol. Sciences, vol 1, part II, p. 632—635 XXII Int. Congr. Leiden 1962.

FOX, R. H., and S. M. HILTON· Bradykinin formation in human skin as a factor in heat vasodilatation. J. Physiol (Lond.) **142**, 219—232 (1958).

FRENCH, J D · Drug actions upon the brain-"psychopharmacology". Ann. Rev. Med. **9**, 333—346 (1958)

FRIEDMAN, U. Blood-brain barrier. Physiol. Rev. **22**, 125—145 (1942).

FUSTER, J. M. Excitation and inhibition of neuronal firing in visual cortex by reticular stimulation. Science **133**, 2011—2012 (1961).

—, and R F. DOCTER Variations of evoked potentials as a function of reticular activity in rabbits with chronically implanted electrodes J. Neurophysiol. **25**, 324—336 (1962).

GADDUM, J H. Push-pull cannulae J. Physiol (Lond.) **155**, 1 P—2P (1961).

GALAMBOS, R.. Suppression of auditory nerve activity by stimulation of efferent fibers to cochlea. J. Neurophysiol. **19**, 424—437 (1956).

GANONG, W F , and P H FORSHAM Adenohypophysis and adrenal cortex. Ann. Rev. Physiol. **22**, 579—614 (1960).

GEORGE, R , W L HASLETT, and D J JENDEN A cholinergic mechanism in the brain stem reticular formation. induction of paradoxical sleep Int J Neuropharmacol. **3**, 541—552 (1964)

GERSCHENFELD, H M , J. H. TRAMEZZANI, and E. DE ROBERTIS Ultrastructure and function in neurohypophysis of the toad Endocrinology **66**, 741—762 (1960).

GLAESSER, A., and P MANTEGAZZINI Action of 5-hydroxytryptamine and of a 5-hydroxy-tryptophan on the cortical electrical activity of the midpontine pretrigeminal preparation of the cat with and without mesencephalic hemisection. Arch. ital. Biol. **98**, 351—366 (1960).

GOLDSTEIN, L., and C. MUNOZ Influence of adrenergic stimulant and blocking drugs on cerebral electrical activity in curarized animals J. Pharmacol. exp Ther. **132**, 345—353 (1961).

GOLLUB, L. R., and J. V. BRADY: Behavioral pharmacology. Ann Rev. Pharmacol. **5**, 235 (1965).

GRANIT, R. Centrifugal and antidromic effects on ganglion cells of retina. J. Neurophysiol. **18**, 388—411 (1955)

— Neural activity in the retina In Handbook of physiology, Sect. II, Neurophysiology, vol 1, p. 693—712 Washington, D C.. Amer. Physiol. Soc 1959.

—, and R. KAADA Influence of stimulation of central nervous structures on muscle spindles in cat. Acta physiol. scand. **27**, 130—160 (1952).

GROSSMAN, S P. Eating and drinking elicited by direct adrenergic or cholinergic stimulation of the hypothalamus Science **132**, 301—302 (1960)

— Direct adrenergic and cholinergic stimulation of hypothalamic mechanisms Amer. J. Physiol. **202**, 872—882 (1962a).

— Effects of adrenergic and cholinergic blocking agents on hypothalamic mechanisms. Amer. J. Physiol. **202**, 1230—1236 (1962b).

— Behavioral effects of direct chemical stimulation of central nervous system structures Int. J. Neuropharmacol. **3**, 45—58 (1964).

GUILLEMIN, R. Sur la nature des substances hypothalamiques qui controlent la secretion des hormones antehypophysaires J. Physiol (Paris) **55**, 7—44 (1963).

HAGBARTH, K. E , and D. E B. KERR· Central influences on spinal afferent conduction. J. Neurophysiol. **17**, 295—307 (1954).

Hardy, J. D.. Physiology of temperature regulation. Physiol Rev. **41**, 521—606 (1961).

Harris, G W. Neural control of the pituitary gland. London E. Arnold 1955.

— The reticular formation, stress, and endocrine activity. In: Reticular formation of the brain, ed. by H. Jasper, L. D. Proctor, R S. Knighton, W. C Noshay anp R. T. Costello, p 207—221 Boston Little, Brown & Co 1958.

— Central control of pituitary secretion. In Handbook of physiology. Neurophysiology, vol 2, p. 1007—1038. Baltimore, Md . Amer. Physiol Soc , Williams & Wilkins Co. 1960.

Haslett jr., W L · The pharmacology of oxotremorine, a tremorogenic agent. Diss University of California, Los Angeles 1963.

— R. George, and D J Jenden Evidence for muscarinic receptors in the midbrain reticular formation. Fed. Proc. **22** (2) 214 (1963a).

— — — Localization of some central sites of action of oxotremorine Proc Western pharmacol Soc 6, 9—10 (1963b).

Hayama, T., and Y Ogura · Site of emetic action of tetrodoxin in dog. J Pharmacol. exp. Ther. **139**, 94—96 (1963).

Henderson, F. G , B. L Martz, and I. H. Slater Blood pressure response to butamoxane derivatives in mammals and man. J Pharmacol exp Ther **122**, 30A (1958).

Hernandez-Peon, R. Central mechanisms controlling conduction along central sensory pathways. Acta neurol lat.-amer. **1**, 256—264 (1955).

— The centrifugal control of afferent inflow to the brain and sensory perception. Acta neurol. lat.-amer. **5**, 279—298 (1959).

— Neurophysiological correlates of habituation and other manifestations of plastic inhibition, internal inhibition Electroenceph. clin. Neurophysiol , Suppl. **13**, 101—114 (1960).

— Reticular mechanisms of sensory control In Sensory communication, p 497—520. New York and London M.I T. Press and John Wiley and Sons 1961.

— Sleep induced by localized electrical or chemical stimulation of the forebrain. Electroenceph. clin. Neurophysiol. **14**, 423—424 (1962).

— Physiological mechanisms in attention. In Modern developments in physiological psychology (ed R Russel) New York Academic Press, Inc 1963a

— Attention, sleep motivation and behavior. In Pleasure, integration and behavior (ed. G Heath) New York P. B Hoeber, Inc. 1963b

— Neurophysiological mechanisms of wakefulness and sleep XVII. Int Congr. Psychology, Washington D C., Aug. 1963c.

— Central neuro-humoral transmission in sleep and wakefulness. In Sleep mechanisms Progress in brain research, vol. 18 (eds K Akert, C Bally and J. P. Schade), p 96—117 Amsterdam Elsevier Publ Co 1965

—, and G. Chavez-Ibarra Sleep induced by electrical or chemical stimulation of the forebrain. Electroenceph clin. Neurophysiol , Suppl. **24**, 188—198 (1963)

— —, and E. Aguilar-Figueroa Somatic evoked potentials in one case of hysterical anesthesia. Electroenceph clin Neurophysiol. **15**, 889—892 (1963)

— — J. P. Morgan, and C. Timo-Iaria Limbic cholinergic pathways involved in sleep and emotional behavior Exp Neurol **8**, 93—111 (1963)

— — — — Induction of sleep by direct cholinergic stimulation of the brain Clin Res **11**, 177 (1963).

—, and H. Scherrer Inhibitory influences of brain stem reticular formation upon synaptic transmission in trigeminal nucleus Fed Proc. **14**, 71 (1955).

— —, and M Jouvet Modification of electrical activity in the cochlear nucleus during "attention" in unanesthetized cats. Science **123**, 331 (1956).

— —, and M Velasco Central influences on afferent conduction in the somatic and visual pathways Acta neurol lat -amer **2**, 8—22 (1956).

Hervey, G R The effects of lesions in the hypothalamus in parabiotic rats. J. Physiol. (Lond) **145**, 336—352 (1959).

Hess, W. R. Stammganglien — Reizversuche uber den Mechanismus des Schlafes Arch. Psychiat Nervenkr. **86**, 287 (1928).

Hess, W. R.: Lokalisatorische Ergebnisse der Hirnreizversuche mit Schlafeffekten. Arch. Psychiat. Nervenkr. **88**, 813 (1929).

— Hypothalamische Adynamie. Helv. physiol pharmacol. Acta **2**, 137 (1944a).

— Das Schlafsyndrom als Folge diencephaler Reizung. Helv. physiol. pharmacol. Acta. **2**, 305 (1944b).

— Das Zwischenhirn Syndrome, Lokalisationen, Funktionen. Basel Benno-Schwabe & Co 1954.

Hetherington, A W., and S W. Ranson· The spontaneous activity and food intake of rats with hypothalamic lesions. Amer. J. Physiol. **136**, 609—617 (1942).

Hiebel, G., M Bonvallet, P. Huvé et P Dell· Analyse neurophysiologique de l'action centrale de la d-amphetamine (maxiton). Sem. Hôp. Paris **30**, 1880 (1954).

Himwich, H. E Prospects in psychopharmacology J nerv ment. Dis. **122**, 413—423 (1955).

—, and F. Rinaldi. The effects of drugs on reticular system. In Brain mechanisms and drug action, ed. by W. E. Fields, p. 15—44. Springfield (Ill.)· Ch. C Thomas 1957.

Hoebel, B G., and Ph Teitelbaum. Hypothalamic control of feeding and self-stimulation. Science **135**, 375—377 (1962).

Holland, R. C., B. A. Cross, and C H. Sawyer Electroencephalographic correlates of osmotic activation of the neurohypophyseal milk ejection mechanism. Amer. J. Physiol. **196**, 796—802 (1959)

— J. W. Sundsten, and C. H. Sawyer: Effects of intracarotid injections of hypertonic solutions on arterial pressure in rabbit Circulat Res **7**, 712—720 (1959)

Hornykiewicz, O , u. W. Kobinger Über den Einfluß von Eserin, Tetraàthylpyrophosphat (TEEP) und Neostigmin auf den Blutdruck und die pressorischen Carotissinusreflexe der Ratte. Naunyn-Schmiedebergs Arch. exp. Path. Pharmak **228**, 493—500 (1956).

Hubel, D. H Electrocorticograms in cats during natural sleep Arch. ital. Biol. **98**, 171—181 (1960)

Hume, D. M. Hypothalamic localization for the control of various endocrine secretions. In The reticular formation of the brain, ed. by H. H. Jasper, p. 231—248 Boston and Toronto Little, Brown & Co. 1958

Hunter, J , and H.H. Jasper. Effects of thalamic stimulation in unanesthetized animals. Electroenceph clin. Neurophysiol. **1**, 305—324 (1949).

Ingvar, D. H. Extraneuronal influences upon the electrical activity of isolated cortex following stimulation of the reticular activating system. Acta physiol. scand. **33**, 169—193 (1955)

— Cortical state of excitability and cortical circulation In· Reticular formation of the brain, ed by H H. Jasper, L. D. Proctor, R. S. Knighton, W. C. Noshay and R. T. Costello, p. 381—408 Boston Little, Brown & Co. 1958

Jacobowitz, D , and G B Koelle Histochemical correlations of acetylcholinesterase and catecholamines in postganglionic autonomic nerves of the cat, rabbit and guinea pig J. Pharmacol. exp. Ther. **148**, 225—237 (1965).

—, P Johnson, I Kitchner, and G B Koelle. The effect of hemicholinium HC-3 on sympathetic transmission at the nictitating membrane of the rabbit. Brit. J. Pharmacol. **25**, 527—533 (1965)

Janowitz, H D , and M T. Grossman Some factors affecting the food intake of normal dogs and dogs with esophagotomy and gastric fistula Amer J Physiol **159**, 143—148 (1949).

Jasper, H Reticular-cortical systems and theories of the integrative action of the brain. In Biological and biochemical bases of behavior (eds. H. F. Harlow and C. N. Woolsey), p 31—61 Madison Wisconsin University Press 1958.

Jeppson, P. G. Studies on the blood-brain barrier in hypothermia. Acta neurol. scand. **38**, Suppl. 160 (1962)

Jewell, P A., and E B Verney. An experimental attempt to determine the site of neurohypophysial osmoreceptors in the dog. Phil. Trans. B **240**, 197—324 (1957)

JOUVET, M. Telencephalic and rhombencephalic sleep in the cat In The nature of sleep. Ciba Foundation Symposium, p 118—208. London Churchill 1961.

— Recherches sur les structures nerveuses et les mécanismes responsables des différentes phases du sommeil physiologique. Arch. ital. Biol. **100**, 125—207 (1962).

—, et J. E. DESMEDT Controle central des messages acoustiques afférents C R Acad. Sci. (Paris) **243**, 1916—1917 (1956)

KANEMATSU, S Effects of intrahypothalamic and intrahypophysial estrogen implants on pituitary LH and Prolactin content in female rabbit. Fed. Proc. **22** (1), 506 (1963).

KARCZMAR, A. G. Pharmacologic, toxicologic and therapeutic properties of anticholinesterase agents In Physiological pharmacology (ed W. S ROOT and F. G HOFFMAN, vol III (in press) NewYork Academic Press 1966

KENNEDY, G C · The role of depot fat in the hypothalamic control of food intake in the rat. Proc roy Soc B **140**, 578—592 (1953)

KEY, B J., and E MARLEY: The effect of the sympathomimetic amines on behavior and electrocortical activity of the chicken Electroenceph. clin. Neurophysiol **14**, 90—105 (1962).

KILLAM, E. K. Drug action on the brain-stem reticular formation. Pharmacol Rev **14**, 175—223 (1962).

KIRPEKAR, S. M., P. CERVONI, and R. F. FURCHGOTT Catecholamine content of the cat nictitating membrane following procedures sensitizing it to norepinephrine. J. Pharmacol exp Ther. **135**, 180—190 (1962).

KNAPP, D. E , and E. F. DOMINO Action of nicotine on the ascending reticular activating system. Int J Neuropharmacol **1**, 333—351 (1962)

KOELLE, G. B.: The histochemical localization of cholinesterases in the central nervous system of the rat. J. comp. Neurol. **100**, 211—236 (1954).

— A proposed dual neurohumoral role of acetylcholine its functions at the pre- and post-synaptic sites Nature (Lond.) **190**, 208—211 (1961)

— A new general concept of the neurohumoral functions of acetylcholine and acetylcholinesterase. J. Pharm Pharmacol **14**, 65—90 (1962).

— Cytological distributions and physiological functions of cholinesterases. In Handbuch der experimentellen Pharmakologie, Erg -Werk, hrsg von O EICHLER u. A FARAH, Bd. XV, S 187—298 (Subeditor G. B. KOELLE). Berlin-Gottingen-Heidelberg Springer 1963.

— C. N. GEESEY, and C F SCHMIDT Localization of acetylcholinesterase in the neurohypophysis and its functional implications. Proc Soc exp Biol (N Y.) **106**, 625—628 (1961).

—, and E. C. STEINER The cerebral distributions of a tertiary and quaternary anticholinesterase agent following intravenous and intraventricular injection J Pharmacol. exp. Ther. **118**, 420—434 (1956).

KOLLER, TH., u. M. MONNIER Über eine humorale Übertragung des experimentellen Schlafes an Tieren mit gekreuzter Zirkulation (quantitativer elektrographischer Nachweis). Helv physiol pharmacol. Acta **20**, C67—C68 (1962)

KORNMULLER, A. E . Die Elemente der nervosen Tatigkeit Stuttgart Georg Thieme 1947.

— D H. LUX u K. WINKEL· EEG-Untersuchungen an Tieren mit gekreuztem Kreislauf. Naturwissenschaften **48**, 381—382 (1961)

— — — u M. KLEE Neurohumoral ausgeloste Schlafzustande an Tieren mit gekreuztem Kreislauf unter der Kontrolle von EEG-Ableitungen. Naturwissenschaften **48**, 503—505 (1961).

KRNJEVIĆ, K., R. LAVERTY, and D. F. SHARMAN. Iontophoretic release of adrenaline, noradrenaline and 5-hydroxytryptamine from micropipettes. Brit. J. Pharmacol. **20**, 491—496 (1963).

—, and J. F. MITCHEL The release of acetylcholine in the isolated rat diaphragm. J. Physiol. (Lond) **155**, 246 (1961)

KRNJEVIĆ, K, and J. W. PHILLIS Sensitivity of cortical neurons to acetylcholine. Experientia (Basel) **17**, 469—470 (1961).

— — Iontophoretic studies of neurones in the mammalian cerebral cortex. J. Physiol (Lond.) **165**, 274—304 (1963a).

— — Acetylcholine-sensitive cells in the cerebral cortex J. Physiol (Lond) **166**, 296—327 (1963b).

— — Pharmacological properties of acetylcholine-sensitive cells in the cerebral cortex J. Physiol. (Lond.) **166**, 328—350 (1963c).

KROGH, A.: Active and passive exchanges of inorganic ions through the surfaces of living cells and through living membranes generally. Proc. roy. Soc. B **133**, 140—200 (1946).

KUHN, W L., H. MAJER, H HEUSSER u. B. ZEN RUFFINEN· Kunstliche Niere und Kapillarsystem fur den Stoffaustausch Experientia (Basel) **13**, 469—471 (1957).

LABORIT, H, J M. JOUANY, J GÉRARD et F FABIANI Resumé d'une étude expérimentale et clinique sur un substrat métabolique à action centrale inhibitrice, le 4-hydroxybutyrate de Na. Presse méd. **68**, 1867 (1960).

LARSSON, S On the hypothalamic organization of the nervous mechanism regulating food intake. Acta physiol scand **32**, Suppl. 115, 1—40 (1954).

LEIMDORFER, A, and W. R T. METZNER Analgesia and anesthesia induced by epinephrine. Amer. J. Physiol **157**, 116—121 (1949)

LESIĆ, R, and V VARAGIĆ: Factors influencing the hypertension effect of eserine in the rat Brit J Pharmacol **16**, 99—107 (1961)

LINDSLEY, D. B. The reticular system and perceptual discrimination In The reticular formation of the brain (ed H H. JASPER, L D. PROCTOR, R S KNIGHTON, W. C HOSHAY and R T COSTELLO), p 513—534 Boston Little, Brown & Co 1958.

LING, G. M, and J G FOULKS A suitable preparation for pharmacological analysis of EEG "activation". Proc. Soc. exp Biol (N Y) **101**, 429—432 (1959)

LISK, R D Estrogen-sensitive centers in the hypothalamus of the rat. J. exp. Zoology **145**, 197—205 (1960).

— Inhibition of castration cell formation in the pituitary of the spayed rat by estradiol implants in the arcuate nucleus. Amer. Zoologist **2** (3) (1962a).

— Diencephalic placement of estradiol and sexual receptivity in the female rat Amer. J. Physiol **203**, 493—496 (1962b)

— Testosterone-sensitive centers in the hypothalamus of the rat. Acta endocr. (Kbh.) **41**, 195—204 (1962c).

— Maintenance of normal pituitary weight and cytology in the spayed rat following estradiol implants in the arcuate nucleus. Anat. Rec. **146**, 281—286 (1963)

—, and M NEWLON Estradiol Evidence for its direct effect on hypothalamic neurons Science **139**, 223—224 (1963)

—, and A J. WEIN Neurological site of action of estradiol in eliciting estrus behavior in the spayed rat Amer Zoologist **2** (3) (1962)

LITTLEJOHN, B. M., and J. DE GROOT. Estrogen-sensitive areas in the rat brain. Fed. Proc. **22** (2), 571 (1963)

LIVINGSTON, R. B Central control of receptors and sensory transmission systems. In: Handbook of physiology. Section 1 Neurophysiology (ed J. FIELD, H. W. MAGOUN and V. E. HALL), p. 741—760. Washington, D.C.· Amer. Physiol Soc. 1959.

LONGO, V. G.: Acetylcholine, cholinergic drugs and cortical electrical activity. Experientia (Basel) **11**, 76—77 (1955)

— Effect of drugs on the electrical activity of the rabbit brain. In· Rabbit brain research, vol. II. Amsterdam: Elsevier Publ. Co. 1962.

—, and B. SILVESTRINI Action of eserine and amphetamine on the electrical activity of the rabbit brain J. Pharmacol exp Ther **120**, 160—170 (1957).

MACHNE, X., I. CALMA, and H. W. MAGOUN. Unit activity of central cephalic brain stem in EEG arousal. J. Neurophysiol **18**, 547—558 (1955).

MACHNE, X , and K. R. UNNA Cholinesterases and anticholinesterase agents. Actions at the central nervous system In Handbuch der experimentellen Pharmakologie, Erg -Werk, Bd. 15, S. 679—700. Berlin-Gottingen-Heidelberg Springer 1963

MACLEAN, P. D.: Chemical and electrical stimulation of hippocampus in unrestrained animals I. Arch Neurol. Psychiat. (Chic.) **78**, 113—142 (1957).

MAGNES, J , G. MORUZZI, and O. POMPEIANO· Electroencephalogram-synchronizing structures in the lower brain stem. In. The nature of sleep Ciba Foundation Symposium (ed G E. W. WOLSTENHOLM and M. O'CONNOR), p 57—78. London Churchill 1960

— — — Synchronization of EEG produced by low-frequency electrical stimulation of the region of the solitary tract Arch ital Biol **99**, 33—68 (1961)

MANTEGAZZINI, P , K. POECK, and H SANTIBAÑEZ The action of adrenaline and noradrenaline on the cortical electrical activity of the "encéphale isolé" cat. Arch ital Biol. **97**, 222—242 (1959).

MARCZYNSKI, T. J Farmakologia i mechanizm działania kilku nowych zwiazkow pochodnych tryptaminy. Folia med cracov **6**, 1—86 (1965)

— A ROSEN, and J HACKETT Orbital cortex and thalamically induced slow wave activity in the brain stem reticular formation Fed Proc **25**, 504 (1966)

—, and N. YAMAGUCHI Central depressant effect of some new tryptamine derivatives Biochem. Pharmacol., Suppl. **12**, 209 (1963)

— —, and G. M. LING: Effects of bufotenine and melatonin applied in crystalline form to the preoptic region or nucleus centralis medialis in unrestrained cats. Pharmacologist **4** (2), 175 (1962)

— — —, and L. GRODZINSKA Sleep induced by the administration of melatonin (5-methoxy-N-acetyltryptamine) to the hypothalamus in unrestrained cats Experientia (Basel) **26**, 435 (1964)

MARLEY, E., and B J KEY Maturation of the electrocorticogram and behavior in the kitten and guinea-pig and the effect of some sympathomimetic amines Electroenceph clin. Neurophysiol. **15**, 620—636 (1963)

MASLAND, R. L , and R. S. WIGTON Nerve activity accompanying fasciculation produced by prostigmin. J. Neurophysiol **3**, 269—275 (1940).

MAYER, J.: Regulation of energy intake and the body weight the glucostatic theory and the lipostatic hypothesis. Ann N Y Acad Sci **63**, 15—43 (1955).

MAYER, S. E., and J. A BAIN Localization of the hematoencephalic barrier with fluorescent quaternary acridones. J Pharmacol exp. Ther. **118**, 17—25 (1956).

McCANN, S. M.. Effect of hypothalamic lesions on the adrenal cortical response to stress in the rat. Amer. J Physiol **75**, 13—20 (1953).

— A hypothalamic luteinizing hormone releasing factor Amer J. Physiol **202**, 395—400 (1962)

McLENNAN, H. Synaptic transmission Philadelphia and London W B Saunders Co 1963.

— The release of acetylcholine and of 3-hydroxytyramine from the caudate nucleus. J Physiol (Lond) **174**, 152—161 (1964)

MEDUKOVIĆ, M., and V VARAGIĆ The effect of eserine and neostigmine on the blood pressure of conscious rats. Brit. J. Pharmacol. **12**, 24—27 (1957).

— — Effect of guanethidine, hemicholinium and mebutamate on the hypertensive response to eserine and catecholamines Brit J Pharmacol **19**, 451—457 (1962).

MICHAEL, R. P Estrogen sensitive neurons and sexual behavior in two female cats. Science **136**, 322—333 (1962).

— Oestrogens in the central nervous system. Brit med Bull. **21**, 87—90 (1965).

MILLER, N. E Experiments on motivation. Science **126**, 1271—1278 (1957)

— Chemical coding of behavior in the brain. Science **148**, 328—338 (1965)

— K. S GOTTESMAN, and N. EMERY Dose response to carbachol and norepinephrine in rat hypothalamus Amer. J. Physiol. **206**, 1384—1388 (1964)

MIRKIN, D L., and T. CERVONI. The adrenergic nature of neurohumoral transmission in the cat nictitating membrane following treatment with reserpine. J. Pharmacol. exp. Ther. **138**, 301 (1962).

MITCHEL, J. F.· The spontaneous and evoked release of acetylcholine from the cerebral cortex J. Physiol (Lond) **165**, 98—116 (1963).

—, and J. C SHERB The spontaneous and evoked release of acetylcholine from the caudate nucleus Abstr. XXII int physiol Congr. 1962, No 819.

MONNIER, M Action de la stimulation électrique du centre somnogene sur l'électro-corticogramme chez le chat. Rev. neurol. **83**, 561—563 (1950).

—, and L HOSLI Humoral regulation of sleep and wakefulness by hypnogenic and activating dialisable factors In Sleep mechanisms (ed. K. AKERT, C. BALLY and J. P. SCHADE). Amsterdam: Elsevier Publ. Co. 1965.

— —, and P. KRUPP Moderating and activating systems in medial thalamus and reticular formation. Electroenceph. clin Neurophysiol , Suppl. **24**, 97—112 (1963).

—, et M. ROMANOWSKI Les systèmes cholinoceptifs cérébraux — action de l'acétylcholine, de la physostigmine, pilocarpine et de GABA Electroenceph. clin. Neurophysiol. **14**, 486—500 (1963).

MORGANE, P. J.. Alterations in feeding and drinking behavior of rats with lesions in globi pallidi Amer. J Physiol **201**, 420—428 (1961a).

— Electrophysiological studies of feeding and satiety centers in the rat. Amer. J. Physiol. **201**, 838—844 (1961b).

— Medial forebrain bundle and "feeding centers" of the hypothalamus. J. comp Neurol **117**, 1—26 (1961c).

— Reinforcing effects of self-injected cholinergic agents into hypothalamic "drinking" areas in rats Fed. Proc **21**, 352 (1962).

MORRISON, S D , and J. MAYER Adipsia and aphagia in rats after lateral subthalamic lesions. Amer. J. Physiol. **191**, 248—254 (1957)

MORUZZI, G. Synchronizing influences of the brain stem and the inhibitory mechanisms underlying the production of sleep by sensory stimulation. Electroenceph. clin. Neurophysiol., Suppl. **13**, 231—256 (1960).

—, and H. W MAGOUN Brain stem reticular formation and activation of the EEG. Electroenceph. clin. Neurophysiol. **1**, 455—473 (1949).

MYERS, R D · An intracranial chemical stimulation system for chronic self-infusion J appl Physiol **18**, 221—223 (1963).

— Emotional and autonomic responses following hypothalamic chemical stimulation. Canadian J. Psychol. **18**, 6—14 (1964).

NAUTA, W. J H. Hypothalamic regulation of sleep in rats. An experimental study. J. Neurophysiol **9**, 285—316 (1946).

— Hippocampal projections and related neural pathways to the mid-brain in the cat. Brain **81**, 319—340 (1958).

—, and G M. M. KUYPERS. Some ascending pathways in the brain stem reticular formation. In The reticular formation of the brain (ed. H H. JASPER, L D. PROCTOR, R S. KNIGHTON, W. C. NOSHAY and R. T. COSTELLO), p 3—30 Boston and Toronto Little, Brown & Co. 1958

NIKITOVITCH-WIENER, M B Induction of ovulation in rats by direct intrapituitary infusion of median eminence extracts. Endocrinology **70**, 350—358 (1962).

OHLIN, P , and B C. R STROMBLAD Observations on the isolated vas deferens. Brit J. Pharmacol **20**, 299—306 (1963).

OLDS, J. Selfstimulation of the brain. Its use to study local effects of hunger, sex and drugs Science **127**, 315—324 (1958).

— Hypothalamic substrates of reward Physiol Rev. **42**, 554—604 (1962).

—, and M. E. OLDS Positive reinforcement produced by stimulating hypothalamus with iproniazid and other compounds. Science **127**, 1175—1176 (1958).

—, A YUWILER, M E OLDS, and CHANG YUN Neurohumors in hypothalamic substrates of reward Amer. J. Physiol. **207**, 242—254 (1964).

OLSZEWSKI, J.. Designated discussion on the paper of MAGDE E. SCHEIBEL and A. SCHEIBEL, Structural substrates for integrative patterns in the brain stem reticular core In Reticular formation of the brain, ed. by H JASPER, L D. PROCTOR, R. S KNIGHTON, W. C. NOSHAY and R T. COSTELLO, p 56—59 Boston Little, Brown & Co. 1958

—, and D. BAXTER Cytoarchitecture of the human brain stem, p. 12—13 Basel and NewYork S. Karger 1954.

PAGE, I. H.   Serotonin (5-hydroxytryptamine); the last four years. Physiol. Rev 38, 277—335 (1958)

PALAY, S L.: The fine structure of the neurohypophysis In H WAELSCH, ed , Ultrastructure and cellular chemistry of neural tissue, Chapt II, p. 31—49. NewYork Hoeber-Harper 1957.

PAPEZ, J. W.· A proposed mechanism of emotion Arch. Neurol Psychiat (Chic.) 38, 725—743 (1937).

PATON, W. D. M · Central and synaptic transmission in the nervous system (pharmacological aspects). Ann. Rev Physiol. 20, 431—470 (1958).

—, and W. L M PERRY The relationship between the depolarization and block in the cat's superior cervical ganglion J. Physiol (Lond.) 119, 43—57 (1953)

PURPURA, D . Neurohumoral mechanisms of reticulo-cortical activation. Amer. J Physiol. 186, 250—254 (1956)

RAAB, W , and W. GIGEE Concentrations and distribution of "encephalin" in the brain of humans and animals. Proc Soc exp Biol (N Y.) 76, 97—100 (1951).

RAKIĆ, L., N. A BUCHWALD, and E J WYERS Induction of seizures by stimulation of the caudate nucleus Electroenceph clin. Neurophysiol. 14, 809—823 (1962)

REAS, H W., and T H. TSAI The antagonism by atropine of the response of the nictitating membrane to sympathetic nerve stimulation J Pharmacol. exp. Ther. 152, 186—196 (1966)

RECH, R , and E F DOMINO Observations on injections of drugs into the brain substance Arch int. Pharmacodyn 121, 429—442 (1959).

RIEHL, J L , J PAUL-DAVID, and K. R UNNA Comparison of the effects of arecoline and muscarine on the central nervous system. Int. J. Neuropharmacol. 1, 393—401 (1962)

RIKER, W. F , J ROBERTS, F. G STANDAERT, and H FUJIMORI The motor nerve terminal as the primary focus for drug-induced facilitation of neuromuscular transmission. J. Pharmacol exp Ther 121, 286—311 (1957).

— G. WERNER, J. ROBERTS, and A. S KUPERMAN Pharmacologic evidence for the existence of a presynaptic event in neuromuscular transmission J Pharmacol exp Ther 125, 150—158 (1959)

RINALDI, F., and H. E. HIMWICH Alerting responses and action of atropine and cholinergic drugs Arch Neurol Psychiat. (Chic.) 73, 387—395 (1955a).

— — Cholinergic mechanisms involved in function of mesodiencephalic activating system Arch Neurol Psychiat (Chic) 73, 396—402 (1955b)

ROBINSON, B W , and M MISHKIN. Alimentary responses evoked from forebrain structures in Macaca mulatta Science 136, 260—262 (1962).

ROBSON, J. M , and R. A. STACEY Pharmacologically active substances in the central nervous system Psychotropic drugs In Recent advances in pharmacology. London Churchill 1962

ROSSI, G. F. Sleep-inducing mechanisms in the brain stem Electroenceph clin Neurophysiol. 14, 428 (1962).

—, and A ZANCHETTI. The brain stem reticular formation Anatomy and physiology. Arch ital. Biol 95, 199—435 (1957).

ROTHBALLER, A. B Studies on the adrenaline-sensitive component of the reticular activating system Electroenceph clin Neurophysiol 8, 603 (1956).

— The effects of catecholamines on the central nervous system Pharmacol. Rev. 11, 494—547 (1959).

SAILER, S., and CH STUMPF. Beeinflußbarkeit der rhinencephalen Tätigkeit des Kaninchens. Naunyn-Schmiedebergs Arch exp. Path Pharmak. 231, 63—77 (1957).

SALMOIRAGHI, G. C. · Electrophoretic administration of drugs to individual nerve cells In Neuropsychopharmacology, vol. 3, p. 219—231. Amsterdam. Elsevier Publ Co. 1964.

— E COSTA, and F. E. BLOOM. Pharmacology of central synapses. Ann Rev. Pharmacol. 5, 213 (1965).

—, and F. A STEINER · Acetylcholine sensitivity of cat's medullary neurons J. Neurophysiol. 26, 581—597 (1963).

SAWYER, CH. H. Reproductive behavior In · Handbook of physiology. Neurophysiology, vol 2 p 1225—1240, ed by Amer. Physiol. Soc Baltimore, Md. Williams & Wilkins Co. 1960.

— Gonadal hormone feed-back and sexual behavior. In Proceedings of the Int. Union of Physiol. Sciences, vol. 1. XXI. Int. Congr, Leiden 1962.

— Mechanisms by which drugs and hormones activate and block release of pituitary gonadotropines. In Proceedings of the First Int. Pharmacol. Meeting, vol 1, p. 27—46 Amsterdam: Pergamon Press 1963.

—, and B V. CRITCHLOW Reproduction. Ann. Rev. Physiol. 19, 467—487 (1957).

—, and B E GERNANDT · Effects of intracarotid and intraventricular injections of hypertonic solutions on electrical activity of the rabbit brain. Amer. J. Physiol. 185, 209—216 (1956).

—, and M. KAWAKAMI. Interactions between the central nervous system and hormones influencing ovulation In Control of ovulation (ed. C. A. VILLEE), p 79—97. Amsterdam Pergamon Press 1961.

SCHAYER, R. W Studies of the metabolism of beta-C^{14}-dl-adrenaline. J. biol. Chem. 189, 301—306 (1951).

SCHLAG, J L'activité spontanée des cellules du system nerveux central. Bruxelles Arscia, S. A. 1959.

—, and F. CHAILLET Thalamic mechanisms involved in cortical desynchronization and recruiting responses Electroenceph. clin Neurophysiol 15, 39—62 (1963).

—, and J. FAIDHERBE Recruiting responses in the brain stem reticular formation Arch. ital. Biol 99, 135—162 (1961)

SCHOOLMAN, A, and E. V EVART Responses to lateral geniculate radiation stimulation in cats with implanted electrodes. J Neurophysiol. 22, 112—129 (1959).

SCHUMANN, H. J.: Probleme der sympathischen Erregungsubertragung. Naunyn-Schmiedebergs Arch. exp. Path. Pharmak 246, 94—101 (1963).

—, and H. GROBECKER · Untersuchungen am isolierten Vas deferens — Hypogastricus Praparat des Meerschweinchens. Naunyn-Schmiedebergs Arch. exp. Path. Pharmak 246, 215—225 (1963).

SEGUNDO, J. P. · The reticular formation A survey. Acta neurol lat.-amer. 3, 245—284 (1956).

SHERRINGTON, C S The integrative action of the nervous system. Cambridge. Cambridge University Press 1906 (1947).

SJOSTRAND, N. O. · Inhibition by ganglionic blocking agents of the motor response of the isolated guinea pig vas deferens to hypogastric nerve stimulation. Acta physiol. scand. 54, 306—315 (1962a)

— Effect of reserpine and hypogastric denervation on the noradrenaline content of the vas deferens and the seminal vesicle of the guinea-pig. Acta physiol. scand. 56, 376—380 (1962b).

SLATER, I. H., and G. T. JONES · Pharmacological properties of ethoxy-butamoxane. J. Pharmacol. exp. Ther. 122, 69A (1958).

SMELIK, P. G., and CH. H. SAWYER · Effects of implantation of cortisol into the brain stem or pituitary gland on the adrenal response to stress in the rabbit. Acta endocrin. (Kbh) 41, 561—570 (1962).

SPEHLMANN, R.: Acetylcholine and prostigmine electrophoresis at visual cortex neurones. J. Neurophysiol 26, 127—139 (1963).

Stein, L , and J Seifter Muscarinic synapses in the hypothalamus. Amer. J. Physiol
    202, 751—756 (1962).
Steiner, F A., u. G C Salmoiraghi Acetylcholin-empfindliche Neurone im Hirn-
    stamm der Katze. Helv. physiol. pharmacol. Acta 20, C87 (1962).
Steriade, M., and M Demetrescu· Unspecific systems of inhibition and facilitation
    of potentials evoked by intermittent light. J. Neurophysiol. 23, 602—617 (1960)
Sterman, M. B., and C. B. Clemente Forebrain inhibitory mechanisms· Cortical syn-
    chronization induced by basal forebrain stimulation. Exp Neurol 6, 91—102 (1962a).
— — Forebrain inhibitory mechanisms. Sleep patterns induced by basal forebrain
    stimulation in the behaving cat Exp. Neurol. 6, 103—117 (1962b).
Straughan, D. W.: The release of acetylcholine from mammalian motor nerve endings
    Brit. J. Pharmacol. 15, 417 (1960).
Stumpf, Ch.· Die Wirkung des Nicotin auf die Hippocampustatigkeit des Kaninchens
    Naunyn-Schmiedebergs Arch. exp Path. Pharmak 235, 421—436 (1959)
— Erzeugung von Krankheitszustanden durch das Experiment Zentralnervensystem
    Pharmakologische Methoden In Handbuch der experimentellen Pharmakologie,
    Erg -Werk, Bd. 16, Teil 7, S 1—105. Berlin-Gottingen-Heidelberg: Springer 1962.
Sundsten, J. W , and C H. Sawyer Osmotic activation of neurohypophyseal milk-
    ejection reflex in rabbits with "diencephalic island". Fed. Proc. 19, 293 (1963)
Takeshige, C., and R. L. Volle Modification of ganglionic responses to cholinomimetic
    drugs following preganglionic stimulation, anticholinesterase agents and pilocarpine
    J. Pharmacol exp. Ther. 146, 335—343 (1964).
Taleisnik, S , and S M McCann Effects of hypothalamic lesions on the secretion and
    storage of hypophyseal luteinizing hormone. Endocrinology 68, 263—272 (1961).
Teitelbaum, Ph.· Sensory control of hypothalamic hyperphagia. J. comp physiol
    Psychol. 48, 156—163 (1955).
— Disturbances in feeding and drinking behavior after hypothalamic lesions Nebraska
    Symposium on Behavior 1961, p. 39—69
—, and E Stellar Recovery from the failure to eat produced by hypothalamic lesions.
    Science 120, 894—895 (1954)
Toman, J. E. P. Some aspects of central nervous system pharmacology Ann. Rev.
    Pharmacol 3, 153—184 (1963).
Traczyk, W., and B Sadowski: Electrical activity of the "cerveau isolé" during caudate
    nucleus stimulation and its modification by eserine and atropine. Acta physiol. pol
    13, 521—533 (1962).
— — Electrical activity of the "cerveau isolé" preparation and its relation to the
    acetylcholine content of the caudate nuclei. Electroenceph. clin. Neurophysiol. 17,
    272—280 (1964).
Trzebski, A The action of adrenaline, noradrenaline and monamine oxidase inhibitors
    injected directly into the reticular formation of the brain stem. Bull Acad. pol. Sci
    Cl 6, 8, 525—528 (1960)
Tschirgi, R. D Blood-brain barrier. In Biology of the mental health and disease,
    p. 34—46. New York Hoeber 1952.
— Blood-brain barrier. In Biology of neuroglia (ed. W. F. Windle), p. 130—138.
    Springfield (Ill) Ch. C Thomas 1958.
— Chemical environment of the central nervous system In Handbook of physiology.
    Neurophysiology, vol. 3, p 1865—1890. Washington, D C.· Amer. Physiol. Soc
    1960
Udenfriend, S., H Weissbach, and D. F Bogdanski. Increase in tissue serotonin
    following administration of its precursor 5-hydroxy-tryptophan. J biol Chem. 224,
    803—810 (1957).
Unna, K. R.· A review of the neurophysiological effects of psychotherapeutic drugs.
    N Y. Acad. Sci. 66, 777—783 (1957).
Varagić, V.· The action of eserine on the blood pressure of the cat. Brit. J. Pharmacol.
    10, 349—353 (1955)

VARAGIĆ, V , R LESIĆ, J. VUCO, and B. STEMENOVIĆ· The effect of eserine on the activity of adrenergic nerves in the rat. Int. J Neuropharmacol. 1, 201—202 (1962).

VARGA, F., A. KOVER, T. KOVACS, and E. HETENYI Changes in cholinesterase activity of striated muscle after denervation Acta physiol Acad Sci. hung. 2, 235—242 (1957)

VELLUTI, R., and R. HERNANDEZ-PEON Atropine blockade within a cholinergic hypnogenic circuit. Exp Neurol 8, 20—29 (1963).

VERHAVE, T., J. E Owen jr., D FADELY, and J R. CLARK Some preliminary observations concerning the behavioral effects of ethoxybutamoxane and chlorethoxybutamoxane. J. Pharmacol exp Ther. 122, 78A (1958).

VERNEY, E. B The antidiuretic hormone and the factors which determine its release. Proc roy. Soc. B 135, 25—105 (1947)

WADA, J. A Behavioral and EEG correlates of modified levels of brain neurohumoral agents. J. Neuropsychiat. 4, 251—254 (1963)

WANG, S. C , and H L. BORISON A new concept of organization of the central emetic mechanisms recent studies on the sites of action of apomorphine, copper sulfate and cardial glycosides Gastroenterology 22, 1 (1952).

WEIL-MALHERBE, H. The passage of catechol amines through the blood-brain barrier. In Adrenergic mechanisms (ed J. R VANE, G E. W. WOLSTENHOLME and M. O'CONNOR). London Churchill 1960.

— J. AXELROD, and R. TOMCHICK Blood-brain barrier for adrenaline. Science 129, 1226—1227 (1959)

— L. G. WHITBY, and J. AXELROD: The uptake of circulating H^3-norepinephrine by the pituitary gland and various areas of the brain J Neurochem. 8, 55—64 (1961).

WEISSCHEDEL, E. Die zentrale Haubenbahn und ihre Bedeutung fur das extrapyramidal-motorische System. Arch Psychiat. Nervenkr. 107, 443—579 (1937).

WHITE, R P., and E A DAIGNEAULT The antagonism of atropine to the EEG effects of adrenergic drugs J Pharmacol. exp Ther 125, 339—346 (1959).

WIKLER, A · The relation of psychiatry to pharmacology Monograph. Baltimore. Williams & Wilkins Co. 1957.

WINTERSTEIN, H The actions of substances introduced into the cerebrospinal fluid and the problem of intracranial chemoreceptors. Pharmacol Rev 13, 71—107 (1961).

--, and N. GOEKHAN Ammoniumchlorid-Acidose und Reaktionstheorie der Atmungs-regulation. Arch int. Pharmacodyn 93, 212 (1953).

WITT, D. M., A. D. KELLER, H. L. BATSEL, and J R LYNCH Absence of thirst and resultant syndrome associated with anterior hypothalamectomy in the dog. Amer. J. Physiol 161, 75—86 (1950).

WYRWICKA, W., and C DOBRZECKA Relationship between feeding and satiation centers of the hypothalamus. Science 132, 805—806 (1960).

YAMAGUCHI, N., G. M LING, and T J. MARCZYNSKI The effects of chemical stimulation of the preoptic region, nucleus centralis medialis, or brain stem reticular formation with regard to sleep and wakefulness. Rec. Advanc. Biol. Psychiat. 6, 9—20 (1964).

— T. J. MARCZYNSKI, and G. M. LING The effects of electrical and chemical stimulation of the preoptic region and some nonspecific thalamic nuclei in unrestrained, waking animals Electroenceph. clin. Neurophysiol. 15, 154 (1963)

ZIEHER, L. M., and E. DE ROBERTIS Subcellular localization of 5-hydroxytryptamine in rat brain Biochem Pharmacol 12, 596—598 (1963).

The Anaphylatoxin-Forming System

WALTHER VOGT *

With 1 Figure

Table of contents

Introduction 160
I Formation of anaphylatoxin Evidence for an enzymic mechanism 162
II The components of the anaphylatoxin system 165
  A. The anaphylatoxin-forming enzymes 165
    1. The cobra enzyme 165
    2 The plasma enzyme . 166
  B. The substrate, anaphylatoxinogen 167
  C. The product, anaphylatoxin . 168
    1 Preparation . 168
    2. Nature and properties . 169
III. Activation and inhibition of the anaphylatoxin-forming enzyme 170
  A. Activation . . . 170
  B Inhibition 171
IV. Nature of the enzymic reaction that leads to the formation of anaphylatoxin . 173
  A. The role of complement . 173
  B Other possible reactions 175
V. Actions of anaphylatoxin 175
  A. Effects on isolated organs and tissues 175
  B In vivo effects . . 177
VI Assay of anaphylatoxin . . . 179
VII Biological significance of anaphylatoxin . . . 180
References . . . 182

Introduction

In 1910 FRIEDBERGER found that mixtures of immune precipitates with guinea pig serum acquired, after incubation at 37° C, a powerful toxicity towards guinea pigs. On intravenous injection the incubates produced signs similar to anaphylactic shock. This finding seemed to prove the theory, put forward earlier by VAUGHAN and WHEELER (1907), that the anaphylactic shock is produced by a toxic compound, which is cleaved from the antigen in the presence of antibody and complement. Therefore, FRIEDBERGER called the toxic principle, the formation of which he had demonstrated in vitro,

* Department of Pharmacology, Max-Planck-Institut fur experimentelle Medizin, Gottingen, Germany.

"anaphylatoxin". During the following 50 years the picture of anaphylatoxin has changed considerably. Many details about its action but also many contradictory data concerning the mechanism of formation have been assembled.

The history of the work on anaphylatoxin can be divided into three phases each of which gave a new stimulus to research. During the first phase, observations about the formation and actions were collected and soon evidence accumulated that anaphylatoxin is neither a split product of antigens (BORDET, 1913) nor the toxic principle of anaphylaxis. Some authors therefore called it "serotoxin", a name which has, however, rarely been used in the more recent literature. The second phase began when anaphylatoxin was recognized as a potent histamine liberator in guinea pigs. This finding again seemed to indicate a relation to anaphylaxis, the relation again being disproved, subsequently. The most recent, third phase which still continues provided the resolution of the anaphylatoxin forming system into an enzyme and its substrate, and new facts about the nature and action of anaphylatoxin. The physiological or pathological significance of this extremely potent substance is still unknown.

Except for the two following paragraphs of the introduction the work of the two earlier phases of research will be mentioned briefly only inasmuch as it seems necessary for the understanding of recent results. The older literature has been reviewed by GIERTZ and HAHN (1957, 1961, 1966). Their most recent review is also the most comprehensive. In the present critical survey, mainly the recent development will be dealt with. In order to keep the reader as up to date as possible, various as yet unpublished new findings will be reported, and even some preliminary results will be mentioned.

From the pioneer work of FRIEDBERGER, BORDET, NOVY and DE KRUIF and others (see GIERTZ and HAHN, 1966) anaphylatoxin can be described as a pharmacologically active compound which forms when serum or plasma of guinea pigs or rats is "activated" by contact with immune aggregates, polysaccharides like starch, inulin, dextran, or simply by dilution. Its main action is to produce an anaphylactoid syndrome in guinea pigs, including severe bronchospasm which may lead to death. Serum from other species does not develop a toxicity of comparable strength, and other animals are not as susceptible to the effects as are guinea pigs. The active compound shows properties of a protein, being precipitated by alcohol or ammonium sulfate destroyed by heat and non-dialysable.

In 1950 HAHN and OBERDORF found that the lethal effect of anaphylatoxin was prevented by the injection of antihistaminics. This suggested the possibility that the effects were mediated by histamine liberation. ROCHA E SILVA and ARONSON (1952) then proved directly that anaphylatoxin liberated histamine from perfused organs of guinea pigs. As a further indication of histamine liberation was taken the finding that anaphylatoxin contracted the isolated guinea pig ileum and that this effect was subject to tachyphylaxis (ROTHSCHILD and ROCHA E SILVA, 1954).

This is a sketch of anaphylatoxin as it can be drawn from the uniform results obtained during the first two phases of research. No such uniform views had developed at that time concerning the formation, and it will be seen that in the light of recent results, even some of the actions of anaphylatoxin will have to be interpreted in another manner.

I. Formation of anaphylatoxin
Evidence for an enzymic mechanism

The now classical way of producing anaphylatoxin is to incubate rat or guinea-pig serum or plasma with a contact agent at 37^0 C for about an hour As the addition of immune aggregates is not essential it is clear — and all investigators have agreed on this, except for FRIEDBERGER in his very first interpretations — that the precursor of anaphylatoxin is a plasma constituent. It is also generally accepted that the activating effect of the "contact agents" is to alter some plasma protein by physical interaction with their surface. This theory goes back to RITZ and SACHS (1911). Whether this physical effect *per se* leads to the formation of anaphylatoxin, by forming toxic aggregates or denatured toxic proteins, or whether the physical interaction activates an enzyme which then produces anaphylatoxin from a substrate, remained unsettled until recently (for details about the various older theories see the review of GIERTZ and HAHN, 1966). GIERTZ et al. (1956; also GIERTZ and HAHN, 1961) separated from rat serum two components both of which were necessary for anaphylatoxin formation. The two components were present in Cohn fractions III-0 and I-III-1,2,3, respectively. Neither of them seemed to act catalytically, but each limited the yield of anaphylatoxin. Therefore the authors considered an enzymic mechanism of formation unlikely. In their opinion anaphylatoxin was rather the result of a stoichiometric reaction of physical or chemical nature between the two components. This view finds support in earlier filtration experiments (cited by GIERTZ and HAHN, 1966, p. 527) of other authors, which seemed to indicate that anaphylatoxin was a larger molecule than its precursor(s).

Other investigators presented facts in favour of an enzymic mechanism of formation of anaphylatoxin. Interest concentrated largely on proteases, which had already been considered by JOBLING and PETERSEN (1914). These authors found that treatment of serum with anaphylatoxin-forming contact agents reduced the antitrypsin titer. ROTHSCHILD and ROCHA E SILVA (1954) showed that the rate of formation of anaphylatoxin is correlated with the temperature of incubation to an extent that suggests an enzymic action. Also the heat-lability of the anaphylatoxin-forming system was taken as indicating the participation of an enzyme. However, all attempts to demonstrate an increase in proteolytic activity accompanying the formation of anaphylatoxin were unsuccessful (ROCHA E SILVA, 1953; ROCHA E SILVA and ARONSON, 1952; ROTHSCHILD and ROCHA E SILVA, 1954). Ungar, on the other hand, observed a parallelism between the development of the activity of a protease and that of anaphylatoxin. The protease investigated was not related to plasmin (UNGAR et al., 1961).

In favour of an enzymic mechanism is also the finding of OSLER et al. (1959) that a correlation exists between the fixation of complement, inactivation of C'3 and formation of anaphylatoxin, in rat serum. Chemicals known to destroy or to inhibit one or the other factor of complement also abolish the ability of the serum to develop anaphylatoxin activity.

Apart from the finding that the formation of anaphylatoxin is highly dependent on temperature — which indicates a chemical process — all other hints as to an enzymic mechanism of formation were up to this stage based only on parallel development or suppression of enzyme activities and anaphylatoxin formation. No direct evidence was obtained that the enzyme under consideration was essential, and attempts to obtain such evidence were largely negative (see section IV). Recently, however, the enzymic formation of anaphylatoxin was demonstrated directly. An enzyme was found that produced anaphylatoxin in samples of rat plasma which had been pretreated thereby abolishing the ability to form the active product on incubation with contact agents alone. The abolition was achieved by treatment of the plasma with ammonia or hydrazine in the manner used for the preparation of R_4-plasma, or by gel filtration. The enzyme was found first in cobra venom, later it was obtained also from rat plasma (VOGT, 1963, 1964; VOGT and SCHMIDT, 1964, 1966a and b).

The active constituent of cobra venom (Naia naia) has been purified (see section II). It is not an activator like contact agents, because it forms anaphylatoxin in rat plasma fractions which represent incomplete systems being unable to form anaphylatoxin on incubation with e.g. dextran or immune precipitates. That it is an enzyme is concluded from the fact that it acts catalytically. The yield of anaphylatoxin depends on the amount of rat plasma (fraction) used, not on the amount of cobra venom (fraction). Thus the plasma contains the substrate, called anaphylatoxinogen, which is acted upon by the enzyme contained in cobra venom. It seems logical further to conclude that in intact plasma anaphylatoxinogen is converted to anaphylatoxin by an endogenous enzyme which corresponds to the cobra venom enzyme. This enzyme should be present in plasma in an inactive state being activated by contact agents. After treatment of rat plasma with ammonia or other carbonyl reagents, this enzyme would be lost.

These conclusions are valid only when it is shown that a corresponding product is formed from the same precursor in both reactions, the formation of

11*

anaphylatoxin by cobra venom and by contact in whole plasma. To check the identity of the product, both anaphylatoxins have been compared biologically and chemically. Qualitatively, their action on the isolated guinea-pig ileum shows the same characteristics: a quickly developing contraction, tachyphylaxis and inhibition of contraction by antihistaminics. Further, there is cross-desensitization between the two anaphylatoxins (VOGT and SCHMIDT, 1964; STEGEMANN, VOGT and FRIEDBERG, 1964). Quantitatively, some difference in activity becomes apparent when the two preparations are compared using various isolated organs. Whereas the index of discrimination (GADDUM, 1955) is near 1 in most pairs of tests, it is 9 in the pair rabbit duodenum/hamster colon (POPPE and VOGT, in preparation). Chemically, no difference between anaphylatoxins formed by contact or by cobra venom was observed as regards their behaviour in chromatography, gel filtration and amino acid composition, this even though for cobra venom another substrate was used — pig plasma (STEGEMANN, BERNHARD and O'NEIL, 1964; STEGEMANN, HILLEBRECHT and RIEN, 1965).

In both processes of anaphylatoxin formation the same substrate is used. When rat plasma has been activated with a contact agent for one hour, further incubation with cobra venom does not produce more anaphylatoxin activity, and *vice versa*. Moreover, no anaphylatoxinogen is found in the plasma of rats which have been injected with purified cobra enzyme fraction 30 min to 24 hours prior to the experiment (VOGT and SCHMIDT, 1966b).

These observations indicate that cobra venom acts on the same anaphylatoxinogen which, in whole plasma, can otherwise be converted to anaphylatoxin by contact activation. The two anaphylatoxins obtained by the conversion are, so far, not chemically distinguishable. They have the same biological actions with some quantitative differences, which may indicate slight differences in structure, if impurities do not account for the variations. It seems justified, then, to conclude that the presumed endogenous plasma constituent which induces anaphylatoxin formation when activated by contact — and which is lost after treatment with ammonia and other carbonyl reagents — acts by the same mechanism as the cobra factor does, i.e. by an enzymic reaction with anaphylatoxinogen. This assumption has been corroborated by later experiments which led to the preparation of the plasma enzyme (see section II. C.).

When the effect of contact agents is to activate a plasma enzyme it should be possible to resolve the process of anaphylatoxin production in contact-activated plasma into two steps: one of enzyme activation followed by the second one, action of the enzyme on anaphylatoxinogen. In the experiments of GIERTZ and HAHN (1961) two plasma fractions and the contact agent had to be present simultaneously to induce anaphylatoxin formation, and attempts to split the process were unsuccessful. This was again taken as indicating a non-enzymic mechanism, e.g. aggregation of two proteins in the presence of suitable surfaces. VOGT and SCHMIDT (1966a and b) recently succeeded in separating the two steps. They incubated rat plasma with Sephadex as contact activator, in the presence of EDTA which prevents the formation of anaphylatoxin (OSLER et al., 1959). After the incubation the Sephadex was removed, then the EDTA was bound by adding $CaCl_2$. New incubation of the mixture for 30 min at 37^0 C in plastic tubings without any contact agent present, led to the development of anaphylatoxin activity. Apparently, the enzyme had been activated by contact during the first incubation, whereas during the second one it acted on its substrate.

II. The components of the anaphylatoxin system

A. The anaphylatoxin-forming enzymes

1. The cobra enzyme. VOGT and SCHMIDT (1964) purified the anaphylatoxin-forming enzyme of cobra venom (Naia naia) by ion exchange chromatography on DEAE-Sephadex A 25, followed by gel filtration on Sephadex G 100. This separated the enzyme from the toxins of low molecular weight, from the directly haemolyzing principle, phospholipase A, as well as from proteolytic enzymes.

Purified enzyme preparations produce anaphylatoxin when incubated with rat plasma at final concentrations equivalent to as low as 5×10^{-7} crude venom, for 15 min at 37^0 C. Such high activities have been obtained from "crystalline" Naia naia samples (air dried), whereas lyophilized venoms are less active, probably because the enzyme is sensitive to freezing. It is not active below pH 5.5. It does not split casein, tosyl-arginine-methylester, lysine-methylester, acetyltyrosine-ethylester nor benzoyl-arginine-nitroanilide, at concentrations up to 100 times higher than are necessary for anaphylatoxin formation. Aniline and hydroxylamine inactivate the enzyme, EDTA causes partial inhibition. Heating of the enzyme solutions to 60^0C for 30 min leaves some residual activity, but at 100^0 C the enzyme is destroyed. It is also inactivated irreversibly, when the solution is acidified to below pH 3 (VOGT and SCHMIDT, 1964, 1966b).

The purified preparations are not markedly toxic. Mice which die after intravenous injection of 35 µg crude venom per 100 g body weight survive without toxic signs an equivalent dose of the purified enzyme fraction. Rats show nothing more than pronounced swelling of the nose and ears after obtaining equivalents to 5 mg crude venom/100 g (VOGT, unpublished observations). Guinea pigs also tolerate well injections of the enzyme fraction. Only when very high doses are given intravenously (corresponding to 10 mg crude venom/100 g) do they die a few minutes later with signs that seem not to be related to anaphylatoxin formation — predominantly lung oedema (BODAMMER, to be published; see also section V.). When injected into rats the presence of the cobra enzyme can be demonstrated in the plasma even after 24 hours (Vogt and Schmidt, 1966 b).

Besides from rat plasma the cobra enzyme releases anaphylatoxin also from the plasma of pig, guinea pig and, not regularly, horse; but not from plasma of man, ox, rabbit, dog, sheep, frog and eel (Vogt and Schmidt, 1964; Schwoerer, 1966). These latter species apparently do not contain measurable amounts of anaphylatoxinogen. They can also not be activated to form anaphylatoxin by contact, which is a further indication that anaphylatoxin formation by the cobra enzyme and by endogenous mechanisms are identical processes.

2. *The plasma enzyme.* Efforts to obtain purified anaphylatoxin-forming enzyme preparations from rat plasma have met with considerable difficulties. A procedure that has been used successfully several times is the following (Vogt and Schmidt, 1966a and b): rat plasma is dialyzed against 0.067 M phosphate buffer, pH 6.0 and is then passed with the same buffer through a column of Biogel P 100. Whereas anaphylatoxinogen appears in the effluent containing the main peak of proteins, the enzyme leaves the column later. There is, however, in most instances some overlap and considerable loss of activity. The fractions which contain the enzyme without substrate contamination are collected and concentrated by ultrafiltration, with agitation (to prevent occlusion of the membrane pores). The concentrate then has a turbid appearance and after centrifugation at $30000 \times g$ the anaphylatoxin-forming activity is found in the residue, not in the supernatant. These preparations contain the enzyme in an already active state. They act catalytically on their substrate like cobra venom (Vogt and Schmidt, 1966a and b).

The (pre-)enzyme behaves like a globulin. When plasma is diluted with water and the pH is lowered to 5.2, the enzyme is precipitated with the euglobulins. However, only rarely is its activity recovered at all. As a rule, it is destroyed, probably because of denaturation.

The interpretation of Jones (1960) that fractionation of rat serum into an euglobulin precipitate and a supernatant separates two factors which are essential for anaphylatoxin formation, is unlikely. Anaphylatoxinogen is also precipitated under the conditions used. His failure to recover a complete anaphylatoxin-forming system from the euglobulins has more likely been due to denaturation of the enzyme (and some loss of substrate as well), than to separation of two components No successful recombination experiments have been reported in the paper of Jones, in the author's own laboratory the supernatant contained neither enzyme nor substrate.

The activated plasma enzyme is more labile to heat than the cobra enzyme, being destroyed at temperatures above 56° C. It is also inactivated by freezing, and when exposed to solutions of pH below 3. In plasma its activity decays in the course of hours. In contrast, the inactive pre-enzyme as present in fresh rat plasma is stable to freezing and does not disappear on standing for days. Heat and acid destroy it, as the active enzyme.

Carbonyl reagents such as ammonia, aniline, hydrazine or hydroxylamine inactivate the plasma enzyme. The inactivation by ammonia can be reversed simply by bringing the pH of the solution down to pH 5—6. EDTA blocks the activity, an effect which can be reversed by addition of $CaCl_2$. Other chelating agents (8-hydroxy-quinoline, o-phenanthroline) are ineffective. Soybean trypsin inhibitor or ε-amino-caproic acid likewise do not inhibit anaphylatoxin formation (VOGT and SCHMIDT, 1966b; to be published). BECKER (1959) found DFP not to interfere with the production of anaphylatoxin in rat plasma. He left open the possibility that the process was activated too quickly to allow DFP to interact. In view of this, VOGT and SCHMIDT (1966b) treated the preactive enzyme with DFP and added substrate only 4 hours later. Also under this condition DFP showed no inhibitory effect. This finding is not in accord with results of BECKER (1962) according to which another alkyl-phosphate ester did block the formation of anaphylatoxin.

The anaphylatoxin-forming enzyme is rather ubiquitous in the animal kingdom. Apart from the plasmas of species known to produce anaphylatoxin by contact activation (guinea pig and rat), it has been found also in human plasma (VOGT, 1963) and in plasma of many animals (ox, dog, sheep, pig, rabbit, chicken, eel) (SCHWOERER, 1966). In fact, horse and frog have been the only species investigated in which the presence of an anaphylatoxin-forming enzyme was not demonstrated. Human plasma induces anaphylatoxin formation in anaphylatoxinogen-containing solutions without any addition of contact agents. However, the presence of, e.g., Sephadex enhances its activity. Apparently, human plasma contains the enzyme partly in a preactive state (VOGT, unpublished).

Of the plasmas investigated by SCHWOERER, only pig and chicken plasmas are able to produce anaphylatoxin on contact activation. The other species do not contain measurable quantities of anaphylatoxinogen. Pig plasma was originally considered to be a pure substrate donor (VOGT and SCHMIDT, 1964). SCHWOERER, however, succeeded in producing anaphylatoxin by incubating pig plasma for at least one hour at 37^0 C with Sephadex as contact agent. Occasional failure has probably to be explained by the presence of an inhibitor in some specimens of pig plasma. The inhibition can be demonstrated by adding pig plasma to rat plasma and incubating the mixture with a contact agent. Anaphylatoxin formation is then sometimes retarded or even prevented (VOGT, unpublished).

B. The substrate, anaphylatoxinogen

The substrate for anaphylatoxin production has been found in the plasma of rat, pig, guinea pig, horse and chicken. Of these, the horse is the only species which does not develop anaphylatoxin activity because of lack of enzyme (SCHWOERER, 1966).

Anaphylatoxinogen has been obtained free of enzyme activity by two different ways — preparative purification or selective destruction of the enzyme in whole plasma. STEGEMANN, VOGT and FRIEDBERG (1964) have prepared anaphylatoxinogen from rat and pig plasma by gel filtration through Sephadex G 100, at 4⁰ C. No activation of the system occurs under these conditions. The substrate is eluted early in the globulin-containing fractions. From these a stable preparation can be obtained by dialysis and lyophilisation. In the early experiments of STEGEMANN et al. (1964) the anaphylatoxinogen thus obtained was free of enzyme activity, i.e. it did not form anaphylatoxin on incubation with contact agents. The enzyme was not recovered from the gel columns. However, in recent experiments the anaphylatoxinogen was contaminated with anaphylatoxin-forming enzyme (VOGT and SCHMIDT, 1966b). The reason for the difference of the results has not been found out. A more reliable purification of anaphylatoxinogen was achieved by gel filtration of rat plasma through Biogel P 100, in 0.067 M phosphate buffer pH 6.0 (VOGT and SCHMIDT, 1966b).

It is possible to inactivate the anaphylatoxin-forming enzyme of rat plasma selectively by preparing R_4 plasma, i.e by treating the plasma with carbonyl reagents (see section II. A. 2.). R_4 plasma is a suitable reagent for the detection of enzyme activity, containing only the substrate. Care must be taken, however, not to reactivate the enzyme. This will happen, when the R_4 plasma has been prepared with the use of ammonia and is adjusted afterwards to pH 5—6 (VOGT and SCHMIDT, 1966b).

Anaphylatoxinogen does not dissolve in distilled water but is soluble in dilute salt solutions (STEGEMANN, VOGT and FRIEDBERG, 1964). It is precipitated from plasma with the euglobulin fraction when the plasma is diluted with water and its pH adjusted to 5.2. This method is not very suitable for the preparation of substrate as considerable losses are encountered. Freezing and lyophilisation do not cause damage to anaphylatoxinogen but after its solution is heated to 56⁰ C for 30 min, it no longer releases anaphylatoxin after incubation with the cobra or plasma enzyme (STEGEMANN, VOGT and FRIEDBERG, 1964; VOGT and SCHMIDT, 1966b).

C. The product, anaphylatoxin

1. Preparation. All former studies of the actions of anaphylatoxin have been undertaken with whole activated serum of plasma. No purified preparations were available. HAHN and LANGE (1955) fractionated anaphylatoxin-containing rat serum by one of the procedures of Cohn. They found the active principle in fraction III-0. The activity per mg protein had increased 3-fold. STEGEMANN, VOGT and FRIEDBERG (1964) purified anaphylatoxin by gel filtration on Sephadex G 100. When activated rat or pig plasma was passed through the columns, the active principle appeared in the eluate shortly after the bulk of proteins. After concentrating the anaphylatoxin-containing fractions, by dialysis and lyophilisation, and repeating the gel filtration, preparations were obtained which contracted the guinea pig ileum in concentrations of 10^{-6} and sometimes less.

Recently another method has been developed which is suitable for large scale preparation of anaphylatoxin. Four liters of pig serum are activated with Sephadex, at 37^{0} C, for about 6 hours. The solution is centrifuged, the supernatant diluted with 4 liters of water, adjusted with formic acid to pH 4.5 and stirred twice with 12 g each time of carboxy-methyl cellulose powder. The exchanger is collected and eluted, at 4^{0} C, by stirring it with 0.5 M ammonium formiate buffer, pH 8.2. Immediately afterwards the eluate is acidified to pH 4—5 and concentrated by dialysis and lyophilisation. It is then passed through a large column of Sephadex G 100 with 0.02 M acetate buffer, pH 5.6. The active material comes off the column well after the bulk of other proteins. When freed from buffer salts and dried, preparations have been obtained which contract the isolated guinea-pig ileum in concentrations of 2 to 4×10^{-7}, and are lethal to guinea pigs when injected intravenously in amounts of 20 µg/100 g body weight. The preparations seem to be fairly pure. Traces of impurities are seen after electrophoresis in Cellogel strips (VOGT, to be published).

2. Nature and properties. The behaviour of anaphylatoxin in ion exchange chromatography reveals that it is a markedly basic compound. Further, there is no doubt that it has a peptide- or protein-like structure. Proteolytic enzymes readily destroy its activity (STEGEMANN, VOGT and FRIEDBERG, 1964; for older literature see GIERTZ and HAHN, 1966). Attempts to estimate the molecular size have given contradictory results. From the inability of anaphylatoxin to pass through dialysis membranes all former investigators have concluded that it is a protein. However, it does dialyze through Visking tubing No. 27/100 though with difficulty (VOGT, unpublished observations). According to gel filtration the molecule seems to be much smaller than the common plasma proteins, including anaphylatoxinogen (VOGT, 1963; STEGEMANN, VOGT and FRIEDBERG, 1964). This suggests a peptide split from its protein precursor. On the other hand, the retardation observed on Sephadex columns could be due to adsorption rather than to small size. In preliminary ultracentrifugation experiments a molecular weight of about 30000 was considered likely (STEGEMANN, HILLEBRECHT and RIEN, 1965). This estimate is in accord with the value found by analysis of terminal amino acids (STEGEMANN, BERNHARD and O'NEIL, 1964).

After acid hydrolysis 17 different amino acids have been identified (STEGEMANN, HILLEBRECHT and RIEN, 1965). For these analyses only a very small quantity of material was available, so the figures obtained may have to be regarded as not yet definitive. As terminal amino acid STEGEMANN et al. (1965) found arginine only. Cystine was not found among the amino acids. Recent experiments indicate, however, that disulfide bridges are essential for the activity of anaphylatoxin. Performic acid as well as thioglycollate destroy more than 90 % of the biological activity on isolated organs. N-ethyl-maleimide has no effect (Vogt, to be published).

The purest preparation of Stegemann had a rather low content of nitrogen. A considerable part of the compound consisted of carbohydrate, probably mannose among other, unidentified sugars (Stegemann et al., 1965). It seems remarkable in this respect that periodic acid does not destroy the biological activity (Lyncker and Vogt, 1966). This indicates that either the carbohydrate is not a component of the molecule, is not essential or that it is linked in such a way as to form no glycol groups.

Former investigators observed already that anaphylatoxin is more stable to heat than its precursor. The stability is remarkable indeed, at acidic pH. In contrast, anaphylatoxin is destroyed even at 40° C when exposed to pH 11 (Stegemann, Vogt and Friedberg, 1964). In whole plasma, *in vitro*, it is stable for at least 24 hours at 37° C. In vivo, in rats, it disappears from the circulation in less than one hour (Vogt and Schmidt, 1966b).

Anaphylatoxin, crude or purified, is precipitated from aqueous solution by addition of 5 to 10 volumes of alcohol (Rocha e Silva, 1954; author's own observations). It is readily soluble in distilled water. Freeze drying, urea or formamide do not inactivate it, suggesting that it has no essential tertiary structure (Stegemann et al., 1965).

There are no gross differences apparent in the chemical structure of ana-phylatoxins from rat and pig plasma (Stegemann, Vogt and Friedberg, 1964; Stegemann et al., 1965). However, slight variations are not unlikely since Poppe and Vogt (to be published) found some quantitative differences in the biological actions of anaphylatoxins prepared by contact activation and by the cobra enzyme. These results will have to be reinvestigated with more purified preparations to exclude the possibility of impurities altering the responses of test organs.

III. Activation and inhibition
of the anaphylatoxin-forming enzyme

A. Activation

Besides immune precipitates and some inorganic compounds like kaolin, predominantly polysaccharides have been used as contact agents to activate the anaphylatoxin formation. Starch, inulin, dextran, Sephadex or cellulose, all being neutral polymers, can be used. Acidic congeners like dextran sulfate or heparin, are inactive as contact agents. Simple dilution of the plasma with water or dialysis against hypotonic solutions also triggers the production. Dilution with saline has no effect, whereas isotonic glucose solution acts like distilled water. Clearly, lowering the ionic strength of plasma is one possibility of activating the system. Some of the contact agents might also have an effect on salt concentration, and it is tempting to assume that the anaphylatoxin-forming enzyme is preactive in plasma, being inhibited only by high ionic strength. ROCHA E SILVA (1954) suggested that negative groups of poly-saccharides would interact with cations of the plasma, thus lowering the ionic strength. This explanation is unlikely in view of the fact that various acidic polysaccharides do not activate the anaphylatoxin-forming system (see GIERTZ and HAHN, 1966, p. 521). Of agar it is the neutral fraction which activates the anaphylatoxin-forming system, the acidic fraction being inactive in this respect (LYNCKER and VOGT, 1966). The experiments of HAINING (1955, 1956a) which have been interpreted as indicating a positive effect of dextran sulfate, do not deal with anaphylatoxin formation as this is absent from rabbit blood (SCHWOERER, 1966).

Some of the contact agents might lower the salt concentration surrounding the enzyme by acting as molecular sieves. However, even the best sieves, e.g. Sephadex, do not show any impairment of their activating effect when they have been equilibrated with physiological saline before addition to the plasma. Under these conditions they cannot take up further ions from the plasma. Moreover, the purified, active plasma enzyme does convert its substrate in a medium of physiological saline or Tyrode solution (VOGT, unpublished). It is therefore more likely that the enzyme is present in fresh plasma as an inactive pre-enzyme, which is converted to the active state by contact with foreign surfaces *or* by lowering the ionic strength of the environment.

The conversion may be a purely physical process, e.g. some sort of denatura-tion. It is remarkable, however, that EDTA not only inhibits the effect of the enzyme but interferes also with its activation, though not as effectively. Whereas contact of rat plasma with Sephadex produces an active enzyme in the course of a few minutes, no or only slight activity is obtained after even 4 hours when EDTA is present. This is evident from the fact that after such incubation and subsequent removal of the contact agent the plasma, on re-calcification, does not readily form anaphylatoxin. In the presence of EDTA it is well activated only when the contact incubation is carried on for about 20 hours (VOGT and SCHMIDT, 1966b). The interference of EDTA with the activation of the enzyme indicates that divalent cations somehow facilitate the process. This may indicate the participation of another enzyme in the activation.

B. Inhibition

Various ions block the formation of anaphylatoxin when present in sufficient concentrations in the rat plasma to be activated (ROTHSCHILD and ROCHA E SILVA, 1954; GIERTZ and HAHN, 1961). NaCl is inhibitory at twice the physio-logical concentration (0.35 M). Whether the inhibiting effect is related to the ionic strength only, in all cases, is doubtful. Citrate at least seems to specifi-cally cause some irreversible damage to the anaphylatoxin-forming system

(ROTHSCHILD and ROCHA E SILVA, 1954), an effect which is not seen with other complexing anions. It is further unknown whether the salt-induced inhibition is directed against the activation or the action of the plasma enzyme. The (preactive) cobra enzyme is inhibited by NaCl at concentrations of 1.1 M.

A peculiar behaviour is shown by EDTA. It blocks anaphylatoxin formation when it is added to rat plasma in amounts sufficient to bind all divalent cations (5.4—10 mM) (OSLER et al., 1959; VOGT, 1963; VOGT and SCHMIDT, 1966b). When plasma containing 5 4 mM EDTA is diluted with isotonic glucose solution (9 volumes) anaphylatoxin formation ensues. This is prevented when the solution used for dilution also contains 5.4 mM EDTA (VOGT, unpublished). Consequently, the inhibiting effect of EDTA is not due to the complexing with free divalent cations, but is a function of its own concentration. This suggests that EDTA forms a complex with the anaphylatoxin-forming enzyme itself, which complex is much more dissociable than are pure metal chelates. The following finding is in support of this conclusion. In EDTA-containing rat plasma the ability to form anaphylatoxin can be restored simply by ultrafiltrating the plasma and washing it with saline. No substitution of any of the divalent cations lost is necessary (VOGT, 1963; VOGT and SCHMIDT, 1966b). Apparently, EDTA combines with the enzyme forming an inactive, loose complex which dissociates on dilution. It has been concluded that the enzyme is a metal protein with which EDTA can combine forming a ternary complex. A similar complex formation has been reported between a zinc containing enzyme, lactic dehydrogenase, and EDTA (CREMONA and SINGER, 1962).

The conclusion that one is dealing with a metal enzyme reconciles the diverse earlier findings regarding the role of Ca-ions. OSLER et al. (1959) using EDTA as Ca-chelating agent considered Ca^{++} to be essential, whereas ROTHSCHILD and ROCHA E SILVA (1954) found no evidence for this, as they removed Ca^{++} from the plasma by ion exchange. The latter procedure would not touch the enzyme-bound metal, indicating clearly that free divalent cations are not essential for anaphylatoxin formation to occur. This is corroborated by results of GIERTZ and HAHN (1961) who were able to induce the formation in rat serum fractions free of Ca^{++}, and by the author's own findings with the purified plasma enzyme.

The chelating agents, o-phenanthroline, dithizone and 8-hydroxy-quinoline do not block the formation of anaphylatoxin (VOGT and SCHMIDT, 1966b). The effect of oxalate is not quite clear. GIERTZ and HAHN (1961) found no inhibition. From their figures given for whole blood one can estimate that they used oxalate at a final concentration of about 37 mM in plasma. ROTHSCHILD and ROCHA E SILVA (1954) on the other hand reported that 0.04 N oxalate (20 mM) did inhibit the generation of anaphylatoxin activity.

The cobra enzyme appears to be less sensitive to the blocking effect of EDTA than is the plasma enzyme. In 5.4 mM EDTA its action is reduced but not prevented (VOGT and SCHMIDT, 1966b). Even in 16 mM EDTA it is not completely blocked.

As has already been mentioned, carbonyl reagents inactivate the anaphyla-toxin forming enzymes (see p. 167). This raises the question whether these enzymes contain pyridoxal-5-phosphate (P-5-P) as a coenzyme. Attempts to cleave the presumed cofactor and to obtain an apoenzyme by dialysis under various conditions have been unsuccessful (VOGT, unpublished). In rats which have been living for several weeks on a diet free of vitamin B^6 and containing the antagonist, desoxy-pyridoxine, anaphylatoxin-forming activity is still demonstrable. In contrast to these negative findings, free P-5-P (4 mM) inhibits the formation of anaphylatoxin (VOGT and SCHMIDT, 1966b). If this effect should be proved to be competitive, then it might indicate that the enzyme contains the same compound as an essential cofactor. The inhibitory effect of P-5-P is not related to its chelating property. It is specific in so far as other, similar substances (pyridoxamine phosphate, pyridoxim phosphate, pyridoxal) are inert (VOGT and SCHMIDT, 1966b).

The well-known protease inhibitors DFP (BECKER, 1959, VOGT and SCHMIDT, 1966b) and soybean trypsin inhibitor (ROTHSCHILD and ROCHA E SILVA, 1954; VOGT, unpublished observations) do not interfere with the formation of anaphylatoxin. Also ε-aminocaproic acid is inactive.

Various unrelated compounds (heparin, basic dyes, hydrocortison; see GIERTZ and HAHN, 1966, p. 521, 523) inhibit the formation of anaphylatoxin by mechanisms not understood.

IV. Nature of the enzymic reaction that leads to the formation of anaphylatoxin

A. The role of complement

Though the original concept of FRIEDBERGER about the origin and develop-ment of anaphylatoxin activity later proved to be wrong, his assumption that complement plays an active part in the formation has been accepted and supported by some recent investigators, in particular by OSLER et al. (1959) and by JONES (1960), though not by others.

ROTHSCHILD and ROCHA E SILVA (1954) as well as GIERTZ and HAHN (1957, 1961) found Ca-ions not to be essential for anaphylatoxin formation. This would rule out complement being involved. The two rat serum fractions of GIERTZ and HAHN which are essential for anaphylatoxin formation were nearly free of complement factors, except for C'3. It is this component, however, which OSLER et al. (1959) considered to be the most important one for anaphylatoxin formation, so that the system of GIERTZ and HAHN may not have been sufficiently free of complement factors to exclude their participation. OSLER et al. demonstrated that many parallels existed between the activation or consumption of complement factors and formation of anaphylatoxin.

Processes which destroyed one or the other component of complement also prevented anaphylatoxin production. *Vice versa*, when anaphylatoxin activity *did* develop, complement factors, particularly C'3, were consumed simultaneously. These findings are, indeed, very suggestive, but the conclusions are not supported by recent results concerning the properties of the anaphylatoxin-forming enzyme and its substrate, obtained by VOGT and SCHMIDT (1966b). It should be kept in mind that neither the reactions used for destruction of single components of complement nor the procedures of activation of complement or anaphylatoxin formation are specific. The parallelism between complement and anaphylatoxin activations observed by OSLER et al. (1959) may well be accidental. Whereas this statement does not exclude the possibility of complement being involved, there are some clear differences between the two reactions. As mentioned earlier, the anaphylatoxin-forming enzyme is a metal protein which does not depend on free divalent cations, being active even in the presence of low concentrations of uncomplexed EDTA. It can be reactivated after having been blocked by ammonia whereas complement is inactivated irreversibly by this treatment.

Rat serum which had been incubated with ammonia for 90 min at 37° C, was acidified afterwards to pH 5, kept at this pH for about half an hour and was neutralized, subsequently. It proved to be fully active, then, when checked for anaphylatoxin formation by contact activation In contrast, when used as source of complement in an immune hemolysis test, the same sample was entirely inactive, even when used undiluted (VOGT, unpublished).

There is one exception, at least, from the parallelism between anaphylatoxin formation and complement consumption: guinea-pig serum forms anaphylatoxin when activated with zymosan at 17° C, without any loss of complement activity (WILLEMS, 1963). Finally, the anaphylatoxin-forming enzyme is clearly not inhibited by DFP, a potent C'1 inhibitor (BECKER, 1956; VOGT and SCHMIDT, 1966b). It seems, therefore, fairly safe to state that complement is not involved in the action of the anaphylatoxin-forming enzyme on its substrate, and that neither of the two principles can be identified with any complement factor.

As regards the activation of the anaphylatoxin-forming enzyme, the possibility of C'1 being involved cannot be rejected with the same certainty. Though the activation may well be a purely physical mechanism, the fact that EDTA inhibits the activation may indicate the participation of the first component of complement*. The finding of ROTHSCHILD and ROCHA E SILVA (1954) that rat serum after passage through a cation exchange column is able to form anaphylatoxin, would speak against this. However, the possibility still exists that in these experiments the enzyme had already been activated during the passage, while Ca-ions were still available. More work is needed to clarify the possible role of C'1 in the activation of the anaphylatoxin-forming enzyme.

* See Note 1 on p 184

B. Other possible reactions

What else could the reaction which leads to anaphylatoxin formation consist of? It does not seem likely that the enzyme splits an ester or peptide bond, comparable to the actions of plasmin, the kallikreins or trypsin — in spite of the increase in proteolytic activity which UNGAR et al. (1961) have observed during anaphylatoxin formation. Synthetic substrates for these hydrolytic enzymes are not utilized by the cobra enzyme, and inhibitors like DFP, soybean trypsin inhibitor or ε-aminocaproic acid are ineffective. On the other hand, a proteolytic action cannot be denied on these grounds as there are various proteases and peptidases known the action of which depends on metal groups and is inhibited by chelating agents, but not by the inhibitors mentioned before. E.g. the proteases of *Naia naia* do not split tosyl-arginine-methylester (MURATA, SATAKE and SUZUKI, 1963, TU, PASSEY and TU, 1966), they are inhibited by EDTA (author's own observations). The anaphylatoxin-forming enzyme of cobra venom is, however, different even from these proteases (VOGT and SCHMIDT, 1964).

As the anaphylatoxin-forming enzyme contains essential carbonyl-groups — being inhibited by carbonyl reagents — and bound metal, the idea is suggestive that it is a pyridoxal-phosphate enzyme, like decarboxylases etc. Decarboxylation of a protein or peptide would give rise to a product which is more basic than its precursor. In fact this is true for anaphylatoxin. However, at present this is mere speculation. Attempts to clearly demonstrate the participation of pyridoxal phosphate in the reaction have, so far, failed (see p. 173).

NETZER and VOGT (1964) put forth the hypothesis that the formations of anaphylatoxin and of endogenous pyrogen are related or identical reactions. However, later these two processes were separated and recognized as being independent from each other (LYNCKER and VOGT, 1966).

V. Actions of anaphylatoxin

During the first two phases of anaphylatoxin research the actions of this compound were studied only by using whole activated plasma or serum, mostly of rats or guinea pigs. Only recently, purified preparations have become available in sufficient quantities for pharmacological experiments. No major pharmacological differences between purified and non-purified samples are apparent, so far.

A. Effects on isolated organs and tissues

Apart from occasional observations of other authors, ROTHSCHILD and ROCHA E SILVA (1954) were the first to give a detailed description of the actions of anaphylatoxin on the isolated guinea-pig ileum. Hitherto, this is the most sensitive organ, contracting to 2×10^{-3} of activated rat plasma or 2×10^{-7} purified anaphylatoxin. The dose-response curve is extremely steep; often doubling the dose will increase the response from nil to maximal. Two further features characterize the effect on the guinea-pig ileum: tachyphylaxis and blockade by antihistaminics.

Other smooth muscle-containing organs such as intestines of rabbit, rat and hamster, uteri of guinea pig and rat, chicken amnion, or arterial strips are also contracted, though higher doses are necessary (FERLUGA, 1960; FRIEDBERG, ENGELHARDT and MEINEKE, 1964; VOGT and ZEMANN, 1964; POPPE and VOGT, to be published). They also show tachyphylaxis, though less in most cases than the guinea-pig ileum. It is noteworthy to mention that among these are several tissues which do not respond to histamine. The isolated atrium of rats and guinea pigs reacts to anaphylatoxin with an increase in contraction height and frequency (GREEFF et al., 1959; FRIEDBERG et al., 1964).

When perfused through isolated guinea-pig tissues, notably lung, anaphylatoxin liberates histamine (ROCHA E SILVA and ARONSON, 1952; FRICK, HALPERN and LIACOPOULOS, 1962). It does so, too, when added to a suspension of guinea-pig mast cells (MOTA, 1959). Surprisingly, rat tissues and mast cells do not respond in the same way (MOTA, 1957). In fact, anaphylatoxin is the only histamine liberator showing this preference for guinea-pig tissues.

ROTHSCHILD and ROCHA E SILVA (1954) concluded from their findings that anaphylatoxin contracts the guinea-pig ileum indirectly liberating histamine, and that the tachyphylaxis is the consequence of exhaustion of histamine stores. This explanation cannot be extended to organs like rat uterus or intestine as these tissues do not respond to histamine with contraction. The same basic mechanism could, however, apply also to histamine-insensitive organs when defined in a more general way: anaphylatoxin might liberate smooth muscle-contracting amines; in the case of rat organs the release of 5-hydroxytryptamine might be involved, in the case of guinea-pig organs it might be histamine. Both amines occur in the respective mast cells. However, apart from the insensitivity of rat mast cells to anaphylatoxin there are several findings which indicate that anaphylatoxin is not or not predominantly acting on smooth muscle *via* release of amines. First, rat intestine reacts to anaphylatoxin even when the donor animal has been pretreated with reserpine or when reserpine is present in the bath, i.e. when 5-hydroxytryptamine is excluded from participating in the effect (VOGT and ZEMANN, 1964). Further, the tachyphylaxis is reversible. Pieces of guinea-pig gut which have been made entirely unresponsive to even high doses of anaphylatoxin, given frequently, may well recover after intervals of 15—30 min (RANDALL, TALBOT, NEU and OSLER, 1961, FRIEDBERG et al., 1964, VOGT and ZEMANN, 1964). In other organs reactivity may be regained even more quickly (FRIEDBERG et al., 1964). It seems highly unlikely to the author and has been considered unlikely by FRIEDBERG et al. (1964) that during the short periods of rest histamine could be resynthetized in sufficient quantities [according to FRICK et al. (1962) refilling of stores takes days in guinea-pig lungs at least]. In the author's opinion anaphylatoxin acts rather directly as a smooth muscle stimulant of its own, especially because it is active in the guinea-pig uterus even when this organ is desensitized to histamine (VOGT and ZEMANN, 1964) or is pretreated with antihistaminic drugs (CONARD and MUTSAARS, 1949; VOGT and ZEMANN, 1964). In fact, the antagonism of antihistaminics to anaphylatoxin is restricted to the guinea-pig ileum, among all isolated organs investigated. Even here, the antagonism does not indicate necessarily the participation of histamine in the contraction. The dipeptide, histidyl-proline, which contracts the isolated guinea-pig ileum without producing any tachyphylaxis, i.e. certainly without acting by liberating histamine, is also antagonized by low doses of tripelenn-amine (VOGT and ZEMANN, 1964).

The phenomenon of tachyphylaxis then has to be explained in a different manner — certainly in rat organs and guinea-pig uterus, but probably also in the isolated guinea-pig ileum. FRIEDBERG et al. (1964) suggest a blockade of receptors by the drug which would dissociate from its point of attack only slowly. In view of the fact that disulfide groups are essential in the molecule of anaphylatoxin (see p. 169) one may consider that anaphylatoxin affects the oxydo-reduction potential in the tissue. Recovery from tachyphylaxis may require the reestablishment of the original status In fact, treatment of guinea-pig ileum with thioglycollate enhances the effects of anaphylatoxin and speeds up recovery from tachyphylaxis.

A few antagonists have been found, apart from antihistaminics. According to MOUSSATCHÉ and DANON (1956) p-chloro-mercuribenzoate blocks the contraction of the guinea-pig ileum. Slightly higher doses than are necessary for this effect also inhibit histamine-induced contractions. HAINING (1956b) found salicylic and 2-phenyl-cinchoninic acids inhibitory. The action on mast cells is antagonized by p-chloromercuribenzoate, iodoacetate and by phenol (MOTA, 1959). It remains to be determined, how far these compounds act specifically.

B. In vivo effects

The predominant effect seen when anaphylatoxin is injected intravenously into guinea pigs, is a pronounced bronchospasm. At autopsy the lungs appear inflated. Lung oedema has been observed occasionally; it is probably due to the injection of big volumes of fluid rather than to an action of anaphyla-toxin itself (KUMAGAI, 1913; for further literature see GIERTZ and HAHN, 1966). Purified preparations of anaphylatoxin from pig serum have never produced lung oedema (BODAMMER, unpublished).

BODAMMER and VOGT (1967) have recorded the effects of highly purified pig serum anaphylatoxin on circulation and respiration of guinea pigs. Under urethane anaesthesia a biphasic effect on blood pressure is observed with 100 µg/kg (about half the lethal dose): a short-lasting moderate fall followed by a rise. The fall is reproducible many times but the rise shows tachyphylaxis, it disappears entirely after a few injections. The second effect can be prevented by dihydro-ergotamine, and is apparently due to liberation of catecholamines. The initial fall in blood pressure is unaffected by tripelennamine, it probably represents a direct action of anaphylatoxin.

The actions on blood pressure do not appear to threaten the life of the animal, and the term "shock" is not really adequate to describe the actions of anaphylatoxin. Far more dramatic are the effects on respiration. A few seconds after the injection — during the hypotensive phase — the flow of air through the respiratory tract is moderately reduced during inspiration but practically stopped during the expiratory phase. This event is caused by strong bronchial constriction. It leads to progressive inflation of the lungs. When this has reached a certain degree, respiratory movements stop entirely for about half a minute. With release of the bronchospasm the lungs empty and respiration recovers unless the animal died during the period of obstruction.

The bronchial effect of anaphylatoxin is similar to that of histamine but unlike the latter it is subject to tachyphylaxis. Tripelennamine protects the animal from death. Recording the respiration has revealed, however, that tripelennamine is not able to entirely block the respiratory actions of anaphylatoxin but only reduces them, even if given in doses which completely suppress the effects of larger than lethal doses of histamine.

A similar antihistamine-resistant action has been observed by GIERTZ (cited by GIERTZ and HAHN, 1966, p. 534). This again indicates a direct component in the action of anaphylatoxin, independent of histamine liberation. The direct effect is reinforced by histamine liberation and, for reasons yet unknown, is subject to tachyphylaxis itself.

When tachyphylaxis to the bronchoconstrictor action has developed, another respiratory effect becomes visible: an increase in frequency and volume of breathing which is reproducible with each injection and which, each time, lasts for about one minute. This effect occurs also after the first injection of anaphylatoxin but is followed by bronchospasm so rapidly that it can hardly be observed.

A few experiments performed in guinea pigs under local anaesthesia have given the same results. On slow infusion of anaphylatoxin bronchospasm does not develop, but only the stimulation of respiration is seen.

From the earlier work on anaphylatoxin it is known that other laboratory animals do not react to anaphylatoxin with impressive signs (see GIERTZ and HAHN, 1966). MARQUARDT and HEDLER (1965) reported a hypotensive action in cats which was tachyphylactic. They used whole, Sephadex-activated rat serum as a preparation of anaphylatoxin. Non-activated control serum had, however, a similar, though less strong effect. Detailed investigations of anaphylatoxin in other animals are lacking.

VI. Assay of anaphylatoxin

Two actions have been used for assay of anaphylatoxin, the lethal effect in guinea pigs and the contraction induced in the isolated guinea-pig ileum. The estimation of the lethal dose requires rather large amounts of material and numbers of animals. Further, the effect depends very much on the speed

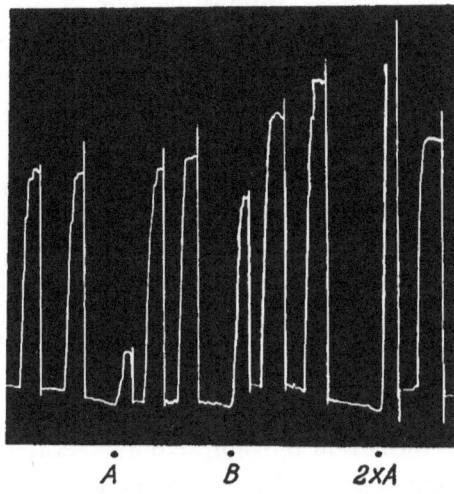

Fig 1 Comparative assay of two preparations of anaphylatoxin on an isolated guinea-pig ileum. *B* is stronger than *A*, but not twice as strong. (The other contractions are evoked by acetylcholine)

of injection. If injected slowly animals may tolerate quantities of anaphylatoxin which otherwise would be several times lethal. This method thus seems not to be highly recommendable.

The assay on the guinea-pig ileum has the advantage that the organ is highly sensitive and that samples can be compared with a standard preparation on the same piece of gut. A serious drawback is the tachyphylaxis. ROCHA E SILVA and ROTHSCHILD (1956) as well as RANDALL et al. (1961) have studied optimal conditions and have designed procedures for assay which overcome this difficulty. ROCHA E SILVA uses four pieces of gut from the same animal, on each of which a four point assay is conducted. By varying the sequence of doses and compounds in the four organs the effect of tachyphylaxis is eliminated. The contraction heights, recorded isotonically, are arranged in a latin square for evaluation of the relative activities. The procedure used by RANDALL et al. (1961) is similar except that contractions are recorded isometrically.

In the author's laboratory the method of ROCHA E SILVA et al. adapted to isometric recording has been found to be optimal as regards ease and accuracy.

12*

A standard preparation has been established which consists of a dry preparation of partially purified anaphylatoxin from Sephadex-activated pig serum. The unit is defined as 100 µg of this sample. The preparation contracts the guinea-pig ileum maximally at concentrations of 2×10^{-5} g/ml. The lethal dose for guinea-pigs is about 15 mg/kg*.

If it is sufficient to check which of two anaphylatoxin preparations is definitely stronger a quick though rough estimation can be made as shown in Fig. 1.

VII. Biological significance of anaphylatoxin

This section should be most interesting but is, in fact, disappointing. Nothing is known about physiological functions of anaphylatoxin and only little about its participation in pathological reactions.

The original assumption that anaphylatoxin is the toxic principle of anaphylactic shock was disputed already a few years after its publication. The findings that immune aggregates are not essential for anaphylatoxin formation (BORDET, 1913) and that anaphylactic reactions can be induced without the presence of plasma (SCHULTZ, 1910; DALE, 1913) only demonstrate that anaphylatoxin can occur independently of anaphylaxis and that the latter can proceed without anaphylatoxin being involved Both statements do not exclude the possibility that anaphylatoxin normally contributes to the symptoms of anaphylaxis. Likewise, the fact that heparin and glucocorticoids inhibit anaphylatoxin formation but not anaphylaxis (GIERTZ, HAHN, JURNA and LANGE, 1958) does not give more evidence against a participation. However, GIERTZ et al. (1958) have clearly demonstrated that anaphylatoxin is not implied in anaphylaxis of the guinea pig, simply because it is not formed. No loss of anaphylatoxin-forming capacity was found in the blood of actively sensitized animals after challenge. In contrast, when a reversed shock (Forssman shock) was elicited, anaphylatoxin-forming capacity was considerably reduced. The difference is well understandable. Anaphylatoxin is produced after contact activation of the plasma enzyme. Immune precipitates possibly do not form in the blood during true anaphylactic shock [the level of circulating antibodies is extremely low in actively sensitized guinea pigs (KABAT and MAYER, 1961, p. 270)], but the reactions take place at cellular surfaces containing sessile antibodies. In reverse shock the antigen is probably everywhere, including blood, and immune complexes may be precipitated in plasma or at the vessel walls. This would activate the enzyme, *in vivo*. Even in Forssman shock, however, anaphylatoxin is not the main lethal factor, for this type of reaction cannot be prevented by antihistaminics (ARBESMAN, NETER and BECKER, 1950; HAHN and OBERDORF, 1950).

A difference between the true anaphylactic reaction and the response to anaphylatoxin is also apparent from the finding that there is no cross-

* See Note 2 on p. 184

desensitization (see GIERTZ and HAHN, 1966, p. 543), at least in isolated organs. Results obtained *in vivo* are contradictory. HALPERN and LIACOPOULOS (1956; also FRICK, HALPERN and LIACOPOULOS, 1962) suppressed passive anaphylaxis in guinea pigs by prolonged pretreatment with anaphylatoxin. FRIEDBERG et al. (1964) on the other hand did not see any desensitizing effect of anaphylatoxin on subsequent responses to challenge in actively sensitized animals. This is not due to a difference between active and passive anaphylaxis, for even passively sensitized guinea pigs reacted with lethal shock to the antigen in 34 out of 40 cases after pretreatment with anaphylatoxin (FRIEDBERG and BAUER, 1965). As GIERTZ and HAHN (1966, p. 544; see there also for further literature on cross-desensitization) have pointed out already, even if there were a cross-desensitization, this would only indicate a participation of mast cells in both reactions but not necessarily an involvement of anaphylatoxin in anaphylactic shock.

A further argument against a role of anaphylatoxin in anaphylaxis is the fact that it would have to be formed first, to reach more or less slowly a pharmacologically active level. Under this condition tachyphylaxis is likely to develop. Even when high doses of the preactive anaphylatoxin-forming enzyme of cobra venom are injected, i.e. when anaphylatoxin is produced at the highest speed experimentally possible, bronchospasm does not develop but only a stimulatory effect on the respiration is seen, as it is observed after slow infusion of preformed anaphylatoxin (BODAMMER, unpublished).

It seems justified to conclude that anaphylatoxin is not likely to be formed during anaphylactic shock unless precipitates appear in the plasma. In general, if formed *in vivo*, it is not likely to produce severe bronchospasm, not even in guinea-pigs.

The hypothesis that anaphylatoxin is related to endogenous pyrogen and could thus play some role in temperature regulation (NETZER and VOGT, 1964), has not been verified by the experiments of LYNCKER and VOGT (1966).

Further search for a possible role of anaphylatoxin will have to deal with the following questions. Is it true that many species contain an anaphylatoxin-forming enzyme, but no anaphylatoxinogen? Or is there only less of the substrate which has hitherto escaped detection? If there is really no anaphylatoxinogen in these species what else is the anaphylatoxin-forming enzyme capable of catalyzing? Can the enzyme be activated *in vivo* without the formation of precipitates in the blood? (In man it is preactive.)

A specific antagonist of anaphylatoxin or an inhibitor of the enzyme which could be used *in vivo* would considerably help to study possible signs of deficiency. It is certainly wise to think of anaphylatoxin not only in terms of toxicity produced by massive amounts. At lower concentrations it may have physiological functions not yet known; and it may occur also in other species in which it has not been found, so far.

References

Arbesman, C. E., E. Neter, and C. F. Becker The effect of pyribenzamine, neo-hetramine, and rutin on reversed anaphylaxis in guinea pigs. J. Allergy 21, 25—33 (1950).

Becker, E. L.: Concerning the mechanism of complement action. I. Inhibition of complement activity by diisopropyl-fluorophosphate. II. The nature of the first component of guinea pig complement. J. Immunology 77, 462—468, 469—478 (1956).

— In vitro models for the allergic reaction. In. Mechanisms of hypersensitivity, ed. by J H. Shaffer, G. A. Lo Grippo and M. W. Chase, p 305—315 Boston 1959

— The possible significance of a particular group of serine esterases in pathological permeability changes In Mechanisms of cell and tissue damage produced by immune reactions, p. 37—50. Basel 1962

Bodammer, G , and W. Vogt Int Arch. Allergy. In preparation (1967).

Bordet, J. Le mécanisme de l'anaphylaxie. C.R Soc. Biol. (Paris) 74, 225—227 (1913).

Conard, V., and W. Mutsaars· Effet antagoniste des colorants basiques sur la genèse de l',,Anaphylatoxine". C.R Soc. Biol. (Paris) 143, 129—130 (1949).

Cremona, T., and T. P. Singer. Non-dissociability of the metal component of D-α-hydroxy acid dehydrogenase by chelating agents. Nature (Lond.) 194, 836 (1962).

Dale, H H.· The anaphylactic reaction of plain muscle in the guinea pig. J. Pharmacol exp. Ther. 4, 167—223 (1913)

Ferluga, J.: Die Wirkung des durch Polysaccharide aktivierten Rattenblutserums auf den isolierten Darm und Uterus der Ratte. Allergie u. Asthma 6, 229—232 (1960).

Frick, O. L., B N. Halpern, and P. Liacopoulos The mechanism of protection from anaphylactic shock in the guinea-pig by pre-treatment with anaphylatoxin. J. Physiol. (Lond.) 163, 191—199 (1962).

Friedberg, K. D., u. U. Bauer: Anaphylaktischer Schock nach passiver Sensibilisierung am anaphylatoxinvorbehandelten Meerschweinchen. Naunyn-Schmiedebergs Arch. exp. Path. Pharmak. 250, 171—172 (1965).

— G. Engelhardt u F. Meineke Untersuchungen uber die Anaphylatoxin-Tachyphylaxie und uber ihre Bedeutung fur den Ablauf echter anaphylaktischer Reaktionen. Int. Arch. Allergy 25, 154—181 (1964).

Friedberger, E.· Weitere Untersuchungen über Eiweißanaphylaxie. IV. Mitt. Z. Immun.-Forsch. 4, 636—689 (1910).

Gaddum, J. H.· Discussion. In: Polypeptides which stimulate plain muscle, ed. by J. H. Gaddum, p. 130—136. Edinburgh 1955.

Giertz, H., u. F. Hahn. Über die Reaktionsbedingungen der Anaphylatoxinbildung. Naturwissenschaften 44, 467 (1957).

— — Weitere Untersuchungen uber die Bildung und die Natur des Anaphylatoxins. Int. Arch. Allergy 19, 94—111 (1961)

— — Makromolekulare Histaminliberatoren. C. Das Anaphylatoxin. In: Heffters Handbuch der Pharmakologie, Erg -Band 18/1, S. 517—545. Springer Berlin-Heidelberg-New York 1966.

— — I. Jurna u. A. Lange· Zur Frage der Beteiligung des Anaphylatoxins im anaphylaktischen Schock. Int. Arch. Allergy 13, 201—212 (1958).

— — u. A. Lange· Über die Natur und Bildung von Rattenserumanaphylatoxin. Naunyn-Schmiedebergs Arch. exp Path Pharmak. 229, 366—373 (1956)

Greeff, K , B G. Benfey u A Bokelmann Anaphylaktische Reaktionen am isolierten Herzvorhofpraparat des Meerschweinchens und ihre Beeinflussung durch Antihistaminica, BOL, Dihydroergotamin und Reserpin. Naunyn-Schmiedebergs Arch. exp. Path Pharmak 236, 421—434 (1959).

Hahn, F, u. A. Lange Versuche zur Isolierung von Rattenserum-Anaphylatoxin. Naunyn-Schmiedebergs Arch exp. Path. Pharmak. 227, 12—22 (1955).

—, u. A. Oberdorf. Antihistaminica und anaphylaktoide Reaktionen. Z. Immun.-Forsch 107, 528—538 (1950).

HAINING, C G   Histamine release in rabbit blood by dextran and dextran sulphate. Brit. J. Pharmacol. **10**, 87—94 (1955).

— Activation of rabbit serum protease by dextran sulphate. Brit J Pharmacol. **11**, 107—110 (1956a).

— Inhibition of histamine release by sodium salicylate and other compounds. Brit. J. Pharmacol. **11**, 357—363 (1956b).

HALPERN, B. N., et P. LIACOPOULOS   Protection du cobaye contre le choc anaphylactique mortel par l-anaphylatoxine et son mécanisme. C.R. Soc. Biol. (Paris) **150**, 108—112 (1956).

JOBLING, J. W., and W. PETERSEN   The mechanism of anaphylatoxin formation. Studies on ferment action XV. J. exp. Med. 20, 37—51 (1914).

JONES, J. H.: Studies on the activation of anaphylatoxin. Int. Arch. Allergy **17**, 99—105 (1960).

KABAT, E. A., and M. M. MAYER·   Experimental immunochemistry, 2nd ed Springfield 1961.

KUMAGAI, T..   Über Anaphylaxie. XXXVI. Mitt. Die Lungenblahung bei der Anaphylatoxinvergiftung und bei einigen ahnlich wirkenden Giften. Z Immun.-Forsch. **17**, 607—638 (1913).

LYNCKER, J., u. W. VOGT· Differenzierung zwischen dem pyrogenbildenden und dem anaphylatoxinbildenden System in Plasma. Naunyn-Schmiedebergs Arch. exp. Path. Pharmak **253**, 71 (1966).

MARQUARDT, P., u. L. HEDLER· Anaphylatoxin und DAS („Fruhgift"). Arzneimittel-Forsch. **15**, 1261—1265 (1965).

MOTA, I.· Action of anaphylactic shock and anaphylatoxin on mast cells and histamine in rats Brit. J Pharmacol. **12**, 453—456 (1957).

— The mechanism of action of anaphylatoxin. Its effect on guinea pig mast cells. Immunology **2**, 403—413 (1959).

MOUSSATCHÉ, H , and A. P DANON   Inhibition by sulfhydril blocking agents of the guinea-pig gut reaction to anaphylaxis and to anaphylatoxin. Naturwissenschaften **43**, 227 (1956).

MURATA, Y., M. SATAKE, and T. SUZUKI   Studies on snake venom. XII. Distribution of proteinase activities among Japanese and Formosan snake venoms. J. Biochem. (Tokyo) **53**, 431—437 (1963).

NETZER, W , u. W. VOGT: Anaphylatoxinbildung durch pyrogenes Lipopolysaccharid. Naunyn-Schmiedebergs Arch exp. Path. Pharmak. **248**, 261—268 (1964).

OSLER, A. G , H. G. RANDALL, B. M. HILL, and Z OVARY. Studies on the mechanism of hypersensitivity phenomena. III. The participation of complement in the formation of anaphylatoxin J. exp. Med. **110**, 311—339 (1959).

RANDALL, H. G , S L. TALBOT, H C NEU, and A. G. OSLER Studies on the mechanism of hypersensitivity phenomena. IV. An isometric smooth muscle assay system. Immunology **4**, 388—400 (1961).

RITZ, H., u H SACHS Über das Anaphylatoxin. Berl. klin. Wschr **48**, 987—992 (1911).

ROCHA E SILVA, M.: Activation by polysaccharides of a histamine liberator (anaphylatoxin) in blood plasma. In· Symposium on the mechanism of inflammation, p. 325—334 Montreal 1953.

— Anaphylatoxin and histamine release. Quart Rev. Allergy **8**, 220—238 (1954).

—, and M. ARONSON. Histamine release from the perfused lung of the guinea pig by serotoxin (anaphylatoxin). Brit. J. exp. Path. **33**, 577—586 (1952).

—, and A. M. ROTHSCHILD   Experimental design for bioassay of a material inducing strong tachyphylactic effect (anaphylatoxin). Brit J. Pharmacol. **11**, 252—262 (1956).

ROTHSCHILD, A M., and M. ROCHA E SILVA. Activation of a histamine-releasing agent (anaphylatoxin) in normal rat plasma. Brit. J. exp Path **35**, 507—518 (1954).

SCHULTZ, W. H.   Physiological studies in anaphylaxis. I. The reaction of smooth muscle of the guinea-pig sensitized with horse serum. J. Pharmacol exp. Ther **1**, 549—567 (1910).

SCHWOERER, D Untersuchungen uber das Vorkommen des anaphylatoxinbildenden Systems im Plasma verschiedener Tierarten Thesis Gottingen 1966

STEGEMANN, H , G. BERNHARD u. J. A O'NEIL Endgruppen von Anaphylatoxin Einfache Sequenzanalyse von carboxyl-endstandigen Aminosauren Hoppe-Seylers Z physiol Chem **339**, 9—13 (1964).

— R HILLEBRECHT u W. RIEN Zur Chemie des Anaphylatoxins Hoppe-Seylers Z physiol. Chem. **340**, 11—17 (1965)

— W VOGT u K. D. FRIEDBERG. Über die Natur des Anaphylatoxins Hoppe-Seylers Z. physiol. Chem **337**, 269—276 (1964)

TU, A. T , R. B PASSEY, and T. TU: Proteolytic enzyme activities of snake venoms Toxicon **4**, 59—60 (1966)

UNGAR, G , T YAMURA, J. B ISOLA, and S KOBRIN Further studies on the role of proteases in the allergic reaction J. exp Med. **113**, 359—380 (1961).

VAUGHAN, V. C , and S. M. WHEELER· The effects of egg-white and its split products on animals, a study of susceptibility and immunity. J. infect. Dis **4**, 476—508 (1907)

VOGT, W. Anaphylatoxinbildung durch ein Metallferment Naunyn-Schmiedebergs Arch. exp. Path Pharmak. **246**, 31—32 (1963).

— Weitere Untersuchungen zur Fermentnatur der Anaphylatoxinentstehung Naunyn-Schmiedebergs Arch. exp Path. Pharmak **247**, 327 (1964).

—, u. G. SCHMIDT Abtrennung des anaphylatoxinbildenden Prinzips aus Cobragift von anderen Giftkomponenten. Experientia (Basel) **20**, 207—208 (1964)

— — Einige Eigenschaften des anaphylatoxinbildenden Fermentes aus Rattenplasma Naunyn-Schmiedebergs Arch exp Path. Pharmak **253**, 91 (1966a).

— — Formation of anaphylatoxin in rat plasma, a specific enzymic process Biochem. Pharmacol **15**, 905—914 (1966b).

—, u. N. ZEMANN. Analyse der erregenden Wirkung von Anaphylatoxin auf glatte Muskulatur. Naunyn-Schmiedebergs Arch exp. Path Pharmak. **247**, 328—329 (1964).

WILLEMS, G.· Properdine et anaphylatoxine Revue Immunol. (Paris) **27**, 285—290 (1963)

Note 1 added in proof In a recent publication DIAS DA SILVA and LEPOW (J Immunol. **95**, 1080—1089 (1966)] report that anaphylatoxin is formed when C'1 esterase is added to serum or plasma. This may indicate the participation of complement in the activation of the anaphylatoxin-forming enzyme The interpretation that the complement system itself effects the release, is unlikely in view of the data presented above.

Note 2 added in proof This dry preparation proofed to be unstable, even at —20° C. The stability of anaphylatoxin preparations stored in slightly acidic solution at —20° C is now under investigation

Magnesium Metabolism*

MACKENZIE WALSER**

With 6 Figures

Table of contents

I. Introduction 189
II. Distribution of magnesium in the organism 190
  A Whole animals 190
    1. Normal values 190
    2 Physiological variations 191
      a) Age and sex 191
      b) Pregnancy and lactation . . 191
  B Plasma 191
    1. Normal values 191
    2 Physiological variations 197
      a) Age and sex . . . 197
      b) Effects of eating 198
      c) Effect of temperature . 198
        i Environmental (in homeothermic animals) . . 198
        ii. Hibernation 198
        iii Hypothermia 198
        iv Role of magnesium in thermoregulation 198
      d) Pregnancy 199
      e) Acid-base balance 199
      f) Ethnic and regional 199
    3. Physicochemical state of magnesium in plasma . 199
      a) Protein-binding 199
      b) Free and complexed magnesium 200
  C. Erythrocytes 201
    1. Normal values 201
    2 Physiological variations 201
    3 Physicochemical state of red cell magnesium 201
  D Soft tissues 202
    1. Normal values 202
    2. Physiological variations . . . 202
    3. Physicochemical state of soft tissue magnesium . . . 202
      a) Binding in the extracellular space 202
      b) Binding to subcellular structures 203
      c) Free and bound magnesium in the cell . . . 203
      d) Isotopic exchangeability 203

    * These studies have been supported by the National Institutes of Health (AM-02306 and GM-K3-2583).

    ** From the Department of Pharmacology and Experimental Therapeutics and the Department of Medicine, Johns Hopkins University School of Medicine, Baltimore, Md. 21205.

E. Hard tissues 204
   1. Normal values 204
   2. Physiological variations 204
   3. Physicochemical state. 204

F. Transcellular fluids 204
   1. Cerebrospinal fluid 204
   2. Digestive secretions 204
   3 Milk 205
   4. Other transcellular fluids 205

III. Intake and excretion in normal animals. 205
  A. Intake and dietary requirements 205
  B. Excretion 207
    1. Fecal 207
     a) Normal values 207
     b) Physiological variations in intestinal absorption 207
      i Flow 207
      ii. Calcium 207
      iii. Phosphate 208
      iv. Lactose 208
      v. Potassium 208
     c) Endogenous fecal excretion 208
    2 Urine 209
     a) Normal values 209
     b) Physiological variations 209
      i Diurnal variation 209
      ii. Meals 210
      iii. Exercise 210
      iv. Intake of other electrolytes 210
    3 Dermal losses 210
  C. Response of excretion to dietary restriction 210

IV. Effects of magnesium 211
  A. On cell-free systems 211
    1. Enzymatic reactions 211
    2. On subcellular structures 212
    3 In body fluids 212
  B Isolated tissues 212
    1. Neural tissue 212
    2. Smooth muscle 213
    3. Skeletal muscle 213
    4. Heart 214
  C. In intact animals following parenteral administration 214
    1. Distribution and fate of injected magnesium 214
    2 Renal effects 214
    3. Cardiovascular effects 215
    4. Effects on the nervous system 216
    5. Other effects 216
  D. In intact animals following oral administration 216
    1. Retention of orally administered magnesium 216
    2 Effects of oral magnesium 217

V. Magnesium deficiency 217

  A. General considerations. 217

  B. The syndrome of magnesium deprivation during growth 218

    1. Symptoms 218
      a) Reduced growth. 218
      b) Neuromuscular and central nervous system malfunction 218
      c) Anorexia 219
      d) Trophic symptoms 219
    2. Magnesium metabolism 219
      a) Plasma. 219
      b) Body 220
      c) Bone 220
      d) Soft tissues 221
      e) Excretion. 222
    3. Calcium and phosphorus metabolism 222
      a) Plasma 222
      b) Carcass 223
      c) Bone 223
      d) Soft tissues 223
      e) Excretion. 224
    4. Potassium metabolism 224
    5. Effects on specific organ functions 225
      a) Thyroid 225
      b) Cardiovascular system 225
      c) Neurological changes. 225
      d) Kidney 225
      e) Liver 226
      f) Gut 226
    6. Pathology 226
    7. Modification by dietary and hormonal factors 227
      a) Calcium and phosphorus 227
      b) Other dietary factors 228
      c) Hormonal factors 228

  C. Magnesium depletion 228
    1. Criteria 228
    2. Causes 229
      a) Dietary 229
      b) Physiological losses 230
      c) Pathological losses 230
        i. Intestinal disease 230
        ii. Drugs 230
        iii Diabetic acidosis 230
        iv Renal magnesium-wasting 230
        v. Hypoparathyroidism 231
    3. Magnesium metabolism 231
      a) Plasma 231
      b) Body 231
      c) Bone 232
      d) Soft tissues 232
      e) Excretion. 233
    4. Calcium and phosphorus metabolism 234
    5. Potassium metabolism 234
    6. Signs and symptoms 234

D. Hypomagnesemia 235
  1. In patients 235
    a) Occurrence . . . 235
      1. Alcoholism and delirium tremens . 235
      ii Malabsorption . . 235
      iii Post-operative 235
      iv Others 235
    b) Symptoms . . 235
  2. In animals . . 236
    a) Symptoms . 237
    b) Magnesium metabolism 237
      i Intestinal absorption . . . 237
      ii Urinary excretion 237
      iii Buffering of plasma magnesium by bone . 237
    c) Calcium metabolism . 237

VI. Magnesium transport 238
  A. Renal . . . 238
    1 Free and complexed magnesium 238
    2. Active and passive transport and exchange 238
    3 Dependence on flow 240
      a) Osmotic diuresis . . 240
      b) Water diuresis 241
    4. Dependence on the transport of other solutes 241
      a) Sodium and calcium 241
      b) Anions 243
  B. Intestinal 243
    1. Mechanism 243
    2. Site 245
  C Other organs 245
    1. Salivary glands 245
    2. Gall bladder . . 246
    3. Stomach 246
    4. Erythrocytes . . 246
    5. Other cells . . 246
    6. Subcellular organelles 247

VII. Hormonal effects on magnesium metabolism . . 247
  A. Calcium-regulating hormones . . 247
    1. Parathyroid hormone 247
    2 Vitamin D . 249
    3. Calcitonin 249
  B. Adrenocortical hormones . 249
  C. Thyroid 250
  D. Other hormones 251

VIII. Effects of drugs on magnesium metabolism . 252
  A. Ethanol 252
  B. Diuretics and cardiac glycosides . . 252
  C. Other drugs 253

IX. Magnesium metabolism in specific disorders . . . 253
  A. Renal failure 253
  B. Alcoholism 255
  C. Atherosclerosis 255

    D. Hypertension.   256
    E. Anemia 257
    F. Central nervous system disorders   257
    G. Muscular dystrophy   257
    H. Other diseases 258

  X. Therapeutic uses of magnesium   .   . . 258
    A. Vascular and renal disorders   . 258
    B. Atherosclerosis   258
    C. Other therapeutic uses   .   259

 XI Discussion 260
    A. Physiological role of magnesium .   . . 260
    B. Antagonism and synergism with calcium   .   260
    C Magnesium transport   .   260
      1. Within the organism   260
      2. Between the organism and the environment 261
    D Possible regulatory mechanisms. .   .   .   261

References   262

I. Introduction

The metabolism of magnesium is far less well understood than that of the other major cations of the body, sodium, potassium, and calcium. Although analytical difficulties accounted for much of this lag in the past, these have been largely overcome during the last decade. Complexometric titration, flame emission spectrophotometry, fluorometry, and atomic absorption spectrophotometry have been applied to magnesium analysis with considerable success, and large analytical errors are now rare, except perhaps in complexometric methods (e.g., see SMITH, FISCHER and ETTELDORF, 1962, for documentation of some of the problems). Nevertheless, the fundamental role of magnesium in the organism and the homeostatic mechanisms which regulate its metabolism are still understood only dimly.

There is a considerable body of evidence, reviewed below, which indicates that the intracellular magnesium concentration is of major importance, while the amount in the extracellular fluids and in bone is of lesser significance. It is perhaps not surprising that a homeostatic mechanism more concerned with intracellular concentration than extracellular should prove elusive. Furthermore, the major regulatory mechanisms concerned with the other predominant cations in body fluids all affect magnesium metabolism. Thus parathyroid hormone, whose principal function is certainly calcium homeostasis, has similar effects upon magnesium in bone and kidney; changes in aldosterone secretion in response to the demands of sodium and potassium metabolism may modify magnesium balance; and the mechanisms (other than aldosterone) involved in regulating the volume of the extracellular fluid and the rate of sodium excretion alter magnesium excretion in a parallel fashion. This interdependence between the metabolism of magnesium and that of the other major cations has been a source of confusion in the search for the magnesium-regulating mechanism, but may indeed be the only regulation which exists.

That magnesium metabolism is of great importance is becoming increasingly evident. In addition to its role in enzymatic processes, where it is often required and may perform a regulatory function, magnesium has also been implicated in the pathogenesis and treatment of arteriosclerosis, among other disorders. Two major questions, closely related, to which the answers are still far from unanimous, are first, the optimal magnesium intake of the adult, and second, how to assess magnesium deficiency.

This review is chiefly concerned with magnesium metabolism in mammals, with particular attention to man. Emphasis is laid on studies which shed light on fundamental mechanisms and on homeostasis, and largely empirical work is given only passing attention. For example, a large body of data reporting growth or mortality rates of animals fed various salt mixtures and adjuvants has been omitted. Similarly, an even larger number of papers in which the effects of omission of magnesium or addition of large amounts of magnesium upon various cell-free systems has been referred to only in passing. As WACKER and VALLEE (1964) have noted, it is uncertain whether these *in vitro* findings have relevance to the living cell.

Particular attention has been given to the most recent references, because several comprehensive reviews, including earlier work, have recently appeared The most extensive are those of WACKER and VALLEE (1958a and b, 1964), MacINTYRE (1959, 1960, 1963a and b), HANZE (1962c, in German), ROOK and STORRY (1962b), AIKAWA (1963, 1965a and b) and BACQ (1964, in French). Other reviews are those of BLAXTER and McGILL (1956), ELKINTON (1958), HASSELMAN and VAN KAMPEN (1958), MAURAT (1959), O'DELL (1960), BERTRAND (1960), WILSON (1960, 1964), PARLIER, HIOCO and LEBLANC (1963), STEWART and FRAZER (1963), SMITH (1964b), FISHMAN (1965), ROGERS (1965), MANITIUS (1965), and WELT and GITELMAN (1965).

II. Distribution of magnesium in the organism

A. Whole animals

1. Normal values

In Table 1 are gathered some recent values for total body magnesium, mostly taken from WIDDOWSON and DICKERSON (1964). Normal adults contain approximately 20 mmoles per kg of fat-free tissue. In an average subject containing 20% fat, and weighing 70 kg, this would amount to 1120 mmoles.

The availability of radioactive magnesium has prompted many studies of the disappearance of isotope from the blood following intravenous injection (for references, see review articles by AIKAWA, 1963 and 1965a, and ROGERS, 1965). Mathematical analysis of such curves leads to estimates of discrete pools of exchangeable magnesium characterized by single rate constants, and estimates of total exchangeable magnesium at various intervals after injection. Since a considerable proportion of body magnesium is unexchangeable before the isotope becomes unmeasurable, total body magnesium cannot be inferred from such data. Furthermore, the existence of a small number of discrete pools is clearly refuted by studies of exchange with individual tissues (see below). In this reviewer's opinion, no useful information can be obtained from such curves in our present state of knowledge.

2. Physiological variations

a) *Age and sex*. WIDDOWSON and DICKERSON (1964) give values of 11 and 15 mmoles per kg fat-free tissue for infants at birth and at 4 years, respectively. The human fetus contains even less early in pregnancy (op. cit.). A similar increase with age is reported in rabbits (DAVIES, WIDDOWSON and McCANCE, 1964), pigs, dogs, cats, rats, and mice (WIDDOWSON and DICKERSON, 1964).

b) *Pregnancy and lactation*. SPRAY (1950) made the interesting observation that the mother rat, after delivery, contains 20 % more magnesium than a non-pregnant control. Similarly, an early study in dogs (TOVERUD and TOVERUD, 1931) indicated an increase of about 30 mmoles, despite the absence of calcium or phosphate retention. In pregnant women, these same authors observed positive magnesium balances during the last trimester, and also during lactation. COONS et al (1935) and HUMMEL et al. (1936, 1937) both reported positive magnesium balances of 2.5 mmoles per day in pregnant women, despite retention of only 1 to 2 grams of nitrogen daily.

Although exchange of isotopic magnesium between mother and fetus is slow in the cow (ROGERS et al., 1964), it is considerably more rapid in the rabbit (AIKAWA and BRUNS, 1960). Studies of fetal and maternal magnesium during dietary deprivation do not appear to have been reported.

B. Plasma

1. Normal values

Table 2 lists mean values for plasma or serum magnesium reported by various authors during recent years. The range of variation among these means within a given species, particularly man, is considerably wider than for other common inorganic constituents. The use of crude or doubtful analytical methods can no longer be cited as an explanation for these discrepancies, because highly sophisticated techniques continue to yield values differing as much as 20 % from one laboratory to another (WACKER and VALLEE, 1964; ALCOCK, MacINTYRE and RADDE, 1965). The highest mean, 1.07 mM, is credited by its authors with "absolute analytical accuracy", but the majority of observers have found considerably lower values. The most reliable criterion of analytical accuracy would appear to be the determination of small amounts

Table 1. *Tissue magnesiu*

Tissue	Cat	Man	Dog	Sheep	Cattle
Body	19.0 [117]	20.0 [117]	—	—	—
Skeletal muscle	11.8 [68]	6 9 [63], 8 4 [117] 8.9 [15], 8.15 [14]* 40 [89]** 35 [40]*** 36 [67]*** 30 [12]† 67 [105A]†††	9 0 [68], 9 8 [117] 10 [20] 11 3 [15] 50 [113, 114]***	11.5 [117]* 45.4 [35]**	7.9 [17]
Whole kidney	—	4.3 [15, 117] 50 [105A]†††	2 85 [117] 26.3 [30]**	6 6 [117] 40 [35]**	5.4 [17] 34.6 [35]**
Cortex	—	—	4 7—5 9 [106] 7 85 [116B] 7 95 [20]	—	—
Medulla	—	—	6 7 [20]	—	—
Liver	—	7.3 [15], 8 2 [117] 54 [105A]†††	7.3 [117], 7 5 [20] 8 [68] 40 [113, 114]***	31 [35]**	4.2 [17] 28 [35]**
Pancreas	—	58 [105A]†††	12.1 [20] 45 5 [113, 114]***	—	—
Heart	—	6.6 [117], 7 2 [15] 71 [105A]†††	8 2 [68], 44 [30]** 52 [114]***	46 [35]**	50 [35]**
Ventricle	—	—	8.5 [117], 9.8 [20]	—	—
Auricle	—	—	7.0 [20]	—	—
Aorta	—	4 6 [75A] 6 2 [105A]†††	4.2 [20] 14.2 [30]** 16 [113, 114]***	—	2.1 [17]
Lung	—	2 4 [117], 3 1 [15] 3 4 [105B] 38 [105A]†††	4 96 [15] 5.1 [68] 5 95 [20]	28 [35]**	1.5 [17] 27 [35]**
Spleen	—	6.5 [117] 42 [105A]†††	7 4 [68]	39 [35]**	7 1 [17] 34 [35]**
Skin	—	1.6 [117] 5.0 [65]** 36 [105A]†††	3 0 [117] 3 3 [20] 3.0 [68]*	—	—
Brain	—	5.7 [117] 5 8 [15] 38 [105A]†††	5.0 [68] 5 6 [8, 117] 6.2—7 2 [20] 48 [113, 114]***	31 3 [35]**	4 2 [17] 27 [35]**
Nerve	—	—	6 6—9.7 [20] 20 5 [113, 114]***	—	—

in mammals

Guinea pig	Rabbit	Rat	Mouse	Pig	Horse
—	20 8 [117]	16.6 [117], 55 [29]*** 58 [82]***	17.9 [117]	18.8 [117]	—
50 [73]***	10 8 [1] 11.4 [68] 11 9 [117] 40.5 [118]***	10.6 [54A] 12.1 [117], 11 0 [34]* 38 [39, 97]** 45 [35]**, 31 [85]*** 50 [27, 66]*** 51 [26]***, 54 [28]*** 56 [96]***	—	11.6 [117] 42 [35]**	41 [35]**
43 [73]***	6 1 [1]	7 4 [117], 25 4 [97]** 29 5 [39]** 38 8 [29]** 41.3 [19, 35]** 42 [66]***, 29 [52]††	—	4 5 [117]	—
—	39 [54 B]**	—	—	—	—
—	—	—	—	—	—
—	6 2 [1] 8 4 [68]	8 6 [117], 32 1 [35]** 38 7 [39]**, 40 [97]** 45 [19]**, 41 [66]***	—	10 2 [117]	—
—	—	—	—	—	—
43 [73]***	6 5 [1]	31 [97]** 37 1 [39]**, 58 [35]** 39 [27]***	—	9 6 [117]	—
—	—	—	—	—	—
—	—	—	—	—	—
—	—	—	—	—	—
—	5.7 [68] 5 9 [1]	24 [35]**	—	—	—
—	—	8 2 [117], 56 [35]**	—	—	—
—	3.1 [1] 14 4 [68]**	—	—	2.2 [117]	—
—	5 6 [8, 87] 5 7 [117] 5 9 [68] 6.1—6 5 [48] 29 [97]**	6 8 [8], 29 5 [39]** 41 [35]** 38 [66]***	—	6 1 [117]	—
—	—	—	—	—	—

Table 1

Tissue	Cat	Man	Dog	Sheep	Cattle
Spinal cord	—	15.8 [117]	—	—	—
Uterus	—	6.4 [117]	—	—	—
		28 [40]***			
	֊	46 [105 A]†††			
Bone	—	41.7 [15]	113 [113, 114]***	2.5 [35]†††	2 3 [35]†††
		233 [71]**			3 0 [96]†††
Dentin	—	321 [71]**	—	—	—
		346 [56]**			
Enamel	—	138 [71]**	—	—	—
		225 [56]**			
CSF	—	1.04 [21]	0.79 [59]	—	—
		1 11 [55]	1.0 [91]		
		1.16 [41]	1.08 [79]		
		1.20 [102]			
		1.25 [50]			
Semen	—	2.7 [83]	0.75 [83]	3 2 [83]	3 7 [83]
Sperm	—	6 38 [83]	7.13 [83]	5 54 [83]	5 75 [83]

Key to Table 1

Values given as mmoles per kg of wet tissue except as follows·

* mmoles per kg fat free wet tissue.   † mmoles per 100 g NCN.
** mmoles per kg dry tissue.   †† mmoles per 100 g N.
*** mmoles per kg fat free dry tissue   ††† mmoles per 100 g ash.

Numbers in brackets refer to the following references.

References to Tables 1 and 2

[1] AIKAWA (1960a)
[2] AIKAWA (1960b)
[3] AIKAWA and REARDON (1965)
[4] AIKAWA, REARDON and HARMS (1962)
[5] ALBERT, MORITA and ISERI (1958)
[6] ALCOCK, MacINTYRE and RADDE (1960)
[7] ALLCROFT (1947a)
[8] AMES and NESBITT (1958)
[9] AMES and SAKANOUE (1964)
[10] AMES, SAKANOUE and ENDO (1964)
[11] AXELROD and BASS (1956)
[12] BALDWIN et al. (1952)
[13] BANG and ØRSKOV (1939)
[14] BARNES, GORDON and COPE (1957)
[15] BERTRAND (1960)
[16] BEST and PICKLES (1965)
[17] BLAXTER, ROOK and MacDONALD (1954a)
[18] BOELLNER et al. (1965)
[19] BUNCE et al. (1963)
[20] BURCH, LAZZARA and YUN (1964)
[21] CANELAS, DE ASSIS and DE JORGE (1965)
[22] CARR and FRANK (1956)
[23] CARR and SCHLOERB (1959)
[24] CARUBELLI, SMITH and HAMMARSTEN (1958)
[25] CHARBON and HOEKSTRA (1962)
[26] CHEEK, GRAYSTONE and CASS (1963)
[27] CHEEK et al (1962)
[28] CHEEK and TENG (1960)
[29] CHEEK and WEST (1956)
[30] CHIEMCHAISRI and PHILLIPS (1963)
[31] CHUTKOW (1965)
[32] CLARKSON et al. (1965)
[33] COPELAND and SUNDERMAN (1952)
[34] COTLOVE et al. (1951)

(continued)

Guinea pig	Rabbit	Rat	Mouse	Pig	Horse
—	—	—	—	—	—
6.0 [16]	4.7 [16]	6 7 [117]	—	—	—
37 [16]**	19 [16]**				
292 [73]***	149 [1]	167 [97]**	—	—	—
		171 [39]**, 201 [28]**			
		200 [66]***			
		1 9 [35]†††			
—	—	18.3—80.0 [71]	—	—	—
—	—	—	—	—	—
—	—	—	—	—	—
—	10 5 [83]	—	—	—	—
—	13.2 [83]	—	—	—	—

[35] CUNNINGHAM (1936)
[36] DAHL (1950)
[37] DE DONCKER and ROSSELLE (1959)
[38A] DE JORGE et al. (1965)
[38B] DREUX and GIRARD (1961)
[38C] DUNN and WALSER (1966)
[39] FIELD and SMITH (1964)
[40] FLOWERS (1965)
[41] FRIEDMAN and RUBIN (1955)
[42] GARTNER, RYLEY and BEATTIE (1965)
[43] GERBRANDY et al. (1960)
[44] GINSBURG et al. (1962)
[45] GREENBERG et al. (1933)
[46] HALL (1957)
[47] HAMBURGER (1957)
[48] HANIG and APRISON (1966)
[49] HANNA (1961a)
[50] HARRIS and SONNENBLICK (1955)
[51] HEATON (1964b)
[52] HEATON and ANDERSON (1965)
[53] HERRING et al (1960)
[54A] HINGERTY (1957)
[54B] HOFER and KLEINZELLER (1963)
[55] HUNTER and SMITH (1960)
[56] JOHANSEN (1963)
[57] JONES et al. (1966)
[58] JONES and FOURMAN (1966)
[59] KEMÉNY, BOLDIZSÁR and PETHES (1961)
[60] KEYL et al. (1965)
[61] KLEEMAN et al. (1958)
[62] KNIPPERS and HEHL (1965)
[63] KOHN, KEYE and ROLLERSON (1961)
[64] LEWIS (1960)
[65] LIPKIN, MARCH and GOWDEY (1964)
[66] MacINTYRE and DAVIDSSON (1958)
[67] MacINTYRE et al. (1961)
[68] MANERY (1954)
[69] MAURAT (1964)
[70] MAXWELL, ELLIOTT and BURNELL (1965)
[71] McCANN (1959)
[72] MERTZ (1957)
[73] MORRIS and O'DELL (1961)
[74] MUNDAY and MAHY (1964)
[75A] NAKAMURA et al. (1965a)
[75B] NIELSEN (1964a)
[76] NIELSEN (1964b)
[77] NORDBO (1939a)
[78] OGASAWARA (1953)
[79] OPPELT, MacINTYRE and RALL (1963)
[80A] ORANGE and RHEIN (1951)
[80B] OUTA (1963)

13*

[81] PRASAD, FLINK and McCOLLISTER (1961)
[82] QUINN, BASS and LEEMAN (1953)
[83] QUINN, WHITE and WIRRICK (1965)
[84] RAY and MULLICK (1965)
[85] RICHARDSON and WELT (1965)
[86] RIZEK, DIMICH and WALLACH (1965)
[87] ROMENSKI (1935)
[88] ROSSELLE and DE DONCKER (1961)
[89] SAVILLE and LIEBER (1965)
[90] SCHACHTER (1959)
[91] SCHAIN (1964a)
[92] SCHWAB, PORYALI and AMES (1964)
[93] SELLER et al. (1966)
[94] SILVERMAN and GARDNER (1954)
[95] SMITH (1957)
[96] SMITH (1959b)
[97] SMITH and FIELD (1963)
[98] SMITH et al. (1962)
[99] SMITH and HAMMARSTEN (1958)
[100] SMITH and HAMMARSTEN (1959)
[101] STILES, BATSAKIS and HARDY (1965)
[102] STUTZMAN and AMATUZIO (1952)
[103] STUTZMAN and AMATUZIO (1953)
[104] TAKAYASU et al. (1962)
[105A] TIPTON and COOK (1963)
[105B] TIPTON and SHAFER (1964)
[106] ULLRICH and JARAUSCH (1956)
[107A] VALBERG et al (1965a)
[107B] VALBERG et al (1965c)
[108] VALLEE, WACKER and ULMER (1960)
[109] VAN LEEUWEN (1964) (see LEEUWEN)
[110] VAN LEEUWEN, THOMASSE and KAPTEYN (1961) (see LEEUWEN et al)
[111] WACKER, IIDA and FUWA (1964)
[112] WACKER and VALLEE (1964)
[113] WALLACH et al (1964)
[114] WALLACH et al. (1965)
[115] WALLACH et al. (1962)
[116A] WALSER (1961b)
[116B] WALSER and NAHMOD (1966)
[117] WIDDOWSON and DICKERSON (1964)
[118] ZUCKERMAN and MARQUARDT (1963)

Table 2 *Blood magnesium in normal animals*[1]

Subject	Plasma or serum, mM	Per cent ultrafiltrable	Erythrocytes (packed), mM
Man	0 62 [40], 0 74 [99] 0.75 [47], 0.77 [72] 0.78 [82], 0 80 [5, 23, 32] 0.81 [18], 0.82 [38C] 0 83 [6, 51, 74, 117] 0 84 [109], 0.85 [53, 92] 0.86 [22, 104, 107b] 0 87 [42, 66], 0 88 [33, 64, 94] 0 89 [31, 61], 0 90 [38B, 76, 81, 100], 0.91 [103] 0 92 [80B], 0.93 [57, 110] 0 94 [46], 0 95 [36] 0 96 [52, 78, 93, 116A] 0.97 [102], 1.00 [86, 108, 112, 115], 1.02 [90], 1.04 [21, 38A, 50], 1 06 [58] 1 07 [9, 111]	55 [42], 59 [77] 62 [74], 64 [81] 65 [33, 80B, 94, 110] 68 [116A], 71 [82] 72 [109], 74 [61] 84 [9], 85 [92, 25]	2 13 [18], 2 15 [38C] 2 20 [93], 2 40 [13] 2 43 [64], 2.53 [107A] 2.55 [25], 2.56 [16] 2 58 [101], 2 60 [69] 2 62 [80A], 2 64 [24, 57, 86], 2.65 [86, 38B, 100, 115], 2 70 [107] 2 71 [36], 2 72 [45] 3 1 [53]
Dog	0.66 [62], 0.68 [91] 0.75 [23, 59, 117], 0 76 [114] 0 78 [104, 113], 0 80 [11, 60, 79, 112], 0.90 [68] 0 92 [70], 0.96 [25]	61 [70], 66 [113, 114]	2.91 [25], 3.07 [107A]
Cat	0 95 [25]	70 [10]	2.80 [25]
Horse	0.65 [25], 0 67 [78] 0.75 [112], 1.0 [117]	44 [78]	3.19 [25]
Cattle	0 80 [112], 0 83 [78] 0.93 [25], 1.00 [7, 41, 84, 95]	65 [78], 72 [77] 79 [95], 38 [78]	1.18 [25]

Table 2 (continued)

Subject	Plasma or serum, mM	Per cent ultrafiltrable	Erythrocytes (packed), mM
Sheep	0 85 [112], 0 88 [78] 1.2 [117]	38 [78]	—
Goat	0.95 [112], 0 96 [78]	43 [78]	—
Monkey	—	—	2.00 [107 A]
Pig	0 65 [117], 0 67 [78] 0 85 [25]	38 [78]	5.79 [25]
Rabbit	0 86 [2], 0 90 [88] 0 98 [75 B], 1.00 [4] 1.05 [112], 1.08 [37, 78] 1 10 [117], 1 20 [3] 1 26 [25], 1 30 [118]	35 [78]	3.57 [107 A], 3 9 [44] 4 63 [25]
Guinea pig	0.94 [88], 0.96 [78] 0.97 [25]	52 [78]	3.2 [25]
Rat	0 67 [78], 0.72 [49] 0.80 [112], 0 87 [66] 0 88 [39], 0.92 [28] 0 96 [19, 52], 1 43 [25] 1.5 [117]	38 [78]	2.22 [107 A], 2 55 [25]
Mouse	0.55 [112]	—	—
Frog	0.60 [117]	—	—
Terrapin	0.92 [78]	23 [78]	—
Turtle	1 60 [112], 1 63 [78]	33 [78]	—
Laying hen	0.58 [78]	57 [78]	—
Hen	0 80 [112], 0 83 [78]	30 [78]	—
Cock	0 90 [112], 0 92 [78]	36 [78]	—
Carp	0.50 [112], 0 54 [78]	31 [78]	—
Duck	—	—	4 24 [107 A]

[1] Numbers in brackets refer to references p 194—196

of magnesium added to plasma which has first been freed of magnesium, a procedure which has only seldom been applied (WALSER, 1960). But the arguments presented for accuracy of many recent studies are convincing, and the possibility that substantial seasonal, climatic, or regional differences in plasma magnesium exist must now be seriously entertained. Evidence for some sources of variation are reviewed below. In my laboratory a recent series of normals yielded a mean significantly lower than a series analyzed five years ago (Table 2). There are no obvious sources of sampling error: venous compression leads to an increase of only about one per cent per minute (LEEUWEN, 1964). Hemolysis would have to be extensive to cause serious errors, since erythrocyte magnesium is only about two and one-half times as great as plasma magnesium.

2. Physiological variations

a) Age and sex. Men and women do not differ significantly in plasma magnesium (DAHL, 1950). Some increase with age is evident in humans (op. cit.)

as in cattle (GARTNER, 1966). The concentration of magnesium in fetal plasma is virtually identical with maternal (BOGERT and PLASS, 1923; DE JORGE, CANATO and DELASCIO, 1966) as is the concentration in plasma of newborn infants; some increase occurs thereafter, especially in breast-fed children (ANAST, 1964). A similar increase is seen in rats after birth (GREENBERG and TUFTS, 1936).

b) Effects of eating. Although fasting causes a considerable reduction of plasma magnesium in sheep (INGLIS, WEIPERS and PEARCE, 1959) and cows (ROBERTSON *et al.*, 1960), no such change occurs in man (STUTZMAN and AMATUZIO, 1952), nor do meals have any discernible effect (HEATON and HODGKINSON, 1963).

c) Effect of temperature. i. Environmental (in homeothermic animals). In the rat, plasma magnesium is not affected by wide changes in environmental temperature (MUNDAY and MAHY, 1962), nor by season (NOWELL and WHITE, 1963), but cold acclimatization produces an increase (HANNON, LARSON and YOUNG, 1958), whether body temperature remains constant or rises. After a month in the cold, the level returns to normal (NEUBEISER, PLATNER and SHIELDS, 1961). INGLIS, WEIPERS and PEARCE (1959) have presented convincing evidence of climatic fluctuations in plasma magnesium in sheep. QUINN, BASS and LEEMAN (1953) observed no change in magnesium in plasma or its ultrafiltrate in men subjected to low environmental temperature for 90 min., but DANIELS *et al.* (1953) noted a 21 % fall in men during longer cold exposure and acclimatization.

ii. Hibernation. Within an hour of the onset of hibernation, plasma magnesium rises substantially (25—65 %), without concomitant hemoconcentration, in bats, squirrels, hamsters, and hedgehogs (RIEDESEL and FOLK, 1956; MUNDAY and MAHY, 1962). It falls progressively during hibernation (PENGELLEY and CHAFFEE, 1966) and returns to normal an hour after arousal.

iii. Hypothermia. Cooling poikilothermic animals, such as goldfish or turtles, leads to pronounced increases in plasma magnesium and ultrafiltrable magnesium (PLATNER, 1950; PLATNER and HOSKO, 1953; MUNDAY and MAHY, 1962); magnesium in muscle falls, but remains constant in skin. These changes are augmented greatly by anoxia. Similar changes (within the limits of tolerance to hypothermia) are seen in dogs (AXELROD and BASS, 1956, MOUSSA and BOBA, 1960), hamsters, cats (PLATNER and HOSKO, 1953) and rats, especially if the experiments are performed in the winter (NOWELL and WHITE, 1963). Cold-acclimated rats show the same increase as controls (NOWELL and WHITE, 1964).

iv. Role of magnesium in thermoregulation. In nonhibernators, injection of magnesium induces a fall in body temperature (HEAGY and BURTON, 1948), while in hibernators, it may induce hibernation (SUOMALAINEN, 1939). Furthermore, magnesium injection in hypothermic dogs impairs their tolerance to low temperatures (MOUSSA and BOBA, 1960). ROGERS (1965) has suggested that release of magnesium from cooled tissues acts as a trigger to the onset of generalized hypothermia (in poikilotherms) or hibernation (in mammals); in nonhibernators compensatory mechanisms prevent the hypermagnesemia and thus further cooling does not occur. Arousal may be triggered by falling plasma magnesium (PENGELLEY and CHAFFEE, 1966).

d) Pregnancy. Plasma magnesium starts to fall early in pregnancy and reaches a mean value 20 % below normal near term (HALL, 1957; DE JORGE et al., 1965).

e) Acid-base balance. According to MAURAT, POCIDALO and LIPSAC (1964) experimentally induced respiratory acidosis augments plasma magnesium about 30 %; other acid-base disturbances are without effect. However, GEORGE, DOTSON and GRANT (1966) report that either hyperventilation or vigorous exercise for five minutes augments plasma magnesium in man.

f) Ethnic and regional. In Australia, the aborigines have lower plasma levels than the remainder of the population (CHARNOCK, CASLEY-SMITH and SCHWARTZ, 1959). In India, however, the same mean values are found throughout the country (BANERJEE and SAHA, 1964).

3. Physicochemical state of magnesium in plasma

a) Protein-binding. Earlier studies yielded widely discrepant estimates of protein-bound magnesium, but more recent observations in which temperature and pH have been controlled (Table 2), are in fair agreement as to the percentage ultrafiltrable, with a mean value of 70 %. This figure represents the difference between magnesium concentration in whole plasma (or serum) and in ultrafiltrate; it does not take into account the water content of these fluids or the Donnan factor, and is therefore a somewhat misleading estimate of the state of plasma magnesium. A more complete account of magnesium-protein interactions in normal human plasma is given by LEEUWEN (1964), who has revised earlier estimates of the net cationic equivalence of the plasma proteins and has thereby obtained more reliable values for the appropriate Donnan factors. In addition he has shown that estimates of protein-bound magnesium obtained by equilibrium dialysis or by measuring the rise in plasma magnesium which ensues during venous compression are close to those obtained by ultrafiltration.

Abundant evidence indicates that magnesium and calcium occupy the same sites on plasma protein molecules and therefore compete with one another for binding. In pure albumin solutions both cations have approximately the same affinity for the protein (CARR and WOODS, 1955), but the greater proportion of plasma calcium than magnesium bound to protein is presumably a reflection of greater affinity for plasma protein rather than its higher molar concentration

(OUTA, 1963). In hypercalcemia one would expect to find a reduction in protein-bound magnesium and possibly a consequent decrease in plasma magnesium; the converse should also be true. Suggestive evidence has been presented (HOPKINS, HOWARD and EISENBERG, 1952; WALSER, 1962), but further study of this question is needed. Measurements of calcium and magnesium binding in hypomagnesemia are discussed below (V A).

A study of magnesium binding by individual plasma proteins, electrophoretically separated, has been reported (PRASAD, FLINK and ZINNEMAN, 1959), but suffers from lack of control of pH or ionic strength, as does the report by WILLIS and SUNDERMAN (1952).

Table 3. *Physicochemical state of plasma magnesium in man* (WALSER, 1961 b)

	Per cent
Protein-bound	32
Ultrafiltrable	
Free Mg^{2+}	55
$MgHPO_4$	3
$MgCit^-$	4
Other complexes	6
	100

The use of ultracentrifugation as a mean of determining protein-binding (LOKEN et al., 1960) has the advantage of obviating the use of a membrane of uncertain pore characteristics, but has apparently not yet been applied to magnesium.

b) Free and complexed magnesium.
The calculation of quantities of individual complexes in an aqueous solution can be made with some assurance if the free cation concentrations, the total anion concentrations, the ionic strength, and the stability constants of the complexes are known. This procedure is useful when not all of the complexing anions are identified. Another procedure, requiring successive approximations, can be applied when the total cation and anion concentrations are known; this is useful only if all of the ligands have been identified. An example of the latter approach, applied to sea water, is given by GARRELS and THOMPSON (1962).

In body fluids the former technique is more useful because the complexing anions are not generally known in detail. Calcium and magnesium ion concentrations can be measured spectrophotometrically in optically clear body fluids (WALSER, 1960), and sodium and potassium can be assumed to be entirely ionized. The individual magnesium complexes estimated to exist in normal human plasma ultrafiltrate by this method are shown in Table 3. A similar value for free magnesium ion concentration in one human plasma sample was reported earlier by NORDBO (1939a) using a colorimetric method. BAUM and CZOK (1959) have reported results with a technique involving activation of isocitric dehydrogenase; between 93 and 102% of the ultrafiltrable magnesium in human plasma was found to be ionic. A preliminary report citing higher values in human and rat serum has recently appeared, in which an ion-exchange resin was used to measure free magnesium (MACLEOD, 1964).

The assertion of HENDRICKS and HILL (1951) that all of the magnesium in solutions "near neutrality" exists as MgOH⁺ is still quoted occasionally. This was probably a miscalculation: published values for the stability of this complex (see PYTKOWICZ, DUEDALL and CONNORS, 1966) indicate that the ratio of free magnesium ions to $MgOH^+$ at pH 7.4 is at least 1000.

C. Erythrocytes

1. Normal values. Magnesium concentrations in packed erythrocytes of several species, according to recent publications, are summarized in Table 2. The estimates in human subjects are quite consistent.

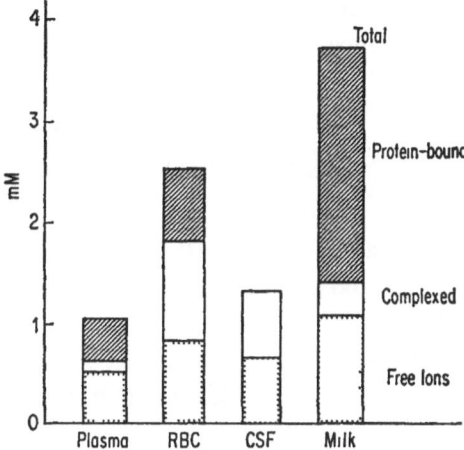

Fig. 1. The physicochemical state of magnesium in body fluids of man Plasma values from WALSER (1961 b) Normal human cerebrospinal fluid and milk samples were analyzed by the same method (WALSER, 1960). Erythrocytes were first hemolyzed by freezing and thawing, and were ultrafiltered overnight at 0°C
Note that the free cation concentration is relatively constant

2. Physiological variations. Young red cells contain much more magnesium than do old. Thus the average erythrocyte magnesium rises considerably following phenylhydrazine (HENRIQUES and ØRSKOV, 1939). In rabbits, the red cell magnesium doubles at a reticulocyte count of 40%, and reaches 14.2 mM at 85% reticulocytes (GINSBURG et al., 1962). By extrapolation, the reticulocytes must contain almost eight times the magnesium concentration of mature cells.

Effects of temperature are considered below (VI C 4).

3. Physicochemical state of red cell magnesium. Ultrafiltration at 0° of packed human erythrocytes, hemolyzed by freezing and thawing, yields a fluid containing 1.8 mM magnesium, of which only 0.8 mM is free magnesium ions (Fig. 1). A portion of red cell magnesium is doubtless bound to the membrane at sites which also bind calcium, but the amount is difficult to estimate (CARVALHO, SANUI and PACE, 1963; GENT, TROUNCE and WALSER, 1964). No amount of washing will remove it all; about 1 μmole per gm dry weight remains (WELT, 1964).

Isotopic magnesium added to plasma or suspension media equilibrates with erythrocyte magnesium remarkably slowly. In sheep, relative specific activity is only 0.2 at 20 hours, and requires 60 hours to reach unity (Care, MacDonald, and Nolan, 1959; MacDonald, Care and Nolan, 1959). Cat and dog erythrocytes equilibrate with a half-time of about 6 hours; human erythrocytes are only 6% equilibrated at this time (Rogers, 1961). Ginsburg *et al.* (1962) could find no appreciable exchange in 24 hours with human erythrocytes or rabbit reticulocytes. Yet, as reported below (VI C), red cell magnesium may change in response to metabolic alterations. Evidently transport processes can exist in the virtual absence of isotopic exchangeability, a finding which may signify that no passive leak or carrier-mediated diffusion is present, despite the existence of an active transport mechanism, or "pump."

D. Soft tissues

1. Normal values

Table 1 lists values from the recent literature for magnesium content of tissues from several mammalian species. All of the values have been converted to mmoles, expressed per 100 gm or kg of referent base, and rounded off. For analyses of other tissues, see Glick, Freier and Ochs (1957), de Azevedo and de Jorge (1965), Tipton and Cook (1963), Burch, Lazzara and Yun (1964), Manery (1954), and Widdowson and Dickerson (1964).

2. Physiological variations

Lowry *et al.* (1946) found no difference in the liver or brain magnesium concentration of rats 600 days old as compared with rats 1000 days old. Mori and Duruisseau (1960), however, observed a progressive decrease in aorta, heart, and liver, in animals 1 to 35 months old; muscle and kidney remained constant. In view of the increase in bone magnesium with age (see below) these results are not in conflict with the increased carcass magnesium with age reported above. Aortic magnesium increased with age (or arteriosclerotic index) in man (Nakamura *et al.*, 1965a).

In an acid medium, Gilbert (1961) noted a decrease in the magnesium content of frog muscle.

No regional differences in soft tissue magnesium were found by Tipton *et al.* (1965) in a survey of 162 adult subjects from various countries.

3. Physicochemical state of soft tissue magnesium

a) Binding in the extracellular space. Connective tissue components in all tissues have negatively charged groups which are associated with cations, including magnesium (Catchpole, Joseph and Engels, 1966). The extent of this binding can be estimated only by indirect methods at present, but it is well to bear in mind that estimates of cellular magnesium from total tissue magnesium, corrected only for extracellular fluid, are subject to considerable uncertainty.

b) Binding to subcellular structures. The distribution of magnesium in subcellular fractions prepared by ultracentrifugation of homogenized tissue has been studied by several authors (GRISWOLD and PACE, 1956; THIERS and VALLEE, 1957; BOROWITZ, FUWA and WEINER, 1965; BRESCIANI and AURICCHIO, 1962; CARVALHO, 1966; WALSER and NAHMOD, 1966); all have noted high affinity of microsomes of magnesium. The last named authors attempted to correct for the trapped supernatant in each step of centrifugation and thus obtained an estimate of the amount of cations bound; by this technique the percentage of tissue magnesium associated with particles of dog renal cortex was as follows: nuclei and cellular debris, 24%; mitochondria, 19%; "heavy" microsomes, 6%; "light" microsomes, 9%. Whether a change in binding occurs in association with energy utilization and membrane transport has yet to be determined. ATP evidently reduces microsomal binding (SANUI and PACE, 1965), but the effect may be attributable to chelation.

c) Free and bound magnesium in the cell. It is well established that cell lipo- and nucleo-proteins, ribonucleic acids, and free adenosine triphosphate are capable of binding magnesium (HAMILTON and PETERMANN, 1959; EDELMAN, TSÒ and VINOGRAD, 1960; CARVALHO, SANUI and PACE, 1965; SILLÉN and MARTELL, 1964). An attempt to estimate the extent of interactions between the major cations and anions within the cell was made by NANNINGA (1961), using the method of calculation described above (II B 3). He concluded tentatively that the free magnesium ion concentration within the muscle cell is 3.4 mM, or 30% of the total. If true, this would indicate a substantial chemical gradient between cell and interstitial fluid. However, the electrochemical gradient is in the opposite direction.

d) Isotopic exchangeability. The use of radioactive magnesium has made possible measurement of the rate at which soft tissue magnesium in various organs exchanges with extracellular magnesium (see AIKAWA, 1963, 1965a). It is generally agreed that liver, heart, and kidney magnesium equilibrates rapidly; pancreas, brain, and nerve more slowly; and muscle even more slowly (ROGERS and MAHAN, 1959; ROGERS et al., 1964; BRANDT, GLASER and JONES, 1958; GLASER and BRANDT, 1959; FIELD and SMITH, 1964). Stimulation increases exchange in muscle (ROGERS and MAHAN, 1959). In small animals equilibration is more rapid; for example, liver and heart have equilibrated with plasma in three hours in the rat (ROGERS and MAHAN, 1959) but not in the dog (WALLACH et al., 1964). In neoplastic cells, exchange was complete in four hours (ROGERS, HAVEN and MAHAN, 1960). Although it is tempting to infer that exchangeable magnesium as measured in this way is "available", i.e., to maintain extracellular concentration, there is abundant evidence, summarized below, that this is not the case. The significance of isotopic exchangeability at our present state of knowledge appears to be wholly negative: the absence of radioactive magnesium in a tissue or fluid can be used to exclude the hypothesis that transfer of magnesium to this site from the extracellular fluid has occurred.

E. Hard tissues

1. Normal values. Table 2 lists bone and tooth magnesium content in several species. According to MORGULIS (1931), the percentage of bone ash contributed by magnesium, 0.5—0.7%, varies little from one species to another. It is also quite constant in different bones of the same animal (R. H. SMITH, 1959b).

2. Physiological variations. The magnesium content of osteons of rats diminishes as they age according to FABRY and CLOOSEN (1961), but BREIBART *et al.* (1960) found an increase in bone magnesium with age. In carious teeth magnesium content is greatly decreased (RITCHIE, 1961; JOHANSEN, 1963).

3. Physicochemical state. Although in principle, magnesium ions could replace calcium ions in the crystalline apatitic structure of bone (NEUMAN and NEUMAN, 1958; POSNER, 1960), the degree to which this occurs is a matter of controversy (see BACQ, 1964). All agree that the majority of skeletal magnesium is adsorbed on the surface of the microcrystals (TAYLOR, 1959). Nevertheless, the extent of exchange of skeletal magnesium with injected radioactive magnesium is very slight (MACINTYRE, 1963b; FIELD and SMITH, 1964; WALLACH *et al.*, 1964; GLASER and GIBBS, 1962). *In vitro*, exchange decreases at low pH, or if metabolic activity increases (AIKAWA, 1965a).

WIDDOWSON and DICKERSON (1964) have calculated that in the adult human, 66% of the body magnesium is found in the skeleton and only 19% in muscle.

F. Transcellular fluids

1. Cerebrospinal fluid. It has been repeatedly demonstrated that the concentration of magnesium in cerebrospinal fluid exceeds that of plasma (see BACQ, 1964), though occasional exceptions have been noted in the rabbit (DAVSON, 1956). The concentration of free magnesium ions in four samples of normal human spinal fluid averaged 0 7 mM, or 55% of the total (WALSER, unpublished observations). The nature of the complexes which account for this difference is unknown. Exchange with plasma magnesium is rapid (OPPELT, MACINTYRE and RALL, 1963).

2. Digestive secretions. Attempts to cite reproducible values for magnesium concentrations in digestive secretions have been uniformly unsuccessful, because of the pronounced variation with secretion rate and other factors. These data are reviewed *in extenso* by BACQ (1964). According to VAGNE,

MARTIN and LAMBERT (1964), human gastric juice contains about 0.6 mM in the fasting state, and less after histamine. The physicochemical state of magnesium in these fluids is not well studied. Some of the magnesium of saliva (GOW, 1965) and small intestinal contents (CARE, 1965) is non-ultrafiltrable. A portion of this magnesium in the gut, in solid form, also fails to exchange with radioactive magnesium entering *via* endogenous secretions — an observation which appears to invalidate the assumption often made, in calculations of endogenous magnesium or calcium secretion, that endogenous and exogenous cation share the same fate in the gut (see section III B 1 c). In human feces, approximately one-third of the magnesium is freely diffusible (WRONG *et al.*, 1965).

3. Milk. BACQ (1964) cites values for milk magnesium concentration culled from the literature varying from 1 mM in humans to 10 mM in cows. The physicochemical state of magnesium in milk has been studied in some detail. Approximately one-fourth is non-ultrafiltrable (BACQ, 1964), According to NORDBO (1938, 1939b), cow's milk ultrafiltrate, containing 2.5 mM magnesium, has only 0.4 mM of ionized magnesium, the remainder being complexed by citrate and phosphate. In 3 samples of normal human milk (Fig. 2), total, ultrafiltrable, and free ionic magnesium averaged 3.8 mM, 1.4 mM, and 1.0 mM, respectively (WALSER, unpublished). Corresponding values in cow's milk are given by CHRISTIANSON, JENNESS and COULTON (1954). The presence of unidentified ligands with strong magnesium- and calcium-binding properties in milk is highly likely. Changes in magnesium activity during processing are described by KREVELD and MINNEN (1955).

4. Other transcellular fluids. Sweat magnesium, according to CONSOLAZIO *et al.* (1963) averages 0.3 mM in human subjects in a hot environment. Values for various ocular fluids are given by DAVSON (1962) and SAKANOUE (1964). For other references see BACQ (1964). Fluids which appear to have the same ionic magnesium concentration as plasma include aqueous humor, amniotic fluid (PRITCHARD, 1955; DE JORGE, CANATO and DELASCIO, 1966), lymph (KEYL *et al.*, 1965), and peritoneal fluid (GROLLMAN, 1954b).

III. Intake and excretion in normal animals

A. Intake and dietary requirements

The intake of magnesium, like that of other common minerals, varies enormously, chiefly because the magnesium content of common foods is so variable. In man, an average figure of 15 mmoles per day is often cited, though normal individuals on self-selected diets have been observed to consume considerably more (SCOULAR, PACE and DAVIS, 1957) or less (DUNN and WALSER, 1966; SEELIG, 1964). The magnesium requirements during growth obviously depend upon the rate of growth of new tissue and its magnesium content;

an amount somewhat in excess of this quantity must be provided (O'DELL, 1960). During lactation, requirements are also high; in sheep 8 to 20 mmoles a day are required to prevent hypomagnesemia (L'ESTRANGE and AXFORD, 1963).

The requirement for magnesium in adults under ordinary circumstances is a matter of considerable importance but widespread disagreement (DUCK-WORTH and WARNOCK, 1942). Some have endeavoured to answer this question by measuring magnesium balance on varying intakes. The basic premises of this approach are first, that the organism will have the wisdom to reject any magnesium beyond its needs, and second, that negative balance indicates that intake is less than the requirement. Both of these may be questioned. As pointed out in the introduction, there is a distinct possibility that no homeostatic mechanism exists in the organism with the primary responsibility of regulating magnesium metabolism. An excessive intake might therefore result in continued positive balance until toxic manifestations ensue. Furthermore, the intake of magnesium is so small in relation to the amount in the body that, if excretion were always proportional to the body content, the establishment of a new steady state following a change in intake would require many months. Thus negative balance persisting for many weeks following a reduction in intake cannot be accepted as evidence that the intake is too low; the preceding intake may have been too high.

Apart from these considerations, there is substantial evidence to suggest that mammalian organisms do not have the capacity rapidly to adjust magnesium excretion to magnesium intake. For example, positive balances were almost as large after four weeks of high magnesium intake as at the onset in a study in normal subjects summarized by SEELIG (1964). Magnesium absorption by the intestine of the rat is proportional to the intake, irrespective of the animal's needs (CHUTKOW, 1964b).

If magnesium requirement is simply defined as the intake necessary to prevent objective evidence of magnesium deficiency, an entirely different value is obtained from the estimate of 6 mg/kg per day arrived at by the balance method (SEELIG, 1964). As indicated below, no more than 1 or 2 mmoles per day is necessary. On the other hand, optimal magnesium intake may be considerably greater than the usual intake in western countries, as suggested by studies of the incidence of atherosclerosis (see Section IX C). Further work on this question is clearly needed.

A number of factors have been observed to augment the magnesium requirement for growth in rats, including a cold environment (HEGSTED, VITALE and McGRATH, 1956), the ingestion of alcohol (GOTTLIEB et al., 1959), and a high protein intake (MENAKER, 1954). Whether these factors also increase the magnesium requirement in adult subjects is uncertain.

B. Excretion

1. Fecal

a) Normal values. The principal loss of magnesium is in the stools, which usually contain about 60% of the ingested magnesium (LEICHSENRING, NORRIS and LAMISON, 1951; HATHAWAY, 1962). In man, 60—88% of an oral dose of radioactive magnesium appears in the feces (AIKAWA, RHOADES and GORDON, 1958). Fecal magnesium output correlates better with fecal nitrogen than with magnesium intake in subjects on hospital diets (HENNEMAN and DEMPSEY, 1956).

b) Physiological variations in intestinal absorption. i. Flow. In section VI B 1 are considered factors which modify the rate of magnesium transport at any point in the gut. The total rate of magnesium reabsorption is the integrated sum of the quantities being reabsorbed at each point. Owing to the progressive changes in the luminal concentration of magnesium (and also of calcium) along the length of the gut, the total reabsorption is not related in a simple way to the local reabsorptive rate. Water removal concentrates luminal magnesium and thereby promotes its reabsorption. Thus the percentage of ingested magnesium absorbed diminishes if water reabsorption decreases. A linear relationship between magnesium reabsorption and transit time to the distal ileum in sheep was demonstrated by R. H. SMITH (1963). The effect of changes in water reabsorption on solute reabsorption from uniform tubular segments has been considered theoretically by WALSER (1966a), and may be applicable to such data. In his later work, R. H. SMITH (1964a) has shown that transit time is decreased by high magnesium intake and increased by propantheline.

ii. Calcium. The percent of entering calcium remaining in gut segments was shown to be a power of the percent of entering strontium remaining (MARCUS and WASSERMAN, 1966). This exponential relationship is seen in mathematical models of uniform segments, when two solutes are reabsorbed passively or by a common "pump" (WALSER, 1966b). The applicability of this relationship to magnesium and calcium remains to be demonstrated, but some such interdependence must occur. Thus a parallelism between calcium and magnesium absorption may exist, whether they share a common "pump" or not, simply because they share the same flow.

Another type of interdependence between calcium and magnesium absorption, independent of transport phenomena, may be expected to result from the fact that both cations are extensively and competitively bound to macromolecules or solid matter in the luminal contents (STORRY, 1961c; SMITH and McALLAN, 1966).

Thus the observations of ALCOCK and MacINTYRE (1964) that reduction in the intake of either calcium or magnesium in rats leads to a corresponding reduction in the fecal excretion of the other cation does not establish the existence of common mechanisms for reabsorption. Such results could reflect altered flow of intestinal contents, or altered binding in the lumen. The increase in fecal magnesium following increased calcium intake in man (HEATON, HODGKINSON and ROSE, 1964), may have a similar explanation. Furthermore, the effects of each of these cations on intestinal absorption of the other are far from consistent (see WILSON, 1964; CLARK, 1965 a).

Evidence of a more direct nature on the question of common transport mechanisms for these two cations is summarized below (VI B).

iii. Phosphate. Although earlier workers (TIBBETTS and AUB, 1937a; MEINTZER and STEENBOCK, 1955) found no inhibition of magnesium absorption by dietary phosphate or phytate, WALKER, FOX and IRVING (1948), and LIKUSKI and FORBES (1965) found them inhibitory, as did HEATON, HODGKINSON and ROSE (1964). According to BUNCE et al. (1965) the effect depends on the level of magnesium intake: at low magnesium intake, phosphate promoted magnesium absorption, but at high intake it was inhibitory; high magnesium intake also diminished phosphate absorption (at high phosphate intake).

iv. Lactose. Addition of lactose to the diet augments magnesium absorption (FEATHERSTON, MORRIS and PHILLIPS, 1963; FORBES, 1964b) and bone uptake of an oral dose of isotopic magnesium (LENGEMANN, 1959). Part of this effect may be due to increased food consumption (SCHOLZ and FEATHERSTON, 1966). Magnesium balance, however, is not increased, because urinary magnesium excretion is stimulated; in fact magnesium balance may be negative (FORBES, 1961) and the magnesium requirement on a low magnesium diet is increased (FORBES, 1964a).

v. Potassium. Increase in dietary potassium augments fecal magnesium in lambs (FONTENOT et al., 1960) and cows (KEMP et al., 1961).

c) Endogenous fecal excretion. Magnesium enters the gut via digestive secretions, in desquamated cells of the gut epithelium, and by transport across the gut wall, with or without an accompanying secretion of *succus entericus.* The amounts contributed by these sources are difficult to estimate. MARCUS and WASSERMAN (1966) have presented evidence that the quantity arising from sloughed cells is relatively small. In sheep, the measured magnesium concentrations of gastrointestinal secretions, multiplied by the volume flow, yield a value of 8 mmoles per day; of this 5 mmoles is contributed by saliva and 2 by gastric secretion (STORRY, 1961a). Using average values for magnesium concentrations in gastrointestinal secretions in man cited by THOREN (1963) or GEERTRUYDEN (1964) and conventional estimates of their volumes, a similar calculation yields a value of about 4 mmoles per day.

In order to estimate the quantity of endogenous magnesium which is excreted in the feces, it is necessary to make one or more simplifying assumptions. The first problem is the relative positions in the gut of the sites of endogenous magnesium secretion and magnesium reabsorption. As noted below (VI B 2), the localization of reabsorption is not well known, and it is clear that magnesium is added to the gut throughout its length. The assumption that endogenous magnesium shares the same fate as exogenous magnesium is therefore questionable at best. Furthermore, endogenous cation may fail to equilibrate with cation in solid intestinal contents (FIELD, 1961). Thus the proportion of endogenous magnesium absorbed may exceed that of exogenous magnesium.

In human subjects given radioactive magnesium intravenously, the proportion of the isotope found in the stool has been reported as 2% (AVIOLI, LYNCH and BASTOMSKY, 1963) or less (ZUMOFF et al., 1958, GRAHAM, CAESAR and BURGEN, 1960; SILVER, ROBERTSON and DAHL, 1960). The first-named authors calculate that 4% of fecal magnesium arose from endogenous secretions, while GRAHAM et al. (1960) estimate fecal magnesium derived from endogenous loss as 1 mmole per day. In sheep, isotopic studies have led to estimates of endogenous fecal loss of about 9 mmoles per day (MACDONALD, CARE and NOLAN, 1959; FIELD, 1959; CARE, 1960), in agreement with the value given above. Similarly, evidence for substantial endogenous loss in rats has been presented (CHUTKOW, 1964a and b). Ingestion of roughage augments the loss in calves greatly (SMITH, 1961a), but not in rats (B. S. W. SMITH, 1966), probably because of differences in salivary flow.

When fecal loss exceeds intake, the difference gives an unequivocal (though minimal) value for endogenous loss. In calves, this difference may amount to 2 mmoles per day (BLAXTER, ROOK and MACDONALD, 1954b), in cows, as much as 20 mmoles (STORRY and ROOK, 1963), and in normal man, 1 mmole (DUNN and WALSER, 1966). In patients with malabsorption, fecal excretion may exceed intake by as much as 6 mmoles per day (PETERSEN, 1963b; OPIE, HUNT and FINLAY, 1964).

In summary, it appears that some 4 mmoles of magnesium per day is added to the gastrointestinal tract in humans, and that more than half of this quantity is normally absorbed. In ruminants, the quantities are approximately twice as great.

2. Urine

a) Normal values. In normal subjects 3 to 6 mmoles of magnesium are excreted daily via the kidneys (LEICHSENRING, NORRIS and LAMSON, 1951; STRIEBEL and BAUR, 1954, HANZE, 1962c, HEATON and PYRAH, 1963; THOREN, 1963). These values correspond to magnesium clearance of 2.5 to 5 ml./min., but a wider range of variation, with a higher mean, is seen in short collection periods (MERTZ, 1957; HAMMARSTEN, ALLGOOD and SMITH, 1957; MCCOLLISTER et al., 1958).

b) Physiological variations. i. Diurnal variation. Conflicting results have been reported. CAMPBELL and WEBSTER (1921) found magnesium excretion to be constant throughout 24 hours. Doe, Vennes and Flink (1960), and Glaubitt and Rausch-Stroomann (1962) observed peak excretion during the night. Wesson (1964) and deVries et al. (1960) found other patterns. Many of these results were doubtlessly obscured by the effects of meals, detected more recently (see below). Heaton and Hodgkinson (1963) observed peak excretion at night in fasting subjects, whether the subjects slept in the daytime or at night. Min, Jones and Flink (1966) fed subjects every three hours and noted an early morning peak in magnesium and calcium excretion, preceding the peak in sodium and potassium excretion, and 180° out of phase with phosphate excretion.

The source of the diurnal variations, as well as the inconsistency in these results is obscure. The adrenal cortex is evidently not responsible (Doe, Vennes and Flink, 1960).

ii. Meals. Ingestion of food is followed by a prompt rise in calcium and magnesium excretion, despite unaltered plasma or ultrafiltrate concentrations (Hodgkinson and Heaton, 1965). Sodium excretion changes little, and filtration rate alterations are not responsible; similar effects can be produced by administering any rapidly metabolizable substrate (Lindeman et al., 1965). The mechanism of this intriguing response remains to be elucidated.

iii. Exercise. Calcium and magnesium excretion decrease following exercise (Heaton and Hodgkinson, 1963).

iv. Intake of other electrolytes. Dietary sodium restriction diminished magnesium excretion according to Hills et al. (1959) but not according to Miller, Faloon and Lloyd (1958). High salt intakes had no effect (Wesson, 1964), nor did KCl (Jabir, Roberts and Womersley, 1957). The latter workers found increased magnesium excretion following NH_4Cl ingestion, as did Hills et al. (1959), but this was not observed by McCollister et al. (1958) or Martin and Jones (1961). $NaHCO_3$ ingestion has variously been reported to lead to an increase in magnesium excretion (op. cit.), a decrease (Hills et al., 1959), and no change (Barker, Elkinton and Clark, 1959). All agree that increasing calcium intake augments magnesium excretion (op. cit.; Mendel and Benedict, 1909; Leichsenring, Norris, and Lamison, 1951; Barker, Elkinton and Clark, 1959). The mechanism is discussed below (VI A).

3. Dermal losses

From earlier literature Seelig (1964) has estimated skin losses as 1 mmole per day. During exercise in a hot, humid environment, normal subjects lost about 0.1 mmole per hour by this route (Conzolazio et al., 1963).

C. Response of excretion to dietary restriction

In normal subjects fed a diet very low in magnesium and calcium, Barnes, Cope and Harrison (1958) noted virtually complete disappearance of magnesium from stools and urine ($< 0 5$ mmoles/day) in three of four subjects; similar results were obtained in a patient tube-fed a magnesium-free diet for 39 days (BARNES, COPE and GORDON, 1960). FITZGERALD and FOURMAN (1956) fed two normal subjects a solid diet containing 0.6 mmoles of magnesium for 3 or 4 weeks. Fecal loss was less than 1 mmole per day, unless an ion-exchange resin was given. Urinary losses became very low. In all of these studies, little or no decrease in plasma magnesium concentration was observed.

Magnesium excretion in growing animals fed low magnesium diets has repeatedly been noted to fall promptly to extremely low values (see section V B 2 e). In fully grown dogs, urinary magnesium excretion on a magnesium-free (but calcium-containing) diet averages 0.325 mmoles per day (WALSER, unpublished observations). In non-lactating cows, urinary excretion falls to unmeasurably low values in a period of four days on a low magnesium diet (STORRY and ROOK, 1963).

IV. Effects of magnesium
A. On cell-free systems

1. Enzymatic reactions. Magnesium is required for a large number of enzymatic reactions, particularly those involving transphosphorylations. This subject has been extensively reviewed by LEHNINGER and WADKINS (1962) and MAHLER (1961) and will be only touched upon here. Not all recent work has reinforced the view that a one-to-one complex of magnesium and adenosine triphosphate (ATP) is the actual substrate of ATP-splitting reactions. The optimum magnesium concentration for transphosphorylating reactions in mitochondrial preparation is that which yields the highest concentration of MgATP$^=$ under the conditions of the experiments (ULRICH, 1964). Similarly, dissociation of actomyosin in solution is produced by a one-to-one complex of magnesium with pyrophosphate but not by magnesium or pyrophosphate alone (GRÁNICHER and PORTZEHL, 1964). However, the substrate for myofibrillar ATPase is free ATP, not MgATP$^=$ (MUHLRAD, KOVACS and HEGYI, 1965), and the course of the curve describing the activation by magnesium does not depend on the concentration of ATP. Similarly, the ATPase activity of erythrocyte membranes (both total and sodium-potassium-activated) varies with magnesium concentration and with ATP concentration but not with MgATP$^=$ (WELT and TOSTESON, 1964; cited by WELT, 1964).

Although the sodium- and potassium-activated, magnesium-requiring, ATP-splitting system discovered by SKOU (1957) usually exhibits some activity with magnesium alone, this activity can be separated from the monovalent-cation-activated portion (RENDI and UHR, 1964; SOMOGYI, 1964; NAKAO *et al.*, 1965). It therefore remains uncertain whether magnesium is specifically involved in enzymatic sodium-potassium exchange across cell membranes, although most suggested mechanisms have assigned it a major role (SKOU, 1964).

Magnesium may also be involved in protein synthesis (REVEL and HIATT, 1965, KRAHL, 1966). Whether variations in intracellular magnesium concentration *in vivo* could exert effects upon enzymatic reaction rates has not been established, but it seems likely that the deleterious consequences of reduced cellular magnesium (v.i.) are related to these processes.

2. *On subcellular structures.* Maximal contraction of myofibrils prepared from skeletal (WATANABE, SARGEANT and ANGLETON, 1964; LEE, TANAKA and YU, 1965), cardiac (EMBRY and BRIGGS, 1966), and smooth muscle (FILO, BOHR and RUEGG, 1965) requires both magnesium and ATP, evidently in nearly equal concentrations. In fibrils from skeletal muscle, these substances are required at two sites, only one being the site of ATP hydrolysis (LEVY and RYAN, 1966). At the latter site, free magnesium increases activity only if the concentration of calcium is high enough to suppress dissociation, according to MUHLRAD, KOVACS and HEGYI (1965); otherwise it is inhibitory. In relaxation, both magnesium and ATP are also required, and magnesium pyrophosphate has a similar effect (GRANICHER-FRICK, 1965); here free magnesium enhances the response. Calcium has little effect on this system (op. cit.) and its requirement for contraction can also be reduced or eliminated (EMBRY and BRIGGS, 1966).

Inhibition of phosphofructokinase by magnesium may be responsible for the accumulation of ATP, diminution of contractile activity, and resulting slowing of glycolysis in muscle to which excess magnesium is added (REGEN et al., 1964). In mitochondria, thyroxin-induced uncoupling of oxidative phosphorylation can be counteracted by addition of magnesium (BAIN, 1954, MUDD, PARK and LIPMANN, 1955; GERSHOFF et al., 1958). It may also play a role in normal calcification (NEUMAN and NEUMAN, 1958; TAVES and NEUMAN, 1964).

3. *In body fluids.* Magnesium increases the apparent solubility of calcium phosphate in urine (VERMEULEN, LYON and MILLER, 1958).

B. Isolated Tissues

1. *Neural tissue.* Magnesium induces neuromuscular block chiefly by inhibition of the release of acetylcholine at the neuromuscular junction (HOFF, SMITH and WINKLER, 1940, ENGBAEK, 1952; DEL CASTILLO and ENGBAEK, 1954; ELMQUIST and QUASTEL, 1965). This effect can be reversed by calcium (HUTTER and KOSTIAL, 1954). Magnesium also increases the stimulus threshold in nerve fibers (FRANKENHAEUSER and MEVES, 1958). The stimulation of catecholamine release from the adrenal medulla by acetylcholine is inhibited by magnesium (DOUGLAS and RUBIN, 1963), as is the stimulation of vasopressin release from the neurohypophysis by potassium (DOUGLAS and POISNER, 1964). All of these effects may reflect alterations in the permeability of cell membranes or subcellular organelles.

When compared with calcium, magnesium is less effective in stabilizing the electrical properties of squid axon (FRANKENHAEUSER and HODGKIN, 1957), but evidently competes with calcium for the same receptors in frog muscle end-plate (JENKINSON, 1957).

In itself, magnesium does not share the stimulatory properties of other divalent cations on adrenal medullary secretion (DOUGLAS and RUBIN, 1964), but isolated chemoreceptors show a pronounced stimulation by magnesium in the concentration range of 0.1 mM to 1 mM; at higher concentrations (5 mM) block appears, which in this case is not reversed by calcium (EYZAQUIRRE and KOYANO, 1965). Other effects are reviewed by ENGBAEK (1952) and BACQ (1964).

2. *Smooth muscle.* As noted by BACQ (1964) and BOHR (1964), magnesium depresses membrane excitability and/or excitation-contraction coupling in smooth muscle. Decreased contractility is seen in blood vessels (SCOTT et al., 1961; BOHR, 1964; HANENSON, 1962; HADDY et al., 1963) and gut (SPERELAKIS, 1962; MEYTS, 1965). The responsiveness of intestinal segments to epinephrine and acetylcholine is inversely related to the concentration of magnesium in the bath (MEYTS, 1965). Again, an effect on permeability may be involved, and TIDBALL (1964) has shown that magnesium can restore normal permeability of rat gut after it has been augmented by EDTA.

In the uterus, it is generally held that magnesium diminishes contractile activity (CLEGG, HOPKINSON and PICKLES, 1963), but augments responsiveness to posterior pituitary hormones. Estradiol and progesterone increase the magnesium content of rabbit and guinea pig myometrium, and estrogen does so in the rat (WALAAS, 1950; COUTINHO, 1961; BEST and PICKLES, 1965); the resulting changes in sensitivity to vasopressin and prostaglandins are mimicked by high external magnesium, and may therefore be attributable to increased intracellular magnesium (BEST and PICKLES, 1963). Similarly, reduction of external magnesium concentration diminishes the vasoconstrictor effects of oxytocin and vasopressin (SOMLYO, WOO and SOMLYO, 1966). Further studies along these lines will be of interest.

3. *Skeletal muscle.* FRATER, SIMON and SHAW (1959) found that magnesium added to toad muscle *in vitro* entered a portion of the tissue water not accessible to inulin, but failed to penetrate all of the chloride space. They inferred that muscle comprises three phases, two of which are intracellular. A single intra-cellular phase inaccessible to magnesium but containing substantial amounts of chloride is also consistent with their data. GILBERT and McGANN (1960) performed rather similar experiments on frog muscle, with results which seem completely opposite (despite the fact that their interpretation is similar). They noted that the increment in tissue magnesium (on soaking the muscles for five hours in magnesium solutions) was even greater (in mmoles/kg.) than the increment in external magnesium concentration (in mM). These results are

contrary to observations in intact animals (see below) as well as to the above *in vitro* data.

4. Heart. In isolated hearts, magnesium induces slowing of the pacemaker (SCHMIDT, SCHMIER and SCHMITZ, 1965).

C. In intact animals following parenteral administration

1. Distribution and fate of injected magnesium. The volume of distribution of magnesium administered intravenously is somewhat greater than that of the extracellular fluid (SMITH, WINKLER and HOFF, 1942). Intracellular magnesium increases little in tissues, if correction for magnesium in extracellular space is made (HINGERTY, 1957; FLOWERS, 1965; WALLACH, BELLAVIA and GAMPONIA, 1965).

Magnesium administered intravenously is largely recovered in the urine (MCCANE and WIDDOWSON, 1939; WOMERSLEY, 1958; FOURMAN and MORGAN, 1962). At least 65% appears in 24 hours (JONES and FOURMAN, 1966). To maintain the plasma concentration at 2.5 mM requires about 100 mmoles per day if glomerular filtration rate is normal (FLOWERS, 1965).

Magnesium concentration in cerebrospinal fluid is only slightly increased during several hours of hypermagnesemia (KEMÉNY, BOLDIZSÁR and PETHES, 1958, 1961; SCHAIN, 1964a), despite rapid exchangeability of this pool (OPPELT, MacINTYRE and RALL, 1963). Studies of cerebrospinal fluid during prolonged hypermagnesemia do not appear to have been reported. Magnesium crosses the placenta more readily (PRITCHARD, 1955) and hypermagnesemia with respiratory depression was observed in a newborn infant whose mother received parenteral magnesium before delivery (FISHMAN, 1965).

In saliva (ESCHLER, OCHS and SCHILLI, 1965), gastric juice, and pancreatic juice, magnesium concentration rises, but apparently not in *succus entericus* (GEERTRUYDEN, 1964). No detectable change in fecal excretion occurs on an ordinary diet, but might well be detectable on a magnesium-free diet; this has not been studied.

2. Renal effects. Glomerular filtration rate, in dogs or humans infused with magnesium salts, may remain constant (WOMERSLEY, 1956; PRITCHARD, 1955; ARDILL et al., 1962; KELLY et al., 1960), rise (ETTELDORF et al., 1952) particularly in hypertensives (KELLY et al., 1960), or fall (PRITCHARD, 1955; HAMMARSTEN, ALLGOOD and SMITH, 1957; SAMIY, BROWN and GLOBUS, 1960; KNIPPERS and HEHL, 1965; BARKER, CLARK and ELKINTON, 1957). It is possible that the reported decreases were the result of the experimental procedures. Renal blood flow and the maximal rate of tubular transport of para-aminohippurate change in the same direction as GFR.

Some have found that increased excretion of sodium, potassium, and calcium occurs (SAMIY, BROWN and GLOBUS, 1960; CHESLEY and TEPPER, 1958; ARDILL et al., 1962; KNIPPERS and HEHL, 1965; WISWELL, 1961). This is by no means invariable, and seems to be pronounced only in subjects with hypertension (KELLY *et al.*, 1960). Others have found variable increments in potassium excretion or a diminution (HELLER, HAMMARSTEN and STUTZMAN, 1953; WOMERSLEY, 1956; HAMMARSTEN, ALLGOOD and SMITH, 1957; JABIR, ROBERTS and WOMERSLEY, 1957; CHESLEY and TEPPER, 1958). The natriuresis may be very transient (PRITCHARD, 1955; CHESLEY and TEPPER, 1958; BARKER, ELKINTON and CLARK, 1959) and is probably in part attributable to the sulfate which is often given concomitantly (WOMERSLEY, 1956). The rise in calcium excretion is perhaps the most consistent finding (besides the hypermagnesuria), and has been shown to be independent of any change in parathyroid activity (MACRAE, KOHLER and PECHET, 1963). Excretion of strontium increases as well as that of calcium (MRAZ and CRAGLE, 1958). It is usually attributed to competition between calcium and magnesium for a common transport mechanism (v.i.). Nevertheless, one report in which calcium excretion fell during intravenous magnesium administration has appeared (WOMERSLEY, 1958). Effects on plasma phosphate and phosphate excretion are variable; usually, increased excretion occurs, independently of parathyroid activity (MACRAE, KOHLER and PECHET, 1963), possibly owing to formation of $MgHPO_4$ ion-pairs in tubular fluid (WALSER and MUDGE, 1960).

3. Cardiovascular effects. In relatively low dosage a modest fall in blood pressure occurs (ENGBAEK, 1952; PRITCHARD, 1955; MAXWELL, ELLIOTT and BURNELL, 1965; KNIPPERS and HEHL, 1965). The fall is greater in hypertensive subjects (KELLY *et al.*, 1960). The mechanism has not been fully elucidated, but blockage of sympathetic impulses to the heart (STANBURY, 1948) apparently has been excluded (MAXWELL, ELLIOTT and BURNELL, 1965). These authors also report that the hypotension is reversed by calcium administration and that coronary flow does not change in intact animals, contrary to earlier reports (STANBURY and FARAH, 1950; SCOTT *et al.*, 1961) and clinical impressions in treating angina pectoris (v.i.). It has evidently not been established that over-all peripheral resistance diminishes, although studies of vascular beds perfused with magnesium-enriched blood (SCHMID *et al.*, 1955; FROHLICH, SCOTT and HADDY, 1962) would suggest that it does.

Heart rate may rise or fall. Atrioventricular conduction is diminished at levels two or three times normal (MILLER and VAN DELLEN, 1938; SMITH, WINKLER, and HOFF, 1939; GRANTHAM, TU and SCHLOERB, 1960; TAKAYASU *et al.*, 1962), even after atropine or vagotomy (DELLEN and MILLER, 1939; SZEKELY and WYNN, 1951). Intraventricular conduction may also become delayed, especially if hyperkalemia is simultaneously present (TAKAYASU *et al.*, 1962). Digitalis-induced ectopic rhythms tend to be suppressed (SZEKELY and WYNN, 1951; STANBURY and FARAH, 1950). However, the delay in atrioventricular conduction may lead to ectopic ventricular arrythmias. Other electrocardiographic alterations are minimal (GARB, 1951) except in the presence of subnormal calcium concentration; under these conditions, the effects of magnesium resemble those of calcium (Surawicz, Lepeschkin and Herrlich, 1961).

4. Effects on the nervous system. Magnesium in increasing dosage induces neuromuscular paralysis, sedation, anesthesia, respiratory depression, coma, and death. The mechanism of the neuromuscular effect has been discussed. Central depression begins to appear at levels above 4 mM (Neuwirth and Wallace, 1929; Smith, Winkler and Hoff, 1942; Flowers, 1965). For surgical anesthesia the optimal level is about 10 mM (Moore and Wingo, 1942). Neostigmine plus an analeptic together counteract all the neural effects (Borglin and Lindsten, 1949). Coutino (1966) has suggested, on the basis of evidence from a model system, that magnesium induces anesthesia by substituting for calcium at phospholipid binding sites in cell membrane. Other evidence is reviewed by Engbaek (1952), Bacq (1964) and Fishman (1965).

5. Other effects. Plasma calcium often falls (Kelly *et al.*, 1960, Kemény, Boldizsár and Pethes, 1961; Knippers and Hehl, 1965; Jones and Fourman, 1966), including free ionic calcium concentration (Gitelman, Kukolj and Welt, 1966). Conceivably, inhibition of parathyroid function is responsible; perfusion of the parathyroids with magnesium-enriched blood diminished parathyroid hormone output (Care *et al.*, 1966). However, plasma phosphate is unaltered (Ardill *et al.*, 1962).

Plasma potassium may also fall (Gower-Smith, 1949; Kemény, Boldizsár and Pethes, 1961; Maxwell, Elliott and Burnell, 1965).

Muscle ATP concentration rises (du Bois, Albaum and Potter, 1943), partly owing to diminished muscular activity (Stoner, 1950).

The contractions of the pregnant uterus at term are diminished (Kumar, Zourlas and Barnes, 1963).

D. In intact animals following oral administration

1. Retention of orally administered magnesium. As noted above (Section III A), increments in magnesium intake are usually accompanied by positive magnesium balance, even in the absence of any indications of magnesium deficiency. This has been demonstrated in the rat (Cunningham, 1936; Chutkow, 1964b), the pig (Bartley *et al.*, 1961; Miller *et al.*, 1965c), and man (Seelig, 1964; Heaton and Parsons, 1961). Although enormous cumulative positive balances have been recorded (see Seelig, 1964), the conditions under which these studies were performed were not sufficiently rigorous to exclude substantial errors. Of special interest is the observation of Heaton and Parsons (1961), who found that the positive balance observed during high magnesium intake was followed by markedly negative balance on reducing the intake to normal; fecal excretion exceeded intake in one subject. These observations suggest that either endogenous magnesium excretion or intestinal absorption of magnesium may possess the capacity to adapt to the magnesium content of the organism, but only slowly.

Plasma magnesium concentration increased in subjects given 0 5 mmoles per kg per day (HAYWOOD and SELVESTER, 1962), and transiently following an oral dose of 6 mmoles $MgSO_4$ (TERKILDSEN, 1952b), or 10 mmoles of $MgCl_2$ (HUNTSMAN, HURN and LEHMANN, 1960). In rabbits, a considerably greater increase occurs (TERKILDSEN, 1952b). In rats on a high magnesium diet, little increase in soft tissue magnesium occurs but bone magnesium increases considerably (CUNNINGHAM, 1936).

These observations fail to support the view that a homeostatic mechanism prevents the magnesium content of the organism from increasing above some optimal level, even though urinary magnesium may be sensitive to dietary intake (FIELD, McCALLUM and BUTLER, 1958). It appears that excess dietary magnesium is to a considerable extent assimilated and stored as osseous tissue.

2. Effects of oral magnesium. Recent work has confirmed that the urinary excretion of calcium is augmented (HART and STEENBOCK, 1931; RANDOIN *et al.*, 1952; HEATON and PARSONS, 1961; MILLER *et al.*, 1965c) as well as that of strontium (CLARK, 1965a). In patients with nephrolithiasis, however, calcium excretion may fall (MOORE and BUNCE, 1964).

Rats exhibit a reduction in fecal calcium, which may overweigh the calciuresis and result in positive calcium balance (CLARK, 1965a and b, CLARK, MOYER and RIVERA-CORDERO, 1965). Fecal strontium may also diminish (CLARK and SMITH, 1962). Phosphorus excretion in the urine remains constant (CARPENTER *et al.*, 1963) or falls (MOORE and BUNCE, 1964; CLARK, 1967). Bone strontium is lessened (CLARK *et al.*, 1964).

When administered with excess dietary fluoride, magnesium inhibits calcification of growing bone (GRIFFITH, PARKER and ROGLER, 1963) and reduces its content of citrate (ibid. 1964).

Metabolic alterations following oral magnesium have also been noted. BROWNE (1964) reports decreases in plasma cholesterol (in arteriosclerotic patients); however, HAYWOOD and SELVESTER (1962) found only a transient increase. Plasma lipoproteins decrease slightly (HAYWOOD and SELVESTER, 1962) and peak thrombin generation time may increase (HUNTSMAN, HURN and LEHMANN, 1960); clotting time is unaltered (SCHIMPF and HARTER, 1957; WILLE, 1960; BROWNE, 1964b). A positive nitrogen balance in response to oral magnesium supplements was noted by HEATON and PARSONS (1961).

V. Magnesium deficiency

A. General considerations

An organism whose total content of magnesium is less than optimal is clearly magnesium deficient, but optimal magnesium content is difficult to determine. The few whole body analyses which have been carried out in normal subjects (Table 1) define an average content; in view of the possibility that magnesium intake in western countries is less than optimal (BERSOHN and OELOFSE, 1957; SEELIG, 1964), bodily content may be likewise. Furthermore, magnesium has been found beneficial in hypomagnesemic states without reduced bodily content (see below), suggesting that extracellular magnesium deficiency, like calcium deficiency, may occur without intracellular deficiency or significant negative balance. Most current knowledge concerning magnesium deficiency is based upon observations of growing animals placed on low magnesium diets. Deficiency develops, not because of negative balance, but because growth must proceed with insufficiently positive magnesium balance. The new tissue formed therefore contains subnormal amounts of magnesium. The manifestations of this disorder may have little in common with the picture seen in an adult on suboptimal magnesium intake, or subjected to large external losses of magnesium. Unfortunately, little attempt has been made to distinguish these disorders, and the term "magnesium deficiency" is now so ambiguous as to be almost useless. Three disorders are defined herein: 1. the syndrome of magnesium deprivation during growth; 2. magnesium depletion, meaning negative magnesium balance of significant degree; and 3. hypomagnesemia without either of the other two disorders, i.e., extracellular magnesium deficiency. Many reports fail to give sufficient data to permit classification of the type of magnesium deficiency under study, and these are referred to only in passing or omitted.

B. The syndrome of magnesium deprivation during growth
1. Symptoms

These have been reviewed extensively by KRUSE, ORENT and McCOLLUM (1932), FOLLIS (1958), SYLLM-RAPOPORT and STRASSBURGER (1958), O'DELL (1960), BACQ (1964), and WACKER and VALLEE (1964). All of the symptoms are more pronounced in younger animals (WATCHORN and McCANCE, 1937; BUNCE, CHIEMCHAISRI and PHILLIPS, 1962; SMITH and FIELD, 1963), as would be anticipated from their higher rate of growth. They fall into several categories:

a) Reduced growth. It was noted many years ago that young animals on low magnesium diets grow poorly (LEROY, 1926). This finding has been confirmed in all species studied. MENAKER and KLEINER (1952) showed that protein synthesis is impaired, as would be expected. The mechanism is conjectural, but may be related to the role of magnesium in protein synthesis referred to above (IV A).

b) Neuromuscular and central nervous system malfunction. Nervousness, hyperirritability, tremors, tetany, spasticity, muscular weakness, manifested notably in the hindlimbs and distal portions of the forelimbs in quadrupeds, with characteristic gaits in each species, are seen, and eventually convulsions, coma, and death. The species variation in these symptoms was well studied by ORENT, KRUSE and McCOLLUM (1932). It is clear that tetany can occur without hypocalcemia or alkalosis, but it is often absent despite severe hypomagnesemia. The convulsions are not prevented by curare (GREENBERG and TUFTS, 1938). As ROOK and STORRY (1962a) point out, extracellular magnesium deficiency could, by inference from physiological studies, produce the neuromuscular symptoms by increasing acetylcholine action at nerve endings, lowering the threshold of the muscle membrane, and inhibiting "relaxing factor". Electromyographic alterations, including fibrillations, repetitive activity, and polyphasic potentials may in fact precede the appearance of symptoms (DE DONCKER and ROSSELLE, 1959).

c) *Anorexia*, which contributes to the deficiency but ceases immediately when magnesium-containing foods are presented (L'ESTRANGE and AXFORD, 1963). The mechanism is unknown.

d) *Trophic symptoms.* Peripheral vasodilation and edema, hair loss, skin lesions, and eosinophilia have been observed in several species. Total leucocyte count may rise too (KASHIWA and HUNGERFORD, 1958). These symptoms are absent in some species, e.g., the pig (MAYO, PLUMLEE and BEESON, 1959).

Endogenous histamine apparently plays a role in the vasodilation, edema, and eosinophilia in rats, as shown by several lines of evidence. A histamine-liberating compound administered to normal rats produces a similar picture transiently; if magnesium deficiency is then induced in these rats, the symptoms do not appear, presumably because tissue stores of histamine have been depleted (BOIS, BYRNE and BÉLANGER, 1960). The eosinophilia in magnesium-deficient rats is prevented by pyribenzamine (HUNGERFORD, 1964). Furthermore, histamine levels rise in urine (BOIS, 1963), plasma, and tissues (BOIS and BEAULNES, 1966) during the first week or two of dietary restriction. The number of tissue mast cells, as well as the extent of granulation in the remaining cells, is reduced (BÉLANGER, VAN ERKEL and JAKEROW, 1957; BOIS, 1963). Histamine decarboxylase activity of tissues increases slightly, probably secondary to histamine liberation (BOIS, 1966). The serum of a magnesium-deficient animal induces histamine liberation when administered to another deficient animal, but not in controls, suggesting that an inhibitor of the histamine-liberating enzyme system may be absent in magnesium-deficient animals (op. cit.).

2. Magnesium metabolism

a) *Plasma.* The rate at which hypomagnesemia and the other manifestations of this syndrome develop depends upon the rate of growth of the animals, the magnesium content of the diet, and the presence of other factors, summarized below, which can modify the magnesium requirement. In one study, the level had fallen 30 % by 16 hours and even further at 24 (CHUTKOW, 1965). Plasma magnesium may fall to one-fourth of normal or even lower before death

(KRUSE, ORENT and McCOLLUM, 1933), but in rats it characteristically falls more rapidly at first (TUFTS and GREENBERG, 1938a), reaching a temporary nadir at one week. Ultrafiltrability is unaltered (HOOBLER, KRUSE and McCOLLUM, 1937).

b) Body. Total carcass magnesium has rarely been measured. MARTIN and WILSON (1960) showed clearly that rats on a magnesium deficient diet sustained no decrease in total carcass magnesium; however, the concentration of magnesium in the carcass fell by half. Similar findings were reported many years ago (TUFTS and GREENBERG, 1938a). Other authors who have reported similar decrements in carcass magnesium concentration (WHANG and WELT, 1963) have unfortunately neglected to report the animals' weights. MACINTYRE and DAVIDSSON (1958) noted a reduction in muscle magnesium concentration of about the same magnitude as the gain in weight. Only two studies of "adult" experimental animals on low magnesium diets have appeared. One of these is summarized below (AIKAWA, REARDON and HARMS, 1962). In another study, 400 gm. rats, 9—12 months old, were used (SMITH and FIELD, 1963); control animals grew 12 % in the 18-day study

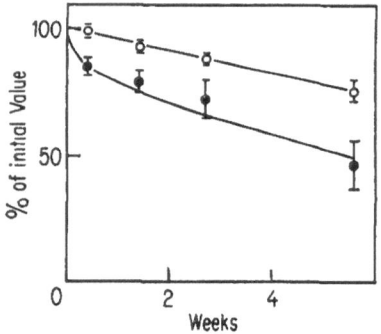

Fig 2 Bone magnesium (open circles) and plasma magnesium (solid circles) in calves on a low magnesium diet, redrawn from SMITH (1961 b), as a function of time Age of calves at the onset was about 2 weeks, and the percentage reduction in bone magnesium concentration was similar in magnitude to the percentage increase in body weight. Therefore the fall in bone magnesium may have been attributable to addition of new bone low in this cation, rather than to mobilization of bone magnesium

but animals on the magnesium-deficient diet did not. Hypomagnesemia was pronounced; nevertheless bone magnesium fell only 10 % and soft tissue magnesium not at all. It follows that total carcass magnesium must have decreased about 6 %, but it is questionable whether this is significant. All of the other studies reported in the literature to date are consistent with the view that carcass magnesium does not decrease in this syndrome, and may in fact increase.

c) Bone. Bone magnesium concentration has been found reduced in every study in which it has been measured. Total magnesium content of the femur has been found unchanged in two studies in young rats (MACINTYRE and DAVIDSSON, 1958; HEATON, 1965a), despite substantial decreases in bone magnesium concentration. The percentage reduction in bone magnesium concentration has been found to be nearly as great as the percentage reduction in plasma magnesium concentration in young animals (Fig. 2), including calves (SMITH, 1959b) and rats (McALEESE and FORBES, 1961; MARTINDALE and HEATON, 1964). Pronounced reductions in bone magnesium concentration were also seen in other studies in rats (CUNNINGHAM, 1936; WATCHORN and McCANCE, 1937; DUCKWORTH, GODDEN and WARNOCK, 1940; CHIEMCHAISRI and PHILLIPS,

1963), young guinea pigs (MORRIS and O'DELL, 1961); puppies (CHIEMCHAISRI and PHILLIPS, 1963; BARNES and MENDELSON, 1963), and calves (KNOOP, KRAUS and HAYDEN, 1939; BLAXTER, ROOK and MACDONALD, 1954a; BLAXTER, 1956). On the other hand, older animals develop considerably greater reductions in plasma magnesium than in bone magnesium concentration (WILSON, 1960; SMITH and FIELD, 1963; AIKAWA, REARDON and HARMS, 1962). Although it is usually stated that reduced bone magnesium concentration indicates that magnesium has been "mobilized" to combat extracellular deficiency, the fact is that significantly reduced total bone magnesium has been demonstrated only rarely (DUCKWORTH, GODDEN and WARNOCK, 1940; AIKAWA, REARDON and HARMS, 1962; SMITH and FIELD, 1963). In the remainder, including those in which total femur magnesium was measured, no mobilization of bone magnesium has been shown to occur. Thus the greater reduction in bone magnesium concentration observed in younger animals may simply reflect the greater rate of bone growth; mobilization of bone stores of magnesium in younger animals may actually be less than in older animals, rather than greater.

A similar reduction in bone magnesium concentration is seen in young rats in whom magnesium loss is induced by diarrhea (FELLERS and HERRERA, 1966).

Magnesium concentration also falls in teeth (WATCHORN and McCANCE, 1937).

The rate of exchange of bone magnesium with plasma radioactive magnesium is decreased in lambs (McALEESE, BELL and FORBES, 1961) and rats (FIELD and SMITH, 1964).

HEATON (1964a, 1965a) has shown that parathyroidectomy (in rats) limits the fall in bone magnesium concentration, even though plasma magnesium is lower; however, it also reduced the rate of bone growth.

Bone magnesium is restored only slowly during repletion (FORBES, 1964a).

d) Soft tissues. Erythrocyte magnesium falls, though usually more slowly than plasma magnesium (BARRON, BROWN and PEARSON, 1949; TUFTS and GREENBERG, 1938a; SHILS, 1966). TUFTS and GREENBERG (1938a) suggested that the magnesium content of red cells reflects the plasma level at the time they are produced, a hypothesis which should be verified by labeling experiments, but seems reasonable. Muscle magnesium concentration remains constant (CUNNINGHAM, 1936; WATCHORN and McCANCE, 1937; BLAXTER, ROOK and MACDONALD, 1954a; PARR, 1957; SMITH, 1957; SMITH and FIELD, 1963; BARNES and MENDELSON, 1963; FELLERS and HERRERA, 1966) or decreases slightly (TUFTS and GREENBERG, 1938a; COTLOVE *et al.*, 1951; MORRIS and O'DELL, 1961; FORBES, 1966); in young rats the decrease may amount to 20% (MACINTYRE and DAVIDSSON, 1958; SMITH *et al.*, 1962). Kidney magnesium concentration is unaltered (CUNNINGHAM, 1936, BLAXTER,

Rook and MacDonald, 1954a; MacIntyre and Davidsson, 1958; Hess et al., 1959; Chiemchaisri and Phillips, 1963; Barnes and Mendelson, 1963) or slightly reduced (Watchorn and McCance, 1937; Tufts and Greenberg, 1938a; Morris and O'Dell, 1961; Heaton and Anderson, 1965), particularly in younger animals (Smith and Field, 1963). Similar results are seen in heart, aorta, liver, spleen, brain, skin and lung (Cunningham, 1936; Tufts and Greenberg, 1938a; Watchorn and McCance, 1937; Blaxter, Rook and MacDonald, 1954a; Suter and Klingman, 1955; MacIntyre and Davidsson, 1958; Morris and O'Dell, 1961; Bunce, Jenkins and Phillips, 1962; Barnes and Mendelson, 1963; Chiemchaisri and Phillips, 1963; Smith and Field, 1963). Unfortunately organ weights are not reported in any of these studies, so that it is uncertain in each instance whether the total magnesium content of these organs changed. However, all of the data are consistent with the hypothesis that soft tissue magnesium is not mobilized in this syndrome.

Cerebrospinal fluid magnesium concentration does not fall (Barnes and Mendelson, 1963).

e) Excretion. Urinary excretion of magnesium falls markedly (Cunningham and Cunningham, 1938; Welt, 1964) and may approach zero (Chutkow, 1965). The dietary calcium intake makes little or no difference (Cunningham and Cunningham, 1938; Welt, 1964). Fecal magnesium excretion falls to a lesser extent, and may exceed intake (Blaxter, Rook and MacDonald, 1954b; Welt, 1964; Smith and Field, 1963). The magnesium absorptive ability of the gut is apparently reduced, rather than augmented (R. H. Smith, 1957), though Kessner and Epstein (1966) found magnesium accumulation by isolated gut sacs to be unaltered.

3. Calcium and phosphorus metabolism

a) Plasma. The concentration of calcium in plasma is variably affected. It may be unaltered in rabbits (de Doncker and Rosselle, 1959), dogs (Vitale et al., 1961; Wener et al., 1964), calves (Smith, 1957), and rats (Ko, Fellers and Craig, 1962; Forbes, 1963, Seta, Hellerstein and Vitale, 1965). In other reports, definite hypocalcemia has occurred in dogs (Bunce, Jenkins and Phillips, 1962; Chiemchaisri and Phillips, 1963; Wener et al., 1964), lambs (McAleese and Forbes, 1959), pigs (Miller et al., 1965), rats (Carillo et al., 1961), and calves (Larvor et al., 1964; Smith, 1957). The fall in plasma calcium can be prevented by vitamin D in calves (Smith, 1958) but not in lambs (McAleese and Forbes, 1959). Magnesium repletion corrects it (Smith, 1961b). It is usually associated with a rise in plasma inorganic phosphate (Bunce, Jenkins and Phillips, 1962; Chiemchaisri and Phillips, 1963). Distinct hypercalcemia has also been noted, particularly in rats (MacIntyre and Davidsson, 1958; Smith et al., 1962; Chiemchaisri and

PHILLIPS, 1963; RICHARDSON and WELT, 1963; WHANG and WELT, 1963; HEATON and ANDERSON, 1965; HEATON, 1965a; SCHNEEBERGER and MORRISON, 1965) and transiently in calves (BLAXTER, ROOK and MACDONALD, 1954a). Ionic calcium increases as well as total (GITELMAN, KUKOLJ and WELT, 1965a). This increase was prevented by restricting intake of calcium as well as magnesium in one study (HEATON and ANDERSON, 1965), but not in another (RICHARDSON and WELT, 1963). It is prevented by parathyroidectomy (HEATON and ANDERSON, 1965; GITELMAN, KUKOLJ and WELT, 1965a). When hypercalcemia occurs, plasma phosphate is usually reduced, providing further evidence for parathyroid overactivity as a mechanism of this response (BUNCE, JENKINS and PHILLIPS, 1962; CHIEMCHAISRI and PHILLIPS, 1963; HEATON, 1965b). Furthermore, plasma citrate rises (LIFSHITZ et al., 1966). Others have reported unchanged (WHANG and WELT, 1963; WENER et al., 1964; BUNCE et al., 1965) or increased (MORRIS and O'DELL, 1963) plasma phosphate, especially in hypocalcemic animals (BUNCE, JENKINS and PHILLIPS, 1962; CHIEMCHAISRI and PHILLIPS, 1963). Alkaline phosphatase is normal (MILLER et al., 1965) or decreased (HEATON, 1965b; LARVOR et al., 1964).

b) Carcass. The concentration of calcium in the whole body is unaffected (WHANG and WELT, 1963) as is the calcium balance (BLAXTER, ROOK and MACDONALD, 1954b).

c) Bone. Bone calcium and phosphate concentration may be normal (MORRIS and O'DELL, 1961), or the calcium:phosphorus ratio may rise (LARVOR et al., 1964).

d) Soft tissues. A pronounced increase occurs in kidney calcium concentration in rats (MACINTYRE and DAVIDSSON, 1958; HESS et al., 1959; MORRIS and O'DELL, 1961; CHIEMCHAISRI and PHILLIPS, 1963; HEATON and ANDERSON, 1965; FORBES, 1965) and guinea pigs (MORRIS and O'DELL, 1961); acid-soluble phosphate rises too. Frank nephrocalcinosis is common (TUFTS and GREENBERG, 1936), and may occur in the absence of hypercalcemia (SETA, HELLERSTEIN and VITALE, 1965), contrary to an earlier report (MACINTYRE and DAVIDSSON, 1958). Parathyroidectomy prevents the rise in kidney calcium (HEATON, 1964a; HEATON and ANDERSON, 1965), and thyroxine (FORBES, 1965) or a low calcium diet (HEATON, 1964a; HEATON and ANDERSON, 1965) ameliorates it. In calves, calcium concentration did not increase in kidney or other soft tissues (BLAXTER, ROOK and MACDONALD, 1954a). Magnesium repletion is only minimally effective in reducing the renal content of calcium in rats (FORBES, 1964a; HEATON and ANDERSON, 1965).

In the other soft tissues, calcium concentration may be normal (MACINTYRE and DAVIDSSON, 1958) or increased (MORRIS and O'DELL, 1961; CHIEMCHAISRI and PHILLIPS, 1963; FORBES, 1965, 1966).

The mechanism of the fall in blood calcium and its deposition in tissues may be a decrease in the solubility of bone salt (or its precursors), occasioned by lowered magnesium concentration (Neuman and Neuman, 1958; Wilson, 1964).

e) Excretion. Urinary excretion of calcium is regularly reduced (Cunningham and Cunningham, 1938, Smith *et al.*, 1962; Alcock and MacIntyre, 1962; Heaton, 1965 a) but restriction of dietary calcium as well as magnesium leads to a further decrease (Welt, 1964). Intestinal absorption of calcium may be increased (Alcock and MacIntyre, 1962), and calcium accumulation by gut sacs from rats with this syndrome is increased (Kessner and Epstein, 1966). Urinary phosphate excretion is increased (Heaton, 1965 a; Whang and Welt, 1963; Lifshitz *et al.*, 1967), as a result of diminished percentage excretion of filtered phosphate; a reduction in phosphate T_m can also be demonstrated (Ginn and Becker, 1962). Citrate excretion is low despite increased blood levels (Lifshitz *et al.*, 1967).

4. Potassium metabolism

Moderate restriction of dietary magnesium in growing rats has no effect on potassium concentration of heart or kidney (Forbes, 1965). More severe restriction, even in older rats (Cotlove *et al.*, 1951), usually leads to diminished tissue potassium concentration (MacIntyre and Davidsson, 1958; Manitius and Epstein, 1963; Whang and Welt, 1963; Schneeberger and Morrison, 1965; Seta, Hellerstein and Vitale, 1965; Forbes, 1966). Total body potassium stops increasing in young rats after two weeks on a magnesium-deficient diet (Martin and Wilson, 1960). Plasma potassium is usually unaffected (Cotlove *et al.*, 1951; Whang and Welt, 1963; Wener *et al.*, 1964; Chutkow, 1965; Schneeberger and Morrison, 1965; Shils, 1966), but has been reported to rise (Forbes, 1966) or fall (Seta, Hellerstein and Vitale, 1965) especially in the later stages (Seta *et al.*, 1966). Carillo *et al.* (1961) report that a high potassium intake minimizes the fall in plasma magnesium in rats on a low magnesium diet, but does not reduce the signs and symptoms of magnesium deficiency. A low potassium diet failed to affect muscle magnesium in older rats (Cotlove *et al.*, 1951) and actually increased it in one study (Gardner, MacLachlan and Berman, 1953) but led to a fall in cardiac magnesium and a rise in plasma magnesium in young rats (Seta, Hellerstein and Vitale, 1965) as well as dogs (Seta *et al.*, 1966). These observations suggest that the new tissue laid down by the growing animal either contains both magnesium and potassium, in proportions close to normal, or else contains neither. Baldwin *et al.* (1952) and Lilienthal *et al.* (1950) have in fact demonstrated that the ratio of muscle magnesium to potassium is constant, being unaffected by potassium deficiency or atrophy, and independent of the ratio of these two cations in plasma. The interrelationships between plasma potassium and magnesium are complicated and require further study.

5. Effects on specific organ functions

a) Thyroid. Thyroid weight and iodine uptake are increased (KLEIBER, BOELTER and GREENBERG, 1941; CORRADINO and PARKER, 1962): however, protein-bound iodine is normal (CORRADINO and PARKER, 1962). According to HONORATA and ROSA (1944), basal metabolic rate is reduced, but RAY and MULLICK (1965) found it to be normal. The increase in total heat production is attributable to increased muscular activity (BLAXTER and ROOK, 1955). The rate of incorporation of inorganic iodide into organic form by thyroid slices from these rats was not affected by the addition of magnesium (CORRADINO and PARKER, 1962).

b) Cardiovascular system. In puppies, electrocardiographic changes in this syndrome include tachycardia, depressed RST segments, and low or inverted T waves. These changes were poorly correlated with hypomagnesemia (WENER et al., 1964), and may reflect concomitant potassium deficiency (SETA et al., 1966). VITALE, NAKAMURA and HEGSTED (1957) and VITALE et al. (1957b) found that mitochondria isolated from the hearts of magnesium-deficient rats exhibited defective oxidative phosphorylation, but BEECHEY et al. (1959) and BEECHEY, ALCOCK and MACINTYRE (1961) reported normal P/O ratios. DI GIORGIO, VITALE and HELLERSTEIN (1962) found normal P/O ratios and ATP levels but diminished respiration in heart sarcosomes from magnesium-deficient ducks. Oxidative phosphorylation by these sarcosomes was abnormally sensitive to reduction of magnesium concentration in the medium. The number of mitochondria in the heart is reduced (MISHRA, 1960b) and they are swollen; this swelling can be reduced by ATP (NAKAMURA et al., 1961). The rate of incorporation of P^{32} into ATP in aorta is depressed (SASAKI, NAKATANI and NAKAMURA, 1965), while the rate of uptake of cholesterol in this tissue increases (HIRANO, 1966).

c) Neurological changes. DE DONCKER and ROSSELLE (1959) reported electromyographic changes in rabbits fed a low magnesium diet, as noted above. The response of muscle to galvanic stimuli is augmented (KRUSE, SCHMIDT and McCOLLUM, 1933). Surprisingly, susceptibility to allergic encephalomyelitis is reduced (McCREARY et al., 1966).

d) Kidney. Severe renal damage from nephrocalcinosis and urolithiasis may occur (CRAMER, 1932; GREENBERG, LUCIA and TUFTS, 1938; BUNCE et al., 1965). But in the absence of frank renal failure, W. O. SMITH et al. (1962) found reduced concentrating ability in rats. MANITIUS and EPSTEIN (1963), however, did not. The effect on phosphate T_m is cited above. In addition, SMITH et al. (1962) found a supranormal increase in titratable acid excretion following ammonium chloride loading. In all probability this is a reflection of the increased quantities of phosphate buffer in the urine. Amino acid excretion rises (BUNCE et al., 1963; MAZZOCCO, FLINK and JONES, 1966). Urine pH is unaltered (CUNNINGHAM and CUNNINGHAM, 1938).

e) Liver. The changes in mitochondrial function seen in the heart are not found in the liver (VITALE, NAKAMURA and HEGSTED, 1957; VITALE *et al.*, 1957b), but swelling is seen (NAKAMURA *et al.*, 1961), and phosphate incorporation into nucleotides is reduced (SASAKI, NAKATANI and NAKAMURA, 1965; YOUNGS and CORNATZER, 1965). Urea cycle enzyme levels are unaltered (LIZARRALDE, MAZZOCCO and FLINK, 1966).

f) Gut. As noted above, calcium transport is increased (ALCOCK and MACINTYRE, 1962; KESSNER and EPSTEIN, 1966). Phosphate transport may be increased (LIFSHITZ, HARRISON and HARRISON, 1967).

6. Pathology

Early workers noted widespread calcifications and degenerative changes in parenchymatous organs (GREENBERG, ANDERSON, and TUFTS, 1936; MOORE, HALLMAN, and SHOLL, 1938), with cardiovascular lesions resembling atherosclerosis. These findings have been amply confirmed in rats (LOWENHAUPT, SCHULMAN and GREENBERG, 1950; MISHRA, 1960a; KO, FELLERS and CRAIG, 1962; MORRIS *et al.*, 1963), dogs (UNGLAUB, SYLLM-RAPOPORT and STRASS-BURGER, 1959; VITALE *et al.*, 1961; WENER *et al.*, 1964), monkeys (VITALE *et al.*, 1963), calves (LARVOR *et al.*, 1964; BLAXTER and SHARMAN, 1955), rabbits (BARRON, BROWN and PEARSON, 1949), chicks (BIRD, 1949), and other species (see reviews referred to in introduction). In some studies (e.g. CARILLO *et al.*, 1961), pathological findings have been virtually absent despite overt symptoms.

Electron microscopy has now made possible a description of the sequential changes in subcellular organelles which eventually lead to these gross pathological alterations. In the kidney, mitochondrial swelling is first seen in the cells of the proximal convoluted tubules according to HESS *et al.* (1959), but SCHNEEBERGER and MORRISON (1964, 1965) and WELT (1964) have presented evidence that the earliest change consists of luminal concretions in the thick ascending limb of Henle's loops; soon thereafter calcium deposits are seen in lysosomes and in the cytoplasm of tubular cells in this segment. Signs of internal hydronephrosis appear in obstructed nephrons. In dilated proximal tubules, the brush border soon disappears (SABOUR, HANNA and MACDONALD, 1964). These lesions differ from those of potassium deficiency and also those of hypercalcemia (SCHNEEBERGER and MORRISON, 1965), although similarities have often been noted (SCHRADER, PRICKETT and SALMON, 1937; HEATON and ANDERSON, 1965).

The reduction in magnesium concentration in hard tissues is associated with histologic evidence of abnormal calcification (BERNICK and HUNGERFORD, 1965).

Swelling is seen also in cardiac sarcosomes, with vacuolization (NAKAMURA *et al.*, 1961; HEGGTVEIT, HERMAN and MISHRA, 1964). This swelling can be reversed by ATP, and thus is probably a manifestation of a defect in transphosphorylations (NAKAMURA *et al.*, 1961). Later, calcium deposits become visible, first on the cristae of the sarcosomes (HEGGTVEIT, 1965b; HEGGTVEIT, HERMAN and MISHRA, 1964). Although the similarity of the changes of magnesium deficiency to those of potassium deficiency has often been noted, these electron microscopic alterations are distinct (op. cit.). Swelling of muscle fibers (HEGGTVEIT, 1965a) also precedes nuclear injury and the deposition of calcium granules intracellularly. In the brain, changes are seen in the Purkinje cells of the cerebellum (MOORE, HALLMAN and SHOLL, 1938; BIRD, 1949; BARRON, BROWN and PEARSON, 1949).

It is clear that pathological changes may develop in tissues whose magnesium concentration is not measurably reduced. This has been demonstrated both with respect to calcific (HEATON and ANDERSON, 1964; BUNCE, JENKINS and PHILLIPS, 1962), and non-calcific (BLAXTER, ROOK and MACDONALD, 1954a) alterations. As noted above, hypercalcemia is not a prerequisite for the development of calcific lesions — though parathyroidectomy and consequent hypocalcemia will prevent them. MORRIS *et al.* (1963) suggest that alterations in tissue mucopolysaccharides may predispose to calcium deposition, but this could not explain the granules which appear to be intracellular (v.s.). The mechanism of the luminal concretions, casts, and urolithiasis is more likely to be directly related to reduced magnesium concentration in the tubular fluid and urine, since the latter is a constant feature of this disorder (v.s.), and since high magnesium intake tends to diminish stone-formation in patients with chronic urolithiasis (see below). This hypothesis will require micropuncture experiments for confirmation.

Both in dogs (VITALE *et al.*, 1961; WENER *et al.*, 1964) and monkeys (VITALE *et al.*, 1963), feeding a low magnesium diet during growth may lead to the appearance of frank atherosclerosis.

Parathyroid enlargement may occur in the absence of hypercalcemia (LARVOR *et al.*, 1964). Other apparently unrelated findings include splenomegaly (HUNGERFORD, 1964) and thymic tumors (BOIS, 1964), swollen gums and hematomas (O'DELL, 1960) and anemia (LARVOR, *et al.*, 1964).

7. Modification by dietary and hormonal factors

a) Calcium and phosphorus. Both elements have repeatedly been observed to aggravate this syndrome when added to the diet in increased amounts (TUFTS and GREENBERG, 1938b; HOGAN, REGAN and HOUSE, 1950; O'DELL, MORRIS and REGAN, 1960; COLBY and FRYE, 1951b; HOUSE and HOGAN, 1955; HEGSTED, VITALE and MCGRATH, 1956; MAYNARD *et al.*, 1958; BUNCE, CHIEMCHAISRI and PHILLIPS, 1962; MORRIS and O'DELL, 1963; FORBES, 1963; NUGARA and EDWARDS, 1963). The effect of each element is independent and additive (MORRIS and O'DELL, 1963) and may be attributable to inhibition by these elements of intestinal absorption of the small amount of magnesium in the diet (O'Dell, Morris and Regan, 1960; Forbes, 1963). However, Welt (1964) found no difference in carcass magnesium or fecal magnesium between rats fed a low magnesium diet and rats fed a low magnesium, low calcium diet.

b) Other dietary factors. Vitamin D did not alter fecal magnesium in calves (Smith, 1957) or fecal or urinary magnesium in young rats (Richardson and Welt, 1963) fed diets deficient in magnesium; it aggravated hypomagnesemia in rats.

A high intake of protein (Colby and Frye, 1951 b; Bunce *et al*, 1963), cholesterol (Olson and Parker, 1964), or lactose (Forbes, 1964a) may also aggravate this syndrome; the explanation in the case of lactose is apparently increased renal losses of magnesium.

Atherogenic diets may lead to the appearance of renal lesions characteristic of magnesium deficiency despite a normal intake of this cation: an 8-fold increase is necessary to prevent their appearance (Hellerstein *et al.*, 1957).

Addition of 25 ppm. of fluoride to the diet largely prevents the development of calcinosis in dogs, but not in rats (Chiemchaisri and Phillips, 1965).

The dietary level of potassium may also modify this syndrome. High intake may aggravate the symptoms (Colby and Frye, 1951a), or fail to affect them; it does lessen the degree of hypomagnesemia (Carillo *et al.*, 1961). Potassium restriction diminishes tissue calcium and the vascular lipoidosis (Forbes, 1966; Vitale, Koide and Hellerstein, 1966); it may increase plasma magnesium (while decreasing cardiac concentration) in rats on a normal magnesium intake (Seta, Hellerstein and Vitale, 1965).

c) Hormonal factors. The role of parathyroid hormone has been considered above. Eliel, Chanes and Hawrylko (1963) have reported that parathyroidectomized rats develop symptomatic hypomagnesemia sooner than controls. Thyrocalcitonin was effective in counteracting the hypercalcemia in rats (Gudmundsson, MacIntyre and Soliman, 1966).

Hypophysectomized rats develop symptoms more slowly (Kashiwa, 1961). On the other hand, cortisone treatment also delays or prevents the appearance of irritability, convulsions, and vasodilation (Farnell, 1966). Serum levels of corticoids are transiently increased (Bois, 1966).

Thyroxine inhibited the appearance of nephrocalcinosis (Gershoff *et al.*, 1958; Forbes, 1965) and reduced the extent of atherosclerotic changes (Hellerstein *et al.*, 1957) but failed to increase the dietary magnesium requirement for normal growth or prevention of hypomagnesemia (Forbes, 1965).

C. Magnesium depletion

1. Criteria

The following types of evidence, in order of diminishing reliability, have been used as criteria of magnesium depletion: 1. diminished magnesium content of the whole body, or sufficient analyses of individual tissues (especially bone) in non-growing subjects to permit this deduction; 2. negative magnesium balance, determined in hospitalized subjects under metabolic ward conditions (constant diet, analyzed for total magnesium and magnesium content of individual food rejects, distilled water to drink, collection and analysis of all excreta, etc.); 3. positive magnesium balance measured under similar conditions in patients presumed to be depleted at the onset of the study; 4. positive magnesium balance during therapy, based only on measurements of oral and parenteral intake and urinary excretion. The first three of these are clearly valid; the fourth becomes invalid only if endogenous fecal excretion of magnesium rises considerably during therapy. As noted above, there is evidence to suggest that this may occur after several days of magnesium loading, although it does not occur acutely (HEATON and PARSONS, 1961; PETERSON, 1963a). A number of other criteria which have been invoked cannot be considered more than suggestive; these include measurement of negative magnesium balance in outpatients on restricted diets (HATHAWAY, 1962); reduced magnesium concentration of one or two soft tissues, such as muscle or erythrocytes, with or without hypomagnesemia; symptomatic improvement during magnesium therapy; urinary excretion of less than the usual fraction of a single parenteral dose; reduced exchangeable magnesium, as measured with Mg^{28}; and various combinations of these. All of these may be questioned, including exchangeable magnesium, which may actually increase (FANKUSHEN et al., 1964)

2. Causes

a) Dietary. In adult man, as noted in section III C, a low magnesium diet does not cause appreciable magnesium depletion (FITZGERALD and FOURMAN, 1956; BARNES, COPE and HARRISON, 1958, BARNES, COPE and GORDON, 1960). Early data by CLARK (1926), who reported enormous cumulative negative balances in prisoners on moderate magnesium intake, have been reproduced without comment by SEELIG (1964) and HATHAWAY (1962). As Clark states in this paper, "The negative balances are the result of cumulative analytical errors."

If magnesium losses are augmented by high dietary calcium, ion exchange resins by mouth, or infusions of sodium sulfate, negative balance may occur (FITZGERALD and FOURMAN, 1956; SHILS, 1964, DUNN and WALSER, 1966). In adult (400 g) rats, pronounced hypomagnesemia occurred, but there was only a questionable reduction of carcass magnesium (SMITH and FIELD, 1963). In rabbits weighing 1.7 kg, fed a diet containing 3 mmoles magnesium per kg for four weeks, moderate hypomagnesemia (0.6 mM) and a 15% reduction in bone magnesium concentration occurred (AIKAWA, REARDON and HARMS, 1962); since weight decreased, body magnesium must have fallen at least 10%. Urinary magnesium was less than 0.4 mmoles per day. It therefore appears that endogenous magnesium loss in the rabbit must be enormous — approximately 3 mmoles per kg per day. This inference is, however, based solely on the finding of a 15% reduction in bone magnesium. In dogs $1^1/_2$ to 7 years old, a low magnesium diet induced weight loss, hypomagnesemia, and hyper-cholesterolemia (BUNCE, CHIEMCHAISRI and PHILLIPS, 1962): plasma calcium did not change, and no atherosclerotic lesions developed. Neither balance data nor tissue analysis were reported, so that the extent of magnesium depletion is conjectural.

In summary, it appears that dietary restriction alone produces neither hypomagnesemia nor magnesium depletion in adult man, but causes hypo-magnesemia accompanied by some degree of reduction in bone magnesium in animals, whose obligatory endogenous losses are greater.

b) Physiological losses. During lactation, the dietary requirement for mag-nesium may be considerably increased (RITCHIE *et al.*, 1962), and depletion could occur on a low intake.

c) Pathological losses i. Intestinal disease. Gastrointestinal losses of magne-sium in patients with malabsorption or diarrhea may be considerable (MELLING-HOFF and VAN LESSEN, 1949; HEATON and FOURMAN, 1965; MELLINGHOFF, 1949). As noted in Section III B, 4 to 6 mmoles per day of endogenous magnesium may appear in the stool in such patients. In one report, far larger losses were recorded in a girl with celiac disease (GOLDMAN, VAN FOSSAN and BAIRD, 1962). Although gastric secretion may occasionally lead to loss of as much as 5 mmoles per day, ordinarily the losses are far smaller (LEVEY *et al.*, 1956; THOREN, 1963; MACBETH and MABBOTT, 1964), and could lead to significant depletion only after many weeks on a low intake.

ii. Drugs. Although alcohol increases renal magnesium excretion in normal and alcoholic subjects (Section VIII A), it did not do so in a magnesium depleted normal subject (DUNN and WALSER, 1966) nor in hypomagnesemic alcoholics (SULLIVAN, LANKFORD and ROBERTSON, 1966). There is therefore some question whether it can induce magnesium depletion, even on a low intake.

Several diuretics augment renal excretion of magnesium (VIII B), but this loss, like that of sodium, is probably self-limited on continued use, and the amounts lost could hardly prove significant on the basis of the data currently available.

iii. Diabetic acidosis. During three days of insulin withdrawal, diabetics lost 0.4 mmoles per kg, or about 2% of body stores (BUTLER, 1950). A similar estimate was made by NABARRO, SPENCER and STOWERS (1952) on the basis of magnesium balance during therapy, corrected for nitrogen balance.

iv. Renal magnesium-wasting with accompanying potassium wastage, has been described in one patient by VOSTAL, KUCHEL and PACOVSKY (1963), and by GITELMAN, GRAHAM and WELT (1966) in two sisters. The extent of negative magnesium balance required to produce hypomagnesemia or symp-toms in this disorder has not yet been reported.

v. Hypoparathyroidism. Patients with postoperative hypoparathyroidism appear to retain large amounts of magnesium when given repeated intravenous injections (JONES and FOURMAN, 1966). In one such patient the difference between total magnesium administered and the increment in urinary excretion amounted to over 300 mmoles.

3. Magnesium metabolism

a) Plasma. The concentration of magnesium in the plasma gives no indication of the presence or absence of magnesium depletion. Reductions in plasma magnesium may occur in the absence of appreciable magnesium losses (V D). Conversely, plasma magnesium may be normal in the presence of convincing evidence of magnesium depletion (FITZGERALD and FOURMAN, 1956; MONTGOMERY, 1960, 1961; FOURMAN and MORGAN, 1962; LINDER, HANSEN and KARABUS, 1963; PRETORIUS, WEHMEYER and THERON, 1963).

b) Body. In an average man containing about 1100 mmoles of magnesium, loss of 50 mmoles would represent less than 5 % of his magnesium stores. Magnesium depletion greater than this has been established only rarely in patients (MACINTYRE *et al.*, 1961; GOLDMAN, VAN FOSSAN and BAIRD, 1962; LINDER, HANSEN and KARABUS, 1963; PETERSEN, 1963a; OPIE, HUNT and FINLAY, 1964; HEATON and FOURMAN, 1965) and only once in normal subjects of any species (DUNN and WALSER, 1966). These six studies, including a meager total of nine subjects, must provide the basis of our knowledge of magnesium depletion. In patients in whom deficits have been estimated during repletion, the simultaneous balance of nitrogen has been measured only rarely (PETERSEN, 1963a); such patients are commonly malnourished and a major portion of the retained magnesium may be associated with formation of new tissue. According to BUTLER (1950), each gram of nitrogen is equivalent to 0.3 mmoles of magnesium. AIKAWA *et al.* (1959) observed urinary loss of 31 mmoles per kg of lost weight during starvation in rabbits. The two normal subjects studied by DUNN and WALSER (1966) had normal protein intakes and lost little or no weight; negative magnesium balances cumulated to 85 and 92 mmoles, respectively.

FOURMAN and MORGAN (1962) cite highly suggestive evidence for much larger deficits (220—240 mmoles) in three patients with malabsorption who had normal plasma magnesium and no symptoms indicating magnesium lack. The deficits were estimated on the basis of the urinary excretion (within two days) of repeated intravenous infusions of magnesium salts (42 mmoles); it was assumed that endogenous fecal loss did not increase. As noted above (IV D 1) this assumption may not be valid. Furthermore, the percentage excretion of the intravenous dose, subnormal at the onset, remained subnormal after 40 to 80 days of infusions. These findings suggest that considerable magnesium depletion may occur without hypomagnesemia, but it may be questioned whether the deficits could have been as large as the authors estimate.

Deficits less than 50 mmoles have been recorded by several authors (FITZGERALD and FOURMAN, 1956; BARNES, COPE and GORDON, 1960; McCOL-LISTER, FLINK and DOE, 1960).

c) Bone. The bone analyses in the two studies in adult animals have already been discussed; both suggest a slight reduction in bone magnesium concentration. The only other observation appears to be one analysis of bone from a patient with chronic diarrhea and hypomagnesemia reported by MacINTYRE *et al.* (1961) and by BOOTH *et al.* (1963). Magnesium depletion was established by the observation of a positive balance of 75 mmoles during therapy. Bone magnesium was reduced from 113 to 96 mmoles/kg fat-free dry tissue. As bone contains two-thirds of body magnesium, this reduction is great enough to account for the entire deficit of 75 mmoles. In a second case of malabsorption (BOOTH *et al.*, 1963), a magnesium deficit was not assessed by balances, but bone magnesium was reduced even more; enough to account for loss of 16 % of body magnesium. These few observations suggest that bone magnesium in man, as in grown animals, falls during magnesium depletion, though it is evidently mobilized less readily in ruminants (V D).

d) Soft tissues. Muscle magnesium was unaltered in the two studies in grown animals (v.s.), in one of the two patients reported by BOOTH *et al.* (1963), and in both normal subjects depleted by DUNN and WALSER (1966). In the other patient reported by BOOTH *et al.* (1963), muscle magnesium was reduced 25 %. Since muscle contains about 19 % of body magnesium, this decrease would account for loss of about 5 % of body stores; although bone magnesium concentration was less depressed, the contribution of bone to the total deficit must have been greater (9 % of body magnesium).

In contrast with these results, infants with kwashiorkor regularly exhibit reduced muscle magnesium concentration, even in the absence of hypo-magnesemia (MONTGOMERY, 1960, 1961); during therapy muscle magnesium may rise, but not within the first 7 to 10 days (METCOFF *et al.*, 1960). In a similar subject, LINDER, HANSEN and KARABUS (1963) have shown that a magnesium deficit, above and beyond nitrogen depletion, exists. This disorder may be analogous to the magnesium deprivation syndrome in growing animals, in which muscle magnesium concentration (though not content) is sometimes low. However, it is not clear why plasma magnesium is often normal, in contrast to the findings in animals.

In one alcoholic subject with hypomagnesemia, FANKUSHEN *et al.* (1964) report a surprisingly low muscle magnesium concentration of 5 mmoles/kg wet tissue.

Erythrocyte magnesium was reduced more than body magnesium but less than plasma magnesium (Fig. 3) in the subjects studied by DUNN and WALSER

(1966). In SHILS' (1964) study of two patients depleted by diet and resins, the magnesium deficit was not measured, but erythrocyte magnesium fell one-third while plasma concentration fell two-thirds. Similar results were subsequently reported in rats and dogs (SHILS, 1966). In kwashiorkor, which is probably a form of magnesium or protein deprivation rather than magnesium depletion, erythrocyte magnesium, like plasma magnesium, is usually normal or nearly so (MONTGOMERY, 1960; PRETORIUS, WEHMEYER and THERON, 1963).

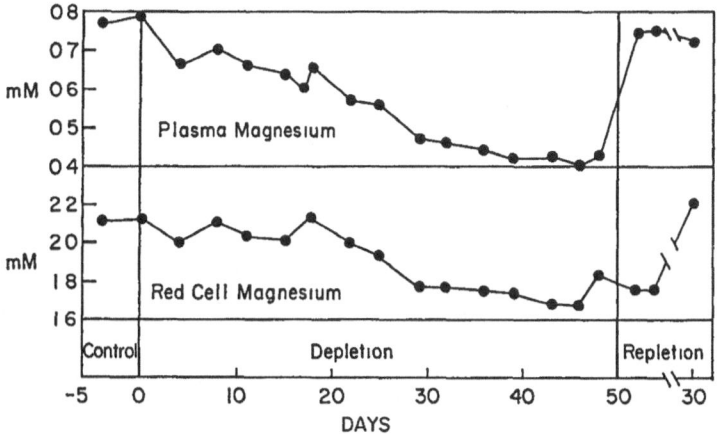

Fig 3 Plasma and erythrocyte magnesium in a normal man during 7 weeks of magnesium depletion. Negative balance amounted to 92 mmoles Note that erythrocyte magnesium fell more slowly, and rose more slowly during repletion, probably because the cells reflect plasma magnesium at the time of synthesis
From DUNN and WALSER (1966) by permission

e) Excretion. Daily renal excretion of magnesium falls to very low levels (FLETCHER and HENLY, 1960; MACINTYRE *et al.*, 1961; GOLDMAN, VAN FOSSAN and BAIRD, 1962; PETERSEN, 1963a, BOOTH *et al.*, 1963, OPIE, HUNT and FINLAY, 1964; SHILS, 1964; DUNN and WALSER, 1966) in patients with magnesium depletion and hypomagnesemia; even in kwashiorkor, with normal plasma magnesium, the urinary excretion of magnesium is low (MONTGOMERY, 1960, 1961; LINDER, HANSEN and KARABUS, 1963; PRETORIUS, WEHMEYER and THERON, 1963).

The amount of magnesium excreted in the feces on a low magnesium diet has been discussed in Sections III C and V C 2 a. There is no evidence that this loss increases or decreases as magnesium depletion progresses. However, R. H. SMITH (1957) has asserted that intestinal absorption of magnesium is impaired in magnesium-deprived calves; an increase in endogenous loss might therefore occur. In support of this possibility is the observation that magnesium deprivation may produce (or aggravate) diarrhea (PETERSEN, 1963 b). In children with kwashiorkor, stool losses may exceed intake, but are poorly correlated with the severity of diarrhea (PRETORIUS, WEHMEYER and THERON, 1963).

4. Calcium and phosphorus metabolism

The plasma concentration of calcium either remains constant (FITZGERALD and FOURMAN, 1956; DUNN and WALSER, 1966) or falls (FLETCHER and HENLY, 1960; PETERSEN, 1963 a; BOOTH et al., 1963; OPIE, HUNT and FINLAY, 1964; SHILS, 1964), despite adequate intake. Calcium balance remains zero (DUNN and WALSER, 1966) despite reduced urinary calcium (op. cit.; OPIE, HUNT and FINLAY, 1964). Magnesium repletion is accompanied by correction of hypocalcemia and hypocalciuria (PETERSEN, 1963 a; SHILS, 1964). Parenteral calcium administration does not alleviate hypomagnesemia; if anything, it aggravates it.

No consistent change in plasma phosphate or renal tubular reabsorption of phosphate was found by DUNN and WALSER (1966), but the tubular transport maximum was not examined.

5. Potassium metabolism

In PETERSEN's (1963 a and b) patient, the development of magnesium depletion was associated with potassium depletion, due to increased fecal loss; this was corrected on magnesium repletion. In SHILS' (1964) subjects, hypokalemia and reduced exchangeable potassium occurred despite large intakes. Similar results occurred in one of DUNN and WALSER's (1966) subjects. During therapy of kwashiorkor, potassium as well as magnesium was retained far in excess of nitrogen, reinforcing the concept of LILIENTHAL and associates (1950) that these two cations are present intracellularly in relatively fixed proportion.

6. Signs and symptoms

It is difficult to distinguish the symptoms of magnesium depletion from those of hypomagnesemia, since the former is often associated with the latter, except in kwashiorkor, where the presence of multiple deficiencies makes it impossible to ascertain which symptoms are due to magnesium lack. As shown below, the association of symptoms with hypomagnesemia is also highly unpredictable. The alleviation of various symptoms during hospitalization and magnesium therapy in a malnourished patient can hardly be taken as evidence that the symptoms were attributable to magnesium depletion (cf. comments by GRAHAM, 1965, on an article by CADDELL, 1965). The only study in which symptoms have been produced in otherwise normal subjects by magnesium depletion is that of SHILS (1964): one subject developed personality changes, anorexia, vomiting, tremors, fasciculations, hyporeflexia, and positive TROUSSEAU and CHVOSTEK signs; as noted above, he was hypocalcemic (5.5 mg %); the other developed paralytic ileus quite suddenly; here potassium deficiency may have been a factor; however, both patients responded to magnesium therapy.

D. Hypomagnesemia

1. In patients

a) Occurrence. i. Alcoholism and delirium tremens. Alcoholics may develop a moderate drop in plasma magnesium following withdrawal of alcohol, without negative magnesium balance (MENDELSON, LA DOU and CORBETT, 1964). Whether this phenomenon accounts for the frequent finding of hypomagnesemia in such individuals, particularly when they are suffering from delirium tremens (MARTIN *et al.*, 1950; STUTZMAN and AMATUZIO, 1953; FLINK *et al.*, 1954, 1957; FLINK, 1956; RANDALL, ROSSMEISL and BLEIFER, 1959; MARTIN, McCUSKEY and TUPIKOVA, 1959; McCOLLISTER, FLINK and DOE, 1960; SMITH, HAMMARSTEN and ELIEL, 1960; PRASAD, FLINK and McCOLLISTER, 1961; NIELSEN, 1963; SULLIVAN *et al.*, 1963; MENDELSON *et al.*, 1965) is uncertain. Such patients retain modest amounts of magnesium during therapy (McCOLLISTER, FLINK and DOE, 1960; MENDELSON *et al.*, 1965) but are doubtlessly also in positive nitrogen balance; in some patients hypomagnesemia persists (op. cit.; SULLIVAN *et al.*, 1963). A part of the decrease in plasma levels may be attributable to reduced protein-binding (PRASAD, FLINK and McCOLLISTER, 1961) Erythrocyte magnesium is less frequently reduced (HEATON *et al.*, 1962).

ii. Malabsorption. As noted above, magnesium depletion both with and without hypomagnesemia has been observed in patients with malabsorption. In some such cases, the drop in plasma concentration is too rapid to be attributed to external losses (BACK, MONTGOMERY and WARD, 1962). Almost half of a group of 24 patients studied by BOOTH *et al.* (1963) had low plasma levels. Whether such individuals are magnesium-depleted is uncertain; they are usually asymptomatic.

iii. Post-operative. Although HEATON (1964b) found that a majority of patients exhibit a fall in plasma magnesium on the first day following surgery, accompanied by negative magnesium balance, THOREN (1963) and MACBETH and MABBOTT (1964) found no change.

iv. Others: Mild degrees of hypomagnesemia have been observed in a large variety of disorders, in which both the mechanism and the significance of this finding is unknown (MARTIN, MEHL and WERTMAN, 1952; STUTZMAN and AMATUZIO, 1953; HANNA, 1961b).

b) Symptoms. Tetany is sometimes associated with hypomagnesemia and may respond to magnesium administration (MILLER, 1944; DURLACH and LEBRUN, 1960; BALINT and HIRSCHOWITZ, 1961; BACK, MONTGOMERY and WARD, 1962; WACKER *et al.*, 1962; GERST, PORTER and FISHMAN, 1964; SHILS, 1964). The presence of hypocalcemia or alkalosis has been excluded in only a minority of such cases (WACKER *et al.*, 1962); furthermore, tetany with normal plasma magnesium often responds to magnesium therapy. The role of hypomagnesemia in producing tetany is therefore uncertain; doubtless it is at least a contributory factor. Sometimes the tetany is aggravated by calcium but relieved by magnesium (George and Chambless, 1962). Many patients exhibit profound hypomagnesemia with no symptoms of neuromuscular dysfunction (Fletcher and Henly, 1960; Booth et al., 1963). As noted above, hypomagnesemia when due to magnesium depletion or deprivation is often associated with hypocalcemia, which may revert to normal on giving magnesium.

Cardiovascular symptoms evidently do not occur, and would not be anticipated on experimental grounds (Grantham, Tu and Schloerb, 1960; Haddy et al., 1963).

A number of central nervous system symptoms have been attributed to hypomagnesemia with or without magnesium depletion, on the basis of their association with hypomagnesemia and subsidence with magnesium administration. These include confusion, delirium, convulsions, tremors, twitching, hyperirritability, and athetoid movements (Flink et al., 1954, 1957; Hammersten and Smith, 1957; Baron, 1960; Smith, 1963; Gerst, Porter and Fishman, 1964; Fishman, 1965). It is of interest in this connection that the magnesium concentration of cerebrospinal fluid is normal in hypomagnesemic alcoholics (Glickman et al., 1962) or in infantile tetany with hypomagnesemia (Pallis, MacIntyre and Anstall, 1965). It is not known whether brain or spinal cord magnesium falls in magnesium depletion or hypomagnesemia.

French and Belgian workers have described a syndrome of "spasmophilia" or latent tetany, associated with an array of nonspecific complaints and with modest hypomagnesemia (Durlach and Lebrun, 1960; Misson and Schirardin, 1965). Electromyographic changes are often seen, similar to those which can be produced in experimental animals by low magnesium diets (Rosselle and de Doncker, 1961). Oral magnesium is beneficial.

2. In animals

On being put out to pasture on lush grass in the spring, cattle may develop hypomagnesemia very rapidly, associated with ataxia and tetany (see reviews by Sjollema, 1932; Blaxter and Sharman, 1955; Allcroft, 1956; Blaxter and McGill, 1956; Rook and Storry, 1962a; Wilson, 1964). That this hypomagnesemia is the result of diminished magnesium intake, without large negative balance, is shown by the observation that plasma magnesium may fall to very low levels in cows (Rook and Balch, 1958; Rook, 1961, Kemp et al., 1961; Care and Ross, 1962) or lactating sheep (L'Estrange and Axford, 1963, 1964a) on artificial magnesium-deficient diets within two days. Nonlactating animals respond more slowly (Ritchie et al., 1962; Storry and Rook, 1963). Milk magnesium does not fall (L'Estrange and Axford, 1964a). Even during lactation, the external losses of magnesium during two days could hardly amount to more than a minute fraction of the bodily content.

Thus this syndrome is a clear example of extracellular deficiency without significant depletion. Magnesium repletion corrects it promptly in calves (SMITH, 1961), but not in lactating sheep (RITCHIE and HEMINGWAY, 1963a and b).

a) Symptoms. Early symptoms include anorexia, restlessness, tetany, and "staggers." Later, convulsions occur. However, marked hypomagnesemia may exist with no symptoms (STORRY, 1961d).

b) Magnesium metabolism. i. Intestinal absorption. Magnesium absorption definitely diminishes on changing from dry feed to grass (KEMP *et al.*, 1961; CARE and ROSS, 1962; CARE, ROSS and WILSON, 1965). Use of ammonium-containing fertilizers was shown to increase the incidence of hypomagnesemia in cattle by BARTLETT *et al.* (1954) and in sheep by HEAD and ROOK (1957). The latter authors also showed that the contents of the small intestine contained less ultrafiltrable magnesium under these conditions, presumably due to the formation of magnesium ammonium phosphate. Lush grass feeding led to a high ruminal ammonium content, and addition of ammonium salts to dry feed led to reduced plasma and urine magnesium (HEAD and ROOK, 1955). However, a high urea intake failed to induce hypomagnesemia in sheep (EVERED, 1961), and CARE (1965) found no change in the ultrafiltrability of small intestinal contents on changing from hay to grass. Furthermore, ruminal and abomasal pH were both shown to be unaltered, despite the rise in ammonia content (SIMESEN, 1963). High citrate intake also aggravates hypomagnesemia (BURT and THOMAS, 1961), as may high potassium intake (KUNKEL, BURNS and CAMP, 1953; KEMP *et al.*, 1961). Both of these factors may play a role. The role of these and other dietary factors is reviewed by L'ESTRANGE and AXFORD (1964b) and DISHINGTON (1965).

ii. Urinary excretion. There is evidently no defect in the renal conservation of magnesium (KEMP *et al.*, 1961; STORRY and ROOK, 1963; L'ESTRANGE and AXFORD, 1964a).

iii. Buffering of plasma magnesium by bone. Adult ruminants evidently do not share with other animals the capacity to stabilize plasma magnesium by calling upon bone reservoirs of this element. Fasting, for example, causes a rapid fall in the plasma level (INGLIS, WEIPERS and PEARCE, 1959; ROOK and STORRY, 1962b). Cows with more severe hypomagnesemia than others in the same herd show the least response to oral magnesium but the greatest response to parenteral magnesium administration (ROOK and STORRY, 1962a). The inference that bone magnesium is less available is supported by the finding of diminished isotopic exchangeability (FIELD, 1960).

c) Calcium metabolism. Plasma calcium may or may not fall (ALLCROFT, 1947a and b; SMITH, 1961; ROOK and STORRY, 1962a). Clearly tetany is more prevalent when it does (HEMINGWAY and RITCHIE, 1965), but may develop at normal calcium concentrations.

The hypomagnesemia which develops in calves fed milk diets (DUNCAN, HUFFMAN and ROBINSON, 1935) is probably an example of magnesium deprivation during growth and is discussed therein (V B).

VI. Magnesium transport

A. Renal

1. Free and complexed magnesium

The glomerular filtrate contains magnesium in two forms, free ionic magnesium, and complexes with ligands small enough to pass through the glomerular membranes. The proportions of these two forms in plasma are given above (Table 3 p. 200).

The physicochemical state of magnesium in the urine has been studied very little. RAAFLAUB (1963) reports a few measurements of ionic magnesium in the urine: the fraction complexed is slightly less than that of calcium in the same urine samples. This is to be expected, since the major ligands in the urine (citrate, phosphate, sulfate and various organic acids) bind calcium with equal or greater affinity (SILLÉN and MARTELL, 1964). Ligands with high selectivity for magnesium relative to calcium have apparently not been identified in body fluids or tissues, although a number with high calcium selectivity are known. The concentration of magnesium and calcium as well as the concentrations of urinary anions are subject to enormously wide variations under normal conditions, so that no generalizations are possible as to the proportions free or complexed. But clearly, if complexes play a role in determining reabsorption under normal conditions, their effect will be far greater during antidiuresis than during diuresis, when magnesium, as well as calcium (WALSER, 1961a), probably exists almost entirely as free ions. The effect of water diuresis on magnesium excretion is considered below.

2. Active and passive transport and exchange

The concentration of magnesium in the urine has repeatedly been observed to be lower than in plasma ultrafiltrate. This establishes that the tubular epithelium is capable of reabsorbing a solution containing a higher magnesium concentration than the tubular fluid. Micropuncture data are not yet available, but it is clear that, if there is any tendency for magnesium to diffuse along its electrochemical gradient, it is more than offset by reabsorptive processes. The Nernst concentration ratio of a divalent cation, at equilibrium, across a 60 millivolt potential difference (as between the distal tubule and peritubular space) is 100:1. Thus passive diffusion, if unopposed, should lead to the appearance in the urine of more magnesium than is filtered.

Some transtubular flux of magnesium from interstitial fluid into the tubule occurs, as shown by the observation that radioactive magnesium injected intra-arterially with inulin during temporary ureteral obstruction appears in the post-obstruction urine before inulin appears (GINN *et al.*, 1959; MURDAUGH and ROBINSON, 1959, 1960; SAMIY *et al.*, 1960). Later studies, however, have indicated little (BRONNER and THOMPSON, 1961) or no (RAYNAUD, 1962) transtubular flux; the explanation of these discrepancies is not at hand. The amount of excreted magnesium has not yet been convincingly demonstrated to exceed the amount filtered in animals or normal man, although occasional clearance ratios slightly over unity have been observed (KNIPPERS and HEHL,

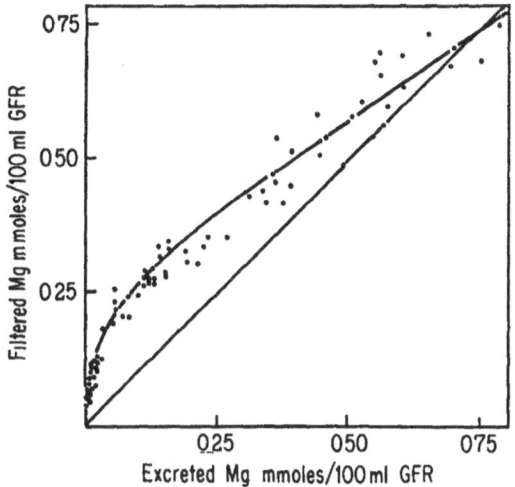

Fig. 4 Tubular reabsorption of magnesium during magnesium loading in dogs, redrawn from KNIPPERS and HEHL (1965). The diagonal line represents equality of amounts filtered and excreted Normal filtered load is about 0 06 mmoles per 100 ml of glomerular filtrate Reabsorption, indicated by the horizontal distance between each point and the diagonal line, increases several fold as magnesium is infused, but later begins to fall A few negative values, suggesting net secretion, are seen, but systematic errors are magnified in this portion of the curve

1965). Highly suggestive evidence for secretion in a patient with a magnesium-wasting disorder has been presented by VOSTAL, KÜCHEL and PACOVSKY (1963). Chickens given magnesium *via* the renal portal circulation show no evidence of secretion (ROBINSON and PORTWOOD, 1962). Transtubular flux, if it occurs, need not signify net movement of magnesium into the tubule at any point, nor that magnesium gradients will be reduced by passive diffusion. In the cerebrospinal fluid, for example, magnesium concentration is hardly affected by marked increases in plasma magnesium, despite the fact that isotopic exchange between the two compartments is extremely rapid (OPPELT, MacINTYRE and RALL, 1963).

The minimum in sodium concentration seen in samples collected after a brief interval of ureteral obstruction coincides with minima in magnesium as well as calcium concentration (WESSON and LAULER, 1959; MURDAUGH and ROBINSON, 1960; SAMIY *et al.*, 1960). Evidently the portion of the nephron capable of most nearly complete reabsorption of all three cations is located distally. However, it is probable that magnesium reabsorption, like that of sodium and calcium (Lassiter, Gottschalk and Mylle, 1963), occurs throughout the nephron, with the proximal tubule accounting for a major fraction.

Little is known of magnesium transport in the loop of Henle. Ullrich and Jarausch (1956) reported a few analyses of magnesium in portions of dog kidney; in each case, magnesium concentration in outer medulla was higher than in cortex or inner medulla; a further rise occurred at the tip of the papilla. The range of variation between the five dogs was disconcertingly large.

Magnesium reabsorption increases with magnesium infusion, but whether a transport maximum exists has not yet been established; fluctuations in renal hemodynamics and in protein-binding make this determination difficult (Barker, Clark and Elkinton, 1957; Knippers and Hehl, 1965). The latter authors' observations suggest that the amount reabsorbed reaches a maximum at plasma levels about four times normal, and then decreases as magnesium infusion is continued (Fig. 4). The assertion by Wilson (1960) and Rook and Storry (1962a) that magnesium reabsorption is fixed at a constant value whenever the filtered load is at least as great as this value is almost certainly incorrect. There is, in fact, no substance known to be handled by the kidney in this manner, and the evidence they present in support of this contention is highly indirect. If a transport maximum for magnesium does exist, it is reached only when excretion is increased far above normal.

3. Dependence on flow

a) Osmotic diuresis. The principal determinant of the rate of magnesium reabsorption is the rate of reabsorption of water in those portions of the nephron proximal to or continuous with the sites where magnesium is reabsorbed. This must be true of any substance which is reabsorbed by a concentration-dependent mechanism, because removal of water increases its concentration in the tubular fluid (Walser, 1966b). If water reabsorption is viewed as being linked with magnesium reabsorption through a form of solute-solvent coupling, the same conclusion follows. Calcium has been shown to be reabsorbed in the same portions of the nephron as sodium (Lassiter, Gottschalk and Mylle, 1963) and the percentage reabsorbed is roughly the same over-all (Walser, 1961a). Water reabsorption, at least in the proximal tubule, follows sodium reabsorption, and it therefore seems likely that all of these cations move together. This may be true whether the mechanisms which transport each are dependent or independent; all three cations share the same tubular fluid flow rates, and their reabsorptive rates are thus inevitably correlated, each being concentrated to the same extent by water reabsorption.

Thus osmotic diuresis augments magnesium clearance profoundly, whether induced by mannitol, sucrose, urea, or saline (WALSER, 1961a, WESSON, 1962a; KUPFER and KOSOVSKY, 1965; BLYTHE, GITELMAN and WELT, 1966; BETTER et al., 1966). When sodium excretion and flow are augmented by drugs which inhibit sodium reabsorption, such as cardiac glycosides or diuretics, the resulting increase in magnesium excretion may be attributable to reduced water reabsorption (see Section VIII B).

b) Water diuresis. Systematic studies of water diuresis have not been made. Water loading in man has been reported by NIELSEN to increase (1962) or decrease (1964) magnesium excretion in separate reports. In nine men, McCOLLISTER et al. (1958) observed no change in magnesium clearance during a day of high water intake (6 liters). HEATON and HODGKINSON (1963) and HEATON et al. (1962) found no effect of acute water diuresis, but MERTZ (1957) found a two-fold increase with an increase of flow of five ml. per minute. The resolution of this question is of some importance, since a lack of change with distal flow would suggest that magnesium reabsorption in this region is independent of concentration.

4. Dependence on the transport of other solutes

a) Sodium and calcium. As noted above, some degree of interdependence in the clearance of reabsorbable solutes is inevitable. The question of whether transport of magnesium at each point in the nephron depends on other transport processes is difficult to answer. WALSER, RAHILL and ROBINSON (1963) have suggested that a common cation transport mechanism may be responsible for the major portion of tubular reabsorption of all of the alkali cations and alkaline earths. Cations larger or smaller than the sodium or calcium ions (in unhydrated radii) appear to be transported less readily. The transport rate of magnesium relative to sodium was estimated to be 0.66. In this scheme, independent distal processes make fine adjustments in cation clearances, which may thus vary almost independently, provided that more proximal reabsorption is nearly complete (see Fig. 5).

Evidence for specific dependence between magnesium transport and that of other solutes may be summarized as follows:

During urea, saline, or mannitol diuresis, excreted/filtered sodium and calcium increase to an approximately equal extent (WALSER, 1961a; WESSON, 1962a; BETTER et al., 1966); excreted/filtered magnesium, normally larger than both the other ratios, rises in a curvilinear fashion, i.e., more rapidly at first. The shape of this curve is consistent with the existence of a common "pump" transporting all three cations, independent pumps, if at least half saturated, should lead to a different relationship (WALSER, 1966b). However, the data are too variable to attach much weight to this inference. Urea diuresis appears to increase divalent cation clearance less, relative to sodium clearance, than does mannitol or saline diuresis in these studies.

Although magnesium salts injected intravenously usually lead to augmented calcium excretion (IV C 2) and, conversely, calcium salts usually, but not always (FREEDMAN, MOULTON and SPENCER, 1958) lead to augmented magnesium excretion (WOLF and BALL, 1949; BARKER, ELKINTON and CLARK, 1959; SAMIY et al., 1960; WALLACH and CARTER, 1961), these observations do

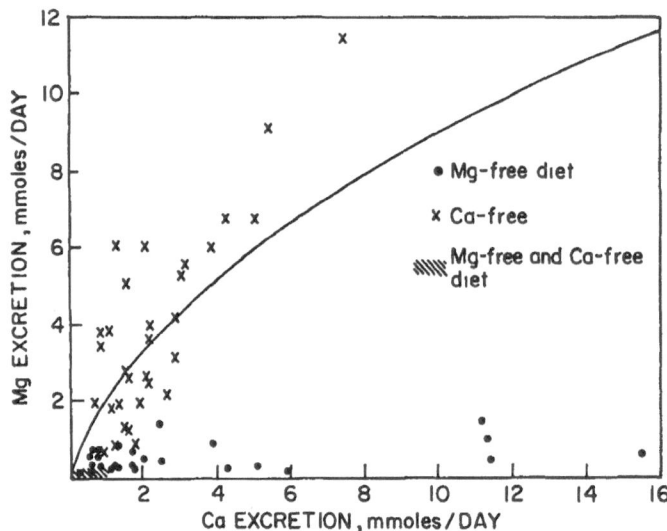

Fig 5 Response of urinary excretion to dietary restriction in dogs. Daily output of magnesium and calcium is related approximately as shown by the curved line in normal dogs (though variation is considerable). On calcium-free milk, little ability to retain calcium without retaining magnesium is seen, but on magnesium-free milk calcium is excreted without magnesium When both cations are absent from the diet, they virtually disappear from the urine (WALSER, unpublished observations)

not establish the existence of a common transport mechanism: sodium excretion, proximal flow, or both, are usually increased by these maneuvers, and the inhibition of cation reabsorption is general, rather than specific for these two divalent cations. Similar data can apparently be obtained for any pair of cations.

More convincing is the demonstration that each cation depresses the excretion rate of the other in the aglomerular fish (BERGLUND and FORSTER, 1958), but the relevance of these findings to mammals is open to question.

When hypertonic saline infusion was combined with artificial reduction of glomerular filtration rate, so that the filtered load of sodium fell, calcium and sodium clearance rose but magnesium clearance did not (BLYTHE, GITELMAN and WELT, 1966). This observation, if confirmed, appears to cast considerable doubt on the existence of common transport pathways for magnesium and either of the other two cations.

On the other hand, abundant evidence exists that magnesium depletion or deprivation reduces both calcium and magnesium clearance without reducing sodium clearance (see Fig. 5). Furthermore, KLEEMAN *et al.* (1966) have reported that magnesium clearance is more closely related to calcium clearance than to filtered magnesium, when hypercalcemia is induced in dogs and filtration rate is simultaneously reduced.

b) Anions. Sulfate augments magnesium clearance markedly (WALSER and BROWDER, 1959). This effect is attributable in part to the resulting natriuresis and diuresis and in part to the formation of $MgSO_4$ ion-pairs, which are evidently less readily reabsorbed. Sulfate given as the salt of an organic cation has little effect, suggesting that the inhibition of fractional sodium reabsorption by sodium sulfate has a directly inhibitory effect on magnesium reabsorption (WALSER, 1961 b).

Lactate has little or no effect (SULLIVAN, LANKFORD and ROBERTSON, 1966), except in hypertonic solution (BARKER, CLARK and ELKINTON, 1957); here the ensuing natriuresis and diuresis is probably the explanation. The same phenomenon probably accounts for reported effects of sodium salts of other organic acids (BRULL and BERNIMOLIN, 1956).

The effect of glucose, galactose or casein (LINDEMAN *et al.*, 1965) is not understood but may be related to a metabolic intermediate excreted in the urine in response to these substrates, or to change in the energy supply to a common cation pump.

B. Intestinal

1. Mechanism. Unlike the renal tubule, the gut has not been demonstrated to possess the capacity to reduce the magnesium concentration of the luminal contents below the concentration in interstitial fluid, although the electro-chemical gradient does suggest active transport. In isolated segments of rat intestine exposed to a magnesium-containing solution on the mucosal side and an initially magnesium-free solution on the serosal surface, HENDRIX, ALCOCK and ARCHIBALD (1963) found that the serosal/mucosal concentration ratio in duodenum, jejunum, ileum and colon after one and one-half hours incubation was close to unity. Ratios somewhat higher than unity (1 3 to 1.5) were seen when extra calcium or strontium (10 mM) was added to the mucosal fluids. A considerably lower ratio (0.4) was observed by SCHACHTER, DOWDLE and SCHENKER (1960). Although Ross and CARE (1962) found that metabolic energy was required for magnesium transport by isolated guinea pig small intestine, no uphill movement has been found in this species nor in the rabbit (ANAST *et al.*, 1964), even when magnesium-deprived. The lumen of the gut is usually somewhat electronegative with respect to the interstitial fluid, and the concentration of ultrafiltrable magnesium in sheep gut *in vivo* is always less than the ultrafiltrable magnesium in plasma, when corrected for this potential difference by the Nernst equation, as was shown by SCOTT (1965), recalculating the data of Storry[1] (1961 b). Although this would indicate active transport in a system otherwise at equilibrium, whether or not it does so in a system in which net movement of water and other solutes is taking place is conjectural.

Scott (1965) was unable to discern any relationship between magnesium transport and the mucosal magnesium concentration in isolated loops of sheep jejunum and ileum; magnesium movement was often inward rather than outward, and was not correlated with the electrical potential difference. Ross (1962), however, found that transport increased with increasing mucosal concentration in rat gut, and approached saturation (Fig. 6). Similar results in the reticulorumen and ileum of intact sheep were obtained by Care and Van't Klooster (1965). Hendrix, Alcock and Archibald (1963) found diminished serosal/mucosal ratios on increasing mucosal magnesium in rat gut. Isotope studies have been hard to interpret because flux of radiomagnesium may occur in the absence of net transport.

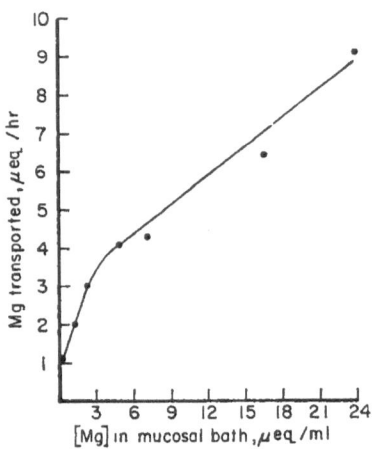

Fig 6 Intestinal transport of magnesium in relation to luminal concentration in the rat redrawn from Ross (1962)

Considerable evidence points to an interdependence between transport of magnesium and other alkaline earths, but as in the kidney, the hypothesis of a single transport mechanism for all is far from proven. That oral calcium tends to depress magnesium absorption from the diet has long been known (IV D). Alcock and MacIntyre showed in 1962 and in 1964 that a magnesium-deficient diet led to augmented calcium absorption in rats while a calcium-deficient diet promoted magnesium absorption. These observations could simply reflect altered water reabsorption (see Section III B 1 b i), but Kessner and Epstein (1966) report that gut segments from magnesium-deficient rats exhibit augmented serosal/mucosal ratios of calcium concentration. In vitro competition between calcium and magnesium for transport by sheep gut was shown by Care and Van't Klooster (1965), in rat gut by Hendrix, Alcock and

---

[1] Care and Van't Klooster (1965) have also compared the chemical gradients of magnesium concentration found between ileum and plasma of sheep with the calculated gradients which would indicate electrochemical equilibrium, based upon simultaneous measurements of the potential difference Unfortunately all of their calculated values are in error, evidently owing to the use of a valence of one instead of two for the magnesium ion They have concluded that magnesium always moves down its electrochemical gradient from ileum to blood, their data clearly show the opposite, as do those of other workers (e g , Scott, 1965)

ARCHIBALD (1963) and in dog jejunum by CRAMER and DUECK (1962), but in none of these studies were the data sufficiently uniform to permit the inference that both share a single "pump". In the study by HENDRIX, ALCOCK and ARCHIBALD (1963), concentration ratios for calcium were roughly equal to those of magnesium though SCHACHTER, DOWDLE and SCHENKER (1960) found much higher calcium ratios. Magnesium addition reduced intestinal transport of strontium (MRAZ, 1962) and of calcium (SCHACHTER and ROSEN, 1959) in rats, but not in sheep gut (SCOTT, 1965). The effects of increasing dietary magnesium on fecal calcium are noted above (III B 1 b *ii*). As MARCUS and WASSERMAN (1965) have shown, a constant ratio between the percentage removal of two divalent cations (e.g., calcium and strontium) from gut lumen leads to percentages of each remaining which diverge progressively along the length of the absorbing segment. Whether this pattern can also be shown to be applicable to magnesium remains to be seen.

Ross (1961) found that sodium stimulated transport of magnesium by rat gut *in vitro:* potassium and ammonium were without effect, as were sulfate, nitrate and acetate.

In summary, magnesium transport by the gut is usually downhill, in a chemical sense, if not an electrochemical one, but does appear to show saturation kinetics; on occasions, net transport may be inward. Calcium and magnesium transport are partially interdependent but strong evidence for a common mechanism is lacking (WILSON, 1964).

2. Site. In rats, magnesium transport apparently occurs throughout the intestine (CHUTKOW, 1964a; HENDRIX, ALCOCK and ARCHIBALD, 1963). In rabbits, indirect evidence indicates that the large intestine is not involved (AIKAWA, 1959). In calves, the large intestine is of greater importance at first, but the small intestine becomes the principal site of transport as the animal grows older (SMITH, 1959a, 1959c, 1962). In sheep, magnesium is absorbed principally from the upper (STEWART and MOODIE, 1956) or middle (FIELD, 1961) portions of the small intestine. In the abomasum and duodenum, the acidity of the intestinal contents and the fraction of luminal magnesium which is ultrafiltrable are relatively high, thus favoring passive reabsorption (STORRY, 1961b). The rumen was impermeable to magnesium according to PHILLIPSON and STORRY (1965), but net absorption was demonstrated by CARE and VAN'T KLOOSTER (1965). No explanation for these discrepancies is at hand. No direct information in human subjects is available.

C. Other organs

1. Salivary glands. CARE (1963) showed that the salivary concentration of magnesium in sheep diminished with increasing flow: the maximal concentration was roughly equal to that of plasma ultrafiltrate. LANGLEY, GRIMES and COCKRELL (1962) found that magnesium concentration in dog parotid saliva decreases during constant flow for up to forty minutes. Concomitantly, the amount of this cation in the gland falls. The ratio of salivary to plasma concentration was near unity, even after an interval of stopped flow. These observations, reinforced by measurements of isotopic magnesium flux, led the authors to conclude that salivary transport is passive. Magnesium infusion increases magnesium concentration in saliva (op.cit), was well as in other gastrointestinal secretions (IV C 1).

2. *Gall bladder.* The rabbit gall bladder *in vitro* elaborates a solution containing magnesium at a concentration somewhat lower than the mucosal fluid; surprisingly, this serosal/mucosal concentration ratio is nearly the same when most of the sodium is replaced by magnesium (PETERS and WALSER, 1966). The authors suggest that alkali cations and alkaline earths may share a common transport mechanism in this organ, with relative transport rate a function of crystal ionic radius, as in the kidney (VI A 4).

3. *Stomach.* GEERTRUYDEN (1964) has reviewed the literature and has presented interesting observations of the relationships between acidity, flow, and cation concentrations of gastric juice in dogs with HEIDENHAIN pouches. Sodium, calcium and magnesium concentration decrease linearly with increasing hydrogen concentration, while potassium concentration remains nearly constant (GEERTRUYDEN and DEJARDIN, 1964). The output of the first three cations is little affected by large increases in secretory rate, their concentrations thus falling; the concentration of potassium, by contrast, is independent of flow. He supports the view that the parietal secretion, rich in acid, contains only traces of sodium, calcium, and magnesium, and varies in flow rate, while the non-parietal secretion, formed at a relatively constant rate, contains substantial amounts of these ions. The minimal magnesium concentration observed at high flow was about 0.25 mM. In man, however, this author found much less change in magnesium concentration with acidity, or flow. He suggests that some mucosal cation may be excreted during rising flow, as in the salivary glands.

4. *Erythrocytes.* HANZE (1962a) has observed that erythrocyte magnesium falls about 0.5 % per hour on incubation at 37^0 but remains constant at 4^0. Incubation with fluoride (0.02 M) led to accumulation of magnesium and loss of potassium; iodoacetate (0.001 M) produced loss of both cations (HÄNZE, 1962b). The significance of these interesting results is as yet uncertain.

5. *Other cells.* Slices of rabbit kidney cortex lose 45 % of their magnesium content on treatment with EDTA. After removing the EDTA, reaccumulation of magnesium occurs in air (HÒFER and KLEINZELLER, 1963). Rat kidney cortex or diaphragm loses little magnesium in a nitrogen atmosphere despite loss of substantial cell potassium (ROGERS, 1965). By observing the rate of exchange of muscle magnesium with isotopic magnesium both *in vivo* and *in vitro*, this author has calculated the Q_{10} of exchange to be 1.8. He suggests

that a major portion of cell transport may be by passive diffusion, and a minor portion active accumulation. These inferences are quite different from those of GILBERT and McGANN (1960), referred to above, who suggest that magnesium is actively extruded from the cell. Reductions in muscle cell magnesium in response to administration of coconut oil, linseed oil, or olive oil in rats have been noted by CHEEK and associates (1962, 1963) suggesting an effect on membrane transport.

Dogs given ouabain into one renal artery exhibit a reduction in cortex magnesium which parallels the rise in cortex calcium (NAHMOD and WALSER, 1966). Magnesium is less effective than calcium in depressing sodium transport in frog skin (CURRAN and GILL, 1962).

Interesting studies of magnesium transport in yeast have been reported by CONWAY and DUGGAN (1956) and ROTHSTEIN (1961), but the relevance of these observations to animal cells is uncertain.

6. *Subcellular organelles.* Mitochondria accumulate magnesium in exchange for hydrogen ions, with or without inorganic phosphate. The process is inhibited by antimycin, cyanide, and uncouplers of oxidative phosphorylation, but not by oligomycin (BRIERLY *et al.*, 1963; BRIERLY, MURER and O'BRIEN, 1964; CARAFOLI, ROSSI and LEHNINGER, 1964; JUDAH *et al.*, 1965; RASMUSSEN, 1966). It has been suggested that ion accumulation by these structures may provide a useful model of membrane transport, and also that transcellular transport of ions may involve mitochondrial accumulation and release.

VII. Hormonal effects on magnesium metabolism

Many of the data in this area are inconsistent or contradictory, but some patterns emerge.

A. Calcium-regulating hormones

1. *Parathyroid hormone.* In hyperparathyroidism plasma magnesium may be depressed (HARMON, 1956; AGNA and GOLDSMITH, 1958), normal, or elevated (WALSER, 1962; HEATON and PYRAH, 1963; WACKER and VALLEE, 1964). Removal of a parathyroid tumor is sometimes, though not always (BASSETT, 1935) followed by retention of modest amounts of magnesium, especially in patients with osteitis fibrosis cystica (BARNES, KRANE and COPE, 1957; HANNA *et al.*, 1961 b). However, parathyroidectomy in such patients (POTTS and ROBERTS, 1958; HANNA *et al.*, 1961 b), in normal animals (in whom the thyroid is generally removed at the same time) (GROLLMAN, 1954 b; WEIL and STATE, 1958; MARTIN, MIKKELSEN and JONES, 1959; MACINTYRE, BOSS and TROUGHTON, 1963; PAYNE and CHAMINGS, 1964; HEATON, 1964 a, 1965 a; HEATON and ANDERSON, 1965), and spontaneous or postoperative hypo-parathyroidism (HOMER, 1961; WALSER, 1962; WALLACH *et al.*, 1965; JONES and FOURMAN, 1966) may be associated with lowered blood levels, though not

invariably (DURLACH *et al.*, 1959; HEATON and ANDERSON, 1965). Administration of parathyroid extract had no effect on plasma magnesium in rats (HEATON, 1965a) or in relatively small doses in dogs (ROBERTS *et al.*, 1954) or man (GILL, BELL and BARTTER, 1962); in larger doses plasma magnesium rises in rats (GREENBERG and MACKEY, 1932; CHEEK and TENG, 1960) and dogs (SCHOLZ, 1931, DURLACH *et al.*, 1959; WALLACH *et al.*, 1965). Tissue magnesium is reduced only in the heart (WALLACH *et al.*, 1965); but bodily content falls (CHEEK and TENG, 1960; ELIEL, CHANES and HAWRYLKO, 1963). In thyroparathyroidectomized sheep (CARE and KEYNES, 1964) and one patient with sprue and hypomagnesemia (HAAS, AFFOLTER and DUBACH, 1965), transient reduction was seen. Hypoparathyroid children given the extract may show an increase in plasma level (NICHOLS, HARRISON and HARRISON, 1964, cited by HARRISON and HARRISON, 1964). When parathyroid extract is given to rats 18 hours after a dose of vitamin D, plasma magnesium rises, but the same dose of extract has no effect in vitamin D-deficient animals (HARRISON and HARRISON, 1964). Ultrafiltrability of plasma magnesium is unaffected by parathyroid extract or parathyroidectomy (WALLACH *et al.*, 1965).

Urinary excretion of magnesium increases following administration of parathyroid extract in man (BULGER and GAUSMANN, 1933; TIBBETTS and AUB, 1937b) though much less than does the excretion of calcium (GILL, BELL and BARTTER, 1962). Similar results are seen in dogs (ROBERTS *et al.*, 1954) and rats (HEATON, 1965a; KOHLER and PECHET, 1966), though diminished renal excretion may be seen with unphysiological doses (CHEEK and TENG, 1960, MACINTYRE, BOSS and TROUGHTON, 1963). Hyperparathyroid patients may exhibit high urinary magnesium, and removal of a parathyroid tumor is followed by reduced excretion (TIBBETTS and AUB, 1937b; HEATON and PYRAH 1963) and modest positive balance (HANNA *et al.*, 1961b), though fecal magnesium may increase enough to overcome urinary retention (HEATON and PYRAH, 1963). The last named authors found renal excretion of magnesium normal in three patients with hypoparathyroidism but increased in three of nine patients with hyperparathyroidism. Parathyroidectomy in animals likewise leads to reduced excretion (HEATON, 1965a). Nevertheless, JONES and FOURMAN (1966) have presented evidence for substantial deficits of magnesium in patients with postoperative hypoparathyroidism, in support of the experimental findings of CHEEK and TENG (1960) and ELIEL, CHANES and HAWRYLKO (1963) cited above.

In thyroparathyroidectomized sheep, reabsorption of magnesium by the mid-ileum was decreased by parathyroid extract (CARE and KEYNES, 1964).

MARTINDALE and HEATON (1965) found that magnesium loss from fresh or boiled bones was increased by parathyroid extract, provided that serum was also present.

These confusing effects of parathyroid hormone on magnesium metabolism can perhaps be summarized by the statement that magnesium is usually (though not always) affected in the same manner as calcium but to a lesser extent.

2. Vitamin D. Vitamin D administration reduces plasma magnesium (SILVERMAN and GARDNER, 1954; HANNA, 1961a; RICHARDSON, HUFFINES and WELT, 1963; GITTELMAN, PINKUS and SCHMERTZLER, 1964; HARRISON and HARRISON, 1964), although it does not usually affect intestinal absorption appreciably (WACKER *et al.*, 1962; HEATON, HODGKINSON and ROSE, 1964; SMITH, 1962). Magnesium balance was found unaltered by RICHARDSON and WELT (1955) in rats on a low magnesium diet given vitamin D, and had no effect in a magnesium-depleted patient (PETERSEN, 1963b). However, ultraviolet irradiation was reported to diminish body magnesium by McHARGUE and ROY (1930), while vitamin D_2 was found to induce positive magnesium balance by MILLER *et al.* (1965a), and its absence led to diminished plasma and bone magnesium (MILLER *et al.*, 1965b). In one study, urinary magnesium rose while fecal magnesium fell (HANNA, 1961a); in the chick, also, intestinal absorption of magnesium is evidently increased by vitamin D_3 (WASSERMAN, 1962). In this species, bone uptake of radiomagnesium given subcutaneously, but not orally, was diminished by vitamin D (WORKER and MIGICOVSKY, 1961). Further study of the effect of vitamin D is clearly needed.

Although there have been reports of improved responsiveness to vitamin D in hypomagnesemic patients following magnesium therapy, it seems likely that the observed rise in plasma calcium was a direct result of the magnesium repletion or was secondary to lessened diarrhea and steatorrhea rather than to a direct relationship between magnesium deficiency and vitamin D resistance.

3. Calcitonin. A calcium-lowering extract of parathyroid glands reduced intestinal absorption of calcium and magnesium (CARE and KEYNES, 1964), but thyrocalcitonin had no effect on plasma magnesium nor its ultrafiltrability (FOSTER *et al.*, 1964, 1966; GITELMAN, KUKOLJ and WELT, 1965b). Urinary excretion of magnesium is decreased by thyrocalcitonin, presumably secondary to inhibition of bone reabsorption (KOHLER and PECHET, 1966).

B. Adrenocortical hormones

Administration of aldosterone increases the renal excretion of magnesium (HANNA and MACINTYRE, 1960; HORTON and BIGLIERI, 1962; SCOTT and DOBSON, 1965); plasma magnesium may remain constant (OYAERT, 1962) or fall (CARE and ROSS, 1963; KAPITOLA and KUCHEL, 1964; SCOTT and DOBSON, 1965). Desoxycorticosterone may produce a fall (WOODBURY *et al.*, 1950), especially in adrenalectomized animals (CARE and ROSS, 1963; MAHY and MUNDAY, 1963). Similarly, plasma magnesium may be low (MADER and ISERI, 1955; MILNE, MUEHRCKE and AIRD, 1957; KAPITOLA and KÜCHEL, 1964) or

normal (HORTON and BIGLIERI, 1962) in primary hyperaldosteronism. Adrenalectomy leads either to a rise in plasma magnesium (HARROP et al., 1933; ZWEM-
MER and SULLIVAN, 1934; CONWAY and HINGERTY, 1946, DA VANZO, CROSS-
FIELD and SWINGLE, 1958; HANNA and MacINTYRE, 1960), no change (HELVE,
1940), or a fall (WALSER, ROBINSON and DUCKETT, 1963). In normal sheep,
aldosterone and desoxycorticosterone augment fecal magnesium (CARE and
ROSS, 1963); a similar increase occurs in adrenalectomized rats given aldosterone (HANNA and MacINTYRE, 1960). CONWAY and HINGERTY (1946) found
increased muscle magnesium in adrenalectomized rats; BUELL and TURNER
(1941) found it decreased, and HANNA and MacINTYRE (1960) found it unaltered, though reduced when aldosterone was given. RUBIN and KRICK (1936)
noted negative magnesium balance after adrenalectomy. BRANDT and GLASER
(1959) reported no alterations in plasma disappearance or tissue uptake of
isotopic magnesium after adrenalectomy, but AIKAWA, HARMS and REARDON
(1960) noted some changes in tissue equilibration. In patients with primary
hyperaldosteronism, the usual nocturnal increase in magnesium excretion is
not seen, and magnesium clearance remains high; spironolactone reduces it;
removal of the tumor is followed eventually by magnesium retention (HORTON
and BIGLIERI, 1962). The effect of aldosterone on the diurnal pattern of
excretion is not shared by glucocorticoids (DOE, VENNES and FLINK, 1960).

Most of these data are consistent with the view that mineralocorticoids have
a potent stimulatory influence on the renal and fecal excretion of magnesium;
whether the effects on plasma magnesium are secondary is uncertain, but they
are in the right direction to be so explained.

The possibility that magnesium deficiency might inhibit aldosterone
secretion was examined by COPE and PEARSON (1963); the results were equivocal.

Effects of glucocorticoids have been less well studied. GROLLMAN (1954a)
observed an increased plasma magnesium following cortisone administration
to a nephrectomized dog maintained by peritoneal dialysis, and MAHY and
MUNDAY (1963) observed a rise in both plasma and urinary magnesium in rats,
as did MERTZ and LUTZ-DETTINGER (1957); but HILLS et al. (1959) observed
a rise in both values on withdrawing cortisone therapy in an Addisonian, and
a fall in excretion on giving hormone. Cushing's syndrome was found to be
associated with hypermagnesuria, but ACTH in normals had no effect (KAPI-
TOLA and KÜCHEL, 1964).

C. Thyroid

As noted above (V B), magnesium-deprived rats show some of the signs
of hyperthyroidism. Following the demonstration (TAPLEY, 1955), that triiodothyronine induced negative balance in patients with hypothyroidism,
TAPLEY and COOPER (1956) went on to confirm and extend the earlier observations of BAIN (1954) that magnesium inhibits thyroxine-induced uncoupling

of oxidative phosphorylation. Further similarities between magnesium deprivation and the effect of thyroid administration in rats were noted by VITALE et al. (1957a), who demonstrated that high magnesium intake could partially overcome thyroxine-induced inhibition of growth. Thyroxine was shown to reduce markedly the increase in kidney calcium in the magnesium deprivation syndrome (HELLERSTEIN et al., 1957; GERSHOFF et al., 1958). Normal rats given thyroxine exhibit reduced oxidative phosphorylation when they are killed and heart mitochondria are isolated minutes later; magnesium injection prevents this reduction (VITALE, NAKAMURA and HEGSTED, 1957; VITALE et al., 1957b). However thyroxine did not affect mitochondrial swelling from hearts and kidneys of such animals (NAKAMURA et al., 1961).

One gram of magnesium intravenously had no effect on basal metabolic rate in patients with hyper- or hypothyroidism, and failed to prevent the metabolic response to triiodothyronine (WISWELL, 1961), but NEGUIB (1963) found distinct improvement in thyrotoxic subjects given 7 g of $MgCl_2$ intramuscularly every day for several weeks, BELLOIU et al. (1965) observed remission of some symptoms, and JONES et al. (1966) observed substantial positive balance during therapy. The latter authors also noted negative balances during treatment of myxedema, but COHEN (1963) did not. A distinct negative correlation between plasma and erythrocyte magnesium and thyroid activity in patients with hyper- and hypothyroidism was noted by RIZEK, DIMICH and WALLACH (1965) and JONES et al. (1966); urinary magnesium was positively correlated. Hypomagnesemia in thyrotoxicosis was also noted by PRASAD, FLINK and McCOLLISTER (1961), and DOE, FLINK and PRASAD (1959) but not by KLEEMAN et al. (1958). More recently, FORBES (1965) has reported that thyroxine did not increase the magnesium requirement for normal growth or prevention of hypomagnesemia in rats on marginally low magnesium diets, although it did counteract the increase in kidney calcium concentration; thyroxine alone reduced positive balances of magnesium, calcium, and phosphorus in growing rats, but not in rabbits, though plasma magnesium fell (AIKAWA, 1960d).

Earlier reports (e.g., SOFFER et al., 1941; DINE and LAVIETES, 1942) of altered protein-binding of magnesium in the plasma of patients with hyper- and hypothyroidism have not been confirmed (COPE and WOLFF, 1942; BISSELL, 1945; SILVERMAN and GARDNER, 1954; KLEEMAN et al., 1958; PRASAD, FLINK and McCOLLISTER, 1961; WACKER and VALLEE, 1964).

D. Other hormones

HANNA et al. (1960, 1961a) have shown that growth hormone promotes the intestinal reabsorption while inhibiting the renal tubular reabsorption of both magnesium and calcium. Plasma magnesium, however, tends to fall while plasma calcium rises.

Insulin and glucose administration tends to lower plasma magnesium (WHANG, WAGNER and RODGERS, 1966) while increasing tissue content (AIKAWA, 1960b). The hypoglycemic response to insulin is apparently augmented by giving 50 micromoles per kg. of magnesium intraperitoneally to mice (PLANCHART, 1965). BHATTACHARYA (1961) has shown that magnesium is required to maintain the sensitivity of the isolated diaphragm to insulin, possibly by regulating its rate of entry.

VIII. Effects of drugs on magnesium metabolism
A. Ethanol

It has repeatedly been observed that alcohol augments renal excretion of magnesium, both in normal and alcoholic individuals (McCOLLISTER et al., 1958; KALBFLEISCH, LINDEMAN and SMITH, 1961; KALBFLEISCH et al., 1963; McCOLLISTER, FLINK and LEWIS, 1963; LINDEMAN et al., 1964; SULLIVAN, ROBERTSON and SHEHAN, 1964). Calcium excretion increases to a comparable extent but sodium excretion does not; filtration rate is unaffected. This appears to be another example of this type of response occurring whenever any rapidly oxidizable substrate is administered in quantity (LINDEMAN et al., 1965).

The inference that magnesium depletion might result from the continued ingestion of alcohol was strengthened by the report that rats fed alcohol require more magnesium for normal growth (GOTTLIEB et al., 1959). However, magnesuria following ethanol is blunted if serum magnesium is low (SULLIVAN, LANKFORD and ROBERTSON, 1966) and urinary magnesium remained close to zero in a magnesium-depleted normal subject despite ingestion of 10 ounces of whiskey daily (DUNN and WALSER, 1966). Furthermore, chronic alcohol administration failed to alter growth, bone density or muscle magnesium in rats (SAVILLE and LIEBER, 1965). Thus the role of ethanol in producing significant magnesium depletion is doubtful at best.

B. Diuretics and cardiac glycosides

As noted above (VI A), any reduction in the reabsorption of sodium and water in the more proximal portions of the nephron tends to be accompanied by a parallel reduction in the reabsorption of magnesium. It is not surprising, therefore, that magnesium excretion is increased markedly by mercurial diuretics (BARKER, ELKINTON and CLARK, 1959; HANZE, 1960a, 1965, WALSER and TROUNCE, 1961; WALSER, 1961c; WESSON, 1962b; KNIPPERS and HEHL, 1965). A less striking increase (McCOLLISTER et al., 1958; ROBINSON, MURDAUGH and PESCHEL, 1959a; HANZE, 1960a; SMITH, KYRIAKOPOULOS and HAMMARSTEN, 1962; GLAUBITT and RAUSCH-STROOMANN, 1962) or no change (KIIL, 1960; POUTSIAKA et al., 1961) is seen following diuretics of the benzothiadiazine group. Acetazolamide, on the other hand, regularly leads to decreased excretion (BARKER, CLARK and ELKINTON, 1955; JABIR, ROBERTS and WOMERSLEY, 1957; BARKER, ELKINTON and CLARK, 1959; HANZE, 1960a). Since proximal sodium reabsorption is considerably inhibited by this drug, the enhancement of magnesium reabsorption must be pronounced, and is evidently highly specific; calcium excretion rises as magnesium excretion falls. This response is not unlike the response to sodium bicarbonate administration (HILLS et al., 1959), and suggests a relationship between hydrogen transport and magnesium transport in the proximal tubule.

Whether diuretics can induce significant magnesium depletion seems doubtful. In all probability, as the natriuretic effect diminishes with repeated administration, the magnesiuretic effect will do likewise. In any case, the amounts involved are small in comparison with body stores. Plasma magnesium shows little or no change (WACKER, 1961; SELLER et al., 1966).

Cardiac glycosides inhibit tubular transport of magnesium as well as that of sodium and calcium (KUPFER and KOSOVSKY, 1965), whether administered into the renal artery or the general circulation (NAHMOD and WALSER, 1966). Magnesium balance is not appreciably altered, and tissue levels remain constant, but equilibration of tissue pools with extracellular magnesium shows subtle changes (AIKAWA, REARDON and HARMS, 1961), and plasma concentration may fall (KLEIBER et al., 1966).

C. Other drugs

NIELSEN (1964b) noted that manic patients being treated with lithium salts exhibit reduced plasma magnesium. However, lithium administration in rabbits increased plasma magnesium (ibid., 1964a).

Salicylate at a dosage of 300 mg/kg daily led to a rise in plasma magnesium and a fall in urinary magnesium in rats (CHARNOCK, OPIT and HETZEL, 1961).

A large dose of acetylcholine increased plasma magnesium within ten minutes in rabbits (TERKILDSEN, 1952a).

Dinitrophenol, pyridoxine and desoxypyridoxine had minimal effects in rabbits except for some increase in plasma level with the last drug (AIKAWA and REARDON, 1965; AIKAWA, 1960c).

IX. Magnesium metabolism in specific disorders
A. Renal failure

It is generally agreed that patients with acute renal failure characteristically exhibit hypermagnesemia (HAURY and CANTAROW, 1942; WACKER and VALLEE, 1957; HAMBURGER, 1957; SMITH and HAMMARSTEN, 1958; LOSSE and KOENIG, 1961), particularly when hyperkalemia is also present (TAKAYASU et al., 1962). Levels as high as 3 mM are seen. A rapid rise occurs in experimental renal failure (YANADORI, 1962, HARRIS, McDONALD and WILLIAMS, 1952). It has not been established that the obtundation or the prolonged QT interval often

seen in this syndrome is in part attributable to hypermagnesemia, although
this possibility has been suggested (HAMBURGER, 1957), and there is evidence
to suggest that ECG changes with hyperkaliemia and hypermagnesemia occur
earlier if renal failure is present (TAKAYASU et al., 1962).

In chronic renal failure variable values have been observed. Hypermagne-
semia occurs (HIRSCHFELDER and HAURY, 1934, HAURY and CANTAROW, 1942;
SILVERMAN and GARDNER, 1954; LOSSE and KOENIG, 1961; DUNN, KEOPLUNG
and KREHL, 1962), but normal or low values have also been reported (HAMMAR-
STEN, ALLGOOD and SMITH, 1957; SMITH and HAMMARSTEN, 1958, WALSER,
1962; CLARKSON et al., 1965). Ultrafiltrability is normal or slightly reduced
(SILVERMAN and GARDNER, 1954; PRASAD, FLINK and McCOLLISTER, 1961;
WALSER, 1962), but the fraction of ultrafiltrable magnesium existing as free
ions is often low (WALSER, 1962). Thus physiologic hypomagnesemia may
exist with normal total plasma magnesium in such patients. In nephrosis, on
the other hand, ultrafiltrability is increased as a result of hypoproteinemia
(PRASAD, FLINK and McCOLLISTER, 1961). Erythrocyte magnesium may also
be increased in chronic renal insufficiency (SMITH and HAMMARSTEN, 1959;
HÁNZE, 1960b; MAURAT, 1964), but is poorly correlated with the plasma
level (HÁNZE and HILLER, 1963). HÁNZE (1962c) has suggested that the
accumulation in cells of organic phosphates with magnesium-complexing ability
may be responsible.

Renal excretion of magnesium loads is impaired, presumably in proportion
to the reduction in glomerular filtration rate (HIRSCHFELDER and HAURY, 1934);
hence hypermagnesemia is more common in patients with severely depressed
filtration rates (ROBINSON, MURDAUGH and PESCHEL, 1959b). Magnesium
clearance tends to remain constant as filtration rate falls, until the percentage
excretion of filtered magnesium has reached a very high value (STEELE et al.,
1966). The mechanism of this homeostatic response is unknown; osmotic
diuresis in the remaining nephrons is evidently not the explanation, since
uremics have higher excreted/filtered magnesium ratios than normals given
saline and urea to achieve comparable osmolar clearance per unit volume of
glomerular filtrate (BETTER et al., 1965). Magnesium infusion does not improve
renal hemodynamics in these patients (HAMMARSTEN, ALLGOOD and SMITH,
1957). Oliguria is said to play a role in the impaired excretion of magnesium
loads (NIELSEN, 1962); this may be because oliguric patients usually have
superimposed acute renal failure, severely depressed filtration rates, or both.
Renal excretion of magnesium is sometimes reduced without resultant elevation
of plasma concentration; in such individuals intestinal absorption of magnesium
is diminished as well (CLARKSON et al., 1965). Symptoms which can be attributed
with some confidence to hypermagnesemia occur in chronic renal failure,
where marked elevations are occasionally seen, especially after the administration of magnesium for hypertension or gastric hyperacidity (SMITH and

HAMMARSTEN, 1958; RANDALL et al., 1964); cardiac conduction defects, are-flexia, hypotension, nausea, vomiting, difficulty in voiding, stupor, and coma are seen. RANDALL et al. (1964) believe that these patients develop symptoms of hypermagnesemia at lower levels than do normal subjects.

B. Alcoholism

The occurrence of hypomagnesemia in alcoholics, especially those with withdrawal symptoms of one kind or another, was summarized above (V D), and the effect of alcohol on magnesium metabolism is discussed in VIII A. Evidence pertinent to the question of whether magnesium deficiency plays a role in the development of alcoholic tremulousness or delirium tremens is equivocal. Some of these patients exhibit reduced erythrocyte magnesium as well as plasma magnesium (SMITH and HAMMARSTEN, 1959) but this is not always the case (HEATON et al., 1962; MAURAT, 1964); some retain moderate amounts of magnesium during treatment, as noted previously (SULLIVAN et al., 1963). However, HEATON and associates (1962) could detect no difference between patients with or without hypomagnesemia in symptomatology or response to conventional therapy, and as WACKER and VALLEE (1964) have noted, complete recovery is often observed without giving extra magnesium.

It has not been shown that raising plasma magnesium to normal (and no further) is associated with improvement: invariably hypermagnesemia is induced at least transiently, and the sedative effects of this procedure are well known.

A role of magnesium in the pathogenesis of these syndromes thus remains to be established, although abnormalities of magnesium metabolism are frequent.

C. Atherosclerosis

Magnesium has been shown to inhibit the development of experimental atherosclerosis in several ways, some of which have already been discussed (V B). Increased magnesium intake in rats inhibits the lipid deposition in the aorta which follows a high cholesterol intake or other atherogenic diets (HELLERSTEIN et al., 1957; VITALE et al., 1957c; VITALE et al., 1959; HELLER-STEIN et al., 1960; NAKAMURA et al., 1960). Although this response is seen only in hypercholesterolemic animals (HELLERSTEIN et al., 1960), it is not associated with a reduction in plasma cholesterol (VITALE et al., 1957c, 1959). Plasma lipoproteins, however, fall somewhat. In rabbits fed cholesterol, no decrease in the resulting deposits of aortic lipid is seen with high magnesium intake, but plasma and tissue cholesterol may fall (BHATTACHARYYA and MULLICK, 1963); a low magnesium diet leads to an increase (NAKAMURA et al., 1965) in tissue levels. Triparanol inhibits lipid deposition on a high magnesium intake but not on a low one (NAKAMURA et al., 1963).

There is also evidence that atherogenic diets aggravate the magnesium deprivation syndrome. Such diets lead to the appearance of this syndrome in rats on magnesium intakes which fail to produce the syndrome on non-atherogenic diets; with increasing dietary magnesium, these manifestations do not occur (Hellerstein et al., 1957; Vitale et al, 1957c). Furthermore, a high cholesterol intake tends to reduce plasma magnesium despite normal magnesium intake (Hellerstein et al, 1960; Bhattacharyya and Mullick, 1963). The skin lesions of the magnesium deprivation syndrome are aggravated if cholesterol is added to the diet (Olson and Parker, 1964). Maize prevents both the hypercholesterolemia and the hypomagnesemia that result from feeding rats a diet high in saturated fats (Rademeyer and Booyens, 1965); furthermore, plasma magnesium was considerably higher and plasma cholesterol lower in normal men during a month on maize supplements than during preceding or following periods on maize-free diets (Booyens et al., 1966). Maize is known to be rich in magnesium, but whether this accounts for its effects is uncertain. High magnesium intake inhibits the development of coronary thromboses following the intravenous administration of certain polysaccharides (Shimamoto et al., 1959) and the vascular and cardiac lesions which follow dihydrotachysterol plus high phosphate intake (Selye, 1958; Mishra, 1960c); potassium chloride is, however, equally effective in the latter syndrome (Selye, 1958).

In man, the evidence for an association between atherosclerosis and magnesium intake is less impressive. Bersohn and Oelofse (1957) noted that ethnic or national groups exhibiting high plasma magnesium values tend to show low plasma cholesterol, and conversely, but Brown et al. (1958) found no correlation between these plasma constituents in a group of middle-aged men. Hughes and Tonks (1965) maintain that plasma magnesium is slightly reduced in patients with myocardial infarctions, or histories thereof, and can be restored to normal by as little as 0 7 mmoles per day extra dietary magnesium supplied in a proprietary mixture. They have presented evidence that platelet aggregation, promoted by ADP, is inhibited by magnesium (Hughes and Tonks, 1966).

D. Hypertension

Mild hypomagnesemia in patients with essential hypertension was reported by Albert, Morita and Iseri (1958), but Seller, Ramirez-Muxo and Brest (1965) found both plasma and erythrocyte levels normal before thiazide treatment was instituted. Toxemia of pregnancy was associated with levels no lower than normal pregnancy (Hall, 1957). A role of hypomagnesemia in augmenting peripheral resistance seems unlikely, since limb perfusion with blood containing subnormal amounts did not lead to vasoconstriction unless concentrations of other ions (potassium, calcium or hydrogen) were also abnormal (HADDY et al., 1963). However, RIGÓ and SZELÉNYI (1963) report that a fivefold augmentation of magnesium intake in rats inhibits the development of hypertension following a cardiopathic diet or noxious stimuli.

E. Anemia

It has long been known that erythrocyte magnesium concentration is increased in the presence of increased blood destruction, whether spontaneous (BANG and ØRSKOV, 1939) or experimentally induced (HENRIQUES and ØRSKOV, 1939). Reticulocytes contain a much higher concentration of magnesium than do mature erythrocytes (GINSBURG et al., 1962). In pernicious anemia, the increase in intracellular magnesium may be related to higher concentrations of phosphorylating enzymes (VALBERG et al., 1965 b). In iron deficiency anemia, red cell magnesium is normal (VALBERG et al., 1965 b), as is intestinal absorption of this metal (POLLACK et al., 1965).

F. Central nervous system disorders

Spinal cord magnesium was found to increase in experimental rabies (ROMENSKI, 1935). More recently, HARRIS and BEAUCHEMIN (1956) reported reduced cerebrospinal fluid magnesium in patients with organic and functional psychoses. SCHAIN (1964b) found normal values in 62 mentally defective children, and normal values have also been reported in epilepsy (GREENBERG and AIRD, 1938), PARKINSON's disease (SCHWAB, PORYALI and AMES, 1964) and the Guillain-Barré syndrome (PALLIS, MACINTYRE and ANSTALL, 1965). However, CANELAS, DE ASSIS and DE JORGE (1965) found high cerebrospinal fluid magnesium and hypomagnesemia in epileptics examined soon after seizures; electroshock therapy also increased the spinal fluid concentration.

JONES, DESPER and FLINK (1965) studied blood and spinal fluid magnesium, as well as magnesium balance, in HUNTINGTON's chorea. They observed no abnormalities and suggest that a group of patients found to have abnormal concentrations by KENYON and HARVEY (1963), all of whom had strong family histories, might be etiologically distinct.

G. Muscular dystrophy

LILIENTHAL et al. (1950) noted that the ratio of muscle magnesium to potassium was normal in atrophic muscle, but ZUCKERMAN and MARQUARDT (1963) found muscle magnesium reduced and muscle potassium normal in rabbits with muscular dystrophy induced by feeding a vitamin E deficient diet. In children with muscular dystrophy, BOELLNER et al. (1965) report that red cell magnesium is increased; plasma magnesium is normal (SMITH, FISCHER and ETTELDORF, 1962).

H. Other diseases

In acute pancreatitis, plasma magnesium may fall briefly, but less than does plasma calcium (EDMONDSON et al., 1952). In rabbits made diabetic with alloxan, skin magnesium fell, bone magnesium rose, and the balance of magnesium remained normal (AIKAWA, 1960a). Diabetic acidosis is associated with a considerable magnesium deficit (ATCHLEY et al., 1933; BUTLER, 1950; NABARRO, SPENCER and STOWERS, 1952). MARTIN and WERTMAN (1947) have emphasized that plasma magnesium may nevertheless be high; after treatment is initiated it falls over a period of several days.

Abnormalities of magnesium metabolism in endocrine disorders are considered in section VII.

X. Therapeutic uses of magnesium

A. Vascular and renal disorders

The vasodilatory action of magnesium has been employed in the treatment of a variety of diseases. HARRIS and DE MARIA (1953) observed that parenteral magnesium improved renal hemodynamics in acute glomerulonephritis, particularly during the hypertensive stage, and showed later (DE MARIA and HARRIS, 1955) that the fall in glomerular filtration rate and renal blood flow which ensued on transfusing hemolyzed blood into dogs could be prevented or reversed. No effect occurred in normal dogs (HARRIS and DE MARIA, 1951). ETTELDORF and TUTTLE (1952), however, found filtration rate unaffected by $MgSO_4$ in acute nephritis, despite a fall in blood pressure and rise in effective renal plasma flow. The sedative action on convulsions in nephritis had been noted earlier by WINKLER, SMITH and HOFF (1942). DAUR (1955) and others (HÁNZE, 1962c) have advocated the use of intravenous magnesium thiosulfate in the treatment of hypertension, and HOZDRX and ATANASSOWA (1956) have documented temporary increments in circulation in the lower extremities in vascular disease. The use of magnesium in eclampsia is reviewed by FLOWERS (1965).

The optimum plasma level in these disorders is 2—3 mM; no respiratory depression is seen at this level. Careful monitoring of the reflexes should provide a warning of impending toxicity, which usually responds promptly to calcium when it does occur. The amount required to maintain this blood level is about 100 mmoles per day if renal function is normal, but may be far less if it is impaired.

B. Atherosclerosis

The use of magnesium supplements to treat atherosclerosis, either peripheral or coronary, has been advocated by MALKIEL-SHAPIRO (1958), PARSONS et al. (1960), and by BROWNE (1964a and b), who gives 2 to 4 ml of 50% $MgSO_4$

once or twice weekly. Such therapy is said to reduce platelet agglutination (HEINRICH, 1957) and to improve the abnormal lipoprotein pattern seen in the plasma of patients with coronary heart disease (MALKIEL-SHAPIRO, BERSOHN and TERNOR, 1956 cited by BROWNE, 1964b). HAYWOOD and SELVESTER (1962) gave oral supplements of 0.5 mmoles per kg. daily for several months to 35 patients: plasma cholesterol rose initially then returned to the control value; α- and β-lipoproteins diminished by ten per cent. Side effects included nausea, cramps, diarrhea and weakness. This preliminary report does not indicate whether any clinical improvement occurred. A more extensive trial of magnesium salts in coronary heart disease is certainly warranted, comparable to the many studies of new drugs promoted for the long term therapy of this condition.

C. Other therapeutic uses

In digitalized animals MILLER and VAN DELLEN (1941) noted further slowing of atrioventricular conduction and ectopic beats, but SZEKELY and WYNN (1951) found suppression of ectopic foci caused by this drug in dogs. ENSELBER, SIMMONS and MINTZ (1950) studied the use of parenteral magnesium in patients with cardiac arrythmias, one-third of whom were intoxicated with digitalis. Ventricular premature beats were suppressed, but they concluded that magnesium has "an ephemeral action and occasional undesirable side effects", chiefly attributable to suppression of atrioventricular conduction.

The use of oral magnesium supplements to counteract renal stone formation has been advocated by HAMMARSTEN (1956), SAUBERLICH et al. (1964) and MOORE and BUNCE (1964). Ten mmoles is given daily. Urinary output of calcium and phosphate in these patients falls, as it may in animals given supplemental magnesium (IV D), which may account in part for the improvement.

An additional mechanism of reduced stone formation may be enhancement of the solubility of calcium salts (NEUMAN and NEUMAN, 1958). GERSHOFF and ANDRUS (1961) showed that oxalate deposition is reduced in rats by high magnesium intake. In their experiments, oxalate excretion was unaffected, but DE ALBUQUERQUE and TUMA (1962) found it reduced in patients with recurrent oxalate lithiasis given magnesium supplements. The tendency of urine from patients with nephrolithiasis to calcify rachitic rat cartilage at lower calcium and phosphate concentrations than normal can be overcome by adding magnesium to the urine (MUKAI and HOWARD, 1963).

In rats, oral magnesium supplements lessen the occurrence of caries (McCLURE and McCANN, 1960). Magnesium content in carious teeth has been noted to be low; oral supplements increase magnesium content of enamel (RITCHIE, 1961).

XI. Discussion

A. Physiological role of magnesium

The major function of magnesium appears to lie in maintaining the appropriate physicochemical milieu within the cell. The state of aggregation of cell proteins and their association with nucleotides are vital aspects of chemical anatomy of the cell, particularly in reactions involving transphosphorylations. To what extent magnesium plays a regulatory role remains to be seen, but it may possibly do so in cell growth and division. Few of these functions can be performed by calcium, though its presence is often essential. At the cell membrane, however, the role of magnesium is usually less striking than that of calcium, whose role in regulating permeability is prominent. In the extracellular fluid, magnesium seems to be far less critical. Much wider changes in its concentration can be tolerated than in the concentration of the other major cations, and in some instances its nearly complete removal is without discernible effect on the organism.

B. Antagonism and synergism with calcium

It is usually said that these two divalent cations have similar effects on certain functions and opposing effects on others. While this may be good pedagogy, it is usually bad chemistry. Unless a process can be isolated and the effects of altering the free ionic concentration of each cation can be assessed while holding constant the free ionic concentration of the other (as well as the ionic strength and pH), such statements are only of empirical value. Whenever multiple binding sites are present, addition of either cation may increase or decrease the amount of the other bound to the reactive site, or effector. The permutations have been ably described by ARIËNS (1965) in general terms. As a corollary, however, any observations of magnesium metabolism become more meaningful if parallel observations of calcium metabolism are also made. Measurement of free cation concentrations are often essential to complete interpretation.

C. Magnesium transport

1. *Within the organism.* Does cell magnesium change when extracellular magnesium changes? Evidently not, despite numerous references to "mobilization" of tissue magnesium in the literature. Neither hypomagnesemia nor hypermagnesemia has been convincingly shown to alter intracellular concentration, at least *in vivo*. The observed reductions in tissue concentrations are all attributable to continued growth of tissue; the reported increments are probably extracellular, not intracellular. Thus cell magnesium, like many of the organic constituents of cells, is almost independent of the extracellular milieu. When growth must occur in the presence of subnormal extracellular concentrations, cells low in magnesium may be formed; more commonly, the new tissue either

contains normal amounts of magnesium or else growth is inhibited. The implication that this cation is an integral component of the chemical anatomy of the cell is strong. The most outstanding exceptions are the erythrocyte, which can evidently be constructed with subnormal magnesium content, without evident disturbance of function, and the muscle cell in kwashiorkor, where the deficiency of protein precursors may be responsible for the formation of defective cells, low in magnesium.

The entry of magnesium into growing cells may be exclusively in conjunction with the uptake of organic substances required for growth. If the cation enters only in the form of complexes with such substances, the question of active transport into or out of the cell of free magnesium becomes obscured. Isotopic exchangeability may merely indicate exchange diffusion, and have no relevance to the net movement of magnesium through its transport pathways. This supposition is borne out by the fact that no physiological correlations with isotopic exchangeability have been clearly drawn.

Bone magnesium has been held to be a reservoir, readily available to buffer changes in plasma concentration. This inference was based upon measurements of the proportions of magnesium in bone ash in growing animals deprived of this element. Later, the rapid fall in plasma magnesium in ruminants very mildly depleted led to the suggestion that bone magnesium is almost totally unavailable, the earlier data being explained by dilution, as new bone low in magnesium was laid down. The truth may lie between these extreme views. There seems little doubt that bone is the site of the extra magnesium retained by adults on high intakes, despite some suggestions to the contrary. Magnesium-depleted adult subjects appear to mobilize some bone magnesium, but the ability to do so varies considerably from one species to another. Further study of animals depleted of more than a tenth of their body content is sorely needed.

2. *Between the organism and the environment.* Are there active mechanisms capable of magnesium transport? The evidence to date shows clearly that net magnesium movement against electrochemical gradients can occur in gut and renal tubule, but the inference that specific mechanisms exist is unwarranted. The most impressive evidence is the tubular capacity to absorb magnesium much more completely than the simultaneous reabsorption of sodium or calcium. If a specific magnesium "pump" is not present in the distal portions of the nephron, then the non-specific cation pump must be subject to considerable alterations in its selectivity.

D. Possible regulatory mechanisms

Cell magnesium content may be determined exclusively by synthetic processes, as noted above. The regulation of extracellular magnesium is also conjectural.

Since hypomagnesemia tends to induce hypocalcemia by lowering the solubility of bone salt, it appears that a basic parallelism between extracellular calcium and magnesium concentration might be observed in the absence of other regulatory influences. But when either cation falls, the other also tends to be retained, owing to the parallelism between transport rates of the two ions in gut and kidney. Such regulation is inefficient because it places extracellular homeostasis at the mercy of dietary intake and the flow of water and other solutes through intestine and renal tubule. Endocrine influences on magnesium so far elucidated seem to be incidental to calcium homeostasis, but it is possible that hormones exist whose primary action is on extracellular magnesium metabolism; the adrenal cortex is a possible source.

Acknowledgments. I am greatly indebted to Miss Abbey Frank, Miss Esther Massengill and Miss Carol Larson for bibliographic and clerical assistance, and to Dr. H. E. Harrison for a critical review of the manuscript.

References

Agna, J. W., and R. E Goldsmith Primary hyperparathyroidism associated with hypomagnesemia. New Engl. J Med **258**, 222 (1958)

Aikawa, J K.. Gastrointestinal absorption of Mg[28] in rabbits Proc Soc. exp Biol. (N.Y.) **100**, 293 (1959).

— Effect of alloxan-induced diabetes on magnesium metabolism in rabbits. Amer. J. Physiol. **199**, 1084 (1960a)

— Effect of glucose and insulin on magnesium metabolism in rabbits A study with Mg[28]. Proc Soc exp Biol. (N Y) **103**, 363 (1960b).

— Effects of pyridoxine and desoxypyridoxine on magnesium metabolism in the rabbit. Proc. Soc. exp Biol (N Y.) **104**, 461 (1960c).

— Effect of thyroxine and propylthiouracil on magnesium metabolism in the rabbit. Study with Mg[28]. Proc Soc exp. Biol (N.Y.) **104**, 594 (1960d).

— The role of magnesium in biologic processes. Springfield (Ill.)· Ch. C. Thomas 1963.

— Mg[28] studies in magnesium metabolism. In Radioisotopes in animal nutrition and physiology, p. 705. Vienna. International Atomic Energy Agency 1965(a)

— The role of magnesium in biologic processes. A review of recent developments. In Electrolytes and cardiovascular diseases (E. Bajusz, ed.), p. 9—27. New York (N.Y). S Karger 1965(b).

—, and P. D. Bruns Placental transfer and fetal tissue uptake of Mg[28] in the rabbit Proc. Soc exp. Biol. (N.Y.) **105**, 95 (1960).

— D. R Harms, and J. Z. Reardon Effect of cortisone on magnesium metabolism in the rabbit. Amer. J. Physiol. **199**, 229 (1960).

—, and J Z. Reardon Effect of 2,4-dinitrophenol on magnesium metabolism. Proc Soc exp Biol. (N.Y.) **119**, 812 (1965).

— —, and D. R. Harms Effect of digitoxin on exchangeable and tissue contents of magnesium. Proc Soc. exp Biol. (N.Y.) **108**, 684 (1961).

— — — Effect of magnesium-deficient diet on magnesium metabolism in rabbits. A study with Mg[28]. J. Nutr. **76**, 90 (1962).

— E. L Rhoades, and G. S. Gordon Urinary and fecal excretion of orally administered Mg[28]. Proc Soc exp Biol (N.Y) **98**, 29 (1958).

— — D R. Harms, and J. Z. Reardon: Magnesium metabolism in rabbits using Mg[28] as a tracer. Amer. J Physiol. **197**, 99 (1959)

Albert, D. G., J Morita, and L. T. Iseri Serum magnesium and plasma sodium levels in essential vascular hypertension. Circulation **17**, 761 (1958).

ALCOCK, N., and I. MacINTYRE· Inter-relation of calcium and magnesium absorption. Clin Sci **22**, 185 (1962).

— — Some effects of magnesium repletion on calcium metabolism in the rat. Clin. Sci. **26**, 219 (1964).

— —, and I RADDE· The determination of magnesium in biological fluids and tissues by flame spectrophotometry J. clin. Path. **13**, 506 (1960).

ALCOCK, N. W., I MacINTYRE, and I. RADDE: Concentration of magnesium in human plasma. Nature (Lond.) **206**, 89 (1965).

ALLCROFT, R.· Hypomagnesemia of cattle and sheep in Britain. J. Brit. Grassld Soc. **2**, 119 (1956).

ALLCROFT, W M.· Observations on some metabolic disorders of cows, evidenced by chemical analysis of samples of blood from clinical cases. I. Condition of occurrence. Vet. J. **103**, 3 (1947a).

— Observations on some metabolic disorders of cows as evidenced by chemical analysis of samples of blood from clinical cases. 2. Some aspects of blood chemistry. Vet. J. **103**, 30 (1947b).

AMES, A III, and F. B. NESBETT· A method for multiple electrolyte analyses on small samples of nervous tissue. J. Neurochem. **3**, 116 (1958).

—, and M SAKANOUE: A new method for preparing a plasma ultrafiltrate distribution of 5 electrolytes. J. Lab. clin. Med. **64**, 168 (1964).

— —, and S. ENDO· Na, K, Ca, Mg and Cl concentrations in choroid plexus fluid and cisternal fluid compared with plasma ultrafiltrate. J. Neurophysiol. **27**, 672 (1964).

ANAST, C., R. KENNEDY, G. VOLK, and L. ADAMSON: Studies of active transport of calcium, magnesium, and sulfate by the small intestine. J. Pediat. **65**, 1105 (1964).

ANAST, C. S. Serum magnesium levels in newborn. Pediatrics **33**, 969 (1964).

ARDILL, B. L., J. A. HALLIDAY, J D. MORRISON, H. C. MULHOLLAND, and R. A. WOMERSLEY Interrelations of magnesium, phosphate and calcium metabolism. Clin. Sci. **23**, 67 (1962).

ARIENS, E. J (ed.) Molecular pharmacology. New York: Academic Press 1964.

ATCHLEY, D W, R. F. LOEB, D. W. RICHARDS jr., E. M. BENEDICT, and M. E. DRISCOLL On diabetic acidosis — a detailed study of electrolyte balance following the withdrawal and reestablishment of insulin therapy. J. clin. Invest **12**, 297 (1933).

AVIOLI, L, T. LYNCH, and C. BASTOMSKY: Tracer-kinetic analysis of Mg^{28} in man. Clin. Res **11**, 40 (1963).

AXELROD, D. R, and D. E. BASS Electrolyte and acid-base balance in hypothermia. Amer J. Physiol. **186**, 31 (1956).

BACK, E. H., R D MONTGOMERY, and E. E. WARD. Neurological manifestations of magnesium deficiency in infantile gastro-enteritis and malnutrition. Arch. Dis. Childh. **37**, 106 (1962).

BACQ, Z. (ed.)· Heffter's Handbuch der experimentellen Pharmakologie, Erganzungswerk, Bd. 17, Teil 2. Ions alcalin-terreux. Tome 2. Organismes entiers. Berlin, Heidelberg, New York. Springer 1964.

BAIN, J. A. The effect of magnesium upon thyroxine inhibition of phosphorylation. J Pharmacol exp. Ther. **110**, 2 (1954).

BALDWIN, J., P. K. ROBINSON, K. L. ZIERLER, and J. L. LILIENTHAL jr · Interrelations of magnesium, potassium, phosphorus and creatine in skeletal muscle of man. J. clin. Invest **31**, 850 (1952).

BALINT, J. A, and B. I. HIRSCHOWITZ: Hypomagnesemia with tetany in nontropical sprue New Engl. J Med. **265**, 631 (1961).

BANERJEE, B., and N. SAHA: Serum magnesium content of Indians. Indian J. Physiol. Pharmacol. **8**, 156 (1964).

BANG, O, and S. ØRSKOV. The magnesium content of the erythrocytes in pernicious and some other anemias. J. clin. Invest **18**, 497 (1939).

BARKER, E. S., J. K CLARK, and J. R. ELKINTON: The renal excretion of magnesium. Fed. Proc. **14**, 8 (1955).

Barker, E. S., J. K. Clark, and J. R. Elkinton: Renal response to magnesium loading in the dog Fed Proc **16**, 6 (1957).
— J. R. Elkinton, and J. K. Clark Studies of the renal excretion of magnesium in man. J. clin Invest **38**, 1733 (1959).
Barnes, B. A., O. Cope, and E. B. Gordon Magnesium requirements and deficits an evaluation in two surgical patients. Ann Surg **152**, 518 (1960).
— —, and T. Harrison Magnesium conservation in the human being on a low magnesium diet J clin. Invest **37**, 430 (1958)
— E. B. Gordon, and O. Cope. Skeletal muscle analyses in health and in certain metabolic disorders 1. The method of analysis and the values in normal muscle. J clin. Invest. **36**, 1239 (1957).
— S. M. Krane, and O. Cope Magnesium studies in relation to hyperparathyroidism. J. clin. Endocr **17**, 1407 (1957)
—, and J. Mendelson The measurement of exchangeable magnesium in dogs. Metabolism **12**, 184 (1963).
Baron, D. N. Magnesium deficiency after gastro-intestinal surgery and loss of secretions. Brit. J Surg **48**, 344 (1960).
Barron, G. P., S. O. Brown, and P. B. Pearson Histological manifestations of a magnesium deficiency in the rat and rabbit Proc Soc. exp Biol (N Y) **70**, 220 (1949).
Bartlett, S., B. B. Brown, A. S. Foot, S. J. Rowland, R. Allcroft, and W. H. Parr The influence of fertilizer treatment of grassland on the incidence of hypomagnesemia in milking cows. Brit vet. J. **110**, 3 (1954).
Bartley, J. C., E. F. Reber, J. W. Yusken, and H. W. Norton Magnesium balance study in pigs 3 to 5 weeks of age J. Anim. Sci **20**, 137 (1961).
Bassett, S. H. Mineral exchanges of man. J. Nutr **9**, 323 (1935).
Baum, P., u. R. Czok Enzymatische Bestimmung von „ionisiertem" Magnesium im Plasma Biochem Z **332**, 121 (1959).
Beechey, R. B., N. Alcock, S. Hanna, and I. MacIntyre Magnesium deficiency and oxidative phosphorylation. Biochem. J. **71**, 18p (1959).
— N. W. Alcock, and I. MacIntyre Oxidative phosphorylation in magnesium and potassium deficiency in the rat. Amer J Physiol. **201**, 1120 (1961).
Bélanger, L. F., G. A. van Erkel, and A. Jakerow Behavior of the dermal mast cells in magnesium-deficient rats Science **126**, 29 (1957).
Belloiu, D., Z. Rodin, F. Cherciulesco et W. Demetresco L'action therapeutique de l'ion magnésium dans un cas d'hyperthyroidie associée à des crises de paralysis periodique. Psychiat et Neurol. (Basel) **149**, 28 (1965).
Berglund, F., and R. P. Forster Renal tubular transport of inorganic divalent ions by the aglomerular marine teleost, *Lophius americanus*. J gen Physiol **41**, 429 (1958)
Bernick, S., and G. H. Hungerford· Effect of dietary magnesium deficiency on the bones and teeth of rats. J dent Res **44**, 1317 (1965)
Bersohn, I., and P. J. Oelofse Correlation of serum-magnesium and serum-cholesterol levels in South African Bantu and European subjects. Lancet **1957 I**, 1020.
Bertrand, D. Le magnésium et la vie. Paris Presses Universitaires de France, Coll Que sais-je, 1960
Best, F. A., and V. R. Pickles A myometrial effect of estradiol, imitated by high Mg++ concentration J. Physiol. (Lond) **166**, 12P (1963).
— — Effects of treatment with oestradiol and progesterone on the calcium and magnesium content of the guinea-pig and rabbit myometrium, and on its sensitivity to stimulants. J. Endocr. **32**, 121 (1965).
Better, O., H. C. Gonick, L. Chapman, and C. R. Kleeman Renal handling of divalent ions in uremia Clin. Res. **13**, 301 (1965).
Better, O. S., H. C. Gonick, L. C. Chapman, P. D. Varrady, and C. R. Kleeman Effect of urea-saline diuresis on renal clearance of calcium, magnesium, and inorganic phosphate in man Proc.Soc. exp. Biol. (N.Y.) **121**, 592 (1966).

BHATTACHARYA, G · Effect of metal ions on the utilization of glucose and on the influence of insulin on it by the isolated rat diaphragm. Biochem J. **79**, 369 (1961).

BHATTACHARYYA, N. K., and D. N. MULLICK. Effect of magnesium sulphate injection in experimental hypercholesteremia in rabbits Part I· on the serum and tissue cholesterol Ann Biochem. **23**, 515 (1963).

BIRD, F. A Magnesium deficiency in the chick. J. Nutr. **39**, 13 (1949).

BISSELL, G. W.. The magnesium partition in hyperthyroidism with special reference to the effect of thiouracil Amer. J med Sci. **210**, 195 (1945)

BLAXTER, K L The magnesium content of bone in hypomagnesemic disorders of livestock. Ciba Foundation Symposium. Bone structure and metabolism Boston Little, Brown & Co. 1956.

—, and R. F. McGILL Magnesium metabolism in cattle Vet. Rev. Annot **2**, 35 (1956).

—, and J A F. ROOK Energy and carbohydrate metabolism in magnesium-deficient calves. Brit. J. Nutr. **9**, 121 (1955).

— —, and A. M. MACDONALD· Experimental magnesium deficiency in calves 1. Clinical and pathological observations. J. comp Path. **64**, 157 (1954a)

— — — Experimental magnesium deficiency in calves. 2 The metabolism of calcium, magnesium and nitrogen and magnesium requirements J. comp Path **64**, 176 (1954b).

—, and B. A. M. SHARMAN Hypomagnesaemic tetany in beef cattle. Vet. Rec. **67**, 108 (1955).

BLYTHE, W B, H. J GITELMAN, and L. G. WELT The effect of expansion of the extracellular space on the rate of urinary calcium excretion Clin Res. **14**, 372 (1966).

BOELLNER, S W, E J OLSON, D. FREDRICKSON, and E. R HUGHES Plasma and erythrocyte magnesium in muscular dystrophy. Amer J. Dis Child. **110**, 172 (1965).

BOGERT, L. J., and E. C. PLASS The calcium and magnesium content of fetal and maternal blood serum. J. biol. Chem. **56**, 297 (1923).

BOHR, D F Electrolytes and smooth muscle concentration. Pharmacol. Rev. **16**, 85 (1964)

BOIS, P. Effect of magnesium deficiency on mast cells and urinary histamine in rats Brit J. exp. Path. **44**, 151 (1963)

— Tumor of the thymus in magnesium-deficient rats. Nature (Lond) **204**, 1316 (1964).

— Rôle du magnésium dans le métabolisme de l'histamine. Un méd. Can. **95**, 313 (1966).

—, and A. BEAULNES Histamine magnesium deficiency, and thymic tumors in rats Canad. J. Physiol. Pharmacol. **44**, 373 (1966).

— E. H. BYRNE, and L. F. BÉLANGER Effect of magnesium deficiency on the regeneration of mast cells after treatment with 48/80 in rats. Canad J. Biochem. **38**, 585 (1960).

BOOTH, C C, S HANNA, N. BABOURIS, and I. MACINTYRE· Incidence of hypomagnesaemia in intestinal malabsorption. Brit. med. J **1963 II**, 141.

BOOYENS, J., M. DE WAAL, L. J. RADEMEYER, and F. CALITZ The effect of dietary maize meal supplementation on the levels of serum cholesterol and magnesium. S. Afr. med. J. **40**, 237 (1966)

BORGLIN, N. E., and T. LINDSTEN On combined effect of neostigmine and pentamethylenetetrazol or β-phenyl-isopropylamine on magnesium anesthesia Acta pharmacol. (Kbh) **5**, 309 (1949).

BOROWITZ, J. L, K. FUWA, and N WEINER Distribution of metals and catecholamines in bovine adrenal medulla sub-cellular fractions. Nature (Lond) **205**, 42 (1965).

BRANDT, J. L., and W. GLASER Effect of corticosteroid excess and of bilateral adrenalectomy on metabolism of Mg^{28}. Proc. Soc. exp. Biol. (N.Y.) **101**, 823 (1959).

— —, and A. JONES· Soft tissue distribution and plasma disappearance of intravenously administered isotopic magnesium with observations on uptake in bone. Metabolism **7**, 355 (1958)

BREIBART, S, J. S LEE, A. McCOORD, and G. B. FORBES Relation of age to radiomagnesium exchange in bone. Proc. Soc exp. Biol. (N.Y.) **105**, 361 (1960).

BRESCIANI, F., and F. AURICCHIO: Subcellular distribution of some metallic cations in the early stages of liver carcinogenesis Cancer Res. **22**, 1284 (1962).

BRIERLEY, G P , E. MURER, E. BACKMANN, and D. E. GREEN Studies on ion transport II. The accumulation of inorganic phosphate and magnesium ions by heart mitochondria. J. biol. Chem. **238**, 3482 (1963).

— —, and R. A. O'BRIEN Studies on ion transport. VI. The accumulation of Mg++ by heart mitochondria in the absence of inorganic phosphate. Biochim. biophys. Acta (Amst) **88**, 645 (1964)

BRONNER, F , and D. D THOMPSON: Renal transtubular flux of electrolytes in dogs with special reference to calcium. J. Physiol (Lond) **157**, 232 (1961).

BROWN, D. F., R. B. McCANDY, E. GILLIE, and J. T. DOYLE Magnesium-lipid relations in health and in patients with myocardial infarction. Lancet **1958 II**, 933.

BROWNE, S. E. · Magnesium for atherosclerosis. Brit. Med J **1964 II**, 5409(a).

— Parenteral magnesium sulphate in arterial disease. Practitioner **192**, 791 (1964b).

BRULL, L , and J. BERNIMOLIN Physico-chemical condition of calcium and magnesium in plasma and their renal excretion. Archs int. Pharmacodyn. **108**, 330 (1956).

BUELL, M. V., and E. TURNER. Cation distribution in the muscles of adrenalectomized rats. Amer. J. Physiol **134**, 225 (1941).

BULGER, H. A., and F. GAUSMANN Magnesium metabolism in hyperparathyroidism. J. clin. Invest. **12**, 1135 (1933).

BUNCE, G E , Y. CHIEMCHAISRI, and P. H. PHILLIPS The mineral requirements of the dog. 4. Effect of certain dietary and physiologic factors upon the magnesium deficiency syndrome. J. Nutr. **76**, 23 (1962a).

— K. J. JENKINS, and P. H. PHILLIPS The mineral requirements of the dog. 3. The magnesium requirement J. Nutr **76**, 17 (1962b).

— P. G REEVES, T. S. OBA, and H. E. SAUBERLICH Influence of the dietary protein level on the magnesium requirement. J Nutr. **79**, 220 (1963).

— H E. SAUBERLICH, P. G. REEVES, and T. S. OBA. Dietary phosphorus and magnesium deficiency in the rat. J. Nutr **86**, 406 (1965).

BURCH, G. E., R. K LAZZARA, and T. K YUN Concentration of magnesium in tissues of the dog. Proc. Soc. exp. Biol. (N.Y.) **118**, 581 (1964).

BURT, A. W. A., and D. C. THOMAS Dietary citrate and hypomagnesemia in the ruminant. Nature (Lond.) **192**, 1193 (1961)

BUTLER, A. M.. Diabetic coma New Engl. J Med **243**, 648 (1950).

CADDELL, J. L.. Magnesium in the therapy of protein-calorie malnutrition of childhood. J. Pediat **66**, 393 (1965).

CAMPBELL, J. A., and T. A. WEBSTER. Day and night urine during complete rest, laboratory routine, light muscular work and oxygen administration Biochem. J. **15**, 660 (1921).

CANELAS, H. M., L. M. DE ASSIS, and F. B. DE JORGE Disorders of magnesium metabolism in epilepsy. J. Neurol. Neurosurg. Psychiat. **28**, 378 (1965)

CARAFOLI, E., C. S. ROSSI, and A. L LEHNINGER Cation and anion balance during active accumulation of Ca++ and Mg++ by isolated mitochondria. J. biol. Chem. **239**, 3055 (1964).

CARE, A. D. The kinetics of magnesium distribution within the sheep Res. Vet Sci **1**, 338 (1960).

— Secretion of magnesium and calcium in parotid saliva of sheep. Nature (Lond.) **199**, 818 (1963).

— Factors which affect the availability of magnesium. Proc Nutr Soc. **24**, 99 (1965).

—, and W M KEYNES The role of parathyroid hormones in the absorption of calcium and magnesium from the small intestine. Proc. roy. Soc. Med. **57**, 867 (1964).

— D. C. MacDONALD, and B. NOLAN. Equilibration of labelled magnesium between sheep plasma and red cells. Nature (Lond.) **183**, 1265 (1959).

—, and D. B. ROSS Gastro-intestinal absorption of ²⁸Mg in sheep Proc Nutr. Soc. **21**, ix (1962).

CARE, A D , and D B Ross: The role of the adrenal cortex in magnesium homeostasis and in the aetiology of hypomagnesaemia Res. Vet. Sci. **4**, 24 (1963).

— —, and A. A. WILSON· The distribution of exchangeable magnesium within the sheep. J. Physiol (Lond) **176**, 284 (1965).

— L M. SHERWOOD, J. T. POTTS jr., and G. D. AURBACH Perfusion of the isolated parathyroid gland of the goat and sheep. Nature (Lond.) **209**, 9 (1966).

—, and A TH. VAN'T KLOOSTER *In vivo* transport of magnesium and other cations across the wall of the gastrointestinal tract of sheep. J. Physiol. (Lond.) **177**, 174 (1965)

CARILLO, B J., W G POND, L. KROOK, F. E LOVELACE, and J. K. LOOSLI. Response of growing rats to diets varying in magnesium, potassium and protein content. Proc. Soc. exp. Biol. (N.Y.) **107**, 793 (1961).

CARPENTER, B. E , M. KEOPLUNG, C. D. ARNAUD, and N H. ENGBRING. Effects of magnesium infusion on renal clearance of phosphorus. Clin. Res. **11**, 238 (1963).

CARR, C. W., and K P. WOODS· Studies on the binding of small ions in protein solutions with the use of membrane electrodes. V. The binding of magnesium ions in solutions of various proteins. Arch. Biochem. **55**, 1 (1955).

CARR, M. H , and H. A. FRANK. Improved method for determination of calcium and magnesium in biologic fluid by EDTA titration. Amer. J. clin. Path. **26**, 1157 (1956).

—, and P. R. SCHLOERB· Blood components in the dog: normal values. J. Lab. clin. Med. **53**, 646 (1959).

CARUBELLI, R., W. O. SMITH, and J F. HAMMARSTEN. Determination of magnesium in erythrocytes. J. Lab. clin. Med. **51**, 964 (1958).

CARVALHO, A. P.: Binding of cations by microsomes from rabbit skeletal muscle. J. cell. comp. Physiol. **67**, 73 (1966).

— H. SANUI, and N. PACE Calcium and magnesium binding properties of cell membrane materials. J. cell. comp. Physiol. **62**, 311 (1963).

— — — Binding of Ca and Mg by lipoprotein and nucleoprotein subfractions of rat liver cell microsomes. J. cell. comp. Physiol. **66**, 57 (1965).

CATCHPOLE, H R., N R. JOSEPH, and M. B ENGEL. Thermodynamic relations of polyelectrolytes and inorganic ions of connective tissue. Fed. Proc. **25**, 1124 (1966).

CHARBON, G A., and M. H. HOEKSTRA· Mineral content in plasma and blood cells of various species. Acta physiol. pharmacol. neerl. **11**, 209 (1962).

CHARNOCK, J. S., J. CASLEY-SMITH, and C. J. SCHWARTZ. Serum magnesium-cholesterol relationship in the central Australian aborigine and in Europeans with and without ischaemic heart disease. Aust J. exp. Biol. med. Sci **37**, 509 (1959)

— L J. OPIT, and B. S. HETZEL. Electrolyte distribution in rats following salicylate. Metabolism **10**, 874 (1961).

CHEEK, D. B., J E. GRAYSTONE, and M. H. CASS. The effect of elevation of plasma triglyceride on muscle electrolytes. Bull. Johns Hopkins Hosp. **113**, 261 (1963).

— — J. B. WILLIS, and A. B. HOLT· Studies on the effect of triglycerides, glycerophosphate, phosphatidyl ethanolamine on skeletal and cardiac muscle composition. Clin. Sci. **23**, 169 (1962).

—, and H. C. TENG Changes in tissue composition of the rat in parathyroid intoxication (as produced by large doses of commercial extract), and the effect of magnesium loading. Clin. Sci. **19**, 195 (1960).

—, and C. D. WEST Alterations in body composition with sodium loading and potassium restriction in the rat. the total body sodium, nitrogen, magnesium and calcium. J clin. Invest. **35**, 763 (1956).

CHESLEY, L. C., and I. TEPPER. Some effects of magnesium loading upon renal excretion of magnesium and certain other electrolytes. J clin. Invest. **37**, 1362 (1958).

CHIEMCHAISRI, Y., and P. H. PHILLIPS Effect of dietary fluoride upon the magnesium calcinosis syndrome. J. Nutr. **81**, 307 (1963)

— — Certain factors including fluoride which affect magnesium calcinosis in the dog and rat. J. Nutr. **86**, 23 (1965).

CHRISTIANSON, G., R. JENESS, and S T COULTON Determination of ionized calcium and magnesium in milk. Analyt. Chem. **26**, 1923 (1954)

CHUTKOW, J. G. Sites of magnesium absorption and excretion in the intestinal tract of the rat J Lab. clin Med. **63**, 71 (1964a).

— Metabolism of magnesium in the normal rat J. Lab. clin. Med. **63**, 80 (1964b).

— Studies on the metabolism of magnesium in the magnesium-deficient rat. J. Lab. clin. Med **65**, 912 (1965).

CLARK, G. W. Studies in the mineral metabolism of adult man. Univ Calif. Publs. Physiol **5**, 17, 195—287 (1926).

CLARK, I.· Relation of magnesium ions to calcium and phosphate absorption. Nature (Lond.) **207**, 982 (1965a).

— Magnesium, calcium, phosphorus and strontium interrelations. In· Radioisotopes in animal nutrition and physiology Vienna Internat. Atomic Energy Agency 1965(b).

— Studies in inorganic metabolism. II. The effects of magnesium ions on phosphorus metabolism. J. exp. Med. (in press) (1967).

— M CHARLES, and F. RIVERA-Cordero. The effect of magnesium on calcium and phosphate absorption and excretion. Fed Proc. **24**, 372 (1965).

— E A. GUSMANO, R. NEVINS, and S H. COHN Effect of magnesium on uptake and retention of radioactive strontium. Proc. Soc. exp Biol. (N.Y) **116**, 984 (1964)

—, and M. R SMITH· Effect of magnesium ions on removal of radiostrontium from rats Proc Soc. exp Biol (N.Y) **109**, 135 (1962)

CLARKSON, E. M., S J McDONALD, H E DE WARDENER, and R. WARREN Magnesium metabolism in chronic renal failure. Clin. Sci **28**, 107 (1965).

CLEGG, P. C., P HOPKINSON, and V. R. PICKLES Some effects of calcium and magnesium ions on guinea-pig uterine muscle. J Physiol (Lond) **167**, 1 (1963)

COHEN, R. D. Water and electrolyte metabolism during the treatment of myxoedema. Clin. Sci. **25**, 293 (1963)

COLBY, R. W, and C. M. FRYE Effect of feeding various levels of calcium, potassium and magnesium to rats. Amer. J. Physiol. **166**, 209 (1951a).

— — Effect of feeding high levels of protein and calcium in rat rations on magnesium deficiency syndrome. Amer. J. Physiol. **166**, 408 (1951b).

CONSOLAZIO, C F., L O MATOUSH, R C. NELSON, R. S HARDING, and J. E CANHAM: Excretion of sodium, potassium, magnesium and iron in human sweat and the relation of each to balance and requirements. J. Nutr. **79**, 407 (1963).

CONWAY, E J., and P. F DUGGAN: A general cation carrier in the yeast cell wall. The general carrier and its amount per Kg yeast. Nature (Lond) **178**, 1043 (1956).

—, and D HINGERTY The influence of adrenalectomy on muscle constituents Biochem. J **40**, 561 (1946)

COONS, C. M., A T. SCHIEFELBUSCH, G. B. MARSHALL, and R R COONS Metabolism during pregnancy Okla. agric exp Station, Bull **223**, 9 (1935).

COPE, C L., and J. PEARSON. Aldosterone secretion in magnesium deficiency. Brit. med. J **1963 II**, 1385

—, and B WOLFF The ultrafilterable serum magnesium in hyperthyroidism. Biochem. J **36**, 413 (1942).

COPELAND, B. E., and F. W. SUNDERMAN Studies in serum electrolytes. XVIII. Magnesium-binding property of serum proteins J. biol. Chem. **197**, 331 (1952).

CORRADINO, R. A, and H. E. PARKER Magnesium and thyroid function in the rat. J. Nutr **77**, 455 (1962)

COTLOVE, E, M. A HOLLIDAY, R SCHWARTZ, and W. W. WALLACE Effects of electrolyte depletion and acid-base disturbance on muscle cations Amer. J. Physiol. **167**, 665 (1951).

COUTINHO, E M Interação dos esteróides ovarianos ao nível cellular. Thesis, Faculty of Pharmacy, Univ. of Bahia, Brazil, 1961.

COUTINO, E. M · Calcium, magnesium and local anesthesia. J. gen Physiol. **49**, 845 (1966).

CRAMER, C. F., and J. DUECK *In vivo* transport of calcium from healed Thiry-Vella fistulas in dogs. Amer. J Physiol. **202**, 161 (1962)

CRAMER, W. · Experimental production of kidney lesions by diet. Lancet **1932 II**, 174.

CUNNINGHAM, I. J The distribution of magnesium in the animal organism and the effect of dietary magnesium. N Z. J. Sci. Technol. **18**, 419 (1936).

—, and CUNNINGHAM, M. M. Dietary magnesium and urinary calculi. N Z J. Sci. Technol. **19**, 529 (1938)

CURRAN, P. F , and J. R. GILL jr.. The effect of calcium on sodium transport by frog skin. J gen Physiol **45**, 625 (1962).

DAHL, S.. Serum magnesium in normal men and women Acta haemat. (Basel) **4**, 65 (1950).

DANIELS, F , M. QUINN, C. R. KLEEMAN, and J. MARINO: Plasma magnesium changes during cold acclimatization in man Fed. Proc. **12**, 31 (1953)

DAUR, W. Magnesium und Hypertonie. Arzneimittel-Forsch. **5**, 719 (1955).

DA VANZO, J. P., H. C CROSSFIELD, and W. W. SWINGLE: Effect of various adrenal steroids on plasma magnesium and the electrocardiogram of adrenalectomized dogs. Endocrinology **63**, 825 (1958).

DAVIES, J S , E M WIDDOWSON, and R. A. McCANCE The intake of milk and the retention of its constituents while the newborn rabbit doubles its weight Brit. J. Nutr. **18**, 835 (1964).

DAVSON, H Physiology of the ocular and cerebrospinal fluids. London J. & A. Churchill Ltd. 1956.

— The eye, vol 1 Vegetative physiology and biochemistry. New York Academic Press 1962.

DE AZEVEDO, M. L , and F B. DE JORGE Some mineral constituents of normal human eye tissue (Na-K-Mg-Ca-P-Cu). Ophthalmologica (Basel) **149**, 43 (1965).

DE ALBUQUERQUE, P. F , and M TUMA Investigations on urolithiasis. II Studies on oxalate J Urol. (Baltimore) **87**, 504 (1962)

DE DONCKER, K., et N. ROSSELLE. Etude de l'activité myoélectrique dans la déficience magnésique expérimentale. J Physiol (Paris) **51**, 39 (1959).

DE JORGE, F B., C. CANATO, y D. DELASCIO Aspectos bioquímicos do líquido amniótico, da placenta e do sangue materno e fetal durante o parto Matern. Inf. (Sao Paulo) (in press) (1966).

— D. DELASCIO, A B D. ULHÔA CINTRA, and M L. ANTUNES Magnesium concentration in the blood serum of normal pregnant women. Obstet. and Gynec. **25**, 253 (1965)

DEL CASTILLO, J , and L ENGBAEK The nature of the neuromuscular block produced by magnesium J. Physiol. (Lond.) **124**, 370 (1954).

DELLEN, T. R VAN, and J R. MILLER Electrocardiographic changes following the intravenous administration of magnesium sulfate. II An experimental study on dogs. J. Lab. clin. Med. **24**, 840 (1939).

DEMARIA, W. J. A., and J. S. HARRIS Effect of magnesium sulfate on the alterations in renal dynamics induced by intravenous hemoglobin. Amer. J. Physiol. **182**, 251 (1955)

DE VRIES, L. A., S. P. TEN HOLT, J. J VAN DAATSELAAR, A. MULDER, and J G. G BORST. Characteristic renal excretion patterns in response to physiological, pathological and pharmacological stimuli. Clin chim. Acta **5**, 915 (1960).

DI GIORGIO, J , J. J. VITALE, and E. E. HELLERSTEIN. Sarcosomes and magnesium deficiency in ducks. Biochem. J. **82**, 184 (1962)

DINE, R F., and P. H. LAVIETES Serum magnesium in thyroid disease. J. clin. Invest. **21**, 781 (1942).

DISHINGTON, I W Changes in serum magnesium levels of ruminants, as influenced by abrupt changes in the composition of the diet. Acta vet. scand. **6**, 150 (1965)

DOE, R P., E B. FLINK, and A. S. PRASAD Magnesium metabolism in hyperthyroidism J. Lab clin. Med **54**, 807 (1959).

Doe, R P , J. A. Vennes, and E B Flink Diurnal variation of 17-hydroxycorticosteroids, potassium, magnesium and creatinine in normal subjects and in cases of treated adrenal insufficiency and Cushing's syndrome J clin. Endocr 20, 253 (1960).

Douglas, W. W., and A. M. Poisner· Stimulus-secretion coupling in a neurosecretory organ: the role of calcium in the release of vasopressin from the neurohypophysis J. Physiol. (Lond.) 172, 1 (1964).

—, and R. P Rubin The mechanism of catecholamine release from the adrenal medulla — the role of calcium in stimulus-secretion coupling. J. Physiol. (Lond.) 167, 288 (1963).

— — The effects of alkaline earths and other divalent cations on adrenal medullary secretion J Physiol. (Lond) 175, 231 (1964).

Dreux, C., et M. Girard Magnésium plasmatique et globulaire. Techniques de dosage complexonométrique. Ann. Biol. clin. 19, 627 (1961)

Du Bois, K. P., H G. Albaum, and V. R Potter· Adenosine triphosphate in magnesium anesthesia. J. biol. Chem. 147, 699 (1943).

Duckworth, J., W. Godden, and G. M Warnock The effect of acute magnesium deficiency on bone formation in rats. Biochem. J 34, 97 (1940).

—, and G M Warnock The Mg requirements of man in relation to Ca requirements, with observations on the adequacy of diets in common use Nutr. Abstr. Rev. 12 (2), 167 (1942)

Duncan, C W., C. F Huffman, and C S. Robinson Magnesium studies in calves. I. Tetany produced by a ration of milk or milk with various supplements. J. biol. Chem 108, 35 (1935).

Dunn, M., M Keoplung, and W A Krehl· Alteration in citrate and magnesium levels in chronic renal disease Clin Res 10, 247 (1962).

Dunn, M J , and M Walser Magnesium depletion in normal man. Metabolism 10, 884 (1966).

Durlach, J., et R. Lebrun Importance de la forme hypomagnésienne de la spasmophilie Ann. Endocr. (Paris) 21, 244 (1960).

— M. Stoliaroff, J. Gauduchon, R. Lebuc et T Cong-Trieu Effects de la thyroparathyroidectomie et de la charge en extrait parathyroidien sur la magnésémie du chien. C R Soc. Biol. (Paris) 153, 1976 (1959).

Edelman, I. S , P O. P Ts'o, and J Vinograd· The binding of magnesium to microsomal nucleoprotein and ribonucleic acid. Biochim. biophys. Acta (Amst.) 43, 393 (1960).

Edmondson, H. A , C. J. Berne, R E. Homann jr , and M. Wertman Calcium, potassium, magnesium and amylase disturbances in acute pancreatitis Amer. J. Med 12, 34 (1952).

Eliel, L. P., R Chanes, and J Hawrylko Influence of parathyroid activity on ion exchange in various tissues. Trans Amer. clin Climat Ass 74, 130 (1963).

Elkinton, J. R The role of magnesium in the body fluids Clin Chem. 3, 319 (1958)

Elmquist, D I., and D M J Quastel A quantitative study of end-plate potentials in isolated human muscles. J. Physiol. (Lond.) 178, 505 (1965).

Embry, R , and A H Briggs Factors affecting contraction and relaxation in dog glycerinated cardiac fiber. Amer J Physiol 210, 826 (1966)

Engbaek, L The pharmacological actions of magnesium ions with particular reference to the neuromuscular and the cardiovascular system Pharmacol Rev 4, 396 (1952).

Enselber, C. D , H. G Simmons, and A. A Mintz The effects of magnesium upon cardiac arrhythmias Amer. Heart J. 39, 703 (1950).

Eschler, J , G Ochs u. W Schilli Die Veränderung des Magnesiumgehaltes des menschlichen Speichels nach Injektion von Magnesiumsalzen unter verschiedenen Bedingungen. Z. ges. exp. Med. 139, 150 (1965).

Etteldorf, J N., G W. Clayton, A H Tuttle, and C R Houck Renal function studies in pediatrics. II. Influence of magnesium sulfate on renal hemodynamics in normal children. Amer. J Dis. Child 83, 301 (1952).

ETTELDORF, J. N , and A. H. TUTTLE The effects of magnesium sulfate on renal function in children with acute glomerulonephritis. J. Pediat. **41**, 524 (1952)

EVERED, D. F.: Magnesium absorption in sheep Nature (Lond) **189**, 228 (1961).

EYZAGUIRRE, C., and H. KOYANO: Effects of some pharmacological agents of chemoreceptor discharges J. Physiol. (Lond.) **178**, 410 (1965).

FABRY, C., et J. CLOOSEN. La composition des sels osseux dans les tissus en voie d'ossification. II. Essais de fractionnement du tissu osseux compact par broyage. Bull Soc Chim. biol. (Paris) **43**, 253 (1961).

FANKUSHEN, D , D. RASKIN, A DIMICH, and S WALLACH The significance of hypomagnesemia in alcoholic patients. Amer. J. Med **37**, 802 (1964).

FARNELL, D R.. Cortisone treatment in magnesium deficiency of rats Amer J vet. Res. **27**, 415 (1966).

FEATHERSTON, W. R., M L MORRIS jr., and P. H. PHILLIPS. Influence of lactose and dried skim milk upon the magnesium deficiency syndrome in the dog 1 Growth and biochemical data J. Nutr. **79**, 431 (1963).

FELLERS, F. X., and C. HERRERA: Studies on magnesium· effect of induced malnutrition in rats. Fed. Proc **25**, 609 (1966).

FIELD, A. C.: Balance trials with magnesium-28 in sheep. Nature (Lond.) **183**, 983 (1959).

— Uptake of magnesium-28 by the skeleton of a sheep. Nature (Lond) **188**, 1205 (1960).

— Studies on magnesium in ruminant nutrition 3 Distribution of Mg[28] in the gastrointestinal tract and tissues of sheep Brit. J. Nutr. **15**, 349 (1961).

— J. W. McCALLUM, and E. J. BUTLER Studies on magnesium in ruminant nutrition. Balance experiments on sheep with herbage from fields associated with lactation tetany and from control pastures. Brit. J Nutr. **12**, 433 (1958).

—, and B. S. W. SMITH Effect of magnesium deficiency on the uptake of Mg[28] by the tissues in mature rats. Brit. J. Nutr. **18**, 103 (1964).

FILO, R. S., D. F. BOHR, and J. C. RUEGG Glycerinated skeletal and smooth muscle. calcium and magnesium dependence. Science **147**, 1581 (1965).

FISHMAN, R A.· Neurological aspects of magnesium metabolism Arch Neurol (Chic) **12**, 562 (1965).

FITZGERALD, M. G , and P FOURMAN. An experimental study of magnesium deficiency Clin. Sci **15**, 635 (1956).

FLETCHER, R. F , and A A. HENLY. A case of magnesium deficiency following massive intestinal resection. Lancet **1960 I**, 522

FLINK, E. B · Magnesium deficiency syndrome in man. J. Amer. med. Ass. **160**, 1406 (1956).

— R. McCOLLISTER, A. S PRASAD, J. C. MELBY, and R. P. DOE Evidences for clinical magnesium deficiency. Ann. intern. Med. **47**, 956 (1957).

— F. L STUTZMAN, A R. ANDERSON, T. KONIG, and R. FRASER. Magnesium deficiency after prolonged parenteral fluid administration and after chronic alcoholism complicated by delirium tremens. J. Lab clin Med **43**, 169 (1954).

FLOWERS, C E. jr . Magnesium sulfate in obstetrics. A study of magnesium in plasma, urine, and muscle. Amer. J. Obstet. Gynec **91**, 763 (1965).

FOLLIS, R. H. jr.. Deficiency disease, p 35—41. Springfield (Ill.): Ch. C. Thomas 1958

FONTENOT, J P , R W. MILLER, C. K. WHITEHAIR, and R. MacVICAR Effect of a high-protein, high-potassium ration on the mineral metabolism of lambs. J. Animal Sci. **19**, 127 (1960)

FORBES, R M Excretory patterns and bone deposition of zinc, calcium and magnesium in therapy as influenced by zinc deficiency, EDTA and lactose J. Nutr. **74**, 194 (1961).

— Mineral utilization in the rat. 1. Effects of varying dietary ratios of calcium, magnesium and phosphorus. J. Nutr. **80**, 321 (1963).

— Mineral utilization in the rat 2. Restoration of normal tissue levels of magnesium and calcium following magnesium deficiency. J. Nutr. **83**, 44 (1964a).

FORBES, R. M.: Mineral utilization in the rat III. Effects of calcium, phosphorus, lactose and source of protein in zinc-deficient and in zinc-adequate diets. J. Nutr. **83**, 225 (1964b).

— Mineral utilization in the rat. V. Effects of dietary thyroxine on mineral balance and tissue mineral composition with special reference to magnesium nutriture. J. Nutr. **86**, 193 (1965).

— Effects of magnesium, potassium and sodium nutriture on mineral composition of selected tissues of the albino rat. J. Nutr **88**, 403 (1966).

FOSTER, G. V., A. BAGHDIANTZ, M. A. KUMAR, E. SLACK, H. A. SOLIMAN, and I MAC-INTYRE Thyroid origin of calcitonin. Nature (Lond.) **202**, 1303 (1964).

— G. F. JOPLIN, I. MACINTYRE, K. E. W. MELVIN, and E. SLACK Effect of thyro-calcitonin in man. Lancet **1966 I**, 107.

FOURMAN, P, and D. B. MORGAN Chronic magnesium deficiency. Proc Nutr. Soc. **21**, 34 (1962)

FRANKENHAEUSER, B., and A. L. HODGKIN The action of calcium on the electrical properties of squid axon. J. Physiol. (Lond) **137**, 218 (1957).

FRANKENHAEUSER, B, and H. MEVES The effect of magnesium and calcium on the frog myelinated nerve fiber J Physiol. (Lond) **142**, 360 (1954) ·

FRATER, R., S. E SIMON, and F. H. SHAW Muscle a three phase system; the partition of divalent ions across the membrane J. gen Physiol **43**, 81 (1959).

FREEDMAN, P., R. MOULTON, and A G SPENCER The effect of intravenous calcium gluconate on the renal excretion of water and electrolytes. Clin. Sci. **17**, 247 (1958).

FRIEDMAN, H S, and M A RUBIN Clinical significance of magnesium: calcium ratio, technic for the determination of magnesium and calcium in biologic fluids. Clin. Chem. **1**, 125 (1955).

FROHLICH, E D, J B SCOTT, and F J HADDY Effect of cations on resistance and responsiveness of renal and forelimb vascular beds. Amer. J. Physiol. **203**, 583 (1962)

GANTT, C. L, and W. J. CARTER Acute effects of angiotensin on calcium, phosphorus, magnesium and potassium excretion. Can. med. Ass. J. **90**, 287 (1964).

GARB, S · The effects of potassium, ammonium, calcium, strontium and magnesium on the electrogram and myogram of mammalian heart muscle. J. Pharmacol. exp Ther. **101**, 317 (1951)

GARDNER, L. I, E. A MACLACHLAN, and H. BERMAN Effect of potassium deficiency on carbon dioxide, cation, and potassium content of muscle. J. gen. Physiol. **36**, 153 (1953).

GARRELS, R. M., and M E. THOMPSON. A chemical model for sea water at 25° and one atmosphere total pressure. Amer. J. Sci. **260**, 57 (1962).

GARTNER, R J. W. Values and variations of blood constituents in grazing Hereford cattle Res Vet. Sci. **7**, 424 (1966)

— J. W. RYLEY, and A. W. BEATTIE The influence of degree of excitation on certain blood constituents in beef cattle. Aust J. exp. Biol. med. Sci. **43**, 713 (1965).

GEERTRUYDEN, J VAN Les ions alcalin-terreux dans les sécrétions digestives Heffter's Handbuch der experimentellen Pharmakologie Erganzungswerk, Bd. 17, Teil 2, S. 736. Springer Berlin-Heidelberg-New York 1964

—, et N. DEJARDIN Lois de sécrétion des ions Ca++ et Mg++ par la muqueuse gastrique. Acta gastro-ent belg **27**, 408 (1964)

GENT, W L. G, J R. TROUNCE, and M. WALSER The binding of calcium ion by the human erythrocyte membrane. Arch. Biochem. **105**, 582 (1964).

GEORGE, W. K., and W. S. CHAMBLESS Tetany associated with simultaneous hypo-magnesemia and hypocalcemia in patient with celiac disease. Tex. St. J. Med. **58**, 812 (1962).

— D A. DOTSON, and W. W. GRANT A comparison of the changes of serum calcium and magnesium during exercise and hyperventilation. Clin Res. **14**, 62 (1966).

GERBRANDY, J, A M VAN LEEUWEN, M. B A HELLENDOORN, and L. A. DE VRIES. The binding between electrolytes and serum proteins calculated from an *in vivo* filtration method Clin Sci **19**, 181 (1960).

GERSHOFF, S N, and S. B. ANDRUS: Dietary magnesium, calcium and vitamin B_1 and experimental nephropathies in rats. calcium oxalate calculi, apatite nephrocalcinosis. J. Nutr. **73**, 308 (1961).

GERSHOFF, S N, J J. VITALE, I. ANTONOWICZ, M NAKAMURA, and E. E. HELLERSTEIN: Studies of the interrelationships of thyroxine, magnesium and vitamin B_{12} J. biol. Chem. **231**, 849 (1958).

GERST, P. H, M. R. PORTER, and R A FISHMAN Symptomatic magnesium deficiency in surgical patients. Ann. Surg **159**, 402 (1964).

GILBERT, D. L.. Effect of pH on muscle calcium and magnesium Proc. Soc exp Biol. (N Y) **106**, 550 (1961)

—, and J McGANN. Magnesium equilibrium in muscle. J. gen. Physiol. **43**, 1103 (1960).

GILL, J. R. jr., N. H BELL, and F. C. BARTTER: The effect of parathyroid extract on magnesium excretion in man Clin. Res **10**, 405 (1962).

GINN, E, and E. L. BECKER Urinary phosphate excretion in magnesium deficient rats. Clin. Res. **10**, 67 (1962).

GINN, H E, W. O. SMITH, J. F. HAMMARSTEN, and D. SNYDER Renal tubular secretion of magnesium in dogs. Proc. Soc. Exp Biol. (N.Y.) **101**, 691 (1959).

GINSBURG, S., J. G. SMITH, F. M. GINSBURG, J. Z. REARDON, and J. K. AIKAWA. Magnesium metabolism of human and rabbit erythrocytes. Blood **20**, 722 (1962).

GITELMAN, H J, J. B. GRAHAM, and L. G. WELT A new familial disorder characterized by hypokalemia and hypomagnesemia Clin Res **14**, 108 (1966).

— S. KUKOLJ, and L. G WELT Effect of thyrocalcitonin on the partition of plasma calcium. Clin Res **13**, 323 (1965).

— — — Inhibition of parathyroid gland function by hypermagnesemia. Fed. Proc. **25**, 495 (1966).

GITTELMAN, I F, J B. PINKUS, and E. SCHMERTZLER Interrelationship of calcium and magnesium in the mature neonate Amer. J. Dis. Child. **107**, 119 (1964).

GLASER, W, and J. L BRANDT· Localization of magnesium-28 in the myocardium. Amer. J. Physiol. **196**, 375 (1959).

—, and W. D. GIBBS· Localization of radiomagnesium in puppies radioautographic study of heart and bone Amer. J. Physiol. **202**, 584 (1962)

GLAUBITT, D., and J. G. RAUSCH-STROOMANN Magnesium-, Calcium- und Phosphorbilanzen bei essentieller Hypertonie und Herzinsuffizienz unter der Behandlung mit Hydrochlorothiazid Klin. Wschr. **40**, 143 (1962).

GLICK, D, E F. FREIER, and M. J. OCHS Studies in histochemistry. XLVII Microdetermination of magnesium and its histological distribution in the adrenal in various functional states. J. biol. Chem. **226**, 77 (1957).

GLICKMAN, L. S., V. SCHENKER, S. GROLNICK, A. GREEN, and A. SCHENKER Cerebrospinal fluid cation levels in delirium tremens with special reference to magnesium. J. nerv. ment. Dis. **134**, 410 (1962).

GOLDMAN, A. S., D D. VAN FOSSAN, and E. E. BAIRD. Magnesium deficiency in celiac disease Pediatrics **29**, 948 (1962).

GOTTLIEB, L S, S A. BROITMAN, J. J VITALE, and N. ZAMCHECK. The influence of alcohol and dietary magnesium upon hypercholesterolemia and atherogenesis in the rat J. Lab clin Med. **53**, 433 (1959).

GOW, B. S. Non-ultrafiltrable calcium and magnesium in human saliva. Arch. oral Biol. **10**, 15 (1965).

GOWER-SMITH, S Magnesium-potassium antagonism. Arch. Biochem. **20**, 473 (1949).

GRANICHER, D., and H. PORTZEHL. The influence of magnesium and calcium pyrophosphate chelates, of free magnesium ions, free calcium ions, and free pyrophosphate ions on the dissociation of actomyosin Biochim. biophys. Acta (Amst.) **86**, 567 (1964)

GRANICHER-FRICK, D.. Der Einfluß der Erdalkali-Pyrophosphat-Komplexe sowie der freien Erdalkali- und Pyrophosphationen auf die Dissoziation des Aktomyosin. Helv. physiol. pharmacol. Acta 23, 1 (1965).

GRAHAM, G. G.. More on magnesium in the protein-calorie malnutrition of childhood. J Pediat. 67, 338 (1965).

GRAHAM, L. A., J. J. CAESAR, and A. S. V. BURGEN. Gastrointestinal absorption and excretion of Mg28. Metabolism 9, 646 (1960).

GRANTHAM, J. J., W. H. TU, and P. R SCHLOERB· Acute magnesium depletion and excess induced by hemodialysis. Amer. J. Physiol. 198, 1211 (1960).

GREENBERG, D. M., and R. B. AIRD Blood and spinal fluid magnesium and calcium levels in epilepsy and convulsive states. Proc Soc exp. Biol (N.Y.) 37, 618 (1938).

— C E. ANDERSON, and E V. TUFTS Pathological changes in tissues of rats reared on diets low in magnesium. J. biol. Chem. 114, xliii (1936).

— S. P. LUCIA, M. A. MACKEY, and E. V. TUFTS. The magnesium content of the plasma and the red corpuscles in human blood. J. biol. Chem. 100, 139 (1933).

— —, and E. V. TUFTS The effect of magnesium deprivation on renal function. Amer. J. Physiol. 121, 424 (1938)

—, and M. A. MACKEY· The effect of parathyroid extract on blood magnesium J. biol Chem. 98, 765 (1932).

—, and E. V. TUFTS Variations in the magnesium content of the normal white rat with growth and development. J. biol. Chem. 114, 135 (1936).

— — The nature of magnesium tetany Amer. J. Physiol. 121, 416 (1938).

GRIFFITH, F. D, H. E PARKER, and J. C ROGLER Observations on a magnesium-fluoride interrelationship in chicks. J. Nutr. 79, 251 (1963).

— — — Effect of dietary magnesium and fluoride on citric acid content on chick bones. Proc Soc. exp Biol (N Y) 116, 622 (1964)

GRISWOLD, R. L., and N PACE The intracellular distribution of metal ions in rat liver. Exp. Cell Res. 11, 362 (1956)

GROLLMAN, A Effect of cortisone on serum calcium, magnesium and phosphate levels in nephrectomized dogs. Proc. Soc. exp. Biol. (N.Y.) 85, 582 (1954a).

— The role of the kidney in the parathyroid control of the blood calcium as determined by studies on the nephrectomized dog Endocrinology 55, 166 (1954b).

GUDMUNDSSON, T V., I. MACINTYRE, and H A SOLIMAN The isolation of thyrocalcitonin and a study of its effects in the rat Proc. roy Soc. B 164, 460 (1966)

HAAS, H. G., H. AFFOLTER, and U. C. DUBACH Evidence for a calcitonin effect in man. Acta endocr. (Kbh.) 48, 132 (1965)

HADDY, F. J, J. B. SCOTT, M. A FLORIO, R. M. DOUGHERTY jr, and J. N. HUIZENGA Local vascular effects of hypokalemia, alkalosis, hypercalcemia and hypomagnesemia. Amer. J. Physiol. 204, 202 (1963)

HANZE, S.· Untersuchungen zur Wirkung verschiedener Diuretica auf die renale Magnesium- und Calcium-Ausscheidung. Klin. Wschr. 38, 1168 (1960a).

— Die intraerythrocytare Magnesium-Konzentration bei Renaler Insuffizienz. Klin. Wschr. 38, 769 (1960b).

— Der erythrocytare Magnesiumstoffwechsel bei intakter und kalte-inhibierter Glykolyse. Experientia (Basel) 18, 45 (1962a).

— Der erythrocytare Magnesiumstoffwechsel bei Glykolyseinhibierung durch Natriumfluorid und Monojodacetat. Naturwissenschaften 49, 39 (1962b).

— Der Magnesiumstoffwechsel. Stuttgart Georg Thieme 1962(c).

— Die diagnostische Bedeutung der Storungen des Magnesiumstoffwechsels. Dtsch. med. Wschr. 90, 1958 (1965).

—, u W. HILLER: Serum- und Erythrocyten-magnesium bei renaler Insuffizienz. Klin. Wschr. 41, 1055 (1963).

HALL, D G.· Serum magnesium in pregnancy. Obstet. and Gynec. 9, 158 (1957)

HAMBURGER, J.: Electrolyte disturbances in acute uraemia. Clin. Chem 3, 332 (1957).

HAMILTON, M G , and M. L. PETERMANN. Ultracentrifugal studies on ribonucleoprotein from rat liver microsomes. J. biol. Chem. **234**, 1441 (1959).

HAMMARSTEN, G.: In: Etiologic factors in renal lithiasis (A. J. BUTT, ed.), p. 56. Springfield (Ill.): Ch. C. Thomas 1956

HAMMARSTEN, J. F , M. ALLGOOD, and W. O. SMITH: Effect of magnesium sulfate on renal function, electrolyte excretion and clearance of magnesium. J. appl. Physiol. **10**, 476 (1957).

—, and W. O. SMITH Symptomatic magnesium deficiency in man. New Engl. J. Med. **256**, 897 (1957).

HANESON, I. B . Effect of electrolytes and hormones on contraction of arterial smooth muscle. Circulation **26**, 727 (1962).

HANIG, R. C., and M H. APRISON Determination of eight cations in five specific brain areas in the rabbit by atomic absorption spectrophotometry. Fed. Proc. **25**, 511 (1966).

HANNA, S.. Influence of large doses of vitamin D on magnesium metabolism in rats. Metabolism **10**, 735 (1961 a).

— Plasma magnesium in health and disease. J. clin. Path. **14**, 410 (1961 b).

— M HARRISON, I. MacINTYRE, and R. FRASER. The syndrome of magnesium deficiency in man. Lancet **1960 II**, 172.

— M. T. HARRISON, I. MacINTYRE, and R. FRASER: Effects of growth hormone in calcium and magnesium metabolism. Brit. Med. J. **1961 II**, 12 (a).

—, and I. MacINTYRE The influence of aldosterone on magnesium metabolism. Lancet **1960 II**, 348.

— K A. K. NORTH, I. MacINTYRE, and R. FRASER: Magnesium metabolism in parathyroid disease Brit. Med. J. **1961 II**, 1253 (b).

HANNON, J P., A. M. LARSON, and D W. YOUNG Effect of cold acclimatization on plasma electrolyte levels. J. appl Physiol **13**, 239 (1958).

HARMON, M · Parathyroid adenoma in a child. Amer. J. Dis. Child. **91**, 313 (1956).

HARRIS, H , I. R. McDONALD, and W WILLIAMS The electrolyte pattern in experimental anuria. Aust. J. exp. Biol. med. Sci. **30**, 33 (1952).

HARRIS, J. S , and W. J. A. DE MARIA· Effects of magnesium sulfate on the renal dynamics of normal dogs. Amer J Physiol. **166**, 199 (1951).

— — Effect of magnesium sulfate on renal dynamics in acute glomerulonephritis in children. Pediatrics **11**, 191 (1953).

HARRIS, W. H., and J A. BEAUCHEMIN· Cerebrospinal fluid calcium, magnesium and their ratio in psychoses of organic and functional origin. Yale J. Biol. Med. **29**, 117 (1956).

—, and E. H. SONNENBLICK A study of calcium and magnesium in the cerebrospinal fluid. Yale J. Biol. Med. **27**, 297 (1955).

HARRISON, H. E., and H C. HARRISON The interaction of vitamin D and parathyroid hormone on calcium phosphorus and magnesium homeostasis in the rat. Metabolism **13**, 952 (1964).

HARROP, G. A., L. J. SOFFER, R. ELLSWORTH, and J. H. TRESCHER: Studies on the suprarenal cortex III. Plasma electrolytes and electrolyte excretion during suprarenal insufficiency in the dog. J exp Med. **58**, 17 (1933).

HART, E B., and H STEENBOCK: The effect of a high magnesium intake on calcium retention by swine. J. biol Chem. **14**, 75 (1913).

HASSELMAN, J J F., and E. J. VAN KAMPEN Magnesium. Clin. chim. Acta **3**, 305 (1958).

HATHAWAY, M. L Magnesium in human nutrition. Home Economics Research Report No 19, Agricultural Research Service, Washington D.C. 1962.

HAURY, V. G., and A. CANTAROW· Variations of serum magnesium in 52 normal and 440 pathologic patients. J. Lab. clin. Med. **27**, 616 (1942).

HAYWOOD, J., and R. SELVESTER· Effects of oral magnesium and potassium on serum lipids. Clin Res **10**, 87 (1962)

HEAD, M. J., and J A. F. ROOK Hypomagnesaemia in dairy cattle and its possible relationship to ruminal ammonia production. Nature (Lond.) 176, 262 (1955).
— — Some effects of spring grass on rumen digestion and the metabolism of the dairy cow Proc. Nutr. Soc. 16, 25 (1957).
HEAGY, F. C., and A. C. BURTON · Effect of intravenous injection of magnesium chloride on the body temperature of the unanesthetized dog, with some observations on magnesium levels and body temperature in man. Amer. J. Physiol 152, 407 (1948).
HEATON, F. W.. The action of the parathyroid glands during magnesium deficiency in the rat. Biochem J. 92, 50P (1964a).
— Magnesium metabolism in surgical patients. Clin chim. Acta 9, 327 (1964b).
— The parathyroid glands and magnesium metabolism in the rat. Clin. Sci. 28, 543 (1965a).
— Effect of magnesium deficiency on plasma alkaline phosphatase activity. Nature (Lond.) 207, 1292 (1965b).
—, and C K. ANDERSON The mechanism of renal calcification induced by magnesium deficiency in the rat. Clin. Sci. 28, 99 (1965).
—, and P FOURMAN Magnesium deficiency and hypocalcaemia in intestinal malabsorption Lancet 1965 II, 50.
—, and A. HODGKINSON External factors affecting diurnal variation in electrolyte excretion with particular reference to calcium and magnesium. Clin. chim. Acta 8, 246 (1963).
— —, and G. A. ROSE. Observations on relation between calcium and magnesium metabolism in man. Clin. Sci. 27, 31 (1964).
—, and F M. PARSONS The metabolic effect of high magnesium intake. Clin. Sci 21, 273 (1961).
—, and L N PYRAH Magnesium metabolism in patients with parathyroid disorders. Clin Sci 25, 475 (1963).
— — C. C. BERESFORD, R. W. BRYSON, and D. F. MARTIN: Hypomagnesaemia in chronic alcoholism. Lancet 1962 II, 802
HEGGTVEIT, H. A.· Myopathy in magnesium deficiency Life Sci 4, 69 (1965a).
— The cardiomyopathy of magnesium-deficiency. In. Electrolytes and cardiovascular diseases, p 204—220 New York. S. Karger 1965(b).
— L. HERMAN, and R. K. MISHRA· Cardiac necrosis and calcification in experimental magnesium deficiency. Amer. J. Path. 45, 757 (1964).
HEGSTED, D. M , J J VITALE, and H. MCGRATH· The effect of low temperature and dietary calcium upon magnesium requirement. J. Nutr. 58, 175 (1956).
HEINRICH, H. G. Prophylaxe und Therapie thrombotischer Zustande mit Magnesium. Z. ges. inn. Med. 12, 777 (1957)
HELLER, B. I., J. F. HAMMARSTEN, and F. I. STUTZMAN Concerning the effects of magnesium sulfate on renal function, electrolyte excretion and clearance of magnesium. J. clin. Invest. 32, 858 (1953).
HELLERSTEIN, E. E., M. NAKAMURA, D. M. HEGSTED, and J. J. VITALE Studies on the interrelationships between dietary magnesium, quality and quantity of fat, hypercholesterolemia and lipidosis. J Nutr. 71, 339 (1960).
— J. J. VITALE, P. L. WHITE, D. M. HEGSTED, N. ZAMCHECK, and M NAKAMURA Influence of dietary magnesium on cardiac and renal lesions of young rats fed on atherogenic diet. J. exp. Med. 106, 767 (1957).
HELVE, O E.. Studien uber den Einfluß der Nebennierenexstirpation auf den tierischen Stoffwechsel Biochem. Z 306, 343 (1940).
HEMINGWAY, R G , and N S RITCHIE The importance of hypocalcemia in the development of hypomagnesemic tetany. Proc. Nutr. Soc. 24, 54 (1965)
HENDRICKS, S. B., and W. L HILL The nature of bone and phosphate rock. Trans of III Conf. JOSIAH MACY, JR. FOUNDA. Metabolic Interrelations 3, 173 (1951)
HENDRIX, J. Z., N. W ALCOCK, and R. M ARCHIBALD Competition between calcium, strontium, and magnesium for absorption in the isolated rat intestine. Clin. Chem. 9, 734 (1963).

HENNEMAN, P. H , and E F. DEMPSEY. Factors determining fecal electrolyte excretion. J. clin. Invest. **35**, 711 (1956)

HENRIQUES, V , u S L ØRSKOV Untersuchungen uber den Magnesium- und den Kalium- gehalt der roten Blutkorperchen bei Anämie Skand. Arch. Physiol. **82**, 86 (1939).

HERRING, W. B., B. S LEAVELL, L M. PAIXAO, and J. H. YOE. Trace metals in human plasma and red blood cells. Amer. J. clin. Nutr. **8**, 846 (1960).

HESS, R., I MACINTYRE, N ALCOCK, and A. G E. PEARSE Histochemical changes in rat kidney in magnesium deprivation. Brit. J. exp Path. **40**, 80 (1959).

HILLS, A G., D W. PARSONS, G D WEBSTER jr , O ROSENTHAL, and H CONOVER. Influence of the renal excretion of sodium chloride upon the renal excretion of magne- sium and other ions by human subjects J. clin. Endocr. **39**, 1192 (1959).

HINGERTY, D. The role of magnesium in adrenal insufficiency. Biochem J. **66**, 429 (1957)

HIRANO, J. Studies on metabolism of cholesterol in magnesium-deficient rabbits Fukuoka Acta med. **57**, 259 (1966)

HIRSCHFELDER, A D , and V. G. HAURY· Clinical manifestations of high and low plasma magnesium Dangers of epsom salt purgation in nephritis. J. Amer. med. Ass. **102**, 1138 (1934)

HODGKINSON, A., and F. W. HEATON The effect of food ingestion on the urinary excretion of calcium and magnesium. Clin. chim. Acta **11**, 354 (1965).

HOFER, M , and A. KLEINZELLER Calcium transport in slices of rabbit kidney cortex the uptake and distribution of calcium. Physiol. bohemoslov. **12**, 405 (1963).

HOFF, H. E , P. K SMITH, and A. W WINKLER Effects of magnesium on the nervous system in relation to its concentration in serum. Amer. J. Physiol **130**, 292 (1940).

HOGAN, A G., W. O. REGAN, and W R HOUSE Calcium phosphate deposits in guinea pigs and the phosphorus content of the diet. J. Nutr. **41**, 203 (1950).

HOMER, L Hypoparathyroidism requiring massive amounts of medication, with apparent response to magnesium sulfate. J. clin. Endocr. **21**, 219 (1961).

HONORATA, C. R., and E. ROSA Importancia del magnesio en la nutrición y su relación con la glándula tiroides. (Importance of magnesium in nutrition and its relation with the thyroid gland.) Rev. méd Chile **72**, 892 (1944).

HOOBLER, S W., H D. KRUSE, and E. V. McCOLLUM Studies on magnesium deficiency in animals VIII. The effects of magnesium deprivation on the total and ultrafilterable Ca and Mg of the serum Amer. J. Hyg **25**, 86 (1937).

HOPKINS, T , J E HOWARD, and H. EISENBERG Ultrafiltration studies on calcium and phosphorus in human serum. Bull Johns Hopkins Hosp. **91**, 1 (1952).

HORTON, R., and E. BIGLIERI Effect of aldosterone on magnesium metabolism. Clin. Res. **10**, 93 (1962).

HOUSE, W B , and A. G. HOGAN Injury to guinea pigs that follows a high intake of phosphorus J. Nutr **55**, 507 (1955).

HOZDRX, T., u L ATANASSOWA Die Kreislaufwirkung der isotonischen und hyper- tonischen Losungen I. Mitt. Die Wirkung von Magnesiumsulfat auf die Blutgefaße der unteren Extremitaten beim Menschen Naunyn-Schmiedebergs Arch exp Path. Pharmak. **227**, 400 (1956).

HUGHES, A , and R. S TONKS Platelets, magnesium and myocardial infarction Lancet **1965 I**, 1044

— — Magnesium, adenosine diphosphate and blood platelets. Nature (Lond.) **210**, 106 (1966)

HUMMEL, F. C., H A HUNSCHER, M. F. BATES, P. BONNER, I. G MACY, and J. A. JOHN- STON. A consideration of the nutrition state in the metabolism of women during pregnancy. J. Nutr. **13**, 263 (1937)

— H R. STERNBERGER, H. A. HUNSCHER, and I. G MACY Metabolism of women during the reproductive cycle. VII. Utilization of inorganic elements (a continuous case study of a multipara) J. Nutr. **11**, 235 (1936).

HUNGERFORD, G. F. Role of histamine in producing the eosinophilia of magnesium deficiency. Proc. Soc. exp Biol. (N Y.) **115**, 182 (1964).

HUNTER, G., and H. V. SMITH. Calcium and magnesium in human cerebrospinal fluid Nature (Lond.) 186, 161 (1960).

HUNTSMAN, R. G., B. A. L. HURN, and H. LEHMANN. Observations on the effect of magnesium on blood coagulation. J clin. Path. 13, 99 (1960).

HUTTER, O. F., and K. KOSTIAL Effect of magnesium and calcium ions on the release of acetylcholine. J. Physiol. (Lond) 124, 234 (1954).

INGLIS, J. S. S., M WEIPERS, and P. J. PEARCE Hypomagnesaemia in sheep. Vet Rec. 71, 755 (1959).

JABIR, F. K., S. D. ROBERTS, and R. A. WOMERSLEY: Studies on the renal excretion of magnesium. Clin. Sci. 16, 119 (1957).

JENKINSON, D. H.. The nature of the antagonism between calcium and magnesium ions at the neuromuscular junction J. Physiol. (Lond.) 138, 433 (1957).

JOHANSEN, E : Ultrastructural and chemical observation on dental caries In Mechanisms of hard tissue destruction (R. F. SOGNNAES, ed), p 187 Washington, D. C. Amer. Ass Advance Sci. 1963

JONES, J. E., P. C. DESPER, S R. SHANE, and E B. FLINK Magnesium metabolism in hyperthyroidism and hypothyroidism J clin. Invest 45, 891 (1966).

— —, and E. B. FLINK. Magnesium metabolism in Huntington's chorea. Metabolism 14, 813 (1965).

JONES, K. H., and P. FOURMAN Effects of infusions of magnesium and of calcium in parathyroid insufficiency. Clin. Sci. 30, 139 (1966).

JUDAH, J. D., K. AHMED, A. E M McLEAN, and G. S CHRISTIE· Uptake of magnesium and calcium by mitochondria in exchange for hydrogen ions Biochim. biophys Acta (Amst.) 94, 452 (1965).

KALBFLEISCH, J. M., R. D. LINDEMAN, H. E. GINN, and W. O SMITH Effects of ethanol administration on urinary excretion of magnesium and other electrolytes in alcoholic and normal subjects. J. clin Invest. 42, 1471 (1963)

— —, and W. O. SMITH The effects of ethyl alcohol administration on urinary excretion of magnesium and other electrolytes in alcoholic and normal subjects J. Lab. clin. Med 58, 833 (1961).

KAPITOLA, J., u. O. KUCHEL. Nebennierenrinde und Magnesiumstoffwechsel. Klin. Wschr. 42, 954 (1964)

KASHIWA, H K. Magnesium deficiency in intact, in adrenalectomized and in hypophysectomized rats. Endocrinology 68, 80 (1961).

—, and G. F HUNGERFORD: Blood leucocyte response in rats fed a magnesium deficient diet. Proc Soc exp. Biol (N.Y) 99, 441 (1958).

KELLY, H. G., H. C. CROSS, M. R TURLON, and J D. HATCHER. Renal and cardiovascular effects induced by intravenous infusion of magnesium sulfate. Canad. med. Ass. J. 82, 866 (1960).

KEMÉNY, A., H. BOLDIZSÁR u. G. PETHES· Angaben uber die Stabilitat der Kationenkonzentration des Liquor cerebrospinalis bei verschiedenen Plasmaspiegeln. Acta physiol. Acad. Sci. hung. 14, 9 (1958).

— — — The distribution of cations in plasma and cerebrospinal fluid following infusion of solutions of salts, potassium, magnesium and calcium. J. Neurochem. 7, 218 (1961).

KEMP, A., O. DEIJS, O. J. HEMKES, and A. J. H. VAN ES Hypomagnesemia in cows Intake and utilization of magnesium from herbage by lactating cows. Netherlands J. agric. Sci. 9, 134 (1961).

KENYON, F. E., and S. M. HARVEY. A biochemical study of Huntington's chorea. J. Neurol. Neurosurg. Psychiat. 26, 123 (1963).

KESSNER, J. M , and F. H. EPSTEIN. Effect of magnesium deficiency on gastrointestinal transfer of calcium. Proc. Soc. exp. Biol. (N Y.) 122, 721 (1966).

KEYL, M. J., J. B SCOTT, J. M DABNEY, F. J HADDY, R B. HARVEY, R D. BELL, and H. E. GINN Composition of canine renal hilar lymph. Amer. J. Physiol. 209, 1031 (1965).

KIIL, F.: Permutation trial of diuretics: chlorothiazide and hydroflumethiazide. Circulation 21, 717 (1960).

KLEEMAN, C. R., F. H. EPSTEIN, D. McKAY, and E. TABORSKY Effects of hypo- and hyperthyroidism on the filterability of serum magnesium. J. clin. Endocr. 18, 1111 (1958).

— S. LING, D. BERNSTEIN, M. H. MAXWELL, and L. CHAPMAN The effect of independent changes in glomerular filtration (GFR) and sodium (Na+) excretion on the renal excretion of calcium (Ca++) and magnesium (Mg++) in acutely hypercalcemic dogs. J. clin Invest. 45, 1032 (1966).

KLEIBER, M., M. D. BOELTER, and D. M GREENBERG Fasting catabolism and food utilization of magnesium deficient rats. J. Nutr 21, 363 (1941).

KLEIBER, R. E., K. SETA, J. J. VITALE, and B. LOWN. Effects of chronic depletion of potassium and magnesium upon the action of acetylstrophanthidin on the heart. Amer J. Cardiol. 17, 520 (1966).

KNIPPERS, R., u. U. HEHL Die renale Ausscheidung von Magnesium, Calcium und Kalium nach Erhöhung der Magnesium-Konzentration im Plasma des Hundes. Z ges. exp. Med. 139, 154 (1965).

KNOOP, C. E., W. E. KRAUSS, and C. C. HAYDEN. Magnesium-vitamin D relationships in calves fed mineralized milk. J. Dairy Sci. 22, 283 (1939).

KO, K. W., F. X. FELLERS, and J. M. CRAIG Observations on magnesium deficiency in the rat. Lab. Invest. 11, 294 (1962).

KOHLER, H F, and M. M. PECHET· The inhibition of bone resorption by thyrocalcitonin. J. clin. Invest. 45, 1033 (1966).

KOHN, R. R, H. KEYE, and E. ROLLERSON Role of magnesium in the variation of swelling ability of human muscle with age. Exp. Cell Res 25, 170 (1961).

KRAHL, M. E.: Insulinlike and anti-insulin effects of chelating agents on adipose tissue. Fed. Proc. 25, 832 (1966).

KREVELD, A. VAN, and G. VAN MINNEN· Calcium and magnesium ion activity in raw milk and processed milk Ned. Melk-en Zuiveltijdschr 9, 1 (1955)

KRUSE, H. D., E. R ORENT, and E. V. McCOLLUM Studies on magnesium deficiency in animals. I. Symptomatology resulting from magnesium deprivation. J. biol. Chem. 96, 519 (1932).

— — — Studies on magnesium deficiency in animals. III. Chemical changes in the blood following magnesium deprivation. J. biol. Chem. 100, 603 (1933).

— M. M. SCHMIDT, and E. V. McCOLLUM Studies on magnesium deficiency in animals. IV. Reaction to galvanic stimuli following magnesium deprivation. Amer. J. Physiol. 105, 635 (1933)

KUMAR, D, P. A. ZOURLES, and A C BARNES· In vitro and in vivo effects of magnesium sulfate on human uterine contractility. Amer. J. Obstet. Gynec. 86, 1036 (1963).

KUNKEL, H. O., K. H BURNS, and B. J. CAMP A study of sheep fed high levels of potassium bicarbonate with particular reference to induced hypomagnesemia. J. Anim. Sci. 12, 451 (1953).

KUPFER, S, and J. D. KOSOVSKY Effects of cardiac glycosides on renal tubular transport of calcium, magnesium, inorganic phosphate, and glucose. J. clin Invest. 44, 1132 (1965).

LANGLEY, L L, O. R. GRIMES jr., and D. F. COCKRELL jr.· Secretion of magnesium by dog parotid gland. Amer. J. Physiol. 202, 707 (1962).

LARVOR, P., A GIRARD, M. BROCHART, A. PARODI et J. SEVESTRE Étude de la carence expérimentale en magnésium chez le veau. I. Observations cliniques, biochemiques, et anatomo-pathologiques Ann. Biol. anim. 4, 345 (1964).

LASSITER, W. E, C. W. GOTTSCHALK, and M. MYLLE Micropuncture study of renal tubular reabsorption of calcium in normal rodents. Amer. J. Physiol. 204, 771 (1963).

LEE, K. S., K. TANAKA, and D. H. YU Studies on the ATPase, calcium uptake and relaxing activity of the microsomal granules from skeletal muscle J. Physiol. (Lond.) 179, 456 (1965).

LEEUWEN, A. M. VAN Net cation equivalency ("base binding power") of the plasma proteins Acta med. scand Suppl **176**, 422 (1964).

— C. M. THOMASSE, and P. C. KAPTEYN: The determination of proteinbound calcium and magnesium by ultrafiltration *in vivo* Clin. chim. Acta **6**, 550 (1961).

LEHNINGER, A L , and C L WADKINS Oxidative phosphorylation Ann. Rev. Biochem **31**, 47 (1962)

LEICHSENRING, J. M , L. M. NORRIS, and S. A. LAMSON Magnesium metabolism in college women Observations on the effect of calcium and phosphorus intake levels. J. Nutr. **45**, 477 (1951).

LENGEMANN, F. Site of action of lactose in enhancement of calcium utilization. J. Nutr. **69**, 23 (1959).

LEROY, J Nécessité du magnésium pour la croissance de la souris. C R. Soc Biol. (Paris) **94**, 431 (1926).

L'ESTRANGE, J L , and R. F E AXFORD: The effects of low magnesium intake on lactating ewes Proc. Nutr. Soc. **22**, 1 (1963).

— — A study of magnesium, and calcium metabolism in lactating ewes fed a semi-purified diet low in magnesium. J. agric. Sci. **62**, 353 (1964a)

— — A study of serum mineral changes in lactating Welsh mountain ewes under different grazing conditions with special reference to hypomagnesaemia. J. agric. Sci. **62**, 341 (1964b)

LEVEY, S , W. E ABBOTT, H. KRIEGER, and J. H DAVIS Metabolic alterations in surgical patients. X. Studies involving iron and magnesium metabolism in patients with gastrointestinal drainage J. Lab clin. Med. **47**, 437 (1956).

LEVY, H. M , and E. M. RYAN· Evidence that the contraction of actomyosin requires the reaction of adenosine triphosphate and magnesium at two different sites (rabbit muscle). Biochem. Z. **345**, 132 (1966).

LEWIS, W. H. A micro-method for the estimation of magnesium J. med Lab Technol. **17**, 32 (1960).

LIFSHITZ, F , H. C HARRISON, E. C. BULL, and H E HARRISON Citrate metabolism and the mechanism of renal calcification induced by magnesium depletion Metabolism (in press) (1967).

— —, and H E HARRISON Intestinal transport of calcium and phosphate in magnesium deficiency. Proc Soc. exp Biol. (N Y) (in press) (1967)

LIKUSKI, H J A , and R M FORBES Mineral utilization in the rat IV Effects of calcium and phytic acid on the utilization of dietary zinc J. Nutr. **85**, 230 (1965).

LILIENTHAL, J L , K L ZIERLER, B. P FOLK, R. BUKA, and M J. RILEY A reference base and system for analysis of muscle constituents J biol Chem **182**, 501 (1950).

LINDEMAN, R. D., S. ADLER, M J. YIENGST, and E. S. BEARD Effect of carbohydrate, protein and fat ingestion on urinary divalent cation excretion Clin Res **13**, 310 (1965).

— H. E. GINN, J M KALBFLEISCH, and W. O. SMITH: Effect of galactose on urinary electrolyte excretion in man Proc. Soc. exp Biol. (N Y) **115**, 264 (1964).

LINDER, G C , J D L HANSEN, and C D KARABUS The metabolism of magnesium and other inorganic cations and of nitrogen in acute kwashiorkor. Pediatrics **31**, 552 (1963).

LIPKIN, G , C. MARCH, and J. GOWDEY Magnesium in epidermis, dermis and whole skin of normal and atopic subjects J. invest. Derm **42**, 293 (1964)

LIZARRALDE, G , V E MAZZOCCO, and E. B FLINK Urea cycle enzymes in magnesium deficiency J. clin. Invest **45**, 1042 (1966).

LOKEN, H F , R. J HAVEL, G. S. GORDAN, and S L WHITTINGTON Ultracentrifugal analysis of protein-bound and free calcium in human serum J. biol. Chem. **235**, 3654 (1960)

LOSSE, H , u. W. KOENIG Das Verhalten des Serum-Magnesiumspiegels bei akuten und chronischen Nierenerkrankungen Dtsch. med. Wschr. **86**, 824 (1961).

LOWENHAUPT, E , M. P. SCHULMAN, and D M. GREENBERG Basic histologic lesions of magnesium deficiency in the rat Arch. Path. 49, 427 (1950).

LOWRY, O. H., A B. HASTINGS, C. M. McCOY, and A. N BROWN Histochemical changes associated with aging. IV. Liver, brain and kidney in the rat. J. Geront. 1, 345 (1946)

MACBETH, R. A , and J D. MABBOTT Magnesium balance in the postoperative patient Surg Gynec. Obstet 118, 748 (1964).

MACDONALD, D. C , A. D CARE, and B. NOLAN Excretion of labeled magnesium by the sheep Nature (Lond.) 184, 736 (1959).

MAHY, B W. J , and K. A. MUNDAY Hypomagnesaemia following aldosterone administration in rats J. Physiol (Lond) 168, 22P (1963).

MACINTYRE, I Some aspects of magnesium metabolism and magnesium deficiency Proc. roy Soc. Med 52, 212 (1959).

— Magnesium metabolism in man and animals Proc. roy. Soc. Med. 53, 1037 (1960).

— Magnesium metabolism Scient. Basis Med. p 216 (1963a)

— An outline of magnesium metabolism in health and disease — a review. J. chron. Dis. 16, 201 (1963b).

— S. Boss, and V. A. TROUGHTON Parathyroid hormone and magnesium homeostasis. Nature (Lond) 198, 1058 (1963).

—, and D. DAVIDSSON The production of secondary potassium depletion, sodium retention, nephrocalcinosis and hypercalcaemia by magnesium deficiency. Biochem. J. 70, 456 (1958)

— S HANNA, C. C. BOOTH, and A. E. READ· Intracellular magnesium deficiency in man Clin Sci. 20, 297 (1961).

MACLEOD, L D Ionized calcium and magnesium in serum. Biochem. J. 91, 29P (1964).

MACRAE, I F., H F. KOHLER, and M. M. PECHET. Interrelationship of Mg, Ca, and P metabolism. Clin. Res. 11, 223 (1963).

MADER, I. J., and L. T. ISERI Spontaneous hypopotassemia, hypomagnesemia, alkalosis and tetany due to hypersecretion of corticosterone-like mineralocorticoid. Amer. J. Med 19, 976 (1955)

MAHLER, H. R.: Interrelationships with enzymes. In· Mineral metabolism (C L COMAR and F. BRONNER, eds.), vol. I, part B. New York. Academic Press 1961.

MAHY, B. W. J., and K. A. MUNDAY Hypomagnesaemia following aldosterone administration in rats J Physiol (Lond.) 168, 22P (1963)

MALKIEL-SHAPIRO, B Further observations on parenteral magnesium sulphate therapy in coronary heart disease a clinical appraisal. S. Afr med. J. 32, 1211 (1958).

MANERY, J. F.: Water and electrolyte metabolism. Physiol. Rev. 34, 334 (1954).

MANITIUS, A Some physiological effects of magnesium deficiency. In. Fundamental aspects. Electrolytes in cardiovascular diseases (E. BAJUSZ, ed), Vol 1, p. 28. New York S. Karger 1965.

—, and F. H EPSTEIN· Some observations on the influence of a magnesium deficient diet on rats with special reference to renal concentrating ability. J clin Invest 42, 208 (1963).

MARCUS, C. S., and R. H. WASSERMAN. Comparison of intestinal discrimination between Ca^{47} and Sr^{85} and Ba^{133} Amer. J. Physiol 209, 973 (1965).

— — Ca and Mg levels in gastrointestinal mucosa of fed, fasted and lactose-treated rats. J. appl. Physiol 21, 1063 (1966).

MARTIN, H. E., H EDMONDSEN, R. HOMANN, and C. F BERNE: Electrolyte problems in the surgical patient, with particular reference to serum Ca, Mg and K levels. Amer. J. Med 8, 529 (1950).

—, and R. JONES The effects of NH_4Cl and $NaHCO_3$ on the urinary excretion of Mg, Ca and phosphate. Amer. Heart J. 62, 206 (1961).

— C. McCUSKEY jr., and N. TUPIKOVA Electrolyte disturbance in acute alcoholism. Amer J. clin. Nutr. 7, 191 (1959)

— J. MEHL, and M WERTMAN Clinical studies of magnesium metabolism. Med Clin. N Amer. 36, 1157 (1952).

MARTIN, H. E , W. P. MIKKELSEN, and R JONES. The effect of parathyroidectomy in the dog on serum and urine magnesium levels Clin. Res 7, 108 (1959).

—, and M. WERTMAN. Serum potassium, magnesium and calcium levels in diabetic acidosis. J. clin Invest 26, 217 (1947).

—, and M. L. WILSON Effect of magnesium deficiency on serum and carcass electrolyte levels in the rat Metabolism 9, 484 (1960).

MARTINDALE, L., and F. W. HEATON Magnesium deficiency in the adult rat. Biochem. J. 92, 119 (1964).

— — The relation between skeletal and extracellular fluid magnesium in vitro. Biochem. J 97, 440 (1965).

MAURAT, J. P.. Le magnésium en pathologie. L'Expansion Scientifique Française, Paris, 1959

— Le dosage du magnésium globulaire· technique et resultats personnels Rev Path. comp. 64, 423 (1964)

— J J. POCIDALO et J. LIPSAC Le magnésium plasmatique au cours des modifications de l'équilibre acido-basique C R Soc. Biol. (Paris) 158, 2303 (1964).

MAXWELL, G. M., R B. ELLIOTT, and R. H. BURNELL: Effects of hypermagnesemia on general and coronary hemodynamics of the dog. Amer. J. Physiol. 208, 158 (1965)

MAYNARD, L. A., D. BOGGS, G. FISK, and D. SEQUIN Dietary mineral interrelations as a cause of soft tissue calcification in guinea pigs J. Nutr 64, 85 (1958).

MAYO, R H., M. P. PLUMLEE, and D M. BEESON· Magnesium requirement of the pig. J. Anim Sci. 18, 264 (1959).

MAZZOCCO, V. E., E. B. FLINK, and J. E. JONES Aminoaciduria in magnesium deficient rats Clin Res. 14, 383 (1966)

McALEESE, D M., M. C. BELL, and R. M FORBES Magnesium-28 studies in lambs. J. Nutr. 74, 505 (1961)

—, and R. M. FORBES Experimental production of magnesium deficiency in lambs on a diet containing roughage. Nature (Lond) 184, 2025 (1959).

— — The requirement and tissue distribution of magnesium in the rat as influenced by environmental temperature and dietary calcium J. Nutr. 73, 94 (1961).

McCANCE, R. A., and E. M WIDDOWSON LXIV. The fate of calcium and magnesium after intravenous administration to normal persons. Biochem. J. 33, 523 (1939).

McCANN, H. G.. Determination of microgram quantities of magnesium in mineralized tissues. Analyt. Chem 31, 2091 (1959).

McCLURE, F. J , and H. G. McCANN Dental caries and composition of bones and teeth of white rats. effects of dietary mineral supplements Arch. oral Biol. 2, 151 (1960)

McCOLLISTER, R , A. S. PRASAD, R. P. DOE, and E. B. FLINK. Normal renal magnesium clearance and the effect of water loading, chlorothiazide and ethanol on magnesium excretion. J. Lab. clin Med. 52, 928 (1958).

McCOLLISTER, R. J , E. B. FLINK, and R P. DOE. Magnesium balance studies in chronic alcoholism. J. Lab. clin. Med. 55, 98 (1960).

— —, and M D. LEWIS Urinary excretion of magnesium in man following the ingestion of ethanol Amer. J. clin. Nutr 12, 415 (1963).

McCREARY, P A , H A. BATTIFORA, G. H LAING, and G. M. HAAS: Protective effect of magnesium deficiency on experimental allergic encephalomyelitis in the rat. Proc. Soc. exp. Biol. (N.Y) 121, 1130 (1966)

McHARGUE, J. S., and W. R. ROY. Effect of ultraviolet irradiation on the magnesium content of rats receiving reflected sunlight and a uniform stock ration. Amer. J. Physiol. 92, 651 (1930).

MEINTZER, R. B., and H. STEENBOCK Vitamin D and magnesium absorption. J. Nutr. 56, 285 (1955).

MELLINGHOFF, K.. Magnesiumstoffwechselstorungen bei Inanition Dtsch. Arch. klin. Med 195, 475 (1949)

—, and W. VAN LESSEN. Magnesium-Calciumbilanz bei Inanition. Dtsch. Arch. klin. Med. 194, 285 (1949).

MENAKER, W. Influence of protein intake on magnesium requirement during protein synthesis. Proc. Soc. exp. Biol. (N.Y.) 85, 149 (1954).

—, and I. S. KLEINER Effect of deficiency of magnesium and other minerals on protein synthesis Proc. Soc. exp. Biol. (N Y) 81, 377 (1952).

MENDEL, L. B , and S. R. BENEDICT: The paths of excretion for inorganic compounds. IV. The excretion of magnesium. Amer. J. Physiol. 25, 1 (1909).

MENDELSON, J. H., B. A. BARNES, C. MAYMAN, and M. VICTOR· The determination of exchangeable magnesium in alcoholic patients. Metabolism 14, 88 (1965)

— J LA DOU, and C. CORBETT· Experimentally induced chronic intoxication and withdrawal in alcoholics. Pt. 9 Serum magnesium and glucose Quart. J Stud. Alcohol, Suppl. No 2, 108 (1964).

MERTZ, D. P Untersuchungen über die physiologischen renalen Ausscheidungsverhältnisse von Magnesium und Calcium Klin Wschr 35, 1171 (1957).

—, u U. LUTZ-DETTINGER: Untersuchungen uber die Wirkung von Prednison auf den Wasser- und Elektrolythaushalt und die Nierenfunktion beim Menschen Z klin. Med. 154, 631 (1957).

METCOFF, J., S FRENK, I. ANTONOWICZ, G. GORDILLO, and E. LOPEZ Relation of intracellular ions to metabolite sequences in muscle in kwashiorkor. Pediatrics 26, 960 (1960).

MEYTS, P. DE· Action des ions magnésium et calcium sur l'intestin isolé du rat. Rev. belge Path. 31, 245 (1965)

MILLER, E. R., D. E. ULLREY, C. L. ZUTAUT, B V. BATTZER, D. A. SCHMIDT, J A. HOEFER, and R. W. LUECKE· Magnesium requirement of the baby pig. J. Nutr. 85, 13 (1965).

— — — J A. HOEFER, and R. W. LUECKE. Mineral balance studies with the baby pig: Effect of dietary vitamin D_2 level upon calcium, phosphorus and magnesium balance. J. Nutr 85, 255 (1965a)

— — — —, and R. L LUECKE Comparison of casein and soy proteins upon mineral balance and vitamin D_2 requirement of the baby pig. J. Nutr. 85, 347 (1965b)

— — — —, and R W. LUECKE· Mineral balance studies with the baby pig. effects of dietary magnesium level upon calcium, phosphorus and magnesium balance. J. Nutr 86, 209 (1965c)

MILLER, J F.· Tetany due to deficiency in magnesium. Its occurrence in a child of six years with associated osteochondroses of capital epiphysis of femur. (Legg-Perth's dis.) Amer. J. Dis. Child. 67, 117 (1944).

MILLER, J. R., and T R. VAN DELLEN: Electrocardiographic changes following the intravenous administration of magnesium sulfate. An experimental study on dogs J. Lab. clin. Med. 23, 914 (1938).

— — Electrocardiographic changes following the intravenous administration of magnesium sulfate. III Combined effect with digitalis. J. Lab. clin Med 26, 1116 (1941).

MILLER, T. R. II, W. W. FALOON, and C. W. LLOYD. Divergence in magnesium, sodium and potassium excretion during stimulation of endogenous aldosterone production. J. clin. Endocr. 18, 1178 (1958)

MILNE, M D., R. C. MUEHRCKE, and I. AIRD· Primary aldosteronism. Quart. J. Med. 26, 317 (1957).

MIN, H K., J. E. JONES, and E. B FLINK Circadian variations in renal excretion of magnesium, calcium, phosphorus, sodium and potassium during frequent feeding and fasting Fed. Proc. 25, 917 (1966)

MISHRA, R. K . Studies on experimental magnesium deficiency in the albino rat. 1. Functional and morphologic changes associated with low intake of Mg. Rev. canad. Biol. 19, 122 (1960a).

— Studies on experimental magnesium deficiency in the albino rat. 2. The influence of Mg-deficient diet on mitochondrial population of heart, kidney and liver. Rev. canad. Biol. 19, 136 (1960b)

— Studies on magnesium deficiency in the albino rat 5. The influence of Mg-deficient regime on "steroid-phosphate-cardiopathy". Rev. canad. Biol. 19, 158 (1960c).

Misson, C , et H. Schirardin Étude de la magnésémie au cours de la tétanie. Strasbourg méd. 16, 267 (1965).

Montgomery, R. D · Magnesium metabolism in infantile protein malnutrition. Lancet 1960 II, 74

— Magnesium balance studies in marasmic kwashiorkor J Pediat. 59, 119 (1961).

Moore, C. A., and G. E. Bunce Reduction in frequency of renal calculus formation by oral magnesium administration. Invest. Urol 2, 7 (1964).

Moore, L A , E. T. Hallman, and L. B. Sholl Cardiovascular and other lesions in calves fed diets low in magnesium. Arch Path 26, 820 (1938).

Moore, R M , and W. J. Wingo Blood level of magnesium ion in relation to lethal, anesthetic, analgesic and antitetanic effects Amer J. Physiol. 135, 492 (1942).

Morgulis, S. Studies on the chemical composition of bone ash. J. biol Chem. 93, 455 (1931)

Mori, K., and J. P. Duruisseau Water and electrolyte changes in aging process with special reference to calcium and magnesium in cardiac muscle Canad. J. Biochem. 38, 919 (1960).

Morris, E R , and B. L. O'Dell Magnesium deficiency in the guinea pig. Mineral composition of tissues and distribution of acid-soluble phosphorus J. Nutr 75, 77 (1961).

— — Relationship of excess calcium and phosphorus to magnesium requirement and toxicity in guinea pigs J. Nutr 81, 175 (1963)

Morris, M. L. jr., W. R. Featherston, P H Phillips, and S. H. McNutt Influence of lactose and dried skim milk upon the magnesium deficiency syndrome in the dog. 2. Pathological changes. J. Nutr. 79, 437 (1963)

Moussa, S L., and A. Boba: Exogenous plasma magnesium increases during hypothermia in dogs Amer J. Physiol. 199, 1090 (1960).

Mraz, F. R.. Intestinal absorption of Ca-45 and Sr-85 as affected by the alkaline earths and pH. Proc Soc. exp. Biol. (N.Y.) 110, 273 (1962)

—, and R. G. Cragle. Fission product metabolism in animals. U.S. Atomic Energy Commission, ORO 177 (1958).

Mudd, S H , J. H. Park, and F. Lipmann Magnesium antagonism of the uncoupling of oxidative phosphorylation by iodo-thyronines. Proc. nat. Acad Sci (Wash) 41, 571 (1955)

Muhlrad, A., M. Kovacs, and G Hegyi The role of Mg^{2+} in the contraction and adenosine triphosphatase activity of myofibrils. Biochim biophys. Acta (Amst.) 107, 567 (1965).

Mukai, T., and J E. Howard Some observations on the calcification of rachitic cartilage by urine. One difference between "good" and "evil" urines, dependent upon content of magnesium Bull Johns Hopkins Hosp. 112, 279 (1963).

Munday, K A , and B W. J. Mahy Plasma electrolyte levels in a mammal and a reptile at various environmental temperatures Proc XXII Int. Physiol Congr 2, 492 (1962).

— — Determination of ultrafiltrable calcium and magnesium on small quantities of plasma Clin. chim Acta 10, 144 (1964).

Murdaugh, H. V., jr , and R R Robinson Magnesium excretion studied by stop-flow analysis. J clin. Invest 38, 1028 (1959).

— — Magnesium excretion in the dog studied by stop-flow analysis Amer. J. Physiol. 198, 571 (1960).

Nabarro, J. D. N., A. G. Spencer, and J. M. Stowers Metabolic studies in severe diabetic ketosis. Quart J. Med. 21, 225 (1952).

Nahmod, V E , and M. Walser The effect of ouabain on renal tubular reabsorption and cortical concentrations of several cations and on their association with subcellular particles Molec. Pharmac. 2, 22 (1966).

Nakamura, M , S Torii, M. Hiramatsu, T. Umezaki, T. Ohta, and J Hirano. Effect of MER 29 (Triparanol) on hypercholesterolemia and atherosclerosis in rabbits. Kyushu J. med Sci. 14, 77 (1963)

NAKAMURA, M , I. CORE, N. YABUTA, S. TORII, Y. ISHIHARA, K. TAMARI, and T. IMAI: Acid mucopolysaccharides, cholesterol, calcium, and magnesium content of Japanese aortas. Jap. Heart J. **6**, 20 (1965a).

— M NAKATINI, M. KOIKE, S. TORII, and M HIRAMATSU· Swelling of heart and liver mitochondria from magnesium deficient rats and its reversal. Proc. Soc. exp. Biol. (N.Y.) **108**, 315 (1961).

— S. TORII, M. HIRAMATSU, J. HIRANO, A. SUMIYOSHI, and K. TANAKA· Dietary effects of magnesium on cholesterol-induced atherosclerosis of rabbits. J. Atheroscler Res **5**, 145 (1965b).

— J. J. VITALE, J. M. HEGSTED, and E. E. HELLERSTEIN: The effect of dietary magnesium and thyroxine on progression and regression of cardiovascular lipid deposition in the rat. J. Nutr. **71**, 347 (1960).

NAKAO, T., Y. TASHIMA, K. NAGANO, and M. NAKAO. Highly specific sodium-potassium-activated adenosine triphosphatase from various tissues of rabbit. Biochem. biophys. Res Comm. **19**, 755 (1965).

NANNINGA, L. B · Calculation of free magnesium, calcium, and potassium in muscle. Biochim. biophys. Acta (Amst) **54**, 338 (1961).

NEGUIB, M A · Effect of Mg on the thyroid. Lancet **1953 I**, 1405.

NEUBEISER, R. E., W. S. PLATNER, and J. L. SHIELDS· Magnesium in blood and tissues during cold acclimation. J. appl. Physiol. **16**, 247 (1961).

NEUMAN, W F., and M. W. NEUMAN Chemical dynamics of bone mineral. Chicago· Univ. of Chicago Press, 1958.

NEUWIRTH, I., and G. B WALLACE. The use of magnesium as an aid in anesthesia. J. Pharmacol. exp. Ther. **35**, 171 (1929).

NIELSEN, B.· Plasma and urinary magnesium concentration in patients with renal insufficiency. Dan. med. Bull. **9**, 235 (1962).

— Correlation between antidiuretic hormone effect and the renal excretion of magnesium and calcium in man. Acta endocr. (Kbh) **45**, 151 (1964).

NIELSEN, J.: Magnesium metabolism in acute alcoholics. Dan. med. Bull. **10**, 225 (1963).

— Magnesium-lithium studies. 2. The effect of lithium on serum magnesium in rabbits. Acta psychiat. scand **40**, 197 (1964a).

— Magnesium-lithium studies. 1. Serum and erythrocyte magnesium with manic states during lithium treatment. Acta psychiat. scand. **40**, 190 (1964b).

NORDBO, R. Die Dissoziation von Magnesiumcitrat. Skand. Arch. Physiol **80**, 341 (1938).

— Bestimmung der Magnesiumionkonzentration im Ultrafiltrat von Blutserum. Skand Arch. Physiol. **81**, 265 (1939a).

— The concentration of ionized magnesium calcium in milk. J. biol. Chem. **128**, 745 (1939b).

NOWELL, N W., and D. C. WHITE Seasonal variation of magnesium and calcium in serum of the hypothermic rat. J. appl. Physiol. **18**, 967 (1963).

— — Note on plasma magnesium in the cold-acclimated rat subjected to hypothermia. Canad. J. Physiol. Pharm. **42**, 679 (1964).

NUGARA, D , and H. M. EDWARDS jr.· Influence of dietary Ca and P levels in the Mg requirement of the chick. J Nutr. **80**, 181 (1963).

O'DELL, B. L.: Magnesium requirement and its relation to other dietary constituents. Fed. Proc. **19**, 648 (1960).

— E. R. MORRIS, and W. O. REGAN. Magnesium requirement of guinea pigs and rats. effects of calcium and phosphorus and symptoms of magnesium deficiency. J. Nutr. **70**, 103 (1960)

OGASAWARA, K.· Inorganic substances in the animal serum I. Total and ionized calcium and total ionized magnesium in the serum. Igaku to Seibutsugaku **29**, 250 (1953).

OLSON, E. J., and H. E. PARKER. Effects of dietary cholesterol on skin lesions of rats with subacute magnesium deficiencies. J. Nutr. **83**, 73 (1964)

OPIE, L. H., B. G. HUNT, and J. M. FINLAY· Massive small bowel resection with malabsorption and negative magnesium balance. Gastroenterology **47**, 415 (1964)

OPPELT, W. W , I. MACINTYRE, and D. P. RALL· Magnesium exchange between blood and cerebrospinal fluid. Amer. J. Physiol. **205**, 959 (1963).

ORANGE, M., and H. C. RHEIN. Microestimation of magnesium in body fluids. J. biol. Chem. **189**, 379 (1951)

ORENT, E R., H. D. KRUSE, and E. V. MCCOLLUM Studies on magnesium deficiency in animals. II Species variation in symptomatology of Mg deprivation. Amer J Physiol. **101**, 454 (1932).

OUTA, T.: Studies on interrelationships between serum magnesium, calcium and cardiovascular disorders. Fukuoka Acta med. **54**, 1208 (1963).

OYAERT, W.. Einfluß von Kalium-Belastung auf den Magnesium-Stoffwechsel. Berl. Munch. tierarztl. Wschr **75**, 323 (1962).

PALLIS, C , I. MACINTYRE, and H ANSTALL Some observations on magnesium in cerebrospinal fluid. J. clin. Path. **18**, 762 (1965).

PARLIER, R , D HIOCO et R. LEBLANC: II. Les troubles du métabolisme magnésien. Symptomes et traitement des carences et des pléthores magnésiennes. Rev. franç. Endocr clin **4**, 335 (1963).

PARR, W. H.· Hypomagnesaemic tetany in calves fed on milk diets Vet Rec. **69**, 71 (1957)

PARSONS, R. S., T. BUTLER, and E. P. SELLARS. Coronary artery disease — further investigation on its treatment with parenteral magnesium sulphate and incorporating minimal doses of heparin. Med Proc. **6**, 479 (1960).

PAYNE, J. M., and J. CHAMINGS. The effect of thyro-parathyroidectomy in the goat with particular respect to clinical effect and changes in the concentrations of plasma calcium, inorganic phosphate and magnesium J. Endocr. **29**, 19 (1964).

PENGELLEY, E. T., and R. R J. CHAFFEE Changes in plasma magnesium concentration during hibernation in the golden-mantled ground squirrel (Citellus lateralis). Comp. Biochem Physiol **17**, 673 (1966).

PETERS, C. J , and M. WALSER Transport of cations by rabbit gall bladder· evidence suggesting a common cation pump. Amer. J Physiol **240**, 677 (1966)

PETERSEN, V. P : Metabolic studies in clinical magnesium deficiency. Acta med. scand. **173**, 285 (1963a).

— Potassium and magnesium turnover in magnesium deficiency Acta med scand. **174**, 595 (1963b)

PHILLIPSON, A. T., and J. E. STORRY. The absorption of calcium and magnesium from the rumen and small intestine of the sheep J. Physiol. (Lond.) **181**, 130 (1965).

PLANCHART, A. Potentiation of insulin action by calcium and magnesium (using the mouse convulsion bioassay method) Diabetes **14**, 430 (1965)

PLATNER, W. S Effects of low temperature on magnesium content of blood, body fluids and tissues of goldfish and turtle Amer. J. Physiol. **161**, 399 (1950).

—, and M G. HOSKO Mobility of serum magnesium in hypothermia. Amer. J. Physiol. **174**, 273 (1953).

POLLACK, S., J. N. GEORGE, R. C. REBA, R M KAUFMAN, and W. H COSBY. The absorption of nonferrous metals in iron deficiency J. clin. Invest. **44**, 1470 (1965).

POSNER, A. S The nature of the inorganic phase in calcified tissues In Calcification in biological systems (R F. SOGNNAES, ed) Washington, D C Amer. Ass Advance. Sci., 1960

POTTS, J T. jr., and B. ROBERTS Clinical significance of magnesium deficiency and its relation to parathyroid disease. Amer. J. med. Sci. **235**, 206 (1958).

POUTSIAKA, J. W., H MADISSOO, L G. MILLSTEIN, and J. KIRPAN Effects of benzydroflumethiazide (Naturetin) on the renal excretion of calcium and magnesium by dogs. Toxicol. appl. Pharmacol. **3**, 455 (1961).

PRASAD, A S , E. B. FLINK, and R MCCOLLISTER Ultrafiltration studies on serum magnesium in normal and diseased states J Lab clin. Med **58**, 531 (1961).

—, —, and H. H ZINNEMAN. The base binding property of the serum proteins with respect to magnesium J. Lab. clin Med. **54**, 357 (1959).

PRETORIUS, P. T., A. S. WEHMEYER, and J. J THERON Magnesium balance studies in South African Bantu children with kwashiorkor. Amer. J. clin. Nutr. **13**, 331 (1963).

PRITCHARD, J. A.: The use of the magnesium ion in the management of eclamptogenic toxemias Surg. Gynec. Obstet **100**, 131 (1955).

PYTKOWICZ, R. M., I. W. DUEDALL, and D N. CONNORS Magnesium ions: activity in seawater Science **152**, 640 (1966).

QUINN, M., D. E BASS, and C R. LEEMAN. Effect of acute cold exposure on serum potassium and magnesium and the electrocardiogram in man. Proc. Soc. exp. Biol. (N.Y.) **83**, 660 (1953).

QUINN, P. J., I G. WHITE, and B. R. WIRRICK. Studies of the distribution of the major cations in semen and male accessory secretions J. Reprod. Fertil. **10**, 379 (1965).

RAAFLAUB, J.: Chemical complex basis of urinary calculus formation. Helv. med. Acta **30**, 724 (1963).

RADEMEYER, L. J., and J. BOOYENS The effects of variations in the fat and carbohydrate content of the diet on the levels of magnesium and cholesterol in the serum of white rats Brit J. Nutr. **19**, 153 (1965).

RANDALL, R. E. jr., M. D. COHEN, C. C. SPRAY jr., and E. C. ROSSMEISL Hypermagnesemia in renal failure etiology and toxic manifestations. Ann. intern. Med. **61**, 73 (1964).

— E. C. ROSSMEISL, and K H. BLEIFER· Magnesium depletion in man. Ann. intern. Med. **50**, 257 (1959)

RANDOIN, L., J. CAUPERET, D. HUGOT, and G. MOREL Influence de divers sels magnésiens administrés par voie orale sur la retention du calcium dans l'organisme. Bull. Soc. chim. Biol. **34**, 1159 (1952).

RASMUSSEN, H Mitochondrial ion transport mechanism and physiological significance Fed Proc. **25**, 903 (1966).

RAY, S. N. jr., and D. N. MULLICK Oxygen consumption and serum magnesium concentration in calves fed milk diet or standard calf ration. Indian J exp Biol. **3**, 121 (1965).

RAYNAUD, C.· Renal excretion of magnesium in the rabbit. Amer. J. Physiol. **203**, 649 (1962).

REGEN, D. M., D. A B. YOUNG, W. W DAVIS, J. JACK jr., and C. R PARK. Adjustment of glycolysis to energy utilization in perfused rat heart. The effect of changes in the ionic composition of the medium on phosphofructokinase activity. J. biol. Chem. **239**, 381 (1964).

RENDI, R, and M. L. UHR Sodium, potassium-requiring adenosine triphosphatase activity Biochim. biophys Acta (Amst) **89**, 520 (1964).

REVEL, M., and H H. HIATT· Magnesium requirement for the formation of an active messenger RNA-ribosome-S-RNA complex J. molec Biol. **11**, 467 (1965).

RICHARDSON, J. A., W. D. HUFFINES, and L. G. WELT: The effect of coincident hypercalcemia and potassium depletion on the rat kidney. Metabolism **12**, 560 (1963).

—, and L G. WELT· The hypercalcemia of magnesium depletion. Clin. Res. **11**, 68 (1963)

— — The hypomagnesemia of vitamin D administration. Proc Soc exp. Biol (N Y) **118**, 512 (1965).

RIEDESEL, M L., and G. E. FOLK· Serum magnesium and hibernation. Fed. Proc **15**, 151 (1956)

RIGÓ, J, u I. SZELÉNYI Die Wirkung von Magnesium auf die neurogene und alimentare Hypertonie bei Ratten. Acta physiol hung **24**, 253 (1963).

RITCHIE, D. B Surface enamel magnesium and its possible relation to incidence of caries. Nature (Lond.) **190**, 456 (1961).

RITCHIE, N. S, and R. G. HEMINGWAY Failure of lactating ewes with low plasma magnesium values to respond to large daily magnesium supplements. J. agric Sci **60**, 305 (1963a).

Ritchie, N S, and R G Hemingway: Effects of conventional daily magnesium supplementation, breed of ewe and continued potassium fertilizer applications on plasma magnesium and calcium levels of ewes. J agric. Sci. **61**, 411 (1963 b)

— — J. S S Inglis, and R M. Peacock Experimental production of hypomagnesaemia in ewes and its control by small magnesium supplement J agric Sci. **58**, 399 (1962).

Rizek, J. A., A Dimich, and S. Wallach. Plasma and erythrocyte magnesium in thyroid disease. J. clin. Endocr. **25**, 350 (1965).

Roberts, B., J. J. Murphy, L. Miller, and O. Rosenthal The effect of parathyroid hormone upon serum levels and urinary excretion of magnesium. Surg. Forum **5**, 509 (1954).

Robertson, A, H Pave, P. Barden, and T. G. Marr Fasting metabolism of the lactating cow. Res. Vet. Sci. **1**, 117 (1960).

Robinson, R R, H. V. Murdaugh jr., and E. Peschel Renal excretion of magnesium and the renal factors responsible for hypermagnesemia of renal disease. Clin. Res. **7**, 162 (1959a).

— — — Renal factors responsible for the hypermagnesemia of renal disease. J. Lab. clin Med **53**, 572 (1959b).

—, and R. M. Portwood Mechanism of Mg excretion by the chicken Amer. J. Physiol. **202**, 309 (1962).

Rogers, T. A · The exchange of radioactive Mg in erythrocytes of several species. J. cell. comp Physiol. **57**, 119 (1961).

— The metabolism of magnesium. In Radioisotopes in animal nutrition and physiology. Vienna International Atomic Energy Agency, 1965.

— F. L. Haven, and P. E. Mahan Exchange of radioactive magnesium in Walker carcinosarcoma 256· A note. J. nat. Cancer Inst **25**, 887 (1960).

—, and P. E. Mahan Exchange of radioactive magnesium in the rat. Proc. Soc exp Biol (N Y.) **100**, 235 (1959).

— M. G. Simesen, T. Lunaas, and J. R. Luick The exchange of radioactive magnesium in the tissues of the cow, calf and fetus. Acta vet scand **5**, 209 (1964).

Romenski, N. W. Sur la composition minérale du système nerveux central chez les lapins, dans la rage expérimentale Ann. Inst. Pasteur, Suppl **189**, 181 (1935).

Rook, J. A. F : Rapid development of hypomagnesaemia in lactating cows given artificial rations low in magnesium Nature (Lond) **191**, 1019 (1961).

—, and C. C. Balch Magnesium metabolism in the dairy cow. II. Metabolism during the spring grazing season. J agric. Sci. **51**, 199 (1958)

—, and J. E. Storry Orally and parenterally administered magnesium in the control of hypomagnesaemia in grazing cows. Proc Nutr. Soc. **21**, xl (1962a).

— — Magnesium in the nutrition of farm animals. Nutr. Abstr. Rev. **32**, 1055 (1962b).

Ross, D. B.. Influence of sodium on the transport of magnesium across the intestinal wall of the rat *in vitro*. Nature (Lond) **189**, 840 (1961).

— *In vitro* studies on the transport of magnesium across the intestinal wall of the rat. J. Physiol (Lond.) **160**, 417 (1962).

—, and A. D. Care The movement of $^{28}Mg^{+2}$ across the cell wall of guinea pig small intestine *in vitro*. Biochem. J. **82**, 21 P (1962)

Rosselle, N., et K De Doncker Étude expérimentale et clinique d'une tétanie larvée Wld Neurol. **2**, 908 (1961).

Rothstein, A. Interrelationships between the ion transporting systems of the yeast cell. In Membrane transport and metabolism (A. Kleinzeller and A. Kotyk, eds), p 270 London Academic Press 1961.

Rubin, M. I., and E. T. Krick The salt and water metabolism of adrenal insufficiency, and partial starvation in rats J. clin. Invest. **15**, 685 (1936)

Sabour, M S, S. Hanna, and M. K. MacDonald The nephropathy of experimental magnesium deficiency an electron microscopic study. Quart. J. exp. Physiol. **49**, 314 (1964)

SAKANOUE, M.. Biochemical studies on the preservation medium for donor eyes. 6. Analysis of Na, K, Ca, Mg, Cl in newly formed ciliary body fluid, posterior chamber fluid, and anterior chamber fluid compared with plasma ultrafiltrate. Jap. J. clin. Ophthal. **18**, 1135 (1964)

SAMIY, A. H. E., J. L. BROWN, and D. L. GLOBUS Effects of magnesium and calcium loading on renal excretion of electrolytes in dogs. Amer. J. Physiol. **198**, 595 (1960).

— — — R. H. KESSLER, and D. D. THOMPSON· Interrelation between renal transport systems of magnesium and calcium. Amer. J. Physiol. **198**, 599 (1960).

SANUI, H., and N. PACE: Mass law effects of adenosine triphosphate on Na, K, Mg and Ca binding by rat liver microsomes. J cell. comp. Physiol. **65**, 27 (1965).

SASAKI, T., M. NAKATANI, and M. NAKAMURA. The rate of respiration and incorporation of P^{32} into adenosine triphosphate by rat aorta tissue. Jap Circulat. J. **29**, 1 (1965).

SAUBERLICH, H. E., G. E. BUNCE, C. A. MOORE, and O. G. STONINGTON Oral magnesium administration in the treatment of renal calculus formation. Amer. J clin. Nutr. **14**, 240 (1964).

SAVILLE, P. D , and C. S. LIEBER. Effect of alcohol on growth, bone density and muscle magnesium in the rat. J. Nutr. **87**, 477 (1965).

SCHACHTER, D. The fluorometric estimation of magnesium in serum and in urine. J. Lab. clin Med. **54**, 763 (1959).

— E. B. DOWDLE, and H. SCHENKER Active transport of calcium by the small intestine of the rat. Amer. J Physiol **198**, 263 (1960)

—, and S M ROSEN· Active transport of Ca^{45} by the small intestine and its dependence on vitamin D^1. Amer. J. Physiol. **196**, 357 (1959).

SCHAIN, R. J.· Cerebrospinal fluid and serum cation levels. Arch. Neurol (Chic) **11**, 330 (1964a).

— Cerebrospinal fluid and blood electrolytes in 62 mentally defective infants and children. J Pediat. **65**, 422 (1964b).

SCHIMPF, K., u. H. HARTER. Zur Frage der Magnesium-Wirkung auf die Blutgerinnung. Klin. Wschr. **35**, 50 (1957)

SCHMID, E , M. V BUBNOFF, U WAGENMANN u. R. RAUGNER Zur Kreislaufwirkung der Magnesiumsalze. Naunyn-Schmiedebergs Arch exp. Path. Pharmak. **224**, 426 (1955).

SCHMIDT, H. D., J. SCHMIER u. S. SCHMITZ. Chrontrope Wirkung von Calcium und Magnesium am isolierten Hundeherzen. Pflugers Arch. ges. Physiol. **284**, 316 (1965).

SCHNEEBERGER, E. E., and A. B. MORRISON. Nephropathy of experimental magnesium deficiency. Clin. Res. **12**, 472 (1964)

— — The nephropathy of experimental magnesium deficiency light and electron microscope investigations. Lab Invest. **14**, 674 (1965).

SCHOLTZ, H. G. Notiz uber die Wirkung des Parathyreoidea-Hormons auf den Magnesiumgehalt des Blutes. Naunyn-Schmiedebergs Arch. exp. Path. Pharmak. **159**, 233 (1931).

SCHOLZ, R W , and W. R FEATHERSTON. Influence of lactose on magnesium utilization by the chick. Fed. Proc. **25**, 609 (1966).

SCHRADER, C. A., C. O. PRICKETT, and W. D. SALMON· Symptomatology and pathology of potassium and magnesium deficiencies in the rat. J. Nutr. **14**, 85 (1937).

SCHWAB, R. S., A. PORYALI, and A. AMES III Normal serum magnesium levels in Parkinson's disease. Neurology (Minneap.) **14**, 855 (1964).

SCOTT, D · Factors influencing the secretion and absorption of calcium and magnesium in the small intestine of sheep. Quart. J. exp. Physiol. **50**, 312 (1965).

—, and A DOBSON Aldosterone and the metabolism of magnesium and other minerals in the sheep. Quart. J. exp. Physiol **50**, 42 (1965).

SCOTT, J. B., E. D. FROHLICH, R. A. HARDIN, and F. J. HADDY. Na^+, K^+, Ca^{++}, and Mg^{++} action on coronary vascular resistance in the dog heart Amer. J. Physiol. **201**, 1095 (1961).

SCOULAR, F. I., J. K. PACE, and A. N. DAVIS The calcium, phosphorus and magnesium balances of young college women consuming self-selected diets. J. Nutr. **62**, 489 (1957).

SEELIG, M. S.. The requirement of magnesium by the normal adult. Summary and analysis of published data. Amer. J. clin Nutr **14**, 342 (1964).

SELLER, R. H., O. RAMIREZ, H. SOLLER, A. N. BREST, and J. H. MOVER Serum and erythrocyte magnesium levels in congestive heart failure Pharmacologist **8**, 222 (1966).

— O. RAMIREZ-MUXO, and A. N. BREST. Magnesium metabolism in hypertension. J. Amer. med. Ass. **191**, 654 (1965).

SELYE, H.. Prophylactic treatment of an experimental arteriosclerosis with Mg and K salts. Amer. Heart J. **55**, 805 (1958).

SETA, K., E. E. HELLERSTEIN, and J J. VITALE Myocardium and plasma electrolytes in dietary magnesium and potassium deficiency in the rat. J Nutr. **87**, 179 (1965).

— R KLEIGER, E. E. HELLERSTEIN, B. LOWN, and J. J. VITALE. Effect of potassium and magnesium deficiency on the electrocardiogram and plasma electrolytes of purebred beagles. Amer. J. Cardiol. **17**, 516 (1966).

SHILS, M. E.: Experimental human magnesium depletion. I. Clinical observations and blood chemistry alterations. Amer. J. clin Nutr. **15**, 133 (1964).

— Species differences in electrolytes in magnesium deficiency. Fed. Proc. **25**, 609 (1966).

SHIMAMOTO, T., T. FUJITA, H. SHIMURA, H YAMAZAKI, S. IWAHARA, and G. YAJIMA. Myocardial infarct-like lesions and arteriosclerosis induced by high molecular substances and prevention by magnesium salt. Amer. Heart J. **57**, 273 (1959)

SILLÉN, L. G., and A E. MARTELL Stability constants of metal-ion complexes. Special Publication No. 17, Chemical Society, London, 1964.

SILVER, L., J. S. ROBERTSON, and L. K. DAHL. Magnesium turnover in the human studied with Mg^{28}. J. clin. Invest **39**, 420 (1960).

SILVERMAN, S. H., and L. I. GARDNER Ultrafiltration studies on serum magnesium. New Eng. J. Med. **250**, 938 (1954).

SIMESEN, M. A.: Proc. Int. Vet Congr. XXII, Hanover, 1963, cited by CARE, 1965.

SJOLLEMA, B.: Nutritional and metabolic disorders in cattle. Nutr. Abstr. Rev. **1**, 621 (1932).

SKOU, J. C · The influence of some cations on an adenosine triphosphatase from peripheral nerves. Biochim. biophys Acta (Amst) **23**, 394 (1957)

— Enzymatic aspects of active linked transport of Na^+ and K^+ through the cell membrane. Progr. Biophys. **14**, 131 (1964).

SMITH, B. S. W.: The relationship between endogenous loss of magnesium and faecal output in the rat. Proc. Nutr. Soc. **25**, xxxvii (1966).

SMITH, H L., R L. FISCHER, and J. N. ETTELDORF Magnesium and calcium in human muscular dystrophy. Amer. J. Dis. Child. **103**, 771 (1962).

SMITH, P. K., A. W. WINKLER, and H E. HOFF. Electrocardiographic changes and concentration of magnesium in serum following intravenous injection of magnesium salts. Amer. J. Physiol. **126**, 720 (1939).

— — — The pharmacological actions of parenterally administered magnesium salts A review. Anesthesiology **3**, 323 (1942).

— —, and B. M SCHWARTZ. The distribution of magnesium following the parenteral administration of magnesium sulfate. J. biol. Chem. **129**, 51 (1939).

SMITH, R. H.. Calcium and magnesium metabolism in calves Plasma levels and retention in milk-fed calves. Biochem. J. **67**, 472 (1957).

— Calcium and magnesium metabolism in calves 2. Effect of dietary vitamin D and ultraviolet irradiation on milk-fed calves. Biochem. J. **70**, 201 (1958).

— Absorption of magnesium in the large intestine of the calf. Nature (Lond.) **184**, 821 (1959a).

— Calcium and magnesium in calves. 4. Bone composition in magnesium deficiency and the control of plasma magnesium. Biochem. J. **71**, 609 (1959b).

— The development and function of the rumen in milk-fed calves. J. agric. Sci. **52**, 72 (1959c).

— Effect of the ingestion of wood shavings on magnesium and calcium utilization by milk-fed calves. J. agric. Sci. **56**, 343 (1961a).

SMITH, R H: Importance of magnesium in the control of plasma calcium in the calf. Nature (Lond.) **191**, 181 (1961 b).

— Net exchange of certain inorganic ions and water in the alimentary tract of the milk-fed calf. Biochem. J. **83**, 151 (1962).

— Small intestine transit time and magnesium absorption in the calf Nature (Lond.) **198**, 161 (1963).

— Passage of digesta through the calf abomasum and small intestine J. Physiol. (Lond.) **172**, 305 (1964a).

— Hypomagnesaemia in calves. Rep. III. Intern. Meeting Diseases of Cattle. Copenhagen 1964(b)

—, and A. B. McALLAN. Binding of magnesium and calcium in the contents of the small intestine of the calf. Brit. J. Nutr. **20**, 703 (1966).

SMITH, R S W , and A C. FIELD· Effect of age on magnesium deficiency in rats Brit. J Nutr. **17**, 591 (1963)

SMITH, W. O.: Magnesium deficiency in the surgical patient. Amer. J. Cardiol. **13**, 667 (1963).

— D. J. BAXTER, A. LINDNER, and H. E. GINN: Effects of magnesium depletion on renal function in the rat J. Lab clin. Med. **59**, 211 (1962).

—, and J. F. HAMMARSTEN· Serum magnesium in renal diseases. Arch intern. Med. **102**, 5 (1958)

— — Intracellular magnesium in delirium tremens and uremia. Amer J med. Sci. **237**, 413 (1959).

— —, and L. P. ELIEL: Clinical expression of magnesium deficiency. J. Amer. med. Ass. **174**, 97 (1960).

— A. A. KYRIAKOPOULOS, and J F. HAMMARSTEN: Magnesium depletion induced by various diuretics. J. Okla med Ass. **55**, 248 (1962).

SOFFER, L. J., C. COHN, E. B GROSSMAN, M. JACOBS, and H. SABOTKA Magnesium partition studies in Grave's disease and in clinical and experimental hypothyroidism. J. clin. Invest **20**, 429 (1941).

SOMLYO, A. V., C. WOO, and A. P. SOMLYO Effect of magnesium on posterior pituitary hormone action on vascular smooth muscle. Amer. J. Physiol. **210**, 705 (1966).

SOMOGYI, J.: Preparation of the Na^+, K^+-activated ATPase system of rat brain free from the Mg^{++}-activated ATP hydrolyzing enzyme. Biochim. biophys. Acta (Amst.) **92**, 615 (1964).

SPERELAKIS, N.: Contraction of depolarized smooth muscle by electric fields. Amer. J. Physiol. **202**, 731 (1962).

SPRAY, C. M. A study of some aspects of reproduction by means of chemical analysis. Brit J. Nutr. **4**, 354 (1950).

STANBURY, J. B.· The blocking action of magnesium ion on sympathetic ganglion. J Pharmacol. exp. Ther. **93**, 52 (1948).

—, and A. FARAH· Effects of the magnesium ion on the heart and on its response to digoxin. J. Pharmacol. exp. Ther. **100**, 445 (1950)

STEELE, T. H., M. A. EVENSON, S. F. WEN, and R. E. REISELBACH: Increased capacity for magnesium excretion by residual nephrons of the chronically diseased kidney. Clin. Res. **14**, 388 (1966).

STEWART, C. P , and S C. FRAZER· Magnesium. Advanc. clin. Chem. **6**, 29 (1963).

STEWART, J., and E. W. MOODIE: The absorption of magnesium from the alimentary tract of sheep. J. comp. Path. **66**, 10 (1956).

STILES, D E , J. G BATSAKIS, and G. C. HARDY Intracellular (erythrocytic) magnesium. estimation by fluorescent analysis. Amer. J. clin. Path. **44**, 82 (1965).

STONER, H B Effect of magnesium on P^{31} and P^{32} partition in muscle. Amer. J. Physiol. **161**, 387 (1950).

STORRY, J. E : Calcium and magnesium content of various secretions entering the digestive tract of sheep. Nature (Lond) **190**, 1197 (1961a).

Storry, J E.: Studies on Ca and Mg in the alimentary tract of sheep. I. The distribution of Ca and Mg in the contents taken from various parts of the alimentary tract. J. agric. Sci 57, 97 (1961 b).
— Studies on Ca and Mg in the alimentary tract of sheep. II. The effect of reducing the acidity of abomasal digesta *in vitro* on the distribution of Ca and Mg. J agric. Sci. 57, 103 (1961 c).
— Changes in blood constituents which occur in dairy cattle transferred to spring pastures Res Vet. Sci. 2, 272 (1961 d).
—, and J. A. F Rook. Magnesium metabolism in the dairy cow. V. Experimental observations with a purified diet low in magnesium. J. agric. Sci. 61, 167 (1963).
Striebel, A , und H Baur. Calcium-Magnesium-Ausscheidung im Urin von Gesunden und Krebskranken. Schweiz. med. Wschr 84, 1082 (1954).
Stutzman, F. L., and D. S. Amatuzio. A study of serum and spinal fluid calcium and magnesium in normal humans. Arch. Biochem. 39, 271 (1952).
—, and D. S. Amatuzio Blood serum magnesium in portal cirrhosis and diabetes mellitus J Lab clin Med 41, 215 (1953)
Sullivan, J., H G Lankford, M J. Swartz, and C. Farrell Magnesium metabolism in alcoholism. Amer J. clin. Nutr. 13, 297 (1963)
— P. Robertson, and J Shehan Magnesium and lactate excretion in alcoholic intoxication. Clin. Res. 12, 358 (1964).
Sullivan, J F , H. G Lankford, and P. Robertson Renal excretion of lactate and magnesium in alcoholism Amer J. clin Nutr 18, 231 (1966).
Suomalainen, P. Hibernation of the hedgehog VI. Serum magnesium and calcium. Artificial hibernation Chemical physiology of diurnal sleep. Ann. Acad. Sci. fenn. A 53, No. 7 (1939).
Surawicz, B., E. Lepeschkin, and H. C Herrlich Low and high magnesium concentrations at various calcium levels Effect on the monophasic action potential, electrocardiogram, and contractility of isolated rabbit hearts Circulat Res. 9, 811 (1961).
Suter, C , and W. D. Klingman Neurologic manifestations of magnesium depletion states. Neurology (Minneap) 5, 691 (1955)
Syllm-Rapoport, I , u. I. Strassburger Über den experimentellen Magnesium-Mangel beim Hund I Mitteilung Allgemeine Symptome, Gewebsverkalkungen und chemische Veranderungen im Blutserum. Acta med. biol. germ 1, 141 (1958).
Szekely, P , and N A. Wynn The effects of magnesium on cardiac arrhythmias caused by digitalis. Clin. Sci. 10, 241 (1951).
Takayasu, H., S. Sato, H. Yanadori, and T Hirata Serum magnesium level in acute renal failure Acta med. biol. (Nugata) 10, 117 (1962)
Tapley, D F.. Magnesium balance in myxedematous patients treated with triiodothyronine Bull Johns Hopkins Hosp 96, 274 (1955)
—, and C. Cooper The effect of thyroxine and related compounds on oxidative phosphorylation J. biol Chem 222, 341 (1956)
Taves, D. R., and W. F. Neuman Factors controlling calcification *in vitro* Fluoride and magnesium. Arch. Biochem 108, 390 (1964)
Taylor, T G The magnesium of bone mineral J. agric Sci 52, 207 (1959)
Terkildsen, T. C The effect of physostigmine, neostigmine and acetylcholine on serum magnesium concentration Acta pharmacol. (Kbh) 8, 385 (1952a).
— Serum magnesium after oral administration of magnesium salts. Acta pharmacol. (Kbh) 8, 374 (1952b)
Thiers, R E , and B. L. Vallee Distribution of metals in subcellular fractions of rat liver. J. biol. Chem 226, 911 (1957).
Thoren, L. Magnesium deficiency in gastrointestinal fluid loss. Acta clin scand., Suppl. 306 (1963).
Tibbetts, D. M , and J. C. Aub Magnesium metabolism in health and disease. I. The magnesium and calcium excretion of normal individuals, also the effects of magnesium, chloride and phosphate ions J. clin. Invest. 16, 491 (1937a).

TIBBETTS, D M , and J. C. AUB: Magnesium metabolism in health and disease. II. The effect of the parathyroid hormone J. clin Invest. **16**, 503 (1937 b).

TIDBALL, C. S.· Magnesium and calcium as regulators of intestinal permeability Amer. J. Physiol. **206**, 243 (1964).

TIPTON, I. H., and M. J. COOK Trace elements in human tissue. Part II. Adult subjects from the United States Hlth Phys. **9**, 103 (1963).

—, and J J SHAFER Statistical analysis of lung trace element levels Arch. environm. Hlth **8**, 58 (1964).

— H. A. SCHROEDER, H. M. PERRY, and M. J. COOK· Trace elements in human tissue. Part III. Subjects from Africa, the Near and Far East, and Europe Hlth Phys. **11**, 403 (1965).

TOVERUD, K U , and G TOVERUD Studies on mineral metabolism during pregnancy and lactation and its bearing on disposition to rickets and dental caries. Acta paediat. (Stockh.) **12**, Suppl II, 1 (1931)

TUFTS, E. V., and D. M. GREENBERG. Calcium involvement in magnesium deficiency. Proc. Soc. exp. Biol. (N.Y.) **34**, 292 (1936).

— — The biochemistry of magnesium deficiency. I. Chemical changes resulting from magnesium deprivation. J. biol. Chem **122**, 693 (1938 a).

— — The biochemistry of magnesium deficiency. II The minimum magnesium requirement for growth, gestation, and lactation, and the effect of the dietary calcium level thereon J biol. Chem. **122**, 715 (1938 b).

ULLRICH, K. J , u K. H. JARAUSCH· Untersuchungen zum Problem der Harnkonzentrierung und Harnverdunnung, uber die Verteilung von Elektrolyten (Na, K, Mg, Ca, Cl, anorganischem Phosphat), Harnstoff, Aminosauren und exogenem Kreatinin in Rinde und Mark der Hundeniere bei verschiedenen Diuresezustanden. Pflugers Arch ges. Physiol **262**, 537 (1956).

ULRICH, F.. Kinetic studies of the activation of mitochondrial adenosine triphosphatase by Mg^{++}. J. biol. Chem. **239**, 3532 (1964).

UNGLAUB, I., I. SYLLM-RAPOPORT u. I. STRASSBURGER· Pathologisch-anatomische Befunde bei experimentellem Magnesiummangel des Hundes. Virchows Arch Path. Anat **332**, 122 (1959).

VAGNE, M., F. MARTIN et R. LAMBERT Variations du taux de calcium et de magnésium dans le suc gastrique humain. C.R. Soc. Biol. (Paris) **158**, 1665 (1964).

VALBERG, L. S , R. T. CARD, E. J. PAULSON, and J. SZIVEK The metal composition of erythrocytes in different species and its relationship to the lifespan of the cells in the circulation Comp Biochem. Physiol. **15**, 347 (1965 a).

— J. M. HOLT, G. M. BROWN, J SZIVEK, and E. PAULSON· Sodium, potassium, calcium, magnesium, copper and zinc composition of erythrocytes in vitamin B_{12} deficiency and iron deficiency. J clin. Invest. **44**, 1225 (1965 b).

— — E. PAULSON, and J SZIVEK: Spectrochemical analysis of sodium, potassium, calcium, magnesium, copper and zinc in normal human erythrocytes J clin. Invest. **44**, 379 (1965 c).

VALLEE, B L., W E C. WACKER, and D D ULMER The magnesium-deficiency tetany syndrome in man New Engl. J. Med. **262**, 155 (1960).

VERMEULEN, C. W., E. S. LYON, and G H MILLER Calcium phosphate solubility in urine as measured by a precipitation test experimental urolithiasis. XIII. J. Urol. (Baltimore) **79**, 596 (1958).

VITALE, J. J., D. M HEGSTED, M NAKAMURA, and P. CONNORS The effect of thyroxine on magnesium requirement. J. biol. Chem. **226**, 597 (1957 a).

— — —, and P. S CONNORS The effect of magnesium deficiency on oxidative phosphorylation. J. biol. Chem **228**, 573 (1957 b).

— E. E. HELLERSTEIN, D M. HEGSTED, M NAKAMURA, and A. FARBMAN Studies on the interrelationships between dietary magnesium and calcium in atherogenesis and renal lesions. Amer. J. clin. Nutr. **7**, 13 (1959).

VITALE, J. J , E E HELLERSTEIN, M. NAKAMURA, and B. LOWN. Effects of magnesium-deficient diet upon puppies. Circulat. Res. 9, 387 (1961).
— H. KOIDE, and E. E. HELLERSTEIN. The effect of magnesium and/or potassium deficiency on the response of Rhesus monkeys to acetylstrophanthidin (AS) and cholesterol feeding. Fed. Proc. 25, 623 (1966).
— M. NAKAMURA, and D M HEGSTED The effect of magnesium deficiency on oxidative phosphorylation. J. biol. Chem. 228, 573 (1957).
— H. VELEZ, C GUZMAN, and P. CORREA Magnesium deficiency in the cebus monkey. Circulat. Res. 12, 642 (1963).
— P. L. WHITE, M. NAKAMURA, D. M. HEGSTED, N. ZAMBRECK, and E. E HELLERSTEIN Interrelationships between experimental hypercholesteremia, magnesium requirement and experimental atherosclerosis. J. exp. Med 106, 757 (1957c)
VOSTAL, J., O. KUCHEL u. V. PACOVSKY. Tubulare Sekretion von Mg beim Menschen. Klin. Wschr. 41, 537 (1963)
WACKER, W E. C.: The effect of hydrochlorothiazide on magnesium excretion J. clin. Invest. 40, 1086 (1961).
— C. IIDA, and K FUWA Accuracy of determinations of serum magnesium by flame emission and atomic absorption spectrometry. Nature (Lond) 202, 659 (1964).
— F. D. MOORE, D. D. ULMER, and B L. VALLEE· Normocalcemic magnesium deficiency tetany. J. Amer. med. Ass. 180, 161 (1962).
—, and B. L VALLEE A study of magnesium metabolism in acute renal failure employing a multichannel flame spectrometer. New Engl J. Med. 257, 1254 (1957).
— — Magnesium metabolism. New Engl. J. Med 259, 431 (1958a).
— — Magnesium metabolism. New Engl. J Med. 259, 475 (1958b).
— — Magnesium. In: Mineral metabolism (C L COMAR and F BRONNER, eds.), vol II, Part A, p 483. New York: Academic Press 1964.
WALAAS, O.. The effect of estrogenic hormone on the content of calcium and magnesium in uterus. Acta physiol scand 21, 27 (1950).
WALKER, A. R P , F W. FOX, and J. T. IRVING· The effect of bread rich in phytate phosphorus on the metabolism of certain mineral salts with special reference to calcium. Biochem. J. 42, 452 (1948).
WALLACH, S , J. V. BELLAVIA, and P. J. GAMPONIA Cellular magnesium transport in hypermagnesemia. Clin. Res. 13, 239 (1965)
— — J. SCHORR, and D L. REIZENSTEIN. Tissue distribution of electrolytes, Ca47, and Mg28 in acute hypercalcemia. Amer. J. Physiol. 207, 553 (1964).
— — —, and A. SCHAFFER: Tissue distribution of electrolytes, Ca47 and Mg28 in experimental hyper- and hypoparathyroidism. Endocrinology 78, 16 (1965).
— L. N. CAHILL, F. H ROGAN, and H L. JONES Plasma and erythrocyte magnesium in health and disease. J. Lab. clin. Med. 59, 195 (1962).
—, and A. C. CARTER. Metabolic and renal effects of acute hypercalcemia in dogs Amer. J. Physiol. 200, 359 (1961).
WALSER, M · Determination of free magnesium ions in body fluids. Improved methods for free calcium ions, total calcium, and total magnesium Analyt. Chem 32, 711 (1960).
— Calcium clearance as a function of sodium clearance in the dog Amer. J. Physiol. 200, 1099 (1961a).
— Ion association. VI. Interactions between calcium, magnesium, inorganic phosphate, citrate and protein in normal human plasma J. clin. Invest 40, 723 (1961b).
— Renal discrimination between alkaline earth cations J. clin. Invest. 40, 1087 (1961c).
— Separate effects of hyperparathyroidism, hypercalcemia of malignancy, renal failure, and acidosis on the state of calcium, phosphate, and other ions in plasma. J. clin. Invest 41, 1454 (1962).
— Mathematical aspects of renal function· the dependence of solute reabsorption on water reabsorption and the mechanism of osmotic natriuresis. J. theor. Biol. 10, 307 (1966a).

WALSER, M : Mathematical aspects of renal function: reabsorption of individual solutes as interdependent processes. J theor. Biol. **10**, 327 (1966b).

—, and A. A. BROWDER Ion association. III. The effect of sulfate infusion on calcium excretion J. clin. Invest **38**, 1404 (1959).

—, and G. H. MUDGE Renal excretory mechanisms In· Mineral metabolism (C. L. CO-MAR and F. BRONNER, eds), vol 1, part A New York Academic Press 1960.

—, and V. E. NAHMOD Association of alkali metal and alkaline earth cations with sub-cellular particles prepared from renal cortex. Molec. Pharmac. **2**, 10 (1966).

— W. J. RAHILL, and B. H. B ROBINSON Interdependence among cations in tubular reabsorption Clin. Res. **11**, 411 (1963)

— B. H B. ROBINSON, and J. W. DUCKETT jr The hypercalcemia of adrenal insuffi-ciency. J. clin. Invest. **42**, 456 (1963).

—, and J. R. TROUNCE. The effect of diuresis and diuretics upon the renal tubular trans-port of alkaline earth cations. Biochem. Pharmacol. **8**, 157 (1961)

WASSERMAN, R. H. Studies on vitamin D_3 and intestinal absorption of calcium and other ions in rachitic chick. J Nutr **77**, 69 (1962).

WATANABE, S., T. SARGEANT, and M. ANGLETON Role of magnesium in contraction of glycerinated muscle fibers. Amer. J Physiol. **207**, 800 (1964).

WATCHORN, E., and R. A. McCANCE Subacute magnesium deficiency in rats. Biochem J. **31**, 1379 (1937).

WEIL, P., and D STATE The effect of the removal of major endocrine glands on the serum level of magnesium in dogs Surg Gynec. Obstet. **107**, 483 (1958).

WELT, L. G Experimental magnesium depletion Yale J. Biol. Med. **36**, 325 (1964).

—, and H. GITELMAN Disorders of magnesium metabolism. Disease-a-Month May (1965).

WENER, J , K. PINTAR, M. A. SIMON, R MOTOLA, R. FRIEDMAN, A. MAYMAN, and R. SCHUCHER· The effects of prolonged hypomagnesemia on the cardiovascular system in young dogs. Amer. Heart J. **67**, 221 (1964).

WESSON, L G , jr.· Magnesium, calcium, and phosphate excretion during osmotic diuresis in the dog. J Lab. clin. Med **60**, 422 (1962a)

— Organic mercurial effects on renal tubular reabsorption of calcium and magnesium and on phosphate excretion in the dog J. Lab. clin. Med. **59**, 630 (1962b).

— Electrolyte excretion in relation to diurnal cycles of renal function. Plasma electrolyte concentration and aldosterone secretion before and during salt and water balance changes in normotensive subjects. Medicine (Baltimore) **43**, 547 (1964).

—, and D. P. LAULER· Nephron reabsorptive site for calcium and magnesium in the dog. Proc. Soc. exp. Biol (N.Y.) **101**, 235 (1959).

WHANG, R., D WAGNER, and D. RODGERS The effect of intravenous insulin and glucose on serum Mg and K concentration. Clin Res. **14**, 390 (1966).

—, and L. G. WELT Observations in experimental magnesium depletion. J. clin. Invest. **42**, 305 (1963).

WIDDOWSON, E. M , and J. W T. DICKERSON. Chemical composition of the body. In. Mineral metabolism (C. L COMAR and F. BRONNER, eds), vol. 2, part A. New York. Academic Press 1964.

WILLE, P.. Über den Einfluß von Magnesium auf die Gerinnungsfunktionen während und nach der Operation Munch. med Wschr **102**, 1195 (1960).

WILLIS, M J , and F W. SUNDERMAN: Studies in serum electrolytes. XIX. Nomograms for calculating magnesium ion in serum and ultrafiltrates. J. biol. Chem. **197**, 343 (1952).

WILSON, A. A.: Magnesium homeostasis and hypomagnesemia in ruminants. Vet Rev. Annot. **6**, 39 (1960).

— Hypomagnesemia and magnesium metabolism. Vet. Rec. **76**, 1382 (1964).

WINKLER, A. W., K. SMITH, and H. E. HOFF. Intravenous magnesium sulfate in the treatment of nephritic convulsions in adults. J. clin. Invest. **21**, 207 (1942).

WISWELL, J. G.: Some effects of magnesium loading in patients with thyroid disorders. J. clin. Endocr. **21**, 31 (1961).

Wolf, A. V., and S. M. Ball· Effect of intravenous calcium salts or renal excretion in the dog. Amer. J. Physiol. **158**, 205 (1949).

Womersley, R. A.. Studies on the renal excretion of magnesium and other electrolytes. Clin. Sci. **15**, 465 (1956).

— Metabolic effects of prolonged intravenous administration of magnesium lactate to the normal human. J. Physiol. (Lond) **143**, 300 (1958).

Woodbury, D. M , C. Cheng, G. Sayers, and L. S. Goodman: Antagonism of adreno-corticotrophic hormone and adrenal cortical extract to desoxycorticosterone electrolytes and electroshock threshold. Amer J. Physiol **160**, 217 (1950).

Worker, N. A., and B. B. Migicovsky. Effect of vitamin D on the utilization of beryllium, magnesium, calcium, strontium and barium in the chick. J. Nutr. **74**, 490 (1961).

Wrong, O., A. Metcalfe-Gibson, R. B. I. Morrison, S. T. Ng, and A. V. Howard: *In vivo* dialysis of faeces as a method of stool analysis. I Technique and results in normal subjects Clin. Sci. **28**, 357 (1965).

Yanadori, H.: Clinical and experimental study of renal insufficiency changes of serum electrolyte composition, with special reference to magnesium III. Studies on anuria. Jap J. Urol. **53**, 367 (1962).

Youngs, J. N., and W. E. Cornatzer· Effect of inhibition of oxidative phosphorylation on incorporation of inorganic phosphate (P^{32}) into liver mitochondrial ATP and phospholipids. Proc. Soc. exp. Biol. (N.Y.) **120**, 281 (1965).

Zuckerman, L., and G. H. Marquardt Muscle, erythrocyte and plasma electrolytes and some other muscle constituents of rabbits with nutritional muscular dystrophy. Proc. Soc. exp Biol. (N.Y.) **112**, 609 (1963).

Zumoff, B., E. H. Bernstein, J. J. Imarisio, and L. Hellman Radioactive magnesium (Mg^{28}) metabolism in man. Clin. Res. **6**, 260 (1958).

Zwemmer, R. L., and R. C. Sullivan: Blood chemistry of adrenal insufficiency in cats. Endocrinology **18**, 97 (1934).

Namenverzeichnis

Die gewöhnlich gesetzten Ziffern weisen auf die entsprechende Stelle im Text und die *kursiven* Seitenzahlen auf das Literaturverzeichnis hin.

Abbott, W. E. s. Levey, S. 230, *280*

Abe, K. s Akimoto, H. 127, *143*

Abrahams, V. C., G. B. Koelle u. P. Smart 135, *143*

Acheson, G. H , u. S. A. Pereira 5, 14, *74*

— u. J. Remolina 66, *74*

— u. H. G. Schwarzacher 66, *74*

Adamson, L s. Anast, C. 243, *263*

Adey, W. R., J. P. Segundo u. R. B. Livingston 91, *143*

Adler, S. s Lindeman, R D. 210, 243, 252, *280*

Affani, J., P. L. Marchiafava u. G. Zernicki 114, *143*

Affolter, H. s. Haas, H. G. 248, *274*

Agna, J. W., u. R. E. Goldsmith 247, *262*

Aguilar-Figueroa, E. s. Hernandez-Peon, R. 131, 132, 135, *150*

Ahlquist, R. P , u B. Levy 71, *74*

— s. Levy, B 37, 38, 54, *80*

Ahmed, K. s. Judah, J. D. 247, *278*

Aikawa, J. K. 190, 194, 203, 245, 251, 252, 253, 258, *262*

— u. P. D. Bruns 191, *262*

— D. R Harms u. J. Z. Reardon 250, *262*

— u. J. Z. Reardon 194, 253, *262*

— — u. D. R. Harms 194, 220, 221, 229, 253, *262*

— E. L. Rhoades u. G. S. Gordon 207, *262*

— — D. R. Harms u. J. Z. Reardon 231, *262*

— s. Ginsburg, S. 195, 201, 202, 257, *273*

Aird, I s Milne, M D. 249, *283*

Aird, R. B. s. Greenberg, D. M. 257, *274*

Akert, K , W. Koella u. R. Hess jr. 127, *143*

Akimoto, H., N. Yamaguchi, K. Okabe, T. Nakagawa, I. Nakamura, K. Abe, H. Torii u. K. Masahasi 127, *143*

Albaum, H. G s. Du Bois, K P. 216, *270*

Albert, D. G , J. Morita u. L. T. Iseri 194, 256, *262*

Alcock, N., u. I. MacIntyre 207, 224, 226, 244, *263*

— — u. I. Radde 194, *263*

— s. Beechey, R B. 225, *264*

— s. Hess, R. 222, 223, 226, *277*

Alcock, N. W., I. MacIntyre u. I. Radde 191, *263*

— s. Beechey, R. B. 225, *264*

— s. Hendrix, J. Z. 243, 244, 245, *276*

Allcroft, R. 236, *263*

— s. Bartlett, S. 237, *264*

Allcroft, W. M 194, 237, *263*

Allen jr., M. B. s. Bonvallet, M 115, *144*

Allgood, M s. Hammarsten, J. F. 209, 214, 215, 254, *275*

Alonso de Florida, F. s. Pardo, E. G. 56, *82*

Alphin, R. S. s. Franko, B. V. 21, 37, 38, 39, 40, *77*

Amatuzio, D. S. s Stutzman, F. L. 196, 198, 235, *292*

Ambache, N. 5, 9, 17, *74*

— u. A. W. Lessin 5, 31, *74*

— W. L. M. Perry u. P. A. Robertson 5, 23, *74*

Ames, A. III, u F. B. Nesbett 194, *263*

— u. M. Sakanoue 194, *263*

— — u S. Endo 194, *263*

— s. Schwab, R. S. 196, 257, *289*

Anand, B. D. s. Delgado, J. M. R 110, *147*

Anand, B. K. 93, 110, *143*

— u. J. R. Brobeck 93, 110, *143*

— S. Dua u. C. L. Malhotra 135, *143*

Anast, C , R Kennedy, G. Volk u. L. Adamson 243, *263*

Anast, C. S. 198, *263*

Andén, N. E., J Haggendal, T. Magnusson u. E. Rosengren 125, *143*

Andersen, P. s. Curtis, D. R. 91, *146*

Anderson, A. R. s. Flink, E. B 235, 236, *271*

Anderson, C. E. s. Greenberg, D. M. 226, *274*

Anderson, C. K. s. Heaton, F. W. 195, 222, 223, 226, 227, 247, 248, *276*

Anderson, E. G. s. Schenk, E. A. 39, *83*

Andersson, B. 110, *143*

— C. C. Gale u. J. W. Sundsten 110, *143*

— u. B. Larsson 93, 110, *143*

Andersson, B u S M
McCann 100, 110, *143*
Andrus, S. B s. Gershoff,
S N. 259, *272*
Angleton, M s. Watanabe,
S. 212, *295*
Anstall, H. s Pallis, C. 236,
257, *286*
Antonowicz, I. s. Gershoff,
S. N. 212, 228, 251,
273
— s. Metcoff, J. 232, *283*
Antunes, M. L s De Jorge,
F. B 195, 199, *269*
Aprison, M H 132, *143*
— s. Hanig, R. C. 195, *275*
Arbesman, C. E., E. Neter
u C. F. Becker 180, *182*
Archibald, R. M. s. Hen-
drix, J Z. 243, 244, 245,
276
Ardill, B. L., J. A. Halliday,
J. D. Morrison, H. C.
Mulholland u. R A.
Womersley 214, 216, *263*
Ariens, E. J. 260, *263*
Armengol, V, W. Lifschitz
u. M. Palestini 117, *144*
Arnaud, C. D. s Carpenter,
B. E. 217, *267*
Aronson, M. s. Rocha e
Silva, M. 161, 163, 176,
183
Assenmacher, J s. Benoit,
J. 105, *144*
Atanackovic, D, A. F. de
Schaepdryver, G. R. de
Vleeschhouwer u. C.
Heymans 60, *74*
Atanassowa, L. s. Hozdrx,
T. 258, *277*
Atchley, D W., R. F. Loeb,
D. W Richards jr.,
E. M Benedict u. W. E.
Driscoll 258, *263*
Aub, J. C s. Tibbetts, D. M
208, 248, *292, 293*
Aurbach, G. D. s. Care,
A. D. 216, *267*
Auricchio, F. s. Bresciani,
F 203, *266*
Avioli, L., T. Lynch u.
C. Bastomsky 209, *263*
Axelrod, D. R, u. D. E.
Bass 194, 198, *263*

Axelrod, J. s Weil-Mal-
herbe, H. 94, *159*
Axford, R F E s.
L'Estrange, J. L. 206,
219, 236, 237, *280*

Babouris, N s. Booth, C. C
232, 233, 234, 235, 236,
265
Back, E H, R. D. Mont-
gomery u. E. E. Ward
235, *263*
Backmann, E. s. Brierley,
G P 247, *266*
Bacq, Z. 190, 204, 205, 213,
216, 218, *263*
Bacq, Z. M, u. A. Simonart
5, *74*
— s. Coppée, G. 61, 62, *75*
Baghdiantz, A s. Foster,
G. V. 249, *272*
Bain, J A 212, 250, *263*
— s Mayer, S E. 97, *154*
Bainbridge, J. G., u. D. M.
Brown 43, *74*
Baird, E. E s Goldman,
A. S. 230, 231, 233, *273*
Balch, C. C s. Rook, J A F.
236, *288*
Baldwin, J., P. K. Robin-
son, K L Zierler u.
J. L Lilienthal jr. 194,
224, *263*
Balint, J. A., u. B I. Hir-
schowitz 235, *263*
Ball, S. M. s Wolf, A V.
242, *296*
Banerjee, B., u N Saha
199, *263*
Bang, O., u. S. Ørskov 194,
257, *263*
Bárány, E H 62, *74*
Barden, P. s. Robertson, A.
198, *288*
Barker, E. S, J. K. Clark u.
J R Elkinton 210, 214,
215, 240, 242, 243, 252,
253, *263, 264*
Barnes, A. C. s Kumar, D.
216, *279*
Barnes, B. A., O Cope u.
E. B. Gordon 194, 211,
229, 232, *264*
— — u. T. Harrison 210,
229, *264*

Barnes, B A, S M Krane
u O Cope 247, *264*
— u. J. Mendelson 221,
222, *264*
— s. Mendelson, J. H 235,
283
Baron, D N. 236, *264*
Barron, G P, S O. Brown
u. P. B. Pearson 221,
226, 227, *264*
Bartlett, S, B. B. Brown,
A. S. Foot, S. J Row-
land, R Allcroft u.
W. H. Parr 237, *264*
Bartley, J C, E F Reber,
J. W Yusken u. H W.
Norton 216, *264*
Bartter, F. C. s. Gill jr.,
J R 248, *273*
Bass, D E. s Axelrod,
D R. 194. 198, *263*
— s Quinn, M 196, 198,
287
Bassett, S H. 247, *264*
Bastomsky, C. s. Avioli, L
209, *263*
Bates, M. F. s. Hummel,
F C. 191, *277*
Batini, C., G. Moruzzi, M
Palestini, G F. Rossi u.
A Zanchetti 114, 126,
144
— M. Palestini, G. F. Rossi
u. A Zanchetti 114,
126, *144*
Batsakis, J G. s. Stiles,
D. E. 196, *291*
Batsel, H. L. s. Witt, D. M
110, *159*
Battifora, H. A. s.
McCreary, P. A 225,
282
Battzer, B V. s. Miller,
E. R. 222, 223, *283*
Bauer, U s Friedberg,
K. D. 181, *182*
Baum, P., u R Czok 200,
264
Baur, H. s. Striebel, A.
209, *292*
Baust, W, u. H. Niemczyk
93, *144*
— — u. J. Vieth 93, *144*
Baxter, D. s. Olszewski, J.
113, *156*

Baxter, D. J. s. Smith,
W O 196, 221, 222, 224,
225, *291*
Beard, E S s Lindeman,
R. D. 210, 243, 252, *280*
Beattie, A. W s Gartner,
R. J. W. 195, *272*
Beauchemin, J. A. s. Harris, W. H 257, *275*
Beaulnes, A. s Bois, P 219,
265
— s. Cordeau, J. P. 102,
103, 118, 119, 120, 121,
122, 123, 124, 126, 132,
135, 138, *146*
Bechterew 129
Becker, C. F. s Arbesman,
C E. 180, *182*
Becker, E. L. 167, 173, 174,
182
— s. Ginn, E. 224, *273*
Beechey, R. B , N Alcock,
S Hanna u. I. MacIntyre 225, *264*
— N. W. Alcock u.
I MacIntyre 225, *264*
Beeson, D. M. s. Mayo,
R H 219, *282*
Bein, J H , u R Meier 7,
74
Bélanger, L. F., G. A. van
Erkel u. A. Jakerow 219,
264
— s. Bois, P 219, *265*
Beleslin, D , B Radmanović u. V. Varagić 43, *74*
— u V Varagić 43, *74*
Bell, M. C. s McAleese,
D. M 221, *282*
Bell, N H s. Gill jr., J. R.
248, *273*
Bell, R D. s. Keyl, M. J.
195, 205, *278*
Bellavia, J V. s Wallach,
S 196, 203, 214, 247,
248, *294*
Belloiu, D , Z. Rodin, F.
Cherciulesco u. W. Demetresco 251, *264*
Benedict, E. M. s Atchley,
D. W. 258, *263*
Benedict, S R. s Mendel,
L B. 210, *283*
Benelli, G , D Della Bella
u A Gandini 39, *74*

Benfey, B G , u. D. R.
Varma 14, *74*
— s. Greeff, K. 176, *182*
Benoit, J., u J Assenmacher 105, *144*
Beraldo, W. T. 8, *74*
— u A. Zanotto 8, *74*
Beresford, C. C. s. Heaton,
F. W. 235, 241, 255, *276*
Berglund, F., u. R. P. Forster 242, *264*
Berman, H s Gardner,
L. I 224, *272*
Berne, C. F. s Martin,
H E. 235, *281*
Berne, C. J. s. Edmondson,
H A 258, *270*
Bernhard, G. s. Stegemann,
H. 164, 169, *184*
Bernick, S , u G. H. Hungerford 226, *264*
Bernimolin, J. s. Brull, L
243, *266*
Bernstein, D. s. Kleeman,
C. R 243, *279*
Bernstein, E. H. s. Zumoff,
B. 209, *296*
Bersohn u. Ternor 259
Bersohn, I., u P. J. Oelofse
218, 256, *264*
Bertler, O., B. Falck u. Ch
Owman 125, *144*
— — u. E Rosengren 90,
144
Bertrand, D 190, 194, *264*
Bessman, S. P., u W. N.
Fishbein 134, *144*
— u. S. J. Skolnik 134, *144*
Best, F A , u. V. R. Pickles
194, 213, *264*
Better, O , H C Gonick,
L Chapman u. C. R
Kleeman 254, *264*
— — — L. C Chapmann,
P. D. Varrady u. C. P.
Kleeman 241, *264*
Bhattacharya, G. 252, *265*
Bhattacharyya, N. K , u.
D. N. Mullick 255, 256,
265
Bigheri, E. s Horton, R.
249, 250, *277*
Bindler, E s Gyermek, L.
31, 40, 42, *78*
Bird, F. A. 226, 227, *265*

Birks, R , B. Katz u. R.
Miledi 63, *74*
Birmingham, A. T., u. A. B.
Wilson 19, *74*
Bishop, G. H , u P. Heinbecker 69, *74*
Bishop, P. O , G. Field,
B L. Hennesy u. J. R.
Smith 89, *144*
Bissell, G W 251, *265*
Blaxter, K. L. 221, *265*
— u. R. F McGill 190,
236, *265*
— u. J. A. F. Rook 225, *265*
— — u. A. M. MacDonald
194, 209, 221, 222, 223,
227, *265*
— u. B. A. M. Sharman
226, 236, *265*
Bleifer, K. H s Randall jr ,
R E 235, *287*
Bloch, V., u. M. Bonvallet
115, *144*
— s Bonvallet, M. 115, *144*
Bloom, F. E., A. P. Oliver
u. G. C Salmoiraghi 89,
91, 126, *144*
— s. Salmoiraghi, G. C. 88,
157
Blythe, W B , H J. Gitelman, u. L. G. Welt 241,
242, *265*
Bockelmann, A. s. Greeff,
K. 176, *182*
Boda, A. s. Moussa, S. L.
198, 199, *284*
Bodammer, G 165, 177, 181
— u. W Vogt 177, *182*
Boellner, S. W , E J. Olson,
D. Fredrickson u. E. R.
Hughes 194, 257, *265*
Boelter, M. D. s. Kleiber, M.
225, 253, *279*
Bogdanove, E. M , u. H C.
Schoen 105, *144*
Bogdanski, D F. s. Udenfriend, S. 95, *158*
Bogert, L. J., u. E. C. Plass
198, *265*
Boggs, D s. Maynard, L. A.
227, *282*
Bohr, D. F 213, *265*
— s Filo, R. S. 212, *271*
Bois, P. 219, 227, 228, *265*
— u. A. Beaulnes 219, *265*

Bois, P , E H Byrne u
L. F. Bélanger 219, *265*
Bokri, E , O Fehér u
G Mózsik 45, 64, 65, *74*
— s. Fehér, O. 45, 52, *76*
Boldizsár, H. s. Kemény, A.
195, 214, 216, *278*
Bonner, P s Hummel, F C
191, *277*
Bonvallet, M , u. M B.
Allen jr. 115, *144*
— u. V. Bloch 115, *144*
— — P. Dell u. G. Hiebel
92, 120, *144*
— s. Bloch, V 115, *144*
— s. Dell, P. C. 88, *147*
— s. Hiebel, G. 96, 139, *151*
Booth, C. C , S. Hanna,
N. Babouris u I. MacIn-
tyre 232, 233, 234, 235,
236, *265*
— s. MacIntyre, I. 195,
231, 232, 233, *281*
Booyens, J., M. de Waal,
L. J. Rademeyer u. F.
Calitz 256, *265*
— s. Rademeyer, L. J. 256,
287
Bordet, J. 161, 180, *182*
Borglin, N. E , u. T Lind-
sten 216, *265*
Borison, H. L. s. Wang,
S. C. 93, *159*
Borowitz, J. L , K. Fuwa u.
N. Weiner 203, *265*
Borst, J. G. G. s. de Vries,
L. A. 210, *269*
Boss, S. s. MacIntyre, I.
247, 248, *281*
Boura, A. L. A., F. C. Copp,
W. G. Duncombe, A. F.
Green u. A. McCoubrey
59, *75*
— u A F. Green 38, *75*
Bovet-Nitti, F , R. Kohn,
M. Marotta, W. P.
Scognamiglio u. B. Sil-
vestrini 39, *75*
Bradley, P. B 88, 97, 138,
139, *144*
— B. N. Dhawan u J H.
Wolstencroft 73, *75*, 89,
125, 126, *144*
— u. B. J. Key 138, *144*
— u. A. Mollica 95, *145*

Bradley, P. B. u. J. H Wolstencroft 89, 91, 95, 125,
126, *145*
Brady, J V. s Gollub,
L R 88, *149*
Brandt, J L , u. W. Glaser
203, 250, *265*
— — u. A Jones 203, *165*
— s Glaser, W. *273*
Breibart, S., J. S Lee,
A. McCoord u. G. B.
Forbes 204, *265*
Bremer, F 114, 116, 117,
145
— u. N. Stoupel 116, 117,
145
— — u P. E. van Reeth
117, 118, *145*
Bresciani, F , u F. Auric-
chio 203, *266*
Brest, A. N. s Seller, R H.
196, 253, 256, *290*
Brierley, G. P., E Murer,
E. Backmann u D. E.
Green 247, *266*
— — u. R. A. O'Brien
247, *266*
Briggs, A. H. s. Embry, R.
212, *270*
Brobeck, J. R. 110, *145*
— J. Tepperman u. C. N. H.
Long 110, *145*
— s. Anand, B. K. 93, 110,
143
Broca 127
Brochart, M s Larvor, P.
222, 223, 226, 227, *279*
Brodal, A 119, *145*
— u G. F. Rossi 119, *145*
Brodie, B B , u. P. A.
Shore 125, 132, 135, *145*
— s. Costa, E. 59, 60, *75*
Broitman, S. A s. Gottlieb,
L. S. 206, 252, *273*
Bronner, F., u. D. D.
Thompson 239, *266*
Browder, A. A s Walser, M.
243, *295*
Brown, A. N. s Lowry,
O. H. 202, *281*
Brown, B. B s Bartlett, S
237, *264*
Brown, D. F., R. B.
McCandy, E. Gillie u. J.
T. Doyle 256, *266*

Brown, D M s Bain-
bridge, J. G. 43, *74*
Brown, G. L , u W. Feld-
berg 14, *75*
— u. J. E Pascoe 66, 70,
75
Brown, G. M. s. Valberg,
L. S. 257, *293*
Brown, J L. s. Samiy,
A. H E. 214, 239, 242,
289
Brown, S. O s Barron,
G. P. 221, 226, 227, *264*
Browne, S. E. 217, 258, 259,
266
Brownlee, G , u. J. Harry
32, *75*
Brucke, F. T. von 63, *75*
Bruegger, M. 110, *145*
Brull, L , u J. Bernimolin
243, *266*
Bruns, P. D. s. Aikawa,
J. K. 191, *262*
Bryson, R W. s. Heaton,
F. W 235, 241, 255,
276
Bubnoff, M V s Schmid,
E 215, *289*
Buchwald, N. A. s. Rakić,
L 122, *156*
Bulbring, E. 56, 58, *75*
— u. J H Burn 56, *75*
Buell, M V., u. E. Turner
250, *266*
Buergi, S , u M Monnier
116, *145*
Buka, R. s. Lilienthal, J L.
224, 234, 257, *280*
Bulger, H. A , u. F Gaus-
mann 248, *266*
Bull, E. C. s Lifshitz, F.
223, 224, *280*
Bunce, G E., Y Chiem-
chaisri u. P. H. Phillips
218, 227, 230, *266*
— K. J. Jenkins u. P H.
Phillips 222, 223, 227,
266
— P G Reeves, T. S Oba
u H E Sauerlich 194,
208, 225, 228, *266*
— s. Moore, C A. 217, 259,
284
— s. Sauerlich, H. E. 259,
289

Burch, G. E., R. K Lazzara u T K. Yun 194, 202, *266*

Burgen, A. S. V s Graham, L A. 209, *274*

Burn, J. H. 87, 136, 143, *145*

— u H. H Dale 7, 38, *75*

— J. J. Dromey u. B J. Large 136, 137, 139, 140, *145*

— u H Froede 136, 137, 139, 140, *145*

— u. M. J. Rand 87, 103, 124, 134, 135, 136, 137, 138, 139, *145*

— — u R. Wien 136, 137, 139, 140, *145*

— u. D. F. Weetman 136, 137, 139, 140, *145*

— s. Bulbring, E. 56, *75*

— s. Fleckenstein, A 20, *77*

Burnell, R H s. Maxwell, G M 195, 215, 216, *282*

Burns, K. H s Kunkel, H. O. 237, *279*

Burt, A. W. A, u. D. C. Thomas 237, *266*

Burton, A. C. s. Heagy, F. C. 198, *276*

Butler, A. M. 230, 231, 258, *266*

Butler, E. J s. Field, A C. 217, *271*

Butler, T. s Parsons, R. S. 258, *286*

Byrne, E. H. s. Bois, P. 219, *265*

Caddell, J. L. 234, *266*

Caesar, J. J s. Graham, L. A 209, *274*

Cahill, L. N s. Wallach, S. 196, *294*

Calitz, F. s. Booyens, J. 256, *265*

Calma, I. s. Machne, X 91, *153*

Camp, B J. s. Kunkel, H. O. 237, *279*

Campbell, J. A, u. T. A. Webster 209, *266*

Canato, C. s. De Jorge, F. B. 198, 205, *269*

Canelas, H M, L M. De Assis u. F B De Jorge 194, 257, *266*

Canham, J. E. s. Conzolazio, C F 205, 210, *268*

Cannon 64

Cantarow, A s. Haury, V G. 253, 254, *275*

Capon, A. 95, *145*

Carafoli, E., C. S. Rossi u. A. L Lehninger 247, *266*

Card, R T. s. Valberg, L. S 196, *293*

Care, A. D. 205, 209, 237, 245, *266*

— u. W. M Keynes 248, 249, *266*

— D C. MacDonald u. B Nolan 202, *266*

— u. D. B. Ross 236, 237, 249, 250, *266, 267*

— — u. A. A. Wilson 237, *267*

— L. M. Sherwood, J. T. Potts jr. u. G. D. Aurbach 216, *267*

— u. A Th. Van't Klooster 244, 245, *267*

— s. MacDonald, D. C. 202, 209, *281*

— s. Ross, D B 243, *288*

Carillo, B. J., W. G. Pond, L. Krook, F. E. Lovelace u. J. K. Loosli 222, 224, 226, 228, *267*

Carlsson, A., B Falck, K. Fuxe u N. Å Hillarp 125, *145*

— — N. Å Hillarp u A. Torp 125, *146*

Carpenter, B. E., M. Keoplung, C D Arnaud u. N. H Engbring 217, *267*

Carr, C W., u K P. Woods 199, *267*

Carr, M. H, u. H A Frank 194, *267*

— u. P. R. Schloerb 194, *267*

Carter, A C s. Wallach, S. 242, *294*

Carter, W J. s Gantt, C. L. *272*

Carton, C A. s. Fields, W. S 105, *148*

Carubelli, R., W. O. Smith u J. F. Hammarsten 194, *267*

Carvalho, A. P. 203, *267*

— H. Sanui u. N. Pace 201, 203, *267*

Casley-Smith, J. s. Charnock, J. S 199, *267*

Cass, M. H. s Cheek, D. B. 194, 247, *267*

Castro, F. de 67, 68, *75*

Catchpole, H. R., N. R. Joseph u. M. B Engel 202, *267*

Cato, J. s. Pardo, E. G 56, *82*

Cauperet, J. s. Randoin, L. 217, *287*

Cerletti, A. s. Weidmann, H. 7, *85*

Cervoni, P s. Kirpekar, S. M. 137, *152*

Cervoni, T. s. Mirkin, D. L. 138, *155*

Chaffee, R R. J s Pengelley, E. T 198, 199, *286*

Chaillet, F s Schlag, J. 131, *157*

Chambless, W S. s. George, W. K 236, *272*

Chamings, J s. Payne, J. M. 247, *286*

Chanes, R s Eliel, L. P. 228, 248, *270*

Chang Yun s Olds, J. 102, 103, *155*

Chapman, L. s. Better, O. 241, 254, *264*

— s Kleeman, C. R. 243, *279*

Chapman, L. C. s. Better, O. 241, *264*

Charbon, G. A., u. M. H. Hoekstra 194, *267*

Charles, M s. Clark, I. 217, *268*

Charnock, J. S., J. Casley-Smith u C J. Schwartz 199, *267*

— L J Opit u. B S Hetzel 253, *267*

Chavez-Ibarra, G. s. Hernandez-Peon, R. 99, 103, 122, 123, 128, 129, 130, 131, 132, 135, *150*

Cheek, D. B , J E. Gray-
stone u M H. Cass 194,
247, *267*
— — J. B. Willis u. A. B
Holt 194, 247, *267*
— u. H. C. Teng 194, 248,
267
— u. C D West 194, *267*
Chen, G , R. Portman u.
A. Wickel 19, *75*
Cheng, C. s. Woodbury,
D. M. 249, *296*
Cherciulesco, F. s Bellou,
D. 251, *264*
Chesley, L. C., u. I. Tepper
214, 215, *267*
Chiemchaisri, Y., u. P. H.
Phillips 194, 220, 221,
222, 223, 228, *267*
— s. Bunce, G E. 218, 227,
230, *266*
Chien, S. 64, *75*
Cho, A. K., W. L. Haslett,
u. D. J Jenden 123, *146*
Chou, T. C., u. F. J. de Elio
53, *75*
Chowers, J., S. Feldman u.
J. M. Davidson 106, *146*
Christianson, G., R. Jeness
u. S. T. Coulton 205, *268*
Christie, G. S. s. Judah,
J. D. 247, *278*
Chutkow, J. G. 194, 206,
209, 216, 219, 222, 224,
245, *268*
Clark 16
Clark, G. W. 229, *268*
Clark, I. 208, 217, *268*
— M. Charles u. F. Rivera-
Cordero 217, *268*
— E. A. Gusmano, R. Ne-
vins u. S. H Cohn 217,
268
— u. M. R Smith 217, *268*
Clark, J K s Barker, E. S.
210, 214, 215, 240, 242,
243, 252, 253, *263, 264*
Clark, J. R. s. Verhave, T.
112, *159*
Clarkson, E. M , S. J.
McDonald, H E. De
Wardener u. R. Warren
194, 254, *268*
Clayton, G. W. s. Etteldorf,
J. N. 214, *270*

Clegg, P. C., P. Hopkinson
u. V. R. Pickles 213, *268*
Clemente, C B. s Sterman,
M B. 103, 121, 123, 127,
128, 130, *158*
Clemente, C D., u. M. B.
Sterman 121, 122, 127,
128, 130, 131, *146*
— J. Sutin u J. T. Silver-
tone 92, *146*
Cloosen, J s. Fabry, C. 204,
271
Cockrell jr., D. F. s Lang-
ley, L. L. 245, *279*
Cohen, C. s. Soffer, L. J.
251, *291*
Cohen, L H , u. P B Hagen
52, *75*
Cohen, M D. s. Randall jr.,
R E. 255, *287*
Cohen, R. D. 251, *268*
Cohn, S. H. s. Clark, I. 217,
268
Colby, R. W., u C. M. Frye
227, 228, *268*
Collier, H. O J , G. W. L.
James u. P. J. Piper 34, *75*
Collumbine, H. s. Dirn-
huber, P. 55, *76*
Conard, V., u. W. Mutsaars
177, *182*
Cong-Trieu, T. s. Durlach,
J. 248, *270*
Connors, D. N s. Pytko-
wicz, R M. 200, *287*
Connors, P. s. Vitale, J. J.
251, *293*
Connors, P. S s Vitale, J. J.
225, 226, 251, *293*
Conover, H. s. Hills, A. G.
210, 250, 253, *277*
Consolazio, C. F., L. O.
Matoush, R C Nelson,
R. S Harding u J. E.
Canham 205, 210, *268*
Conway, E. J., u. P. F.
Duggan 247, *268*
— u. D. Hingerty 250, *268*
Cook, M. J. s. Tipton, I. H.
196, 202, *293*
Coon, J. M. s. Salerno, P. R.
54, *83*
Coons, C. M., A. T. Schiefel-
busch, G. B. Marshall u.
R. S. Coons 191, *268*

Coons, R S s Coons, C. M.
191, *268*
Cooper, C. s. Tapley, D. F
250, *292*
Cope, C. L , u. J. Pearson
250, *268*
— u B Wolff 251, *268*
Cope, O. s. Barnes, B A.
194, 210, 211, 229, 232,
247, *264*
Copeland, B. E., u. F W.
Sunderman 194, *268*
Copp, F C. s. Boura, A L A.
59, *75*
Coppée, G., u. Z. M Bacq
61, 62, *75*
Corbett, C. s Mendelson,
J. H 235, *283*
Cordeau, J. P 102, 114,
118, 119, 121, *146*
— u. M. Mancia 115, *146*
— A. Moreau u. A. Beaul-
nes 118, 119, 120, 121,
135, 138, *146*
— — — u C Laurin 102,
103, 119, 120, 122, 123,
124, 126, 132, 135, 138,
146
— s Courville, J. 102, 103,
118, 119, *146*
Core, I. s. Nakamura, M
195, 202, 255, *285*
Cornatzer, W E s. Youngs,
J. N. 226, *296*
Corradino, R. A., u H. E.
Parker 225, *268*
Correa, P. s. Vitale, J. J
226, 227, *294*
Cosby, M. H. s. Pollack, S.
257, *286*
Costa, E., A. M Revzin, R.
Kuntzman, S. Spector u.
B. B. Brodie 59, 60, *75*
— s. Salmoiraghi, G C. 88,
157
Cotlove, E., M. A Holliday,
R. Schwartz u W. W.
Wallace 194, 221, 224, *268*
Coulton, S T. s. Christian-
son, G. 205, *268*
Courville, J., J. Walsh u.
J. P. Cordeau 102, 103,
118, 119, *146*
Coury, J. N. s. Fisher, A E.
104, 112, *149*

Coutinho, E. M. 213, *268*
Coutino, E. M. 216, *268*
Cragg, B. G. 60, *75*
Cragle, R G. s. Mraz, F. R
215, *284*
Craig, J. M. s. Ko, K W.
222, 226, *279*
Cramer, C. F., u. J. Dueck
245, *269*
Cramer, W 225, *269*
Crawford, J M , D. R. Curtis, P. E. Voorhoeve u.
V. J. Wilson 91, *146*
Cremona, T., u. T. P. Singer
172, *182*
Creuzfeldt, O., u. R. Jung
116, *146*
Critchlow, B. V. s. Sawyer,
Ch. H. 105, *157*
Cross, B. A., u. J. D Green
92, *146*
— s. Holland, R. C. 92,
151
Cross, H. C s. Kelly, H. G.
214, 215, 216, *278*
Crossfield, H C. s DaVanzo,
J. P. 250, *269*
Crossland, J. 88, *146*
Cunningham, I. J. 195,
216,
217, 220, 221, 222, *269*
— u. Cunningham, M M.
222, 224, 225, *269*
Cunningham, M. M s. Cunningham, I. J 222, 224,
225, *269*
Curran, P. F., u. J. R.
Gill jr. 247, *269*
Curtis, D. R. 88, *146*
— u. P. Andersen 91, *146*
— J. C. Eccles u. R. M.
Eccles 97, *146*
— u. R. M. Eccles 89, 90,
91, *146*, *147*
— u. K. Koizumi 91, *147*
— u. J. W. Phillis 130,
147
— J. W. Phillis u. J. C.
Watkins 89, 126, *147*
— u. R. W. Ryall 73, *75*
— s. Crawford, J. M. 91,
146
Cuthbert, M. F. 60, *75*
Czok, R. s. Baum, P. 200,
264

Daatselaar, J. J. van s.
de Vries, L A. 210, *269*
Dabney, J M s Keyl,
M. J. 195, 205, *278*
Dahl, L. K. s. Silver, L.
209, *290*
Dahl, S. 195, 197, *269*
Dahlstrom, A , u K. Fuxe
125, *147*
Daigneault, E. A. s. White,
R. P. 139, *159*
Dale, H H 4, 12, *76*, 180,
182
— u. P. P. Laidlaw 4, 5, *76*
— s. Burn, J. H 7, 38, *75*
Daniels, F., M. Quinn, C. R.
Kleeman u. J. Marino
198, *269*
Danon, A. P. s. Moussatché,
H. 177, *183*
Daur, W. 258, *269*
DaVanzo, J. P , H. C.
Crossfield u. W. W.
Swingle 250, *269*
David, J. P , S. Murayama,
X. Machne u. K. R.
Unna 89, *147*
Davidovich, A , u J. V.
Luco 61, *76*
Davidson, J. M., u. S. Feldman 93, 105, 106, *147*
— u Ch. H. Sawyer 100,
106, 108, *147*
— s. Chowers, J. 106, *146*
Davidsson, D. s. MacIntyre, I. 195, 220, 221,
222, 223, 224, *281*
Davies, J. S., E M. Widdowson u R. A. McCance 191, *269*
Davis, A. N. s. Scoular,
F. I. 205, *289*
Davis, J. H. s. Levey, S
230, *280*
Davis, W. W. s Regen,
D. M. 212, *287*
Davson, H. 90, 94, *147*, 204,
205, *269*
— u. E Spaziani 90, *147*
Day, M. D., u. M. J. Rand
136, *147*
De Albuquerque, P F., u.
M. Tuma 259, *269*
De Assis, L. M. s. Canelas,
H. M. 194, 257, *266*

De Azevedo, M. L , u. F. B.
De Jorge 202, *269*
De Doncker, K , u N Rosselle 195, 219, 222, 225,
269
de Groat, W. C., u R L.
Volle 40, 57, 58, 60, *76*
Deijs, O. s. Kemp, A 208,
236, 237, *278*
Dejardin, N. s. van Geertruyden, J. 246, *273*
De Jorge, F. B , C. Canato
u D. Delascio 198, 205,
269
— D. Delascio, A. B. D.
Ulhôa Cintra u. M L.
Antunes 195, 199, *269*
— s. Canelas, H. M. 194,
257, *266*
— s De Azevedo, M L.
202, *269*
Delascio, D. s. De Jorge,
F. B. 195, 198, 199, 205,
269
Del Castillo, J., u L. Engbaek 212, *269*
— u. B Katz 15, *76*
Delgado, J. M. R. 99, *147*
— u. B D Anand 110, *147*
Dell, P. s. Bonvallet, M. 92,
120, *144*
— s. Dumont, S. 117, *148*
— s. Hiebel, G. 96, 139, *151*
Dell, P. C. 88, *147*
— M Bonvallet u. A. Hugelin 88, *147*
Della Bella, D. s. Benelli, G.
39, *74*
Dellen, T. R van, u. J. R.
Miller 215, *269*
— s. Miller, J. R. 215, 259,
283
De Maria, W. J. A., u. J. S.
Harris 258, *269*
— s. Harris, J. S. 258, *275*
Dement, W. 127, *147*
Demetrescu, M s. Steriade,
M. 118, *158*
Demetresco, W. s. Belloiu,
D. 251, *264*
Dempsey, E. F. s. Henneman, P. H. 207, *277*
Desmedt, J. E., u J Schlag
125, *147*
— s. Jouvet, M. 116, *152*

Desper, P. C s. Jones, J. E.
195, 251, 257, *278*
De Vries, L. A., S. P. ten
Holt, J. J. van Daatse-
laar, A Mulder u.
J. G. G. Borst 210, *269*
— s. Gerbrandy, J. 195,
272
De Wardener, H. E. s.
Clarkson, E. M. 194,
254, *268*
Dews, P. B. 88, *147*
— u. W. H. Morse 88, *147*
Dey, F. L. 105, *148*
Dhawa, B N. s. Bradley,
P. B. 73, *75*, 89, 125,
126, *144*
Dickerson, J. W T. s.
Widdowson, E. M 190,
191, 196, 202, *295*
Dickinson, W. L. s. Lang-
ley, J. N. 11, *80*
Di Giorgio, J., J. J. Vitale
u. E E. Hellerstein 225,
269
Dimich, A. s Fankushen, D.
229, 232, *271*
— s. Rizek, J. A. 196, 251,
288
Dine, R. F., u. P. H
Lavietes 251, *269*
Dirnhuber, P., u. H Col-
lumbine 55, *76*
Dishington, I W. 237, *269*
Dixon, W. E., u. F. Ransom
4, *76*
Dobbing, J. 90, *148*
Dobrzecka, C. s. Wyrwicka,
W. 111, *159*
Dobson, A. s. Scott, D. 249,
289
Docter, R. F. s Fuster,
J M. 117, *149*
Doe, R P, E B. Flink u.
A. S Prasad 251, *269*
— J. A. Vennes u. E. B.
Flink 210, 250, *270*
— s. Flink, E B 235, 236,
271
— s McCollister, R. 209,
210, 241, 252, *282*
— s. McCollister, R. J. 232,
235, *282*
Domeij, B. s. von Euler,
U. S. 5, 39, *76*

Domer, F R., u W Feld-
berg 97, 98, *148*
Domino, E. F. 88, *148*
— s. Knapp, D. E 123, *152*
— s. Rech, R 102, *156*
Doncker, K. de s Rosselle,
N. 196, 236, *288*
Dotson, D. A. s. George,
W. K. 199, *272*
Dougherty jr, R. M. s.
Haddy, F. J. 213, 236,
257, *274*
Douglas, W. W, u. A. M.
Poisner 30, 32, 50, *76*,
212, *270*
— u. J. M. Ritchie 40, *76*
— u. R P. Rubin 212, 213,
270
Dowdle, E B s. Schachter,
D. 243, 245, *289*
Doyle, J T s. Brown, D.F.
256, *266*
Dresel, P E. s. Slater, I. H.
7, 37, 38, *83*
Dresse, A. s. Lecomte, J.
8, *80*
Dreux, C, u. M. Girard 195,
270
Driscoll, W. E. s. Atchley,
D. W. 258, *263*
Dromey, J. J. s. Burn, J H.
136, 137, 139, 140, *145*
Dua, S. s Anand, B. K.
135, *143*
Dubach, U. C s. Haas,
H. G. 248, *274*
Du Bois, K. P., H. G. Al-
baum u. V. R. Potter
216, *270*
Duckett jr, J. W s. Walser,
M. 250, *295*
Duckworth, J., W. Godden
u. G. M. Warnock 220,
221, *270*
— u. G. M. Warnock 206,
270
Dudel, J. 40, *76*
Dueck, J. s Cramer, C F.
245, *269*
Duedall, I. W. s. Pytkowicz,
R. M. 200, *287*
Duggan, P. F. s. Conway,
E. J. 247, *268*
Dumont, S, u. P. Dell 117,
148

Duncan, C. W., C. F. Huff-
man u. C S. Robinson
238, *270*
Duncombe, W. G s
Boura, A L. A. 59, *75*
Dunn, M, M. Keoplung u.
W. A. Krehl 254, *270*
Dunn, M. J, u M. Walser
195, 205, 209, 229, 230,
231, 232, 233, 234, 252,
270
Durlach, J., u. R. Lebrun
235, 236, *270*
— M Stol, aroff, J. Gau-
duchon, R. Lebuc u
T. Cong-Trieu 248, *270*
Duruisseau, J P. s. Mori, K.
202, *284*

Eccles, J. C. 69, *76*, 88, 122,
148
— R. M Eccles u. P. Fatt
97, *148*
— P. Fatt u. K Koketsu 90,
91, *148*
— u. B Libet 122, *148*
— s. Curtis, D R. 97, *146*
Eccles, R. M. 12, *76*
— u. B. Libet 47, 48, 49,
58, 59, 71, *76*
— s. Curtis, D. R. 89, 90,
91, 97, *146*, *147*
— s. Eccles, J C 97, *148*
Eckstein, J. W. s Long,
J. P. 54, *81*
Economo, C v. 127, *148*
Edelman, I S, P. O P.
Ts'o u. J. Vinograd 203,
270
Edmondsen, H s. Martin,
H. E. 235, *281*
Edmondson, H. A, C J.
Berne, R E Homann jr.
u. M Wertman 258, *270*
Edwards jr, H M. s.
Nugara, D 227, *285*
Eisenberg, H s. Hopkins, T.
200, *277*
Elfvin, L G. 125, *148*
Eliel, L P, R. Chanes u.
J. Hawrylko 228, 248,
270
— s Smith, W. O 235, *291*
Elio, F. J. de s Chou, T C.
53, *75*

Elkinton, J R 190, *270*
— s. Barker, E S. 210, 214, 215, 240, 242, 243, 252, 253, *263, 264*
Elliott, R B s Maxwell, G. M. 195, 215, 216, *282*
Elliott, R. C 57, *76*
Ellsworth, R s Harrop, G A 250, *275*
Elmquist, D I , u D. M. J. Quastel 212, *270*
Embry, R , u. A. H. Briggs 212, *270*
Emery, N. s. Miller, N. E. 113, *154*
Emmelin, N , u B C R. Stromblad 62, *76*
Endo, S. s. Ames, A III 194, *263*
Engbaek, L. 212, 213, 215, 216, *270*
— s Del Castillo, J. 212, *269*
Engbring, N H s Carpenter, B E 217, *267*
Engel, M B. s. Catechpole, H R. 202, *267*
Engelhardt, G. s. Friedberg, K. D. 176, 177, 181, *182*
Enselber, C. D., H G. Simmons u. A. A. Mintz 259, *270*
Epstein, A. N. 100, 110, *148*
Epstein, F. H. s. Kessner, J M. 222, 224, 226, 244, *278*
— s Kleeman, C R. 195, 251, *279*
— s Manitius, A. 224, 225, *281*
Eranko, O., u. M. Harkonen 59, *76*
Erkel, G A. van s Bélanger, L F 219, *264*
Es, A J. H van s. Kemp, A. 208, 236, 237, *278*
Eschler, J , G. Ochs u W. Schilli 214, *270*
Etteldorf, J N., G. W. Clayton, A. H. Tuttle u. C R. Houck 214, *270*
— u. A. H. Tuttle 258, *271*
— s. Smith, H. L 189, 221, 222, 224, 257

Euler, C v 92, 93, *148*
Euler, U S v , u B Domeij 5, 39, *76*
Evart, E V. s Schoolman, A. 117, *157*
Evarts, E V., W. Landau, W Freygang jr u W. H Marshall 89, *148*
Evenson, M A. s. Steele, T H. 254, *291*
Evered, D F. 237, *271*
Exley, K A. 38, *76*
Eyzaguirre, C , u. H. Koyano 213, *271*

Fabiani, F. s. Laborit, H. 134, *153*
Fabry, C., u. J. Cloosen 204, *271*
Fadely, D. s. Verhave, T. 112, *159*
Faidherbe, J. s Schlag, J. 91, 131, *157*
Falck 59
Falck, B. 125, *148*
— s Bertler, O 90, 125, *144*
— s. Carlsson, A. 125, *145, 146*
Faloon, W. W s Miller, T. R II 210, *283*
Fankushen, D., D. Raskin, A Dimich u. S. Wallach 229, 232, *271*
Farah, A. s. Stanbury, J. B. 215, *291*
Farbman, A. s. Vitale, J J. 255, *293*
Farnell, D. R 228, *271*
Farrell, C. s Sullivan, J. 235, 255, *292*
Fastier, F. N , M A. Mc-Dowall u H. Waal 40, *76*
Fatt, P. s. Eccles, J. C. 90, 91, 97, *148*
Favale, E., C Loeb, G. F. Rossi u. G. Sacco 115, *148*
Featherston, W. R., M. L. Morris jr. u. P. H. Phillips 208, *271*
— s Morris jr M. L. 226, 227, *284*
— s. Scholz, R W. 208, *289*

Fehér, O. u. E. Bokri 45, 52, *76*
— s. Bokri, E 45, 64, 65, *74*
Feldberg, W. 6, 61, 62, *76*, 98, *148*
— u. K. Fleischhauer 98, *148*
— u G. P. Lewis 9, 32, 33, 42, 43, *77*
— B Minz, u H Tsudzimura 5, 44, 51, *77*
— u R D. Myers 98, *148*
— u S L Sherwood 98, *148*
— u. A. Vartiainen 6, 7, 17, 53, *77*
— s Brown, G. L. 14, *75*
— s Domer, F. R. 97, 98, *148*
Feldman, S. s Chowers, J. 106, *146*
— s. Davidson, J. M. 93, 105, 106, *147*
Fellers, F. X , u. C. Herrera 221, *271*
— s. Ko, K. W. 222, 226, *279*
Ferluga, J. 176, *182*
Ferry, C. B. 134, 138, *148*
Fiedl, G. s. Bishop, P. O. 89, *144*
Field, A. C. 209, 237, 245, *271*
— J. W McCallum, u E. J. Butler 217, *271*
— u B. S. W. Smith 195, 203, 221, *271*
— s. Smith, R. S W. 196, 218, 220, 221, 222, 229, *291*
Fields, W. S , R Guillemin u. C. A. Carton 105, *148*
Filo, R. S., D. F. Bohr, u. J. C Ruegg 212, *271*
Finlay, J. M s. Opie, L H. 209, 231, 233, 234, *285*
Fischer, R L s Smith, H. L. 189, 221, 222, 224, 257, *290*
Fishbein, W. N. s. Bessman, S. P. 134, *144*
Fisher, A. E. 94, 99, 102, 108, 109, *149*
— u. J. N. Coury 104, 112, *149*

Fishman, R A. 190, 214, 216, 236, *271*
— s. Gerst, P. H. 235, 236, *273*
Fisk, G s. Maynard, L. A. 227, *282*
Fitzgerald, M. G., u. P. Fourman 211, 229, 231, 232, 234, *271*
Flacke, W., u R. A. Gillis 49, 50, 53, *77*
Fleckenstein, A., u. J. H. Burn 20, *77*
— u. D. Stockle 20, *77*
Fleischhauer, K. s. Feldberg, W. 98, *148*
Flerko, B. 93, 105, *149*
Fletcher, R. F., u. A. A. Henly 233, 234, 236, *271*
Flink, E. B. 235, *271*
— R. McCollister, A S. Prasad, J. C. Melby u. R. P. Doe 235, 236, *271*
— F. L. Stutzman, A R. Anderson, T. König u. R. Fraser 235, 236, *271*
— s. Doe, R. P. 210, 250, 251, *269, 270*
— s. Jones, J E. 195, 251, 257, *278*
— s Lizarralde, G. 226, *280*
— s. Mazzocco, V. E. 225, *282*
— s. McCollister, R. 209, 210, 241, 252, *282*
— s. McCollister, R. J. 232, 235, 252, *282*
— s. Min, H. K. 210, *283*
— s. Prasad, A. S 196, 200, 235, 251, 254, *286*
Florio, M. A. s. Haddy, F. J. 213, 236, 257, *274*
Flowers, C. E. jr. 195, 214, 216, 258, *271*
Folk, B. P. s Lilienthal, J L. 224, 234, 257, *280*
Folk, G. E. s Riedesel, M. L. 198, *287*
Folkow, B, B. Johansson u B. Öberg 69, *77*
Follis, R. H. jr. 218, *271*
Fontenot, J. P., R W. Miller, C. K. Whitehair u. R. MacVicar 208, *271*

Foot, A S s. Bartlett, S 237, *264*
Forbes, G. B. s. Breibart, S 204, *265*
Forbes, R. M. 208, 221, 222, 223, 224, 227, 228, 251, *271, 272*
— s. Likuski, H. J A. 208, *280*
— s. McAleese, D. M. 220, 221, 222, *282*
Forsham, P. H. s. Ganong, W. F. 105, *149*
Forster, R. P. s. Berglund, F. 242, *264*
Fossan, D. D. van s. Goldman, A. S. 230, 231, 233, *273*
Foster, G. V, A. Baghdiantz, M. A. Kumar, E. Slack, H. A. Soliman u. I. MacIntyre 249, *272*
— G. F. Joplin, I. Mac Intyre, K. E. W. Melvin u E. Slack 249, *272*
Foulks, J. G. s. Ling, G M 95, *153*
Fourman, P., u. D. B. Morgan 214, 231, *272*
— s. Fitzgerald, M. G. 211, 229, 231, 232, 234, *271*
— s. Heaton, F. W. 230, 231, *276*
— s. Jones, K. H. 195, 214, 216, 231, 247, 248, 251, *278*
Fox, F. W. s. Walker, A R. P 208, *294*
Fox, R. H, u S. M Hilton 99, *149*
Frank Abbey 262
Frank, H. A. s. Carr, M. H. 194, *267*
Frankenhaeuser, B, u. A L. Hodgkin 213, *272*
— u. H Meves 212, *272*
Franko, B V., J W. Ward u. R S. Alphin 21, 37, 38, 39, 40, *77*
Fraser, R s Flink, E. B. 235, 236, *271*
— s Hanna, S. 247, 248, 251, *275*
Frater, R., S. E Simon u F. H Shaw 213, *272*

Frazer, J W. s Grimson, K S. 53, *78*
Frazer, S C s Stewart, C P. 190, *291*
Fredrickson, D s Boellner, S. W. 194, 257, *265*
Fredricsson, B, u F Sjoqvist 63, 67, *77*
Freedman, P, R. Moulton u. A. G. Spencer 242, *272*
Freier, E F. s. Glick, D. 202, *273*
French, J. D 88 *149*
Frenk, S. s. Metcoff, J. 232, *283*
Freyburger, W. A., C C. Gruhzit, B. R Rennick u. G K. Moe 48, 49, *77*
— s Moe, G. K 48, *81*
Freygang jr. W. s Evarts, E V. 89, *148*
Frick, O L, B N. Halpern u. P. Liacopoulos 176, 177, 181, *182*
Friedberg, K. D, u U. Bauer 181, *182*
— G. Engelhardt u. F. Meinecke 176, 177, 181, *182*
— s. Stegemann, H. 164, 168, 169, 170, *184*
Friedberger, E 160, 161, 162, 173, *182*
Friedman, H. S, u. M A Rubin 195, *272*
Friedman, R. s Wener, J. 222, 223, 224, 225, 226, 227, *295*
Friedman, U 90, *149*
Froede, H. s. Burn, J. H. 136, 137, 139, 140, *145*
Frohlich, E D., J. B. Scott u. F J Haddy 215, *272*
— s. Scott, J B 213, 215, *289*
Frye, C M. s. Colby, R. W. 227, 228, *268*
Furchgott, R F. s. Kirpekar, S M. 137, *152*
Fujimori, H. s. Riker, W. F. 134, *156*
Fujita, T. s. Shimamoto, T. 256, *290*
Fuster, J. M. 116, *149*
— u R. F. Docter 117, *149*

Fuwa, K s Borowitz, J. L. 203, *265*
— s. Wacker, W. E C 196, *294*
Fuxe, K. s. Carlsson, A 125, *145*
— s. Dahlstrom, A. 125, 147

Gaddum, J H. 8, *77*, 99, *149*, *182*
— u N. J. Giarman 42, *77*
— u. M K. Paasonen 42, *77*
— u. Z. P. Picarelli 8, 30, 31, *77*
Galambos, R. 116, *149*
Gale, C C s. Andersson, B. 110, *143*
Gamponia, P. J. s. Wallach, S. 196, 214, 247, 248, *294*
Gandini, A. s Benelli, G. 39, *74*
Ganong, W F , u. P. H. Forsham 105, *149*
Gantt, C L., u. W. J. Carter *272*
Garb, S 215, *272*
Gardner, L I., E. A. Mac-Lachlan u. H Berman 224, *272*
— s Silverman, S. H 196, 249, 251, 254, *290*
Garrels, R. M., u. M E. Thompson 200, *272*
Gartner, R J. W. 198, *272*
— J. W. Ryley u A W. Beattie 195, *272*
Gauduchon, J. s. Durlach, J. 248, *270*
Gausmann, F. s Bulger, H. A. 248, *266*
Gebber, G L., u. R. L. Volle 17, 18, *77*
Geertruyden, J. van 208, 214, 246, *273*
— u N Dejardin 246, *273*
Geesey, C. N s Koelle, G B 135, *152*
Gent, W L. G , J. R. Trounce u M. Walser 201, *272*
George, J. N. s Pollack, S. 257, *286*
George, R., jr. W. L. Haslett u. D. J Jenden 123

George, R. s. Haslett jr. W L. 123, *150*
George, W. K., u. W. S. Chambless 236, *272*
— D. A. Dotson u. W. W. Grant 199, *272*
Gérard, J. s. Laborit, H. 134, *153*
Gerbrandy, J., A. M. van Leeuwen, M. B. A. Hellendorn u. L. A. De Vries 195, *272*
Gernandt, B E. s. Sawyer, Ch H 92, *157*
Gerschenfeld, H. M., J. H. Tramezzani u. E. de Robertis 135, *149*
Gershoff, S. N., u. S. B. Andrus 259, *272*
— J. J. Vitale, I. Antonowicz, M. Nakamura u. E. E. Hellerstein 212, 228, 251, *273*
Gerst, P. H., M. R. Porter u. R. A. Fishman 235, 236, *273*
Gertner, S. B. 41, 60, *77*
— u R. Kohn 24, 42, *77*
— M. K. Paasonen u. N. J. Giarman 42, *77*
— u. A. Romano 59, *77*
Giarman, N. J s. Gaddum, J. H. 42, *77*
— s. Gertner, S. B. 42, *77*
Gibbs, W. D. s. Glaser, W. *273*
Gibson, W C. 61, 67, *77*
Giertz, H., u. F Hahn 161, 162, 164, 169, 171, 172, 173, 177, 178, 180, 181, *182*
— — I. Jurna u. A. Lange 180, *182*
— — u A Lange 162, *182*
Gigee, W s. Raab, W. 94, *156*
Gijón, E. s Pardo, E G. 56, *82*
Gilbert, D. L. 202, *273*
— u. J. MacGann 213, 247, *273*
Gill, jr. J. R., N. H. Bell u. F. C. Bartter 248, *273*
— s Curran, P. F. 247, *269*

Gillie, E. s. Brown, D. F. 256, *266*
Gillis, R A. s. Flacke, W. 49, 50, 53, *77*
Ginn, E., u. E L Becker 224, *273*
Ginn, H E., W. O. Smith, J F Hammarsten u. D. Snyder 239, *273*
— s. Kalbfleisch, J. M. 252, *278*
— s. Keyl, M. J. 195, 205, *278*
— s. Lindeman, R. D. 252, *280*
— s. Smith, W. O. 196, 221, 222, 224, 225, *291*
Ginsborg, B. L. 32, *77*
— u. S Guerrero 12, 15, *77*
Ginsburg, F M. s. Ginsburg, S 195, 201, 202, 257, *273*
Ginsburg, S , J. G. Smith F. M. Ginsburg, J. Z. Reardon u. J. K. Aikawa 195, 201, 202, 257, *273*
Ginzel, K. H., u. S. R. Kottegoda 39, *78*
Girard, A. s. Larvor, P. 222, 223, 226, 227, *279*
Girard, M. s. Dreux, C. 195, *270*
Gitelman, H. s. Welt, L. G 190, *295*
Gitelman, H. J , J. B. Graham u. L. G Welt 230, *273*
— S. Kukolj u L. G. Welt 216, 223, 249, *273*
— s. Blythe, W. B. 241, 242, *265*
Gittelman, I F , J. B. Pinkus u. E Schmertzler 249, *273*
Glaesser, A., u P. Mantegazzini 95, 96, *149*
Glaser, W , u. J. L. Brandt *273*
— u. W. D. Gibbs *273*
— s Brandt, J. L. 203, 250, *265*
Glaubitt, D., u. J. G. Rausch-Strooman 210, 252, *273*
Glick, D., E. F. Freier u. M. J. Ochs 202, *273*

Glickman, L. S , V. Schenker, S Grolnick, A. Green u. A. Schenker 236, *273*

Globus, D L. s Samiy, A. H. E 214, 239, 242, *289*

Godden, W. s. Duckworth, J 220, 221, *270*

Goekhan, N. s Winterstein H. 93, *159*

Gokhale, S. D , O. D. Gulati u N Y. Joshi 55, *78*

Gold, D , u. H Reinert 60, *78*

Goldman, A. S , D. D. van Fossan u. E E Baird 230, 231, 233, *273*

Goldsmith, R E. s Agna, J. W 247, *262*

Goldstein, L., u. C. Munoz 95, *149*

Gollub, L. R., u. J V. Brady 88, *149*

Gomez Alonso de la Sierra, B. s. Jones, A. 17, 37, 38, 39, *79*

Gonick, H C. s. Better, O. 241, 254, *264*

Goodman, L S. s Woodbury, D M. 249, *296*

Gordan, G. S. s. Loken, H. F 200, *280*

Gordillo, G. s Metcoff, J. 232, *283*

Gordon, E. B. s. Barnes, B A. 194, 211, 229, 232, *264*

Gordon, G. S s Aikawa, J. K 207, *262*

Gottesman, K. S. s. Miller, N. E. 113, *154*

Gottlieb, L. S , S. A. Broitman, J J. Vitale u. N. Zamcheck 206, 252, *273*

Gottschalk, C. W. s Lassiter, W. E. 240, *279*

Gow, B S 205, *273*

Gowdey, J s. Lipkin, G. 195, *280*

Gower-Smith, S. 216, *273*

Granicher, D., u. H Portzehl 211, *273*

Graniches-Frick, D. 212, *274*

Graham, G. G. 234, *274*

Graham, J B s Gitelman, H J 230, *273*

Graham, L A , J J. Caesar u A S V. Burgen 209, *274*

Granit, R. 116, *149*

— u R. Kaada 116, *149*

Grant, W. W. s. George, W. K. 199, *272*

Grantham, J. J , W. H Tu u. P. R. Schloerb 215, 236, *274*

Graystone, J. E s. Cheek, D B 194, 247, *267*

Greef, K., B. G. Benfey u. A. Bockelmann 176, *182*

Green, A s. Glickman, L S. 236, *273*

Green, A F. s Boura, A. L A. 38, 59, *75*

Green, D. E. s Brierley, G. P. 247, *266*

Green, J D. s Cross, B A. 92, *146*

Green, J. H. s McCubbin, J. W. 8, 39, *81*

Greenberg, D M., u R. B Aird 257, *274*

— C E. Anderson u. E V. Tufts 226, *274*

— S P Lucia, M. A Mackey u E. V. Tufts 195, *274*

— — u. E. V. Tufts 225, *274*

— u. M. A. Mackey 248, *274*

— u. E V. Tufts 198, 219, *274*

— s. Kleiber, M 225, 253, *279*

— s Lowenhaupt, E. 226, *281*

— s. Tufts, E. V 220, 221, 222, 223, 227, *293*

Griffith, F D , H E. Parker u. J. C Rogler 217, *274*

Grimes, jr., O. R. s. Langley, L. L 245, *279*

Grimson, K S., A. K. Tarazi u J W. Frazer 53, *78*

Griswold, R. L., u. N. Pace 203, *274*

Groat, W. C. de s. Takeshige, C. 21, *83*

Grobecker, H s. Schumann, H J 137, *157*

Grodzinska, L. s Marczynski, T J. 124, *154*

Grollman, A. 205, 247, 250, *274*

Grolnick, S. s Glickman, L S 236, *273*

Gromadzki, C. G., u. G. B. Koelle 63, 67, *78*

Groot, J. de s Littlejohn, B M. 107, *153*

Grossman, E B s. Soffer, L. J. 251, *291*

Grossman, M. T. s. Janowitz, H. D. 110, *151*

Grossman, S P. 99, 100, 101, 103, 104, 111, 112, *149*

Gruhzit, C C. s. Freyburger, W. A. 48, 49, 77

Gudden 128, 129

Gudmundsson, T. V., I. MacIntyre u H. A. Soliman 228, *274*

Guerrero, S. s. Ginsborg, B L. 12, 15, *77*

Guillemin, R 105, 108, *149*

— s. Fields, W. S. 105, *148*

Gulati, O D. s. Gokhale, S. D 55, *78*

Gusmano, E A. s Clark, I. 217, *268*

Guth, L. 68, *78*

Guzman, C s. Vitale, J. J. 226, 227, *294*

Gyermek, L. 30, 31, 41, 42, 78

— u. E. Bindler 31, 42, 78

— E. B. Sigg u. E. Bindler 31, 40, 78

— u L Sztanyik 7, 78

— u. K. R Unna 40, 78

— s. Herr, F. 6, 19, 40, 78

Haas, G. M. s. MacCreary, P. A. 225, *282*

Haas, H. G , H. Affolter u U. C. Dubach 248, *274*

Hackett, J. s. Marczynski, T. J. *154*

Haddy, F. J , J B Scott, M. A. Florio, R M Dougherty jr u J. N. Huizenga 213, 236, 257, *274*
— s. Frohlich, E D 215, *272*
— s. Keyl, M. J. 195, 205, *278*
Haddy, U F. J. s. Scott, J B 213, 215, *289*
Haggendal, J. s. Andén, N. E 125, *143*
Hanze, S. 190, 209, 246, 252, 253, 254, 258, *274*
— u. W. Hiller 254, *274*
Harkonen, M. s. Eranko, O. 59, *76*
Hagbarth, K. E., u D E B Kerr 116, *149*
Hagen, P. B s Cohen, L H 52, *75*
Hahn, F , u A Lange 168, *182*
— u. A. Oberdorf 161, 180, *182*
— s Giertz, H. 161, 162, 164, 169, 171, 172, 173, 177, 178, 180, 181, *182*
Haining, C G 171, 177, *183*
Hall, D. G 195, 199, 256, *274*
Halliday, J. A. s Ardill, B. L. 214, 216, *263*
Hallman, E. T. s Moore, L A. 226, 227, *284*
Halpern, B N., u. P. Liacopoulos 181, *183*
— s. Frick, O. L. 176, 177, 181, *182*
Hamberger, B., u. K. A Norberg 58, 59, 71, *78*
— — u F. Sjoqvist 58, 71, *78*
— — u. U. Ungersted 58, 59, 71, *78*
Hamburger, J. 195, 253, 254, *274*
Hamilton, M. G., u. M. L. Petermann 203, *275*
Hammarsten, G 259, *275*
Hammarsten, J F., M. Allgood u W. O. Smith 209, 214, 215, 254, *275*

Hammarsten, J F u. W. O. Smith 236, *275*
— s. Carubelli, R. 194, *267*
— s. Ginn, H E 239, *273*
— s. Heller, B. I. 215, *276*
— s Smith, W. O 196, 221, 222, 235, 252, 253, 254, 255, *291*
Haneson, I B 213, *275*
Hanig, R C , u. M. H. Aprison 195, *275*
Hanna, C., P B McHugo, u. W. H MacMillan 39, 78,
Hanna, S. 195, 235, 249, *275*
— M. Harrison, I. MacIntyre u R. Fraser 251, *275*
— M T. Harrison, I. MacIntyre u R Fraser 251, *275*
— u. I MacIntyre 249, 250, *275*
— K. A. K. North, I. MacIntyre u R Fraser 247, 248, *275*
— s Beechey, R B 225, *264*
— s. Booth, C C 232, 233 234, 235, 236, *265*
— s MacIntyre, I. 195, 231, 232, 233, *281*
— s Sabour, M. S. 226, *288*
Hannon, J P , A. M. Larson u. D. W. Young 198, *275*
Hansen, J D. L. s Linder, G. C 231, 232, 233, *280*
Hardin, R A. s Scott, J. B. 213, 215, *289*
Harding, R. S. s. Conzolazio, C F. 205, 210, *268*
Hardy, G. C. s Stiles, D. E. 196, *291*
Hardy, J D. 93, *149*
Harmon, M 245, *275*
Harms, D R. s. Aikawa, J. K. 194, 220, 221, 229 231, 250, 253, *262*
Harris, G. W. 94, 100, 105, 108, *149*, *150*
Harris, H , I. R McDonald u. W. Williams 253, *275*

Harris, J. S , u. W. J A. DeMaria 258, *275*
— s. DeMaria, W. J. A 258, *269*
Harris, W. H , u. J. A. Beauchemin 257, *275*
— u. E. H Sonnenblick 195, *275*
Harrison, H. C. s. Harrison, H E 248, 249, *275*
— s. Lifshitz, F. 223, 224, 226, *280*
Harrison, H. E. 262
— u. H. C Harrison 248, 249, *275*
— s. Lifshitz, F 223, 224, 226, *280*
Harrison, M. s. Hanna, S 251, *275*
Harrison, M T. s. Hanna, S. 251, *275*
Harrison, T. s. Barnes, B A. 210, 229, *264*
Harrop, G. A , L. J. Soffer, R. Ellsworth u J. H. Trescher 250, *275*
Harry, J. 32, *78*
— s Brownlee, G. 32, *75*
Hart, E. B., u H. Steenbock 217, *275*
Harter, H. s. Schimpf, K 217, *289*
Harvey, R. B. s. Keyl, M. J 195, 205, *278*
Harvey, S M. s. Kenyon, F. E. 257, *278*
Haslett, W. L s Cho, A. K. 123, *146*
Haslett jr., W. L. 123, *150*
— R George u D. J. Jenden 123, *150*
— s George, R 124
Hasselman, J. J. F , u. E. J. van Kampen 190, *275*
Hastings, A B. s. Lowry, O H 202, *281*
Hatcher, J D s Kelly, H. G. 214, 215, 216, *278*
Hathaway, M. L. 207, 229, *275*
Haury, V. G., u A Cantarow 253, 254, *275*
— s. Hirschfelder, A. D. 254, *277*

Havel, R. J s Loken, H. F. 200, *280*

Haven, F. L. s. Rogers, T. A. 203, *288*

Hawrylko, J. s. Eliel, L. P. 228, 248, *270*

Hayama, T., u. Y Ogura 93, *150*

Hayden, C. C. s Knoop, C. E 221, *279*

Haywood, J., u. R. Selvester 217, 259, *275*

Head, M J , u. J. A F. Rook 237, *276*

Heagy, F. C , u. A C. Burton 198, *276*

Heaton, F. W. 195, 220, 221, 223, 224, 235, 247, 248, *276*

— u. C. K. Anderson 195, 222, 223, 226, 227, 247, 248, *276*

— u. P Fourman, 230, 231, *276*

— u. A. Hodgkinson 198, 210, 241, *276*

— — u. G A. Rose 208, 249, *276*

— u. F. M. Parsons 216, 217, 229, *276*

— u. L. N. Pyrah 209, 247, 248, *276*

— — C. C. Beresford, R. W. Bryson u. D. F. Martin 235, 241, 255, *276*

— s. Hodgkinson, A 210, *277*

— s. Martindale, L 220, 248, *282*

Heaton, T B., u. M. H. MacKeith 5, *78*

Hebb, C. O., u. H. Konzett 29, *78*

— u. G M. H. Waites 62, *78*

Hedler, L. s. Marquardt, P. 178, *183*

Heggtveit, H. A. 227, *276*

— L. Herman u. R. K. Mishra 226, 227, *276*

Hegsted, D. M. s Hellerstein, E. E 228, 251, 255, 256, *276*

— s. Vitale, J. J. 225, 226, 251, 255, 256, *293, 294*

Hegsted, I. M., J. J. Vitale u. H. McGrath 206, 227, *276*

Hegsted, J M. s. Nakamura, M. 255, *285*

Hegyi, G s Muhlrad, A 211, 212, *284*

Hehl, U. s. Knippers, R 195, 214, 215, 216, 239, 240, 252, *279*

Heidenhain 246

Heinbecker, P. s. Bishop, G. H. 69, *74*

Heinrich, H G 259, *276*

Hellendoorn, M. B. A. s Gerbrandy, J. 195, *272*

Heller, B. I., J. F. Hammarsten u. F. I. Stutzman 215, *276*

Hellerstein, E E , M. Nakamura, D. M. Hegsted u. J. J. Vitale 255, 256, *276*

— J. J. Vitale, P. L. White, D. M. Hegsted, N. Zamcheck u. M. Nakamura 228, 251, 255, 256, *276*

— s. Gershoff, S N. 212, 228, 251, *273*

— s. de Giorgio, J. 225, *269*

— s. Nakamura, M. 255, *285*

— s. Seta, K 222, 223, 224, 225, 228, *290*

— s. Vitale, J. J. 222, 226, 227, 228, 255, 256, *293, 294*

Hellman, L. s. Zumoff, B. 209, *296*

Helve, O. E. 250, *276*

Hemingway, R. G., u. N. S. Ritchie 237, *276*

— s. Ritchie, N S. 230, 236, 237, *287, 288*

Hemkes, O. J. s. Kemp, A. 208, 236, 237, *278*

Henderson, F. G., B. L. Martz, u. I. H. Slater 112, *150*

Hendricks, S. B , u. W. L. Hill 201, *276*

Hendrix, J. Z , N. W Alcock u. R. M. Archibald 243, 244, 245, *276*

Henle 240

Henly, A A. s Fletcher, R F 233, 234, 236, *271*

Henneman, P H , u. E F. Dempsey 207, *277*

Hennesy, B. L. s Bishop, P. O. 89, *144*

Henriques, V., u. S L. Ørskov 201, 257, *277*

Herman, L. s. Heggtveit, H A. 226, 227, *276*

Hernandez-Peon, R 116, 117, 127, 128, 129, 130, *150*

— u. G. Chavez-Ibarra 128, 129, 130, 131, 132, 135, *150*

— — u. E Aguilar-Figueroa 131, 132, 135, *150*

— — J. P Morgan u. C. Timo-Iaria 99, 103, 122, 123, 129, 131, 132, 135, *150*

— u. H Scherrer 116, *150*

— — u. M Jouvet 116, *150*

— — u M. Velasco 116, 117, *150*

— s Velluti, R 129, *159*

Herr, F., u L Gyermek 6, 19, 40, *78*

Herrera, C. s Fellers, F X 221, *271*

Herring, W. B , B. S Leavell, L M. Paixao u. J. H. Yoe 195, *277*

Herrlich, H. C. s. Surawicz, B. 216, *292*

Hertzler, E. C 31, *78*

Hervey, G. R. 110, *150*

Hess, R., I. MacIntyre, N. Alcock u. A. G. E. Pearse 222, 223, 226, *277*

Hess jr , R. s. Akert, K 127, *143*

Hess W. R. 88, 127, 132, *150, 151*

Hetenyi, E. s. Varga, F. 135, *159*

Hetherington, A. W., u. S W. Ranson 110, *151*

Hetzel, B. S. s. Charnock, J S. 253, *267*

Heusser, H. s. Kuhn, W. L. 133, *153*

Heuvel-Heymans, G. van den s Heymans, C 7, 8, *79*

Heymans, C., u G. van den Heuvel-Heymans 7, 8, *79*

— s. Atanackovic, D 60, *74*

Hiatt, H. H. s. Revel, M. 212, *287*

Hiebel, G., M. Bonvallet, P. Huvé u P. Dell 96, 139, *151*

— s Bonvallet, M. 92, 120, *144*

Hill, B M s. Osler, A. G. 163, 165, 172, 173, 174, *183*

Hill, W L s. Hendricks, S. B. 201, *276*

Hillarp, N. Å. s. Carlsson, A. 125, *145, 146*

Hillebrecht, R s. Stegemann, H. 164, 169, 170, *184*

Hiller, W. s Hanze, S. 254, *274*

Hills, A. G , D. W. Parsons, G. D. Webster jr., O. Rosenthal u H. Conover 210, 250, 253, *277*

Hilton, J. G. 54, *79*

— u. M. Steinberg 48, 49, *79*

Hilton, S. M. s. Fox, R. H 99, *149*

Himwich, H. E 88, *151*

— u F. Rinaldi *151*

— s. Rinaldi, F. 88, 96, 122, 138, *156*

Hingerty, D. 195, 214, *277*

— s. Conway, E. J. 250, *268*

Hioco, D. s Parlier, R. 190, *286*

Hiramatsu, M. s. Nakamura, M. 225, 226, 227, 251, 255, *284, 285*

Hirano, J. 225, *277*

— s. Nakamura, M 255, *284, 285*

Hirata, T. s. Takayasu, H. 196, 215, 253, 254, *292*

Hirschfelder, A. D , u. V. G. Haury 254, *277*

Hirschmann, L. 11, *79*

Hirschowitz, B. I. s. Balint, J A. 235, *263*

Hodgkin, A. L s Frankenhaeuser, B. 213, *272*

Hodgkinson, A., u. F. W. Heaton 210, *277*

— s Heaton, F. W. 198, 208, 210, 241, 249, *276*

Hoebel, B G., u. Ph. Teitelbaum 110, 111, *151*

Hoefer, J. A. s. Miller, E. R. 216, 217, 222, 223, 249, *283*

Hofer, M , u. A Kleinzeller 195, 246, *277*

Hoekstra, M. H. s. Charbon, G. A. 194, *267*

Hösli, L. s Monnier, M. 115, 127, 133, 140, *155*

Hoet, J. 4, 5, 38, *79*

Hoff, H. E., P. K. Smith u. A. W. Winkler 212, *277*

— s. Smith, P. K. 214, 215, 216, *290*

— s. Winkler, A. W. 258, *295*

Hogan, A G , W. O. Regan u. W. R. House 227, *277*

— s. House, W. B 227, *277*

Hogben, L. T., W Schlapp u. A. D MacDonald 7, *79*

Holaday, D A., K. Kamijo u. G B. Koelle 53, 54, *79*

Holland, R C., B. A. Cross u. C. H. Sawyer 92, *151*

— J. W. Sundsten u. C. H. Sawyer 92, *151*

Holliday, M. A s. Cotlove, E 194, 221, 224, *268*

Holmstedt, B., u. F. Sjöqvist 67, *79*

Holt, A. B. s. Cheek, D. B 194, 247, *267*

Holt, J. M s. Valberg, L. S. 196, 257, *293*

Holt, S. P ten s. De Vries, L. A. 210, *269*

Homann, R. s Martin, H. E. 235, *281*

Homann jr., R. E s. Edmondson, H. A. 258, *270*

Homer, L. 247, *277*

Honorata, C. R., u. E. Rosa 225, *277*

Hoobler, S. W., H. D. Kruse u. E. V. McCollum 220, *277*

Hopkins, T., J. E. Howard u. H. Eisenberg 200, *277*

Hopkinson, P. s. Clegg, P. C 213, *268*

Hornykiewicz, O., u. W. Kobinger 139, *151*

Horton, R., u. E. Bigheri 249, 250, *277*

Hosko, M. G. s. Platner, W. S. 198, *286*

Houck, C. R. s. Etteldorf, J. N. 214, *270*

House, W. B., u. A. G. Hogan 227, *277*

House, W. R. s. Hogan, A. G. 227, *277*

Howard, A. V. s. Wrong, O 205, *296*

Howard, J. E. s. Hopkins, T. 200, *277*

— s. Mukai, T. 259, *284*

Hozdrx, T., u. L. Atanassowa 258, *277*

Hubel, D. H. 121, 127, *151*

Huffines, W. D. s. Richardson, J. A. 249, *287*

Huffman, C. F. s. Duncan, C. W. 238, *270*

Hugelin, A. s. Dell, P. C. 88, *147*

Hughes, A., u. R. S. Tonks 256, *277*

Hughes, E. R s. Boellner, S. W. 194, 257, *265*

Hugot, D. s Randoin, L. 217, *287*

Huizenga, J. N. s. Haddy, F J. 213, 236, 257, *274*

Hume, D. M. 105, *151*

Hummel, F. C., H. A. Hunscher, M. F. Bates, P. Bonner, I. G Macy u. J.A. Johnston 191, *277*

— H. R. Sternberger, H. A. Hunscher u. I. G. Macy 191, *277*

Hungerford, G. F. 219, 227, *277*

— s Kashiwa, H. K. 219, *278*

Hungerford, G. H. s. Bernick, S. 226, *264*

Hunt, B. G s Opie, L H 209, 231, 233, 234, *285*

Hunt, C. C , u P. G. Nelson 61, 62, 63, *79*

Hunt, R. 5, *79*

Hunter, G , u. H V. Smith 195, *278*

Hunter, J , u H. H. Jasper 127, *151*

Hunscher, H. A. s. Hummel F. C. 191, *277*

Huntington 257

Huntsman, R. G., B. A. L. Hurn u. H. Lehmann 217, *278*

Hurn, B. A. L. s. Huntsman, R G. 217, *278*

Hutter, O F., u. K. Kostial 212, *278*

Huvé, P. s Hiebel, G. 96, 139, *151*

Iida, C. s. Wacker, W. E. C. 196, *294*

Imai, T. s Nakamura, M 195, 202, 255, *285*

Imarisio, J. J. s Zumoff, B. 209, *296*

Inglis, J S. S., M Weipers u. P. J. Pearce 198, 237, *278*

— s. Ritchie, N. S. 230, 236, *288*

Ingvar, D H 133, *151*

Iorio, L. C., u. R. J. McIsaac 22, 24, 25, 29, *79*

Irving, J T. s Walker, A. R. P. 208, *294*

Iseri, L. T. s Albert, D. G. 194, 256, *262*

— s. Mader, I J 249, *281*

Ishihara, Y. s. Nakamura, M. 195, 202, 255, *285*

Isola, J. B s. Ungar, G. 163, 175, *184*

Iwahara, S. s. Shimamoto, T. 256, *290*

Jaanus, S. D., u. R. P. Rubin 60, *79*

Jabir, F. K., S. D. Roberts u. R. A Womersley 210, 215, 253, *278*

Jack jr., J s Regen, D. M. 212, *287*

Jacobowitz, D. 60, 71, *79*

— u. G B Koelle 138, *151*

— P. Johnson, I. Kitchner u. G B Koelle 138, *151*

Jacobs, M s Soffer, L. J. 251, *291*

Jakerow, A. s. Bélanger, L. F 219, *264*

James, G W L s. Collier, H O. J 34, *75*

Janowitz, H D , u M T. Grossman 110, *151*

Jarausch, K. H. s Ullrich, K J 196, 240, *293*

Jasper, H 118, *151*

Jasper, H. H. s. Hunter, J. 127, *151*

Jenden, D. J. s Cho, A. K 123, *146*

— s George, R 124

— s. Haslett jr., W. L. 123, *150*

Jeness, R. s. Christianson, G. 205, *268*

Jenkins, K J s Bunce, G E. 222, 223, 227, *266*

Jenkinson, D. H 213, *278*

Jeppson, P. G. 90, *151*

Jewell, P. A., u E B. Verney 92, *151*

Job, C., u A Lundberg 70, *79*

Jobling, J. W , u. W Petersen 162, *183*

Johansen, E. 195, 204, *278*

Johansson, B. s. Folkow, B 69, *77*

Johnson, P. s Jacobowitz, D 138, *151*

Johnston, J A s Hummel, F C. 191, *277*

Jones, A. 21, 22, 23, 24, 25, 26, 27, 29, 31, 32, 40, 45, *79*

— B Gomez Alonso de la Sierra u U Trendelenburg 17, 37, 38, 39, *79*

— s. Brandt, J. L. 203, *265*

— s Trendelenburg, U 22, 23, 27, 45, *84*

Jones, G T. s Slater, I. H. 112, *157*

Jones, H L s Wallach, S. 196, *294*

Jones, J. E., P. C. Desper u. E B. Flink 257, *278*

— — S. R Shane u. E B. Flink 195, 251, *278*

— s Mazzocco, V E 225, *282*

— s Min, H. K. 210, *283*

Jones, J. H. 166, 173, *183*

Jones, K. H., u. P. Fourman 195, 214, 216, 231, 247, 248, 251, *278*

Jones, R s. Martin, H E 210, 247, *281*, *282*

Joplin, G F s. Foster, G. V. 249, *272*

Joseph, N R. s. Catchpole, H R. 202, *267*

Joshi, N. Y. s. Gokhale, S. D 55, *78*

Jouany, J M s. Laborit, H 134, *153*

Jouvet, M 121, 127, 131, *152*

— u J E Desmedt 116, *152*

— s Hernandez-Peon, R 116, *150*

Judah, J. D., K. Ahmed, A E M. McLean u G. S Christie 247, *278*

Jung, R. s Creuzfeldt, O. 116, *146*

Jurna, I. s Giertz, H 180, *182*

Kaada, R. s. Granit, R. 116, *149*

Kabat, E A., u. M M. Mayer 180, *183*

Kalbfleisch, J M , R. D. Lindeman, H E Ginn u. W. O. Smith 252, *278*

— — u W O Smith 252, *278*

— s. Lindeman, R D 252, *280*

Kamijo, K , u. G. B. Koelle 52, 53 , *79*

— s Holaday, D A 53, 54, *79*

Kampen, E. J. van s. Hasselman, J J. F. 190, *275*

Kaneko, Y , J W. McCubbin u I. H Page 8, *79*

Kanematsu, S. 107, *152*

Kapitola, J., u. O. Kuchel 249, 250, *278*

Kapteyn, P. C s van Leeuwen, A. M. 196, *280*

Karabus, C D s. Linder, G C 231, 232, 233, *280*

Karczmar, A. G. 88, 122, 134, 138, *152*

Kashiwa, H. K. 228, *278*

— u. G. F. Hungerford 219, *278*

Katz, B. s. Birks, R. 63, *74*

— s. Del Castillo, J. 15, *76*

Kaufman, R. M. s Pollack, S. 257, *286*

Kawakami, M. s. Sawyer, Ch. H. 105, *157*

Keller, A. D s Witt, D. M 110, *159*

Kelly, H. G , H. C Cross, M. R. Turlon u. J. D. Hatcher 214, 215, 216, *278*

Kemény, A., H. Boldizsár u. G Pethes 195, 214, 216, *278*

Kemp, A , O. Deijs, O J. Hemkes u. A. J. H. van Es 208, 236, 237, *278*

Kennedy, G. C. 110, *152*

Kennedy, R. s Anast, C 243, *263*

Kenyon, F. E , u. S M. Harvey 257, *278*

Keoplung, M s. Carpenter, B. E. 217, *267*

— s Dunn, M. 254, *270*

Kerr, D. E. B. s. Hagbarth, K E. 116, *149*

Kessler, R. H. s. Samiy, A. H. E. 239, 242, *289*

Kessner, J. M., u F. H. Epstein 222, 224, 226, 244, *278*

Kewitz, H. 72, *79*

— u. H. Reinert 60, *79*

Key, B. J., u. E. Marley 98, *152*

— s. Bradley, P. B. 138, *144*

— s. Marley, E. 98, *154*

Keye, H. s Kohn, R. R. 195, *279*

Keyl, M. J , J. B Scott, J. M. Dabney, F. J. Haddy, R. B. Harvey, R. D. Bell u. H. E. Ginn 195, 205, *278*

Keynes, W. M. s Care, A. D 248, 249, *266*

Kibjakow, A. W. 9, *79*

Kiil, F. 252, *279*

Killam, E K. 88, 138, *152*

Kirpan, J. s. Poutsiaka, J. W. 252, *286*

Kirpekar, S M., P. Cervoni u. R. F. Furchgott 137, *152*

Kitchner, I. s Jacobowitz, D. 138, *151*

Klee, M. s. Kornmuller, A E 133, 140, *152*

Kleeman, C. R., F. H Epstein, D McKay u E. Taborsky 195, 251, *279*

— S. Ling, D. Bernstein, M H Maxwell u. L. Chapman 243, *279*

— s. Better, O. 241, 254, *264*

— s. Daniels, F. 198, *269*

Kleiber, M , M. D Boelter u. D. M. Greenberg 225, 253, *279*

Kleiber, R. E., K. Seta, J J. Vitale u. B Lown *279*

Kleiger, R. s. Seta, K 224, 225, *290*

Kleiner, I S s Menaker, W. 218, *283*

Kleinzeller, A s. Hofer, M. 195, 246, *277*

Klingman, W. D. s. Suter, C 222, *292*

Knapp, D. E., u. E F. Domino 123, *152*

Knippers, R , u. U Hehl 195, 214, 215, 216, 239, 240, 252, *279*

Knoop, C. E., W. E Krauss u. C. C Hayden 221, *279*

Ko, K. W , F X. Fellers u. J. M Craig 222, 226, *279*

Kobinger, W. s. Hornykiewicz, O. 139, *151*

Kobrin, S. s Ungar, G. 163, 175, *184*

Kocacs, M. s Muhlrad, A 211, 212, *284*

Koella, W. s. Akert, K. 127, *143*

Koelle, G B 52, 63, *79*, 87, 103, 134, 135, 143, *152*

— C. N Geesey u. C. F. Schmidt 135, *152*

— u E. C. Steiner 97, *152*

— s. Abrahams, V. C. 135, *143*

— s. Gromadzki, C G. 63, 67, *78*

— s. Holaday, D A. 53, 54, *79*

— s Jacobowitz, D. 138, *151*

— s Kamijo, K 52, 53, *79*

— s. Matsumara, M. 68, *81*

— s McKinstry, D. N 20, *81*

— s. Volle, R. L 18, 19, 44, 54, *84*

Koelle, W. A. s. McKinstry, D. N. 20, *81*

Koenig, E. s McKinstry, D. N 20, *81*

Koenig, W. s. Losse, H. 253, 254, *280*

Kohler, H. F., u M M. Pechet 248, 249, *279*

— s Macrae, I. F 215, *281*

Kohn, R s. Bovet-Nitti, F. 39, *75*

— s. Gertner, S B. 24, 42, *77*

Kohn, R. R., H. Keye u. E. Rollerson 195, *279*

Koide, H. s. Vitale, J J. 228, *294*

Koike, M. s. Nakamura, M. 225, 226, 227, 251, *285*

Koizumi, K. s. Curtis, D. R. 91, *147*

Koketsu, K. s. Eccles, J. C. 90, 91, *148*

Koller, Th., u. M. Monnier 133, 140, *152*

Komalahiranya, A., u. R.L. Volle 55, *79*

Konig, s. Flink, E. B. 235, 236, *271*

Konzett, H. 6, 23, 29, 30, 56, *80*
— u. E. Rothlin 39, 43, *80*
— u P. G. Waser 23, 31, *80*
— s. Hebb, C O 29, *78*
Koppányi, Th. 4, 5, 37, *80*
Kornmuller, A. E 133, *152*
— D. H. Lux u K. Winkel 133, 140, *152*
— — — u. M. Klee 133, 140, *152*
Kosovsky, J D. s. Kupfer, S. 241, 253, *279*
Kosterlitz, H. W , u. J. A. Robinson 71, *80*
Kostial, K s. Hutter, O. F. 212, *278*
Kottegoda, S R s. Ginzel, K. H. 39, *78*
Kovacs, T s. Varga, F. 135, *159*
Kover, A. s Varga, F. 135, *159*
Koyano, H. s Eyzaguirre, C 213, *271*
Krahl, M E. 212, *279*
Krane, S. M. s. Barnes, B. A. 247, *264*
Krauss, W. E. s. Knoop, C. E 221, *279*
Krehl, W. A. s. Dunn, M. 254, *270*
Kreveld, A. van, u. G van Minnen 205, *279*
Krick, E. T. s. Rubin, M. I. 250, *288*
Krieger, H. s. Levey, S. 230, *280*
Kristic, M., u. V. Varagić 55, *80*
Krnjević, K., R. Laverty u. D. F. Sharman 89, *152*
— u. J. F. Mitchel 136, *152*
— u. J. W Phillis 89, 91, 97, *153*
Krogh, A. 90, *153*
Krook, L. s Carillo, B. J. 222, 224, 226, 228, *267*
Kruif de s. Novy 161
Krupp, P s. Monnier, M 115, 127, *155*
Kruse, H. D , E R. Orent u. E V. McCollum 218, 220, *279*

Kruse, H. D., M M. Schmidt u. E. V. McCollum 225, *279*
— s. Hoobler, S. W. 220, *277*
— s. Orent, E R 219, *286*
Kuchel, O. s. Kapitola, J. 249, 250, *278*
— s. Vostal, J 230, 239, *294*
Kuhn, W. L., H. Majer, H. Heusser u. B. Zen Ruffinen 133, *153*
Kukolj, S. s. Gitelman, H. J. 216, 223, 249, *273*
Kumagai, T. 177, *183*
Kumar, D., P. A. Zourles, u. A. C. Barnes 216, *279*
Kumar, M. A. s. Foster, G. V. 249, *272*
Kunkel, H O., K. H. Burns u. B. J. Camp 237, *279*
Kuntz, A , u G. Saccomanno 70, *80*
Kuntzman, R. s. Costa, E. 59, 60, *75*
Kuperman, A. S. s. Riker, W. F. 134, *156*
Kupfer, S., u. J. D. Kosovsky 241, 253, *279*
Kuypers, G. M. M. s. Nauta, W J. H. 119, *155*
Kuz'mina, S. V. 70, *80*
Kyriakopoulos, A. A. s. Smith, W O. 196, 221, 222, 224, 252, *291*

Laborit, H , J. M. Jóuany, J. Gérard u. F. Fabiani 134, *153*
La Dou, J. s. Mendelson, J. H. 235, *283*
Laidlaw, P. P. s. Dale, H. H. 4, 5, *76*
Laing, G. H. s. McCreary, P. A. 225, *282*
Lambert, R. s Vagne, M. 205, *293*
Lamson, S. A s Leichsenring, J M 206, 209, 210, *280*
Landau, W. s. Evarts, E V. 89, *148*
Lange, A. s. Giertz, H. 162, 180, *182*
— s. Hahn, F. 168, *182*

Langer, S. Z. 62, 63, *80*
— u. U. Trendelenburg 63, 64, *80*
— s. Smith, C. B 62, *83*
Langley, J. N , u W L. Dickinson 11, *80*
Langley, L L , O R Grimes jr. u. D. F. Cockrell jr 245, *279*
Lankford, H. G. s. Sullivan, J. 235, 255, *292*
— s. Sullivan, J. F. 230, 243, 252, *292*
Large, B. J. s. Burn, J. H 136, 137, 139, 140, *145*
Larson, A. M s. Hannon, J. P. 198, *275*
Larson Carol 262
Larsson, B. s Andersson, B. 93, 110, *143*
Larsson, S 100, 110, *153*
Larvor, P., A. Girard, M. Brochart, A Parodi u J. Sevestre 222, 223, 226, 227, *279*
Lassiter, W. E , C W Gottschalk u. M. Mylle 240, *279*
Lauler, D. P. s. Wesson jr , L G. 239, *295*
Laurin, C. s. Cordeau, J. P. 102, 103, 119, 120, 122, 123, 124, 126, 132, 135, 138, *146*
Laverty, R. s. Krnjević, K. 89, *152*
Lavietes, P. H s Dine, R F 251, *269*
Lazzara, R K. s. Burch, G. E 194, 202, *266*
Leavell, B S. s Herring, W. B. 195, *277*
Leblanc, R. s. Parlier, R. 190, *286*
Lebrun, R s Durlach, J. 235, 236, *270*
Lebuc, R. s. Durlach, J 248, *270*
Lecomte, J 7, 32, *80*
— J. Troquet u. A. Dresse 8, *80*
Lee, F. L , u. U. Trendelenburg 32, 33, 51, *80*
Lee, J. S. s. Breibart, S. 204, *265*

Lee, K. S , K Tanaka u. D. H. Yu 212, *279*

Lee, W. C., u. F. E. Shideman 18, *80*

Leeman, C R. s. Quinn, M. 196, 198, *287*

Leeuwen, A. M van 196, 197, 199, *280*

— C. M. Thomasse u. P. C. Kapteyn 196, *280*

— s. Gerbrandy, J. 195, *272*

Lehmann, H s. Huntsman, R. G. 217, *278*

Lehninger, A. L , u. C. L Wadkins 211, *280*

— s. Carafoli, E. 247, *266*

Lei, B W , u. G. D. Maengwyn-Davies 59, *80*

Leichsenring, J. M., L. M. Norris u. S. A. Lamson 206, 209, 210, *280*

Leimdorfer, A , u. W. R. T. Metzner 98, *153*

Lengemann, F. 208, *280*

Lepeschkin, E s. Surawicz, B. 216, *292*

Leroy, J 218, *280*

Lesić, R., u. V. Varagić *153*

— s Varagić, V. 139, *159*

Lessen, W. van s Mellinghoff, K 230, *282*

Lessin, A. W s. Ambache, N. 5, 31, *74*

L'Estrange, J. L., u. R. F. E Axford 206, 219, 236, 237, *280*

Levey, S., W E Abbott, H Krieger u. J. H. Davis 230, *280*

Levy, B., u R P. Ahlquist 37, 38, 54, *80*

— s. Ahlquist, R P. 71, *74*

Levy, H M , u E. M. Ryan 212, *280*

Lewis, G. P., u. E. Reit 9, 21, 23, 24, 29, *80*

— s. Feldberg, W. 9, 32, 33, 42, 43, *77*

Lewis, M. D. s. McCollister, R. J. 252, *282*

Lewis, W. H 195, *280*

Liacopoulos, P. s. Frick, O. L. 176, 177, 181, *182*

— s Halpern, B N. 181, *183*

Libet, B 48, *80*

— s. Eccles, J. C 122, *148*

— s. Eccles, R. M. 47, 48, 49, 58, 59, 71, *76*

Lieber, C S s Saville, P D 196, 252, *289*

Lifschitz, W. s. Armengol, V. 117, *144*

Lifshitz, F., H. C. Harrison, E. C. Bull u H. E. Harrison 223, 224, *280*

— — u. H. E. Harrison 223, 224, 226, *280*

Likuski, H. J. A , u R M. Forbes 208, *280*

Lilienthal, J. L., K. L. Zierler, B. P. Folk, R. Buka u. M. J. Riley 224, 234, 257, *280*

Lilienthal jr., J. L s Baldwin, J. 194, 224, *263*

Lindell, S. E., H. Westling u. T White 41, *80*

Lindeman, R. D., S. Adler, M. J. Yiengst u E. S Beard 210, 243, 252, *280*

— H. E Ginn, J. M. Kalbfleisch u. W. O Smith 252, *280*

— s Kalbfleisch, J. M 252, *278*

Linder, G C., J D L. Hansen u. C. D Karabus 231, 232, 233, *280*

Lindmar, R. 19, *81*

Lindner, A. s. Smith, W. O. 196, 221, 222, 224, 225, *291*

Lindsley, D. B. 116, *153*

Lindsten, T s. Borglin, N E. 216, *265*

Ling, G. M., u. J. G. Foulks 95, *153*

— s. Marczynski, T. J. 124, *154*

— s. Yamaguchi, N. 99, 101, 103, 104, 115, 119, 120, 121, 122, 124, 127, 130, 131, 132, 135, 138, *159*

Ling, H. W. 19, *81*

Ling, S. s. Kleeman, C R. 243, *279*

Lipkin, G., C. March u. J. Gowdey 195, *280*

Lipmann, F. s Mudd, S.H. 212, *284*

Lipsac, J. s. Maurat, J. P. 199, *282*

Lisk, R. D. 94, 100, 106, 108, 109, *153*

— u. M. Newlon 107, *153*

— u A. J. Wein 109, *153*

Littlejohn, B M., u. J. de Groot 107, *153*

Livingston, R. B 116, *153*

— s. Adey, W. R. 91, *143*

Lizarralde, G , V E. Mazzocco u. E. B Flink 226, *280*

Lloyd, C. W s. Miller, T. R. II 210, *283*

Locke 136

Lockhart 102

Loeb, C s. Favale, E. 115, *148*

Loeb, R F s. Atchley, D W. 258, *263*

Loken, H. F , R. J. Havel, G. S. Gordan u. S. L. Whittington 200, *280*

Long, C. N. H s Brobeck, J. R 110, *145*

Long, J. P., u. J. W. Eckstein 54, *81*

Longo, V G. 88, 95, 96, 139, *153*

— u. B. Silvestrini 88, 95, 97, *153*

Loosli, J. K. s. Carillo, B. J. 222, 224, 226, 228, *267*

Lopez, E. s Metcoff, J. 232, *283*

Losse, H., u. W. Koenig 253, 254, *280*

Lovelace, F. E. s. Carillo, B. J. 222, 224, 226, 228, *267*

Lowenhaupt, E., M. P. Schulman u. D. M. Greenberg 226, *281*

Lown, B s. Kleiber, R. E. 279

— s. Seta, K. 224, 225, *290*

— s. Vitale, J. J. 222, 226, 227, *294*

Lowry, O. H , A. B. Hastings, C. M. McCoy u A N. Brown 202, *281*

Lucia, S P. s. Greenberg, D. M. 195, 225, *274*

Luco, J V. s. Davıdovıch, A 61, *76*

Luecke, R L. s. Mıller, E. R 249, *283*

Luecke, R. W. s. Mıller, E. R. 216, 217, 222, 223, 249, *283*

Luıck, J. R. s. Rogers, T. A 191, 203, *288*

Lunaas, T. s Rogers, T. A. 191, 203, *288*

Lundberg, A 56, *81*

— s. Job, C. 70, *79*

Lutz-Dettınger, U. s Mertz, D P. 250, *283*

Lux, D. H s Kornmuller, A. E 133, 140, *152*

Lynch, J. R s Wıtt, D. M 110, *159*

Lynch, T s. Avıolı, L. 209, *263*

Lyncker, J., u. W. Vogt 170, 171, 175, 181, *183*

Lyon, E S s Vermeulen, C. W. 212, *293*

Mabbott, J. D. s. MacBeth, R A 230, 235, *281*

MacBeth, R. A., u. J. D. Mabbott 230, 235, *281*

MacDonald, A. D. s. Hogben, L. T. 7, *79*

MacDonald, A. M. s Blaxter, K L 194, 209, 221, 222, 223, 227, *265*

MacDonald, D C., A. D. Care u. B. Nolan 202, 209, *281*

— s. Care, A. D 202, *266*

MacDonald, M. K s. Sabour, M S. 226, *288*

Machne, X., I Calma u. H. W. Magoun 91, *153*

— u. K R. Unna 88, *154*

— s Davıd, J. P. 89, *147*

MacIntosh, F. C 20, 61, 62, *81*

MacIntyre, I. 190, *281*

— S. Boss u V. A. Troughton 247, 248, *281*

— u D. Davıdsson 195, 220, 221, 222, 223, 224, *281*

— S. Hanna, C C Booth u. A E. Read 195, 231, 232, 233, *281*

MacIntyre, I s Alcock, N. 194, 207, 224, 226, 244, *263*

— s Alcock, N. W. 191, *263*

— s. Beechey, R B 225, *264*

— s. Booth, C. C. 232, 233, 234, 235, 236, *265*

— s. Foster, G. V. 249, *272*

— s Gudmundsson, T. V. 228, *274*

— s. Hanna, S. 247, 248, 249, 250, 251, *275*

— s Hess, R 222, 223, 226, *277*

— s Oppelt, W W 195, 204, 214, 239, *286*

— s Pallıs, C 236, 257, *286*

MacKeıth, M H s Heaton, T B 5, *78*

Mackey, M A. s. Greenberg, D M. 195, 248, *274*

MacLachlan, E. A s. Gardner, L I. 224, *272*

MacLean, P. D. 99, 100, 101, 102, *154*

MacLeod, L D 200, *281*

Macmillan, W H 39, *81*

— s Hanna, C 39, *78*

Macrae, I. F., H. F Kohler u M M Pechet 215, *281*

MacVıcar, R. s. Fontenot, J. P 208, *271*

Macy, I. G s Hummel, F. C 191, *277*

Mader, I. J., u. L T. Iserı 249, *281*

Madıssoo, H. s Poutsıaka, J. W. 252, *286*

Maengwyn-Davıes, G. D. s. Leı, B. W 59, *80*

Magnes, J., G. Moruzzı u O Pompeıano 115, *154*

Magnusson, T s Andén, N E. 125, *143*

Magoun, H. W. s Machne, X. 91, *153*

— s Moruzzı, G. 114, 115, *155*

Mahan, P. E. s. Rogers, T. A. 203, *288*

Mahler, H. R. 211, *281*

Mahy, B. W. J , u K. A. Munday 249, 250, *281*

Mahy, B W s Munday, K. A. 195, 198, *284*

Majer, H. s Kuhn, W L. 133, *153*

Malhotra, C. L s. Anand, B K. 135, *143*

Malkıel-Shapıro, B. 258, 259, *281*

Malméjac, J. 56, *81*

Mancıa, M s. Cordeau, J. P. 115, *146*

Manery, J. F. 195, 202, *281*

Manıtıus, A 190, *281*

— u. F. H Epsteın 224, 225, *281*

Mantegazzını, P , K Poeck u H. Santıbañez 94, *154*

— s. Glaesser, A. 95, 96, *149*

March, C. s Lıpkın, G 195, *280*

Marchıafava, P. L s Affanı, J. 114, *143*

Marcus, C S , u R H Wasserman 207, 208, 245, *281*

Marczynskı, T. J 122, 124, 130, 131, *154*

— u N. Yamaguchı 124, 131, *154*

— — u. G. M Lıng 124, *154*

— — — u L Grodzınska 124, *154*

— A Rosen u. J Hackett *154*

— s. Yamaguchı, N 99, 101, 103, 104, 115, 119, 120, 121, 122, 124, 127, 130, 131, 132, 135, 138, *159*

Marıno, J. s Danıels, F. 198, *269*

Marley, E., u B. J Key 98, *154*

— s Key, B J 98, *152*

Marotta, M. s. Bovet-Nıttı, F. 39, *75*

Marquardt, G H s Zuckerman, L. 196, 257, *296*

Marquardt, P , u L Hedler 178, *183*

Marr, T G s. Robertson, A. 198, *288*

Marrazzi, A. S. 4, 56, *81*
— u R N Marrazzi 56, *81*
— s. tum Suden, C. 56, *83*
Marrazzi, R. N s Marrazzi,
A. S 56, *81*
Marshall, G. B. s. Coons,
C. M 191, *268*
Marshall, W H. s. Evarts,
E V. 89, *148*
Martell, A E. s Sillén,
L G 203, 238, *290*
Martin, D F. s Heaton,
F. W. 235, 241, 255, *276*
Martin, F. s. Vagne, M. 205,
293
Martin, H E , H. Edmond-
sen, R. Homann u. C F
Berne 235, *281*
— u R. Jones 210, *281*
— C. McCuskey jr. u. N.
Tupikova 235, *281*
— J Mehl u. M Wertman
235, *281*
— W. P. Mikkelsen u R.
Jones 247, *282*
— u. M. Wertman 258, *282*
— u M. L. Wilson 220, 224,
282
Martindale, L , u F W.
Heaton 220, 248, *282*
Martz, B. L. s Henderson,
F. G 112, *150*
Masahasi, K s. Akimoto, H.
127, *143*
Masland, R L , u. R. S
Wigton 134, *154*
Mason, D. F J. 54, *81*
Massengill Esther 262
Matoush, L. O s. Conzo-
lazio, C. F. 205, 210, *268*
Matsumura, M., u. G. B.
Koelle 68, *81*
Matthews jr , R J. 56, *81*
Maurat, J P. 190, 195, 254,
255, *282*
— J. J. Pocidalo u. Lipsac
199, *282*
Maxwell, G M., R. B. Elliott
u R. H. Burnell 195,
215, 216, *282*
Maxwell, M. H. s. Kleeman,
C R 243, *279*
Maxwell, R A., A. J.
Plummer, S D Ross u.
M. W. Osborne 48, *81*

Mayer, J. 110, *154*
— s. Morrison, S D. 110,
155
Mayer, M M s Kabat, E. A
180, *183*
Mayer, S. E., u. J. A. Bain
97, *154*
Mayman, A s. Wener, J.
222, 223, 224, 225, 226,
227, *295*
Mayman, C s. Mendelson,
J. H. 235, *283*
Maynard, L. A., D. Boggs,
G. Fisk u D. Sequin
227, *282*
Mayo, R H , M. P. Plumlee
u D. M. Beeson 219, *282*
Mazzocco, V. E., E. B.
Flink u. J E Jones
225, *282*
— s Lizarralde, G 226,
280
McAleese, D M , M. C Bell
u. R. M Forbes 221, *282*
— u. R. M Forbes 220,
222, *282*
McAllan, A B. s Smith,
R. H 207, *291*
McCallum, J. W. s. Field,
A C. 217, *271*
McCance, R A., u. E. M.
Widdowson 214, *282*
— s. Davies, J. S. 191, *269*
— s. Watchorn, E 218,
220, 221, 222, *295*
McCandy, R. B. s. Brown,
D. F. 256, *266*
McCann, H. G. 195, *282*
— s. McClure, F. J. 259,
282
McCann, S M. 105, 108,
154
— s. Andersson, B. 100,
110, *143*
— s. Taleisnik, S. 105, *158*
McClure, F. J , u. H. G.
McCann 259, *282*
McCollister, R., A. S. Prasad, R P. Doe u. E. B.
Flink 209, 210, 241, 252,
282
— s. Flink, E. B. 235, 236,
271
— s. Prasad, A S 196, 235,
251, 254, *286*

McCollister, R. J., E. B.
Flink u. R P Doe 232,
235, *282*
— — u M D Lewis 252, *282*
McCollum, E V. s Hoob-
ler, S. W. 220, *277*
— s. Kruse, H D. 218,
220, 225, *279*
— s. Orent, E. R. 219, *286*
McCoord, A s. Breibart, S.
204, *265*
McCoubrey, A. s Boura,
A. L. A 59, *75*
McCoy, C. M. s. Lowry,
O. H. 202, *281*
McCreary, P. A , H A.
Battifora, G. H. Laing
u G. M Haas 225, *282*
McCubbin, J. W., J. H.
Green, G C Salmoiraghi
u. I H. Page 8, 39, *81*
— s Kaneko, Y. 8, *79*
— s. Page, I. H 8, *82*
McCuskey jr., C. s. Martin,
H E. 235, *281*
McDonald, I R. s Harris,
H 253, *275*
McDonald, S. J s Clarkson,
E M. 194, 254, *268*
McDowall, M. A s Fastier,
F N 40, *76*
McGann, J. s. Gilbert, D L.
213, 247, *273*
McGill, R. F. s Blaxter,
K L 190, 236, *265*
McGrath, H s. Hegsted,
I. M. 206, 227, *276*
McHargue, J. S , u. W. R.
Roy 249, *282*
McHugo, P. B. s. Hanna, C.
39, *78*
McIsaac, R. J. s. Iorio, L. C.
22, 24, 25, 29, *79*
McKay, D. s. Kleeman,
C. R. 195, 251, *279*
McKinstry, D. N , E. Koenig,
W. A Koelle, u G B.
Koelle 20, *81*
McLean, A. E. M. s. Judah,
J. D. 247, *278*
McLennan, H 88, 91, 122,
125, 136, 140, *154*
— u. J. E. Pascoe 70, *81*
— s. Weir, M. C. L. 56, 58,
59, *85*

McNutt, S. H. s. Morris jr.,
    M L. 226, 227, *284*
Meduković, M., u. V. Vara-
    gić 139, *154*
Mehl, J. s Martin, H. E.
    235, *281*
Meier, R s Bein, J H. 7,
    74
Meineke, F. s. Friedberg,
    K. D. 176, 177, 181,
    182
Meintzer, R B , u. H.
    Steenbock 208, *282*
Melby, J. C. s. Flink, E. B.
    235, 236, *271*
Meledi, R. s. Birks, R. 63,
    74
Mellinghoff, K. 230, *282*
— u. W. van Lessen 230,
    282
Melvin, K. E. W. s. Foster,
    G. V. 249, *272*
Menaker, W 206, *283*
— u. I. S. Kleiner 218,
    283
Mendel, L B , u S. R.
    Benedict 210, *283*
Mendelson, J s Barnes,
    B A. 221, 222, *264*
Mendelson, J. H., B. A.
    Barnes, C. Mayman u.
    M. Victor 235, *283*
— J LaDou u. C. Corbett
    235, *283*
Mendez, R , u. A Ravin 54,
    81
Mertz, D. P., u. U. Lutz-
    Dettinger 250, *283*
Mertz, J. P. 195, 209, 241,
    283
Metcalfe-Gibson, A. s.
    Wrong, O. 205, *296*
Metcoff., S. Frenk, I. Anto-
    wicz, G. Gordillo u.
    E Lopez 232, *283*
Metzner, W. R T. s. Leim-
    dorfer, A. 98, *153*
Meves, H s. Franken-
    haeuser, B 212, *272*
Meyts, P. de 213, *283*
Michael, R. P. 93, 108, *154*
Migicovsky, B. B. s Wor-
    ker, N. A 249, *296*
Mikkelsen, W. P. s. Martin,
    H. E. 247, *282*

Miller, E R , D E. Ullrey,
    C. L. Zutaut, B. V. Batt-
    zer, D A Schmidt, J A.
    Hoefer u. R. W. Luecke
    222, 223, *283*
— — — J A. Hoefer u.
    R. W. Luecke 216, 217,
    249, *283*
— — — — u. R. L.
    Luecke 249, *283*
Miller, G. H. s. Vermeulen,
    C. W. 212, *293*
Miller, J. F. 235, *283*
Miller, J R , u. T. R van
    Dellen 215, 259, *283*
— s. van Dellen, T R. 215,
    269
Miller, L. s Roberts, B.
    248, *288*
Miller, N. E 88, 110, 111,
    113, *154*
— K S Gottesman u N.
    Emery 113, *154*
Miller, R. W. s Fontenot,
    J. P. 208, *271*
Miller, T. R. II, W. W.
    Faloon u. C. W Lloyd
    210, *283*
Millstein, L G. s Poutsiaka,
    J. W. 252, *286*
Milne, M. D., R C. Muehrcke
    u. I. Aird 249, *283*
Min, H. K , J. E. Jones u.
    E. B Flink 210, *283*
Minnen, G. van s. A. van
    Kreveld 205, *279*
Mintz, A. A. s. Enselber,
    C. D. 259, *270*
Minz, B s Feldberg, W.
    5, 44, 51, 77
Mirkin, D L , u. T Cervoni
    138, *155*
Mishkin, M. s. Robinson,
    B W 110, *156*
Mishra, R K. 225, 226, 256,
    283
— s. Heggtveit, H. A. 226,
    227, *276*
Misson, C , u. H Schirardin
    236, *284*
Mitchel, J. F. 136, 140, *155*
— u J. C Sherb 140, *155*
— s Krnjević, K. 136, *152*
Moe, G K., u. W. A Frey-
    burger 48, 81

Moe, G K s. Freyburger,
    W. A. 48, 49, 77
Mollica, A s. Bradley, P. B
    95, *145*
Monnier, M 127, *155*
— u. L. Hosli 133, 140, *155*
— — u P Krupp 115, 127,
    155
— u. M. Romanowski 89,
    155
— s. Buergi, S. 116, *145*
— s. Koller, Th. 133, 140, *152*
Montgomery, R. D. 231,
    232, 233, *284*
— s. Back, E. H. 235, *263*
Moodie, E. W s. Stewart, J.
    245, *291*
Moore, C. A , u. G E Bunce
    217, 259, *284*
— s. Sauberlich, H E. 259,
    289
Moore, F D. s. Wacker,
    W. E C 235, 249, *294*
Moore, L. A , E T Hall-
    man u. L B. Sholl 226,
    227, *284*
Moore, R. M., u. W. J.
    Wingo 216, *284*
Moreau, A. s. Cordeau, J. P.
    102, 103, 118, 119, 120,
    121, 122, 123, 124, 126,
    132, 135, 138, *146*
Morel, G. s. Randoin, L
    217, *287*
Morgan, D B s. Fourman,
    P. 214, 231, *272*
Morgan, J. P. s. Hernandez-
    Peon, R. 99, 103, 122,
    123, 129, 131, 132, 135,
    150
Morgane, P J. 110, *155*
Morgulis, S 204, *284*
Mori, K., u J P. Duruis-
    seau 202, *284*
Morita, J. s Albert, D G.
    194, 256, *262*
Morris, E R , u B L
    O'Dell 195, 221, 222,
    223, 227, *284*
— s O'Dell, B L 227,
    228, *285*
Morris jr., M. L., W. R.
    Featherston, P H
    Phillips u. S. H. McNutt
    226, 227, *284*

Morris jr , M L , s. Featherston, W. R. 208, *271*

Morrison, A. B. s Schnee-
berger, E. E. 223, 224,
226, *289*

Morrison, J. D. s. Ardill,
B. L. 214, 216, *263*

Morrison, R. B. I. s. Wrong,
O. 205, *296*

Morrison, S. D , u. J. Mayer
110, *155*

Morse, W. H. s Dews, P. B.
88, *147*

Moruzzi, G. 114, *155*

— u. H. W. Magoun 114,
115, *155*

— s. Batini, C 114, 126,
144

— s Magnes, J. 115, *154*

Mota, I. 176, 177, *183*

Motola, R s. Wener, J.
222, 223, 224, 225, 226,
227, *295*

Moulton, R. s. Freedman, P.
242, *272*

Moussa, S L , u. A Boda
198, 199, *284*

Moussatché, H , u A. P.
Danon 177, *183*

Mover, J. H. s. Seller, R. H.
196, 253, *290*

Mózsik, G. s. Bokri, E. 45,
64, 65, *74*

Mraz, F. R. 245, *284*

— u. R G. Cragle 215, *284*

Mudd, S H., J. H. Park u.
F Lipmann 212, *284*

Mudge, G H. s. Walser, M.
215, *295*

Muehrcke, R. C s Milne,
M D. 249, *283*

Muller 69

Muhlrad, A , M. Kovacs u
G. Hegyi 211, 212, *284*

Mukai, T , u. J. E. Howard
259, *284*

Mulder, A. s. de Vries, L. A
210, *269*

Mulholland, H. C s. Ardill,
B. L. 214, 216, *263*

Mullick, D. N s. Bhatta-
charyya, N. K. 255, 256,
265

— s Ray jr., S. N. 196,
225, *287*

Munday, K. A., u. B. W. J.
Mahy 195, 198, *284*

— s. Mahy, B. W. J. 249,
250, *281*

Munoz, C s Goldstein, L.
95, *149*

Murata, Y., M Satake u.
T Suzuki 175, *183*

Murayama, S., u. K R.
Unna 21, 24, 25, 27, 29,
31, *81*

— s. David, J. P. 89, *147*

Murdaugh jr , H V., u.
R. R Robinson 239, *284*

— s. Robinson, R. R. 252,
254, *288*

Murer, E s Brierley, G. P.
247, *266*

Murphy, J. J s. Roberts,
B. 248, *288*

Murray, J. G , u. J. W.
Thompson 68, *81*

Mutsaars, W. s. Conard, V.
177, *182*

Myers, R. D. 102, *155*

— s. Feldberg, W. 98, *148*

Mylle, M. s. Lassiter, W. E.
240, *279*

Nabarro, J. D. N., A. G.
Spencer u. J M Sto-
wers 230, 258, *284*

Nagano, K. s. Nakao, T.
211, *285*

Nahmod, V E , u. M.
Walser 247, 253, *284*

— s. Walser, M. 196, 203,
295

Nakagawa, T. s. Akimoto,
H. 127, *143*

Nakamura, I s. Akimoto,
H. 127, *143*

Nakamura, M , I. Core,
N. Yabuta, S Torii,
Y Ishihara, K. Tamari
u. T. Imai 195, 202,
255, *285*

— M. Nakatini, M. Koike,
S Torii u. M. Hiramatsu
225, 226, 227, 251, *285*

— S. Torii, M. Hiramatsu,
J. Hirano, A Sumiyoshi
u. K. Tanaka 255, *285*

— — — T Umezaki, T.
Ohta u. J. Hirano 255, *284*

Nakamura, M , J. J. Vitale,
J M. Hegsted u E. E.
Hellerstein 255, *285*

— s. Gershoff, S. N 212,
228, 251, *273*

— s. Hellerstein, E. E.
228, 251, 255, 256, *276*

— s. Sasaki, T. 225, 226
289

— s. Vitale, J J 222, 225,
226, 227, 251, 255, 256,
293, 294

Nakao, M s Nakao, T.
211, *285*

Nakao, T., Y. Tashima,
K. Nagano u. M. Nakao
211, *285*

Nakatini, M. s. Nakamura,
M. 225, 226, 227, 251,
285

— s. Sasaki, T. 225, 226,
289

Nanninga, L B. 203, *285*

Nauta, W. J. H. 103, 123,
127, 128, 129, 130, 135,
142, *155*

— u. G. M. M. Kuypers
119, *155*

Neguib, M A. 251, *285*

Nelson, P. G. s. Hunt, C. C
61, 62, 63, *79*

Nelson, R. C. s. Conzolazio,
C. F. 205, 210, *268*

Nesbett, F. B. s. Ames, A.
III 194, *263*

Neter, E s Arbesman,
C. E. 180, *182*

Netzer, W., u. W. Vogt
175, 181, *183*

Neu, H. C. s. Randall,
H. G 176, 179, *183*

Neubeiser, R E , W. S.
Platner, u. J. L. Shields
198, *285*

Neuman, M W. s Neuman,
W. F. 204, 212, 259,
285

Neuman, W. F., u. M. W.
Neuman 204, 212, 259,
285

— s. Taves, D. R. 212, *292*

Neuwirth, I., u. G. B.
Wallace 216, *285*

Nevins, R s. Clark, I. 217,
268

Newlon, M s Lisk, R D. 107, *153*

Ng, S. T. s Wrang, O. 205, *296*

Nichols 248

Nielsen, B 195, 241, 254, *285*

Nielsen, J 195, 235, 253, *285*

Niemczyk, H s Baust, W. 93, *144*

Nikitovitch-Wiener, M. B 108, *155*

Nolan, B s Care, A. D 202, *266*

— s. MacDonald, D. C. 202, 209, *281*

Norberg, K. A. 60, 71, *82*

— s. Hamberger, B. 58, 59, 71, *78*

Nordbo, R. 195, 200, 205, *285*

Norris, L. M. s Leichsen-ring, J. M. 206, 209, 210, *280*

North, K A K s. Hanna, S. 247, 248, *275*

Norton, H. W s. Bartley, J. C 216, *264*

Novy u de Kruif 161

Nowell, N. W., u. D. C. White 198, *285*

Nugara, D., u. H. M. Edwards jr 227, *285*

Oba, T. S. s. Bunce, G. E. 194, 208, 225, 228, *266*

Oberdorf, A s. Hahn, F. 161, 180, *182*

O'Brien, R A s. Brierley, G. P. 247, *266*

Ochs, G s Eschler, J. 214, *270*

Ochs, M J s Glick, D. 202, *273*

O'Dell, B L. 190, 206, 218, 227, *285*

— E R. Morris u W O Regan 227, 228, *285*

— s. Morris, E. R. 195, 221, 222, 223, 227, *284*

Öberg, B s Folkow, B. 69, 77

Oelofse, P. J s. Bersohn, I. 218, 256, *264*

Ogasawara, K 195, *285*

Ogura, Y s Hayama, T. 93, *150*

Ohlin, P , u B C R Stromblad 137, *155*

Ohta, T. s Nakamura, M. 255, *284*

Okabe, K s Akimoto, H. 127, *143*

Olds, J 110, 111, *155*

— u. M. E. Olds 102, *155*

— A. Yuwiler, M E Olds u. Chang Yun 102, 103, *155*

Olds, M E. s Olds, J. 102, 103, *155*

Oliver, A. P. s. Bloom, F E. 89, 91, 126, *144*

Olson, E J , u. H. E Parker 228, 256, *285*

— s. Boellner, S. W. 194, 257, *265*

Olszewski, J. 113, *156*

— u. D Baxter 113, *156*

O'Neil, J. A. s. Stegemann, H 164, 169, *184*

Opie, L. H , B. G. Hunt u. J. M Finlay 209, 231, 233, 234, *285*

Opit, L J. s Charnock, J S 253, *267*

Oppelt, W. W., I. Mac-Intyre u. D P. Rall 195, 204, 214, 239, *286*

Orange, M , u. H. C. Rhein 195, *286*

Orent, E R , H D. Kruse u E. V. McCollum 219, *286*

— s Kruse, H D. 218, 220, *279*

Ørskov, S. s Bang, O. 194, 257, *263*

Ørskov, S L s Henriques, V. 201, 257, *277*

Osborne, M W s. Maxwell, R A 48, *81*

Osler, A. G., H G. Randall, B. M. Hill u. Z. Ovary 163, 165, 172, 173, 174, *183*

— s. Randall, H. G. 176, 179, *183*

Outa, T. 195, 200, *286*

Ovary, Z. s. Osler, A. G. 163, 165, 172, 173, 174, *183*

Owen jr , J E. s Verhave, T. 112, *159*

Owman, Ch s. Bertler, O. 125, *144*

Oyaert, W. 249, *286*

Paasonen, M K. s. Gaddum, J. H. 42, 77

— s. Gertner, S B 42, 77

Pace, J. K s Scoular, F I 205, *289*

Pace, N. s Carvalho, A P 201, 203, *267*

— s. Griswold, R L. 203, *274*

— s Sanui, H 203, *289*

Pacovsky, V. s. Vostal, J. 230, 239, *294*

Page, I. H 96, *156*

— u J W. McCubbin 8, *82*

— s. Kaneko, Y. 8, *79*

— s. McCubbin, J. W. 8, 39, *81*

Paixao, L M s Herring, W. B. 195, *277*

Palay, S L 135, *156*

Palestini, M. s. Armengol, V. 117, *144*

— s Batini, C. 114, 126, *144*

Pallis, C., I. MacIntyre u. H. Anstall 236, 257, *286*

Papez, J. W. 113, *156*

Pappano, A. J. s. Take-shige, C. 21, *83*

Pardo, E. G , J Cato, E. Gijón u. F. Alonso de Florida 56, *82*

Park, C. R s. Regen, D. M. 212, *287*

Park, J. H. s. Mudd, S. H. 212, *284*

Parker, H E. s Corradino, R A. 225, *268*

— s Griffith, F D. 217, *274*

— s. Olson, E. J. 228, 256, *285*

Parkinson 257

Parlier, R , D. Hioco u. R. Leblanc 190, *286*

Parodi, A s. Larvor, P.
222, 223, 226, 227, *279*
Parr, W. H. 221, *286*
— s Bartlett, S 237, *264*
Parsons, D. W. s Hills,
A G 210, 250, 253,
277
Parsons, F. M s Heaton,
F. W. 216, 217, 229,
276
Parsons, R S , T Butler u
E P Sellars 258, *286*
Pascoe, J E. 12, *82*
— s. Brown, G. L. 66, 70,
75
— s McLennan, H. 70,
81
Passey, R B. s Tu, A T
175, *184*
Paton, W. D. M 9, 16, 17,
72, *82*, 88, *156*
— u. W. L. M. Perry 12,
14, 15, 17, 18, 25, 26, 47,
54, *82*, 122, *156*
— u J. W. Thompson 57,
82
Paul-David, J s. Riehl,
J. L. 123, *156*
Paulson, E s. Valberg,
L S 257, *293*
Paulson, E. J. s Valberg,
L S 196, *293*
Pave, H. s. Robertson, A.
198, *288*
Payne, J M , u J Cha-
mings 247, *286*
Peacock, R M s Ritchie,
N. S. 230, 236, *288*
Pearce, P. J s Inglis,
J. S S 198, 237, *278*
Pearse, A G E. s Hess, R
222, 223, 226, *277*
Pearson, J. s. Cope, C L
250, *268*
Pearson, P B s. Barron,
G. P. 221, 226, 227, *264*
Pechet, M. M s Kohler,
H. F. 248, 249, *279*
— s. Macrae, I F 215, *281*
Pelikan, E W. 16, *82*
Pengelley, E. T , u R. R J
Chaffee 198, 199, *286*
Peptone 8
Pereira, S A. s. Acheson,
G. H. 5, 14, *74*

Perry, H M s Tipton, I. H
202, *293*
Perry, W L M 62, *82*
— u H. Reinert 64, 65, *82*
— s. Ambache, N. 5, 23, *74*
— s Paton, W D M 12,
14, 15, 17, 18, 25, 26, 47,
54, *82*, 122, *156*
Peschel, E s Robinson,
R. R 252, 254, *288*
Petermann, M L s. Hamil-
ton, M G 203, *275*
Peters, C. J., u. M. Walser
246, *286*
Petersen, W. s. Jobling,
J W 162, *183*
Petersen, V. P. 209, 229,
231, 233, 234, 249, *286*
Pethes, G. s. Kemény, A
195, 214, 216, *278*
Phillips, P H s. Bunce,
G E 218, 222, 223, 227
230, *266*
— s. Chiemchaisri, Y. 194,
220, 221, 222, 223, 228,
267
— s. Featherston, W. R.
208, *271*
— s. Morris jr., M L. 226,
227, *284*
Phillipson, A. T., u J E
Storry 245, *286*
Philis, J. W. s. Curtis,
D R 126, 130, *147*
— s. Krnjević, K. 89, 91,
97, *153*
Picarelli, Z. P. s Gaddum,
J H. 8, 30, 31, *77*
— s Rocha e Silva, M 8,
31, *82*
Pickles, V. R. s. Best, F. A.
194, 213, *264*
— s. Clegg, P C 213, *268*
Pinkus, J. B s. Gittelman,
I F 249, *273*
Pintar, K. s. Wener, J.
222, 223, 224, 225, 226,
227, *295*
Piper, P. J. s. Collier,
H. O J. 34, *75*
Planchart, A. 252, *286*
Plass, E C s Bogert, L J
198, *265*
Platner, W. S. 198, *286*
— u. M. G. Hosko 198, *286*

Platner, W. S. s. Neubeiser,
R. E. 198, *285*
Plumlee, M P. s Mayo,
R. H 219, *282*
Plummer, A. J. s. Maxwell,
R. A. 48, *81*
Pocidalo, J J. s Maurat,
J. P. 199, *282*
Poeck, K. s Mantegazzini,
P 94, *154*
Poisner, A M. s. Douglas,
W W. 30, 32, 50, *76*,
212, *270*
Pollack, S , J N George,
R C. Reba, R M.
Kaufman u. M H. Cosby
257, *286*
Pompeiano, O. s. Magnes,
J 115, *154*
Pond, W. G. s. Carillo,
B. J. 222, 224, 226, 228,
267
Poppe u. Vogt, W. 164,
170, 176
Porter, M R. s Gerst, P. H.
235, 236, *273*
Portman, R s Chen, G 19, *75*
Portwood, R M. s. Robin-
son, R. R. 239, *288*
Portzehl, H. s Gränicher,
D. 211, *273*
Poryali, A. s. Schwab, R S.
196, 257, *289*
Posner, A S 204, *286*
Potter, V R. s Du Bois,
K. P. 216, *270*
Potts, jr., J. T., u. B. Ro-
berts 247, *286*
— s. Care, A. D 216, *267*
Poutsiaka, J W., H. Ma-
dissoo, L. G. Millstein u
J. Kirpan 252, *286*
Prasad, A S , E B Flink
u R McCollister 196,
235, 251, 254, *286*
— — u. H. H. Zinneman
200, *286*
— s Doe, R P. 251, *269*
— s Flink, E. B. 235, 236,
271
— s. McCollister, R. 209,
210, 241, 252, *282*
Pretorius, P. T., A. S.
Wehmeyer u. J. J.
Theron 231, 233, *287*

Prickett, C O. s. Schrader, C. A. 226, *289*

Pritchard, J A 205, 214, 215, *287*

Purpura, D. 133, *156*

Pyrah, L N s. Heaton, F. W. 209, 235, 241, 247, 248, 255, *276*

Pytkowicz, R M., I. W Duedall u D N Connors 200, *287*

Quastel, D. M J s Elmquist, D I 212, *270*

Quinn, M., D E. Bass u. C R Leeman 196, 198, *287*

— s Daniels, F 198, *269*

Quinn, P J, I. G. White u B R Wirrick 196, *287*

Raab, W., u. W. Gigee 94, *156*

Raaflaub, J 238, *287*

Radde, I. s. Alcock, N. 194, *263*

— s. Alcock, N W 191, *263*

Rademeyer, L J, u J Booyens 256, *287*

— s Booyens, J 256, *265*

Radmanović, B s. Beleslin, D. 43, *74*

Rahill, W. J s Walser, M 241, *295*

Rakić, L., N A Buchwald u E. J Wyers 122, *156*

Rall, D P s Oppelt, W W 195, 204, 214, 239, *286*

Ramirez, O s Seller, R H 196, 253, *290*

Ramirez-Muxo, O. s Seller, R H. 256, *290*

Rand, M J s Burn, J H 87, 103, 124, 134, 135, 136, 137, 138, 139, 140, *145*

— s Day, M D 136, *147*

Randall, H G., S L. Talbot, H. C Neu u. A. G Osler 176, 179, *183*

— s. Osler, A G 163, 165, 172, 173, 174, *183*

Randall, jr, R E., M. D. Cohen, C. C. Spray jr. u E C. Rossmeisl 255, *287*

— E. C Rossmeisl u K H Bleifer 235, *287*

Randoin, L., J. Cauperet, D Hugot u G Morel 217, *287*

Ransom, F. s Dixon, W E 4, *76*

Ranson, S W s. Hetherington A W 110, *151*

Raskin, D s Fankushen, D. 229, 232, *271*

Rasmussen, H 247, *287*

Raugner, R s Schmid, E 215, *289*

Rausch-Strooman, J G. s Glaubitt, D 210, 252, *273*

Ravin, A s Mendez, R. 54, *81*

Ray jr S N, u D. N. Mullick 196, 225, *287*

Raynaud, C. 239, *287*

Read, A E s MacIntyre, I 195, 231, 232, 233, *281*

Reardon, J Z s Aikawa, J K 194, 220, 221, 229, 231, 250, 253, *262*

— s Ginsburg, S 195, 201, 202, 257, *273*

Reas, H. W, u U. Trendelenburg 61, 62, *82*

— u T H Tsai 137, *156*

Reba, R C. s. Pollack, S 257, *286*

Reber, E. F s Bartley, J C. 216, *264*

Rech, R, u. E. F. Domino 102, *156*

Reeves, P. G s Bunce, G E 195, 208, 225, 228, *266*

Regan, W O. s Hogan, A G 227, *277*

— s O'Dell, B L. 227, 228, *285*

Regen, D M, D. A B Young, W W. Davis, J. Jack jr. u. C. R. Park 212, *287*

Reid, G 32, *82*

Reinert, H 58, 59, 60, *82*

Reinert, H s. Gold, D. 60, *78*

— s. Kewitz, H. 60, *79*

— s. Perry, W. L. M. 64, 65, *82*

Reiselbach, R E s Steele, T. H. 254, *291*

Reit, E s Lewis, G P 9, 21, 23, 24, 29, *80*

Reizenstein, D L s Wallach, S 196, 203, *294*

Remolina, J. s. Acheson, G H 66, *74*

Rendi, R, u M L Uhr 211, *287*

Rennick, B R s. Freyburger, W. A. 48, 49, *77*

Renshaw 126, *130*

Revel, M, u H H Hiatt 212, *287*

Revzin, A M s. Costa, E. 59, 60, *75*

Rhein, H C s Orange, M 195, *286*

Rhoades, E L s Aikawa, J K 207, 231, *262*

Richards jr., D. W. s Atchley, D W 258, *263*

Richardson, J A, W D Huffines u. L. G Welt 249, *287*

— u L G Welt 196, 223, 228, 249, *287*

Riedesel, M L, u. G E Folk 198, *287*

Riehl, J L., J. Paul-David u. K. R Unna 123, *156*

Rien, W. s. Stegemann, H. 164, 169, 170, *184*

Rigø, J, u I. Szelényi 257, *287*

Riker, W F., J Roberts, F. G Standaert u H Fujimori 134, *156*

— G Werner, J. Roberts u. A S Kuperman 134, *156*

Riley, M J. s Lilienthal, J. L 224, 234, 257, *280*

Rinaldi, F., u. H E Himwich 88, 96, 122, 138, *156*

— s. Himwich, H E *151*

Ritchie, D B 204, 259, *287*

Ritchie, J. M s Douglas,
W. W 40, *76*

Ritchie, N S , u. R. G.
Hemingway 237, *287*,
288

— — J S S Inglis u.
R M. Peacock 230, 236,
288

— s. Hemingway, R. G.
237, *276*

Ritz, H., u H Sachs 162,
183

Rivera-Cordero, F s Clark,
I 217, *268*

Rizek, J A., A Dimich u
S. Wallach 196, 251,
288

Robertis, de E. s. Ger-
schenfeld, H M. 135, *149*
— s Zieher, L. M. 132, *159*

Roberts, B , J J Murphy,
L. Miller u. O. Rosen-
thal 248, *288*

— s. Potts jr , J T. 247,
286

Roberts, J. s. Riker, W F
134, *156*

Roberts, S D. s. Jabir,
F. K. 210, 215, 253, *278*

Robertson, A , H. Pave,
P. Barden u. T. G Marr
198, *288*

Robertson, J. S s. Silver,
L 209, *290*

Robertson, P. s Sullivan, J
252, *292*

— s. Sullivan, J F 230,
243, 252, *292*

Robertson, P. A 8, 31, *82*
— s Ambache, N. 5, 23 *74*

Robinson, B. H B. s. Wal-
ser, M 241, 250, *295*

Robinson, B. W., u. M.
Mishkin 110, *156*

Robinson, C. S s Duncan,
C W 238, *270*

Robinson, J. A s. Koster-
litz, H W 71, *80*

Robinson, P. K. s. Baldwin,
J. 194, 224, *263*

Robinson, R R , H V.
Murdaugh jr u E
Peschel 252, 254, *288*

— u R. M. Portwood 239,
288

Robinson, R R s Murdaugh
jr., H. V
239, *284*

Robson, J. M , u. R. A.
Stacey 88, 125, *156*

Rocha e Silva, M 163, 170,
171, 179, *183*
— u. M. Aronson 161, 163,
176, *183*
— u A M Rothschild 179,
183
— J R. Valle u. Z P.
Picarelli 8, 31, *82*
— s Rothschild, A M.
162, 163, 171, 172, 173,
174, 175, 176, *183*

Rodgers, D s Whang, R.
252, *295*

Rodin, Z s Belloiu, D.
251, *264*

Rogan, F H s Wallach S ,
196, *294*

Rogers, T A. 190, 199, 202,
246, *288*
— F L Haven u. P. E.
Mahan 203, *288*
— u. P E Mahan 203, *288*
— M G Simesen, T Lu-
naas u J. R. Luick 191,
203, *288*

Rogler, J C. s. Griffith,
F. D 217, *274*

Rogoff, J. M s Steward,
G. N 5, *83*

Rollerson, E. s. Kohn, R R
195, *279*

Romano, A s Gertner,
S B 59, *77*

Romanowski, M s. Monnier,
M. 89, *155*

Romenski, N W. 196, 257,
288

Rook, J A. F 236, *288*
— u C. C. Balch 236, *288*
— u. J E. Storry 190, 219,
236, 237, 240, *288*
— s. Blaxter, K. L. 194,
209, 221, 222, 223, 225,
227, *265*
— s. Head, M. J. 237, *276*
— s. Storry, J E 209,
211, 236, 237, *292*

Root, M A 5, 7, 37, 38, *82*

Rosa, E. s. Honorata, C. R
225, *277*

Rose, G. A. s. Heaton,
F. W. 208, 249, *276*

Rosen, A s Marczynski,
T J *154*

Rosen, S. M. s. Schachter,
D. 245, *289*

Rosengren, E s. Andén,
N. E 125, *143*
— s Bertler, O 90, *144*

Rosenthal, O. s. Hills, A. G
210, 250, 253, *277*
— s Roberts, B. 248, *288*

Ross, D B 244, 245, *288*
— u. A D Care 243, *288*
— s Care, A. D. 236, 237,
249, 250, *266*, *267*

Ross, S D s. Maxwell,
R A 48, *81*

Rosselle, N , u K De
Doncker 196, 236, *288*
— s. De Doncker, K. 195,
219, 222, 225, *269*

Rossi, C S. s. Carafoli, E
247, *266*

Rossi, G. F. 115, *156*
— u A. Zanchetti 113,
119, *156*
— s Batini, C. 114, 126,
144
— s. Brodal, A 119, *145*
— s Favale, E 115, *148*

Rossmeisl, E. C s. Randall
jr., R E 235, 255, *287*

Rossum, J. M. van 13, 19,
31, *82*

Roszkowski, A P. 21, 37,
38, 39, 40, *82*

Rothballer, A B 88, 98,
118, 120, *156*
— u. Sharpless 98

Rothlin, E. s. Konzett, H.
39, 43, *80*

Rothschild, A M , u. M
Roche e Silva 162, 163,
171, 172, 173, 174, 175,
176, *183*
— s. Rocha e Silva, M.
179, *183*

Rothstein, A 247, *288*

Rowland, S J s Bartlett,
S. 237, *264*

Roy, W. R. s McHargue,
J. S. 249, *282*

Rubin, M I., u E. T.
Krick 250, *288*

Rubin, M A s. Friedman,
H S 195, *272*

Rubin, R. P. s. Douglas,
W. W. 212, 213, *270*

— s. Jaanus, S. D 60, *79*

Ruegg, J C. s. Filo, R S.
212, *271*

Ryall, R W s Curtis, D R
73, *75*

Ryan, E M s Levy, H M
212, *280*

Ryley, J. W s Gartner,
R. J. W 195, *272*

Sabotka, H s Soffer, L J
251, *291*

Sabour, M. S , S. Hanna u.
M K MacDonald 226,
288

Sacco, G s Favale, E 115,
148

Saccomanno, G s Kuntz,
A. 70, *80*

Sachs, H s Ritz, H 162,
183

Sadowski, B s Traczyk, W
122, *158*

Saha, N s Banerjee, B
199, *263*

Sailer, S , u Ch Stumpf 89,
157

Sakanoue, M 205, *289*

— s. Ames, A III 194, *263*

Sakussow, W W 70, *82*

Salerno, P R , u J. M
Coon 54, *83*

Salmoiraghi, G C 89, 126,
157

— E Costa u. F E Bloom
88, *157*

— u F A. Steiner 89, 91,
92, 125, *157*

— s Bloom, F E 89, 91,
126, *144*

— s McCubbin, J. W 8,
39, *81*

— s Steiner, F. A *158*

Salmon, W D s Schrader,
C A. 226, *289*

Samiy, A. H. E , J L
Brown u. D. L. Globus
214, 239, 242, *289*

— — — R H Kessler u.
D D Thompson 239,
242, *289*

Santibañez, H s Mante-
gazzini, P 94, *154*

Sanui, H , u. N Pace 203,
289

— s Carvalho, A. P. 201,
203, *267*

Sargeant, T s Watanabe,
S 212, *295*

Sasaki, T , M Nakatani u
M Nakamura 225, 226,
289

Satake, M s Murata, Y
175, *183*

Sato, S s Takayasu, H
196, 215, 253, 254, *292*

Sauberlich, H E., G. E
Bunce, C A Moore u
O G Stonington 259,
289

— s Bunce, G E 194, 208,
225, 228, *266*

Saville, P D , u C S. Lieber
196, 252, *289*

Sawyer, Ch H 93, 105, *157*

— u B V Critchlow 105,
157

— u B. E. Gernandt 92,
157

— u M Kawakami 105, *157*

— s Holland, R C 92, *151*

— s Davidson, J. M 100,
106, 108, *147*

— s Smelik, P G 105, *157*

— s Sundsten, J W *158*

Sayers, G s Woodbury,
D M 249, *296*

Schachter, D 196, *289*

— E B Dowdle u H
Schenker 243, 245, *289*

— u S M Rosen 245, *289*

Schaepdryver, A F de s
Atanackovic, D. 60, *74*

Schaffer, A s. Wallach, S
196, 247, 248, *294*

Schain, R J. 196, 214, 257,
289

Schayer, R W 94, *157*

Schenk, E A , u E G.
Anderson 39, *83*

Schenker, A s. Glickman,
L. S. 236, *273*

Schenker, H. s Schachter,
D 243, 245, *289*

Schenker, V s. Glickman,
L S 237, *273*

Scherrer, H. s Hernandez-
Peon, R. 116, 117, *150*

Schiefelbusch, A T s
Coons, C M 191, *268*

Schilli, W. s. Eschler, J.
214, *270*

Schimpf, K , u. H. Harter
217, *289*

Schirardin, H s Misson, C.
236, *284*

Schlag, J 91, *157*

— u F Chaillet 131, *157*

— u. J. Faidherbe 91, 131,
157

— s Desmedt, J E. 125, *147*

Schlapp, W. s Hogben,
L T 7, *79*

Schloerb, P. R. s. Carr,
M H 194, *267*

— s Grantham, J J 215,
236, *274*

Schmertzler, E s Gittel-
man, I. F. 249, *273*

Schmid, E , M V Bubnoff,
U. Wagenmann u. R.
Raugner 215, *289*

Schmidt, C F s Koelle,
G B 135, *152*

Schmidt, D A s Miller,
E R 222, 223, *283*

Schmidt, G s Vogt, W.
163, 164, 165, 166, 167,
168, 170, 171, 172, 173,
174, 175, *184*

Schmidt, H D , J Schmier
u S Schmitz 214, *289*

Schmidt, M. M. s Kruse,
H D 225, *279*

Schmiedeberg, O 11, *83*

Schmier, J s. Schmidt,
H D 214, *289*

Schmitz, S. s. Schmidt,
H D 214, *289*

Schneeberger, E E , u
A B Morrison 223, 224,
226, *289*

Schoen, H C. s. Bogdanove,
E M. 105, *144*

Scholtz, H G 248, *289*

Scholz, R. W , u W. R
Featherston 208, *289*

Schoolman, A , u. E V.
Evart 117, *157*

Schorr, J s. Wallach, S
196, 203, 247, 248, *294*

Schrader, C A , C O. Prickett u. W. D. Salmon 226, *289*

Schroaeder, H. A. s. Tipton, I. H. 202, *293*

Schucher, R s Wener, J. 222, 223, 224, 225, 226, 227, *295*

Schümann, H. J. 134, 137, *157*
— u. H Grobecker 137, *157*

Schulman, M P s. Lowenhaupt, E 226, *281*

Schultz, W H 180, *183*

Schwab, R. S., A. Poryali u A. Ames III 196, 257, *289*

Schwartz, B M s. Smith, P K *290*

Schwartz, C J s Charnock, J. S. 199, *267*

Schwartz, R s Cotlove, E. 194, 221, 224, *268*

Schwarzacher, H G s Acheson, G. H 66, *74*

Schwoerer, D 166, 167, 171, *184*

Scognamiglio, W. P s. Bovet-Nitti, F 39, *75*

Scott, D 243, 244, 245, *289*
— u A. Dobson 249, *289*

Scott, J. B , E D Frohlich, R A Hardin u U F J. Haddy 213, 215, *289*
— s Haddy, F J 213, 236, 257, *274*
— s Frohlich, E D 215, *272*
— s Keyl, M J 195, 205, *278*

Scoular, F I , J K Pace u A. N Davis 205, *289*

Seelig, M S 205, 206, 210, 216, 218, 229, *290*

Segundo, J P 88, *157*
— s Adey, W R. 91, *143*

Seifter s. Stein, L 99, 100, 101, 111, 112, *158*

Sellars, E P s Parsons, R S 258, *286*

Seller, R H , O Ramirez, H Soller, A. N. Brest u J H Mover 196, 253, *290*
— O Ramirez-Muxo u A N Brest 256, *290*

Selvester, R s Haywood, J. 217, 259, *275*

Selye, H 256, *290*

Sequin, D s Maynard, L. A 227, *282*

Seta, K , E. E. Hellerstein u J J Vitale 222, 223, 224, 228, *290*
— R Kleiger, E E Hellerstein, B Lown u. J J Vitale 224, 225, *290*
— s Kleiber, R E *279*

Sevestre, J. s. Larvor, P 222, 223, 226, 227, *279*

Shafer, J J. s. Tipton, I H. 196, *293*

Shand, D G 59, *83*

Shane, S R. s Jones, J. E. 195, 251, *278*

Sharman, B A M s Blaxter, K. L 226, 236, *265*

Sharman, D F s. Krnjević, K 89, *152*

Sharpless s Rothballer, A B 98

Shaw, F. H. s. Frater, R. 213, *272*

Shehan, J s Sullivan, J. 252, *292*

Sherb, J C. s. Mitchel, J F 140, *155*

Sherrington, C S 117, *157*

Sherwood, L M s Care, A D 216, *267*

Sherwood, S L s Feldberg, W 98, *148*

Shideman, F E s Lee, W C 18, *80*

Shields, J L s Neubeiser, R E 198, *285*

Shils, M E 221, 224, 229, 233, 234, 235, *290*

Shimamoto, T , T Fujita, H. Shimura, H Yamazaki, S Iwahara u G Yajima 256, *290*

Shimura, H s Shimamoto, T 256, *290*

Sholl, L B s Moore, L A 226, 227, *284*

Shore, P A s Brodie, B B 125, 132, 135, *145*

Siehe, H J. 6, 7, 8, 33, *83*

Sigg, E B s Gyermek, L 31, 40, *78*

Sillén, L G , u A E Martell 203, 238, *290*

Silver, L , J. S Robertson u L K Dahl 209, *290*

Silverman, S H., u. L I Gardner 196, 249, 251, 254, *290*

Silvertone, J. T. s. Clemente, C D 92, *146*

Silvestrini, B s Bovet-Nitti, F 39, *75*
— s Longo, V G. 88, 95, 97, *153*

Simesen, M A 237, *290*

Simesen, M G s. Rogers, T. A 191, 203, *288*

Simmons, H G. s Enselber, C D 259, *270*

Simon, M A s Wener, J 222, 223, 224, 225, 226, 227, *295*

Simon, S E s Frater, R 213, *272*

Simonart, A s. Bacq, Z M 5, *74*

Singer, T P s. Cremona, T. 172, *182*

Sjoqvist, F. s Fredricsson, B. 63, 67, *77*
— s Hamberger, B 58, 71, *78*
— s. Holmstedt, B 67, *79*

Sjostrand, N O 137, *157*

Sjollema, B 236, *290*

Skolnik, S J s Bessman, S P 134, *144*

Skou, J C. 211, *290*

Slack, E s Foster, G. V. 249, *272*

Slater, I. H , u P. E. Dresel 7, 37, 38, *83*
— u G T Jones 112, *157*
— s Henderson, F. G. 112, *150*

Smart, P s. Abrahams, V. C. 135, *143*

Smelik, P G , u Ch H Sawyer 105, *157*

Smith, B S W. 209, *290*
— s Field, A C. 195, 203, 221, *271*

Smith, C. B , U Trendelenburg, S. Z Langer u. T. H. Tsai 62, *83*

Smith, H L , R L Fischer
  u J N Etteldorf 189,
  221, 222, 224, 257, *290*
Smith, H V s Hunter, G
  195, *278*
Smith, J C 22, 24, 25, 28,
  29, 31, 38, *83*
Smith, J G. s Ginsburg, S
  195, 201, 202, 257, *273*
Smith, J R s Bishop,
  P O 89, *144*
Smith, K s Winkler, A W
  258, *295*
Smith, M. R. s Clark, I.
  217, *268*
Smith, P K , A W Winkler
  u H E Hoff 214, 215,
  216, *290*
— — u B M Schwartz
  290
— s Hoff, H E. 212, *277*
Smith, R H 190, 196, 204,
  207, 209, 220, 221, 222,
  228, 233, 237, 245, 249,
  290, 291
— u A B McAllan 207,
  291
Smith, R. S. W., u A C.
  Field 196, 218, 220, 221,
  222, 229, *291*
Smith, W O 236, *291*
— D J. Baxter, A Lindner
  u. H. E. Ginn 196, 221,
  222, 224, 225, *291*
— u J F Hammarsten
  196, 253, 254, 255, *291*
— — u. L. P Ehel 235, *291*
— A A. Kyriakopoulos u
  J. F. Hammarsten 196,
  221, 222, 224, 252, *291*
— s Carubelli, R 194, *267*
— s Ginn, H E 239, *273*
— s Hammarsten, J F
  209, 214, 215, 236, 254,
  275
— s Kalbfleisch, J. M
  252, *278*
— s. Lindeman, R. D 252,
  280
Snyder, D. s. Ginn, H. E.
  239, *273*
Soffer, L J., C Cohn, E B
  Grossman, M Jacobs u.
  H. Sabotka 251, *291*
— s Harrop, G A 250, *275*

Soliman, H A s Foster,
  G V 249, *272*
— s. Gudmundsson, T. V.
  228, *274*
Soller, H s Seller, R. H
  196, 253, *290*
Somlyo, A P. s Somlyo,
  A V 213, *291*
Somlyo, A V , C Woo u.
  A P Somlyo 213, *291*
Somogyi, J 211, *291*
Sonnenblick, E H. s. Har-
  ris, W H. 195, *275*
Spaziani, E s Davson, H
  90, *147*
Spector, S s Costa, E 59,
  60, *75*
Spehlmann, R 89, 91, 97,
  157
Spencer, A G s Freedman,
  P 242, *272*
— s Nabarro, J. D N. 230,
  258, *284*
Sperelakis, N 213, *291*
Spray jr , C C. s Randall
  jr , R E. 255, *287*
Spray, C M 191, *291*
Stacey, R A. s Robson,
  J M 88, 125, *156*
Stanbury, J. B 215, *291*
— u A Farah 215, *291*
Standaert, F G s Riker,
  W. F 134, *156*
Staszewska-Barczak, J., u
  J R Vane 9, 30, 32, 34,
  83
State, D s Weil, P 247,
  295
Steele, T H , M. A Even-
  son, S F Wen u R E.
  Reiselbach 254, *291*
Steenbock, H s. Hart, E B
  217
— s Meintzer, R B. 208,
  282
Stegemann, H , G Bern-
  hard u J A O'Neil 164,
  169, *184*
— R. Hillebrecht u W
  Rien 164, 169, 170, *184*
— W. Vogt u K D Fried-
  berg 164, 168, 169, 170,
  184
Stein, L , u Seifter 99, 100,
  101, 111, 112, *158*

Steinberg, M s Hilton,
  J G 48, 49, *79*
Steiner, E. C. s Koelle,
  G. B 97, *152*
Steiner, F A , u G C.
  Salmoiraghi *158*
— s Salmoiraghi, G. C. 89,
  91, 92, 125, *157*
Stellar, E s Teitelbaum,
  Ph 110, *158*
Stemenović, B s Varagić,
  V. 139, *159*
Steriade, M., u M Deme-
  trescu 118, *158*
Sterman, M B , u C B
  Clemente 103, 121, 123,
  127, 128, 130, *158*
— s. Clemente, C D 121,
  122, 127, 128, 130, 131,
  146
Sternberger, H. R s Hum-
  mel, F C 191, *277*
Steward, G. N , u J. M.
  Rogoff 5, *83*
Stewart, C P , u S C Fra-
  zer 190, *291*
Stewart, J , u E W. Moo-
  die 245, *291*
Stiles, D E , J G. Batsakis
  u G C Hardy 196, *291*
Stockle, D s Fleckenstein,
  A 20, *77*
Stoliaroff, M s Durlach, J.
  248, *270*
Stoner, H B. 216, *291*
Stonington, O G. s Sauber-
  lich, H E 259, *289*
Storry, J E 207, 208, 237,
  244, 245, *291, 292*
— u J A F Rook 209,
  211, 236, 237, *292*
— s. Phillipson, A. T. 245,
  286
— s Rook, J A F 190,
  219, 236, 237, 240, *288*
Stoupel, N. s Bremer, F.
  116, 117, 118, *145*
Stowers, J M s Nabarro,
  J D. N 230, 258, *284*
Strassburger, I. s Syllm-
  Rapoport, I 218, *292*
— s Unglaub, I 226, *293*
Straughan, D W 136, *158*
Striebel, A , u. H. Baur 209,
  292

Strömblad, B C R s. Emmelin, N. 62, *76*
— s. Ohlin, P 137, *155*
Stumpf, Ch. 123, 143, *158*
— s Sailer, S 89, *157*
Stutzman, F I s. Heller, B. I 215, *276*
Stutzman, F L, u D S. Amatuzio 196, 198, 235, *292*
— s. Flink, E B 235, 236, *271*
Suden, C tum, u A S. Marrazzi 56, *83*
Sullivan, J, H. G Lankford, M. J. Swartz u C Farrell 235, 255, *292*
— P Robertson u J Shehan 252, *292*
Sullivan, J. F, H G Lankford u P. Robertson 230, 243, 252, *292*
Sullivan, R. C. s Zwemmer, R L 250, *296*
Sumiyoshi, A. s. Nakamura, M. 255, *285*
Sunderman, F W s Copeland, B E. 194, *268*
— s Willis, M J 200, *295*
Sundsten, J. W, u. C H. Sawyer *158*
— s Andersson, B. 110, *143*
— s Holland, R C. 92, *151*
Suomalainen, P 198, *292*
Surawicz, B, E. Lepeschkin u H. C. Herrlich 216, *292*
Suter, C, u W. D Klingman 222, *292*
Sutin, J. s Clemente, C D 92, *146*
Suzuki, T s Murata, Y 175, *183*
Swartz, M J s Sullivan, J. 235, 255, *292*
Swingle, W W. s Da Vanzo, J P 250, *269*
Syllm-Rapoport, I, u I Strassburger 218, *292*
— s Unglaub, I 226, *293*
Szczygielski, J. 6, 7, 32, 33, *83*
Szekely, P., u N. A. Wynn 215, 259, *292*
Szelényi, I s Rigø, J. 257, *287*

Szivek, J s Valberg, L S 196, 257, *293*
Sztanyik, L. s Gyermek, L 7, *78*

Taborsky, E s Kleeman, C. R 195, 251, *279*
Takayasu, H, S Sato, H. Yanadori u. T. Hirata 196, 215, 253, 254, *292*
Takeshige, C, A. J. Pappano, W C de Groat u R. L. Volle 21, *83*
— u R. L Volle 23, 29, 44, 45, 46, 47, 53, 54, 55, 64, *83*, 122, *158*
Talbot, S L s Randall, H G. 176, 179, *183*
Taleisnik, S, u S. M. MacCann 105, *158*
Tamari, K s Nakamura, M. 195, 202, 255, *285*
Tanaka, K. s. Lee, K S 212, *279*
— s. Nakamura, M. 255, *285*
Tapley, D F 250, *292*
— u. C Cooper 250, *292*
Tarazi, A. K. s. Grimson, K S. 53, *78*
Tashima, Y. s Nakao, T. 211, *285*
Taves, D R, u. W. F. Neuman 212, *292*
Taylor, T. G. *292*
Teitelbaum, Ph 110, 111, *158*
— u E Stellar 110, *158*
— s Hoebel, B. G 110, 111, *151*
Teng, H C. s Cheek, D B 194, 248, *267*
Tepper, I. s. Chesley, L C 214, 215, *267*
Tepperman, J s Brobeck, J R 110, *145*
Terkildsen, T C 217, 253, *292*
Ternor, s Bersohn 259
Theron, J J. s Pretorius, P T 231, 233, *287*
Thier, R E u B L Vallee 203, *292*
Thomas, D C. s Burt, A. W A. 237, *266*

Thomasse, C M s van Leeuwen, A M 196, *280*
Thompson, D D s. Bronner, F 239, *266*
— s Samiy, A H E 239, 242, *289*
Thompson, J. W. 30, *84*
— s. Murray, J G. 68, *81*
— s Paton, W. D. M. 57, *82*
Thompson, M E. s Garrels, R. M 200, *272*
Thoren, L. 208, 209, 230, 235, *292*
Tibbets, D. M, u. J C Aub 208, 248, *292*, *293*
Tidball, C. S. 213, *293*
Timo-Iaria, C. s. Hernandez-Peon, R 99, 103, 122, 123, 129, 131, 132, 135, *150*
Tipton, I H, u. M. J. Cook 196, 202, *293*
— u J J. Shafer 196, *293*
— H A. Schroaeder, H. M. Perry u M J. Cook 202, *293*
Toman, J. E. P. 88, *158*
Tomchick, R s Weil-Malherbe, H 94, *159*
Tonks, R S. s Hughes, A. 256, *277*
Torii, H. s Akimoto, H. 127, *143*
Torii, S. s. Nakamura, M. 195, 202, 225, 226, 227, 251, 255, *284*, *285*
Torp, A s Carlsson, A. 125, *146*
Tosteson s. Welt 211
Toverud, G. s. Toverud, K U. 191, *293*
Toverud, K. U., u. G. Toverud 191, *293*
Traczyk, W, u B Sadowski 122, *158*
Tramezzani, J H s. Gerschenfeld, H M. 135, *149*
Trendelenburg, P. 70, *84*
Trendelenburg, U. 9, 12, 13, 14, 15, 16, 17, 18, 19, 21, 22, 23, 24, 25, 26, 27, 28, 29, 30, 32, 33, 34, 37, 38, 39, 44, 45, 46, 49, 50, 56, 66, 72, *84*

Trendelenburg, U u A.
Jones 22, 23, 27, 45, *84*
— s. Jones, A 17, 37, 38,
39, *79*
— s Langer, S Z 63, 64,
80
— s. Lee, F. L. 32, 33, 51,
80
— s Reas, H W 61, 62, *82*
— s Smith, C B 62, *83*
Trescher, J H s Harrop,
G. A 250, *275*
Troughton, V A s Mac-
Intyre, I 247, 248, *281*
Trounce, J. R s Gent,
W. L G 201, *272*
— s Walser, M 252, *295*
Trouquet, J. s Lecomte, J
8, *80*
Trousseau u Chvostek 234
Trzebski, A 125, *158*
Tsai, T. H s. Reas, H W
137, *156*
— s. Smith, C B 62, *83*
Tschirgi, R D 90, *158*
Ts'o, P. O P s Edelman,
I S 203, *270*
Tsudzimura, H. s. Feld-
berg, W 5, 44, 51, *77*
Tu, A. T., R. B Passey u
T. Tu 175, *184*
Tu, T s Tu, A T 175, *184*
Tu, W H. s. Grantham,
J. J 215, 336, *274*
Tufts, E V., u. D M.
Greenberg 220, 221,
222, 223, 227, *293*
— s Greenberg, D. M 195,
198, 219, 225, 226, *274*
Tuma, M. s De Albu-
querque, P. F 259, *269*
Tupikova, N s Martin,
H E 235, *281*
Turlon, M. R s Kelly,
H. G 214, 215, 216, *278*
Turner, E s Buell, M. V.
250, *266*
Tuttle, A. H s. Etteldorf,
J N. 214, 258, *270, 271*

Udenfriend, S , H Weiss-
bach u D F Bogdanski
95, *158*
Uhr, M. L. s Rendi, R 211,
287

Ulhôa Cintra, A B D s.
De Jorge, F. B 195, 199,
269
Ullrey, D E. s. Miller, E R.
216, 217, 222, 223, 249,
283
Ullrich, K. J , u K H
Jarausch 196, 240, *293*
Ulmer, D D s Vallee, B. L
196, *293*
— s. Wacker, W E. C. 235,
249, *294*
Ulrich, F 211, *293*
Umezaki, T. s Nakamura,
M. 255, *284*
Ungar, G., T Yamura,
J. B Isola u. S Kobrin
163, 175, *184*
Ungersted, U. s. Hamber-
ger, B. 58, 59, 71, *78*
Unglaub, I , I. Syllm-
Rapoport u I Strass-
burger 226, *293*
Unna, K. R 88, 143, *158*
— s David, J P 89, *147*
— s Gyermek, L. 40, *78*
— s. Machne, X. 88, *154*
— s Murayama, S. 21, 24,
25, 27, 29, 31, *81*
— s Riehl, J L 123, *156*

Vagne, M., F Martin u
R. Lambert 204, 205, *293*
Valberg, L. S , R. T Card,
E J. Paulson u. J.
Szivek 196, *293*
— J. M Holt, G M. Brown,
J. Szivek u E. Paulson
257, *293*
— — E Paulson u J Szi-
vek 196, *293*
Valle, J R s. Rocha e
Silva, M. 8, 31, *82*
Vallee, B L , W E. C.
Wacker u D D Ulmer
196, *293*
— s Thier, R E 203, *292*
— s Wacker, W E C 190,
191, 196, 218, 235, 247,
249, 251, 253, 255, *294*
Vane, J. R s Staszewska-
Barczak, J 9, 30, 32,
34, *83*
Van Reeth, P E s Bremer,
F. 117, 118, *145*

Van't Klooster, A Th. s
Care, A D 244, 245, *267*
Varagić, V. 55, *84*, 139, *158*
— R Lesić, J Vuco u B
Stemenović 139, *159*
— s Beleslin, D 43, *74*
— s Kristic, M 55, *80*
— s Lesić, R. *153*
— s. Meduković, M 139,
154
Varga, F , A Kovér, T
Kovacs u. E. Hetényi
135, *159*
Varma, D R. s Benfey,
B G 14, *74*
Vartiainen, A. 6, 41, *84*
— s Feldberg, W 6, 7, 17,
53, *77*
Varrady, P. D s. Better, O
241, *264*
Vaughan, V. C , u S M.
Wheeler 160, *184*
Velasco, M s. Hernandez-
Peon, R 116, 117, *150*
Velez, H s Vitale, J. J
226, 227, *294*
Velluti, R , u R Hernan-
dez-Peon 129, *159*
Vennes, J A s Doe, R P.
210, 250, *270*
Verhave, T , J E. Owen jr.,
D Fadely u. J R Clark
112, *159*
Vermeulen, C W., E S
Lyon u G H. Miller
212, *293*
Verney, E B 92, *159*
— s Jewell, P A 92, *151*
Victor, M s Mendelson,
J H. 235, *283*
Vieth, J. s. Baust, W. 93,
144
Vinograd, J s Edelman,
I S 203, *270*
Vitale, J. J , D. M Heg-
stedt, M Nakamura u
P Connors 251, *293*
— — — u. P S Connors
225, 226, 251, *293*
— E E Hellerstein, D. M.
Hegstedt, M Nakamura
u A Farbman 255, *293*
— — M Nakamura u.
B Low 222, 226, 227,
294

Vitale, J. J , H Koide u E
E. Hellerstein 228, *294*
— M. Nakamura u D M
Hegstedt 225, 226, 251,
294
— H Velez, C. Guzman u
P Correa 226, 227, *294*
— P L. White, M Naka-
mura, D. M. Hegstedt,
N. Zambreck u E E.
Hellerstein 255, 256, *294*
— s Gershoff, S. N. 212,
228, 251, *273*
— s di Giorgio, J 225, *269*
— s Gottlieb, L. S. 206,
252, *273*
— s. Hegstedt, I. M. 206,
227, *276*
— s. Hellerstein, E. E 228,
251, 255, 256, *276*
— s. Kleiber, R. E 279
— s Nakamura, M 255, *285*
— s Seta, K. 222, 223, 224,
225, 228, *290*
Vleeschhouwer, G. R. de s
Atanackovic, D 60, *74*
Vogt, M. 7, 9, 31, 34, *84*
Vogt, W 163, 165, 167, 169,
170, 171, 172, 173, 174,
184
— u. G Schmidt 163, 164,
165, 166, 167, 168, 170,
171, 172, 173, 174, 175,
184
— u N Zemann 176, 177,
184
— s Bodammer, G. 177,
182
— s. Lyncker, J 170, 171,
175, 181, *183*
— s Netzer, W 175, 181,
183
— s Poppe 164, 170, 176
— s Stegemann, H 164,
168, 169, 170, *184*
Volk, G s Anast, C. 243,
263
Volle, R L 22, 26, 55, 69,
84
— u G B Koelle 18, 19,
44, 54, *84*
— s de Groat, W. C 40,
57, 58, 60, *76*
— s Gebber, G L 17, 18,
77

Volle, R L s Komalahir-
anya, A 55, *79*
— s Takeshige, C 21, 23,
29, 44, 45, 46, 47, 53, 54,
55, 64, *83*, 122, *158*
Voorhoeve, P. E. s Craw-
ford, J M 91, *146*
Vostal, J., O Kuchel u
V Pacovsky 230, 239, *294*
Vuco, J. s Varagić, V
139, *159*

Waal, W. s. Fastier, F N
40, *76*
Waal, M de s Booyens, J
256, *265*
Wacker, W E. C. 253, *294*
— C Iida u K. Fuwa 196,
294
— F. D Moore, D. D
Ulmer u B. L Vallee
235, 249, *294*
— u. B L. Vallee 190, 191,
196, 218, 247, 251, 253,
255, *294*
— s Vallee, B. L 196, *293*
Wada, J A 131, *159*
Wadkins, C L. s. Lehninger,
A. L 211, *280*
Wagenmann, U s Schmid,
E 215, *289*
Wagner, D s Whang, R.
252, *295*
Waites, G. M H s. Hebb,
C. O 62, *78*
Walaas, O. 213, *294*
Walhs, J s Courville, J
102, 103, 118, 119, *146*
Walker, A R P., F W
Fox u J. T. Irving 208,
294
Wallace, G. B s Neuwirth,
I. 216, *285*
Wallace, W W s. Cotlove,
E. 194, 221, 224, *268*
Wallach, S , J V Bellavia
u P J Gamponia 196,
214, 247, 248, *294*
— — J Schorr u D L.
Reizenstein 196, 203, *294*
— — — u A Schaffer
196, 247, 248, *294*
— L N. Cahill, F. H Ro-
gan u. H L Jones 196,
294

Wallach, S u A. C Carter
242, *294*
— s. Fankushen, D. 229,
232, *271*
— s Rizek, J. A. 196, 251,
288
Walser, M 196, 197, 200,
201, 204, 205, 207, 211,
238, 240, 241, 242, 243,
247, 252, 254, *294, 295*
— u A A Browder 243,
295
— u G. H Mudge 215, *295*
— u V E Nahmod 196,
203, *295*
— W J. Rahill u B H. B
Robinson 241, *295*
— B H B. Robinson u
J. W. Duckett jr. 250,
295
— u. J R Trounce 252, *295*
— s Dunn, M J 195, 205,
209, 229, 230, 231, 232,
233, 234, 252, *270*
— s Gent, W. L. G. 201,
272
— s Nahmod, V. E. 247,
253, *284*
— s. Peters, C J. 246, *286*
Wang, S C., u H L Bori-
son 93, *159*
Ward, E E s Back, E. H.
235, *263*
Ward, J. W s Franko,
B V 21, 37, 38, 39, 40,
77
Warnock, G M s Duck-
worth, J 206, 220, 221,
270
Warren, R s Clarkson,
E M 194, 254, *268*
Waser, P. G 6, 40, *85*
— s Konzett, H. 23, 31, *80*
Wasserman, R H 249, *295*
— s. Marcus, C S 207, 208.
245, *281*
Watanabe, S., T. Sargeant u,
M. Angleton 212, *295*
Watchorn, E , u. R A.
McCance 218, 220, 221,
222, *295*
Watkins, J C. s Curtis,
D R 89, 126, *147*
Webster jr , G D s Hills,
A. G. 210, 250, 253, *277*

Webster, T A s Campbell,
J. A 209, *266*

Weetman, D. F. s. Burn,
J. H 136, 137, 139, 140,
145

Wehmeyer, A S s Preto-
rius, P T. 231, 233, *287*

Weidmann, H., u. A. Cer-
letti 7, *85*

Weil, P , u D State 247,
295

Weil-Malherbe, H 94, *159*

— J. Axelrod u R Tom-
chick 94, *159*

— L G Whitby u J
Axelrod 94, *159*

Wein, A J s Lisk, R D
109, *153*

Weiner, N s. Borowitz, J.L
203, *265*

Weipers, M. s Inglis,
J S S 198, 237, *278*

Weir, M C L , u H
McLennan 56, 58, 59, *85*

Weissbach, H. s Uden-
friend, S 95, *158*

Weisschedel, E 130, *159*

Welt u Tosteson 211

Welt, L. G. 201, 211, 222,
224, 226, 228, *295*

— u H. Gitelman 190, *295*

— s Blythe, W B 241,
242, *265*

— s Gitelman, H J 216,
223, 230, 249, *273*

— s Richardson, J A 196,
223, 228, 249, *287*

— s Whang, R 220, 223,
224, *295*

Wen, S F s Steele, T. H
254, *291*

Wener, J , K. Pintar, M A
Simon, R Motola, R.
Friedman, A Mayman
u R. Schucher 222, 223,
224, 225, 226, 227, *295*

Werner, G s. Riker, W. F
134, *156*

Wertman, M. s. Edmond-
son, H A 258, *270*

— s Martin, H E 235,
258, *281*

Wesson jr , L G 210, 241,
252, *295*

— u D. P. Lauler 239, *295*

West, C D s Cheek, D B
194, *267*

Westling, H. s. Lindell,
S E. 41, *80*

Whang, R , D Wagner u
D. Rodgers 252, *295*

— u L G Welt 220, 223,
224, *295*

Wheeler, S W s Vaughan,
V C 160, *184*

Whitby, L G s Weil-
Malherbe, H 94, *159*

White, D C s Nowell,
N W. 198, *285*

White, I G s Quinn, P J
196, *287*

White, P L s Hellerstein,
E E 228, 251, 255, 256,
276

— s Vitale, J J 255, 256,
294

White, R P , u E A
Daigneault 139, *159*

White, T. s. Lindell, S E
41, *80*

Whitehair, C K s Fon-
tenot, J P. 208, *271*

Whittington, S L s Loken,
H. F. 200, *280*

Wickel, A s Chen, G 19,
75

Widdowson, E M , u
J W. T Dickerson 190,
191, 196, 202, *295*

— s Davies, J S 191, *269*

— s McCance, R. A 214,
282

Wien, R s Burn, J H 136,
137, 139, 140, *145*

Wigton, R S s Masland,
R L 134, *154*

Wikler, A 88, *159*

Wille, P 217, *295*

Willems, G 174, *184*

Williams, W s. Harris, H
253, *275*

Willis, J B s Cheek, D B
194, 247, *267*

Willis, M J , u F W Sun-
derman 200, *295*

Wilson, A A 190, 208, 221,
236, 240, 245, *295*

— s Care, A D 237, *267*

Wilson, A. B s Birming-
ham, A. T 19, *74*

Wilson, M L s Martin,
H E 220, 224, *282*

Wilson, V. J. s. Crawford,
J M 91, *146*

Winbury, M M 18, 19, *85*

Wingo, W J. s Moore,
R M 216, *284*

Winkel, K s. Kornmuller,
A E 133, 140, *152*

Winkler, A W , K Smith
u H E Hoff 258, *295*

— s Hoff, H. E 212, *277*

— s Smith, P K 214, 215,
216, *290*

Winterstein, H. 93, 94, *159*

— u N Goekhan 93, *159*

Wirrick, B R s Quinn,
P J 196, *287*

Wispelaere, H de 39, *85*

Wiswell, J. G 214, 251, *295*

Witt, D M , A. D Keller,
H L Batsel u J R
Lynch 110, *159*

Wolf, A V , u. S M Ball
242, *296*

Wolff, B s Cope, C L 251,
268

Wolstencroft, J H. s
Bradley, P. B. 73, *75*, 89,
91, 95, 125, 126, *144*, *145*

Womersley, R A 214, 215,
296

— s Ardill, B L 214, 146,
263

— s Jabir, F K 210, 215,
253, *278*

Woo, C. s Somlyo, A. V
213, *291*

Woodbury, D M , C Cheng,
G Sayers u L S
Goodman 249, *296*

Woods, K P s Carr, C W
199, *267*

Worker, N A , u B B
Migicovsky 249, *296*

Wrong, O , A Metcalfe-
Gibson, R B I Morri-
son, S. T. Ng u A. V.
Howard 205, *296*

Wyers, E J s Rakić, L
122, *156*

Wynn, N A s Szekely, P.
215, 259, *292*

Wyrwicka, W , u C Dobr-
zecka 111, *159*

Yabuta, N. s Nakamura, M.
195, 202, 255, *285*
Yajima, G. s. Shimamoto,
T. 256, *290*
Yamaguchi, N., G. M. Ling
u T J Marczynski 99,
101, 103, 115, 119, 120,
121, 122, 124, 127, 130,
131, 132, 135, 138, *159*
— s Akimoto, H. 127, *143*
— s Marczynski, T J. 124,
131, *154*
Yamazaki, H s Shimamo-
to, T 256, *290*
Yamura, T s Ungar, G
163, 175, *184*
Yanadori, H 253, *296*
— s Takayasu, H 196,
215, 253, 254, *292*
Yiengst, M J. s. Lindeman,
R D 210, 243, 252, *280*
Yoe, J H s. Herring, W B
195, *277*
Young, D A B s Regen,
D M 212, *287*

Young, D. W. s Hannon,
J P 198, *275*
Youngs, J N , u W. E
Cornatzer 226, *296*
Yu, D. H s Lee, K S 212,
279
Yun, T K s Burch, G E
194, 202, *266*
Yusken, J. W s Bartley,
J. C 216, *264*
Yuwiler, A s. Olds, J 102,
103, *155*

Zambreck, N s Vitale,
J J 255, 256, *294*
Zamcheck, N. s. Gottlieb,
L S 206, 252, *273*
— s. Hellerstein, E E 228,
251, 255, 256, *276*
Zanchetti, A s Batini, C
114, 126, *144*
— s Rossi, G F 113, 119,
156
Zanotto, A s. Beraldo,
W. T 8, *74*

Zemann, N s Vogt, W.
176, 177, *184*
Zen Ruffinen, B s. Kuhn,
W L 133, *153*
Zernicki, G s Affani, J
114, *143*
Zieher, L. M., u. E. de
Robertis 132, *159*
Zierler, K L s Baldwin, J.
194, 224, *263*
— s Lilienthal, J L 224,
234, 257, *280*
Zinneman, H H s Prasad,
A S 200, *286*
Zourles, P A s Kumar, D
216, *279*
Zuckerman, L , u. G. H.
Marquardt 196, 257, *296*
Zumoff, B , E H Bernstein,
J J Imarisio u L Hell-
man 209, *296*
Zutaut, C. L. s. Miller, E R
216, 217, 222, 223, 249, *283*
Zwemmer, R L , u. R. C.
Sullivan 250, *296*

Sachverzeichnis

acetylcholine 43—47
—, adrenergic transmission 134
—, anticholinesterases 53, 54
—, atropine 13
—, calcium chloride 29
—, catecholamines 30
—, cocaine 28
—, compound 48/80 39—43
—, denervated ganglia 19, 20, 30, 61, 65, 66
—, depolarization of ganglion cells 14, 15
—, dose-response curve 45
—, electrical stimulation, release of 136, 140
—, endogenous 20, 48, 50, 62
—, injected to carotid artery 96, 97
—, nicotine 13
—, perfused adrenal medulla 7, 32
—, physostigmine 53
—, preoptic region 132
—, range of doses 21
—, reticular formation, injection to 118—126
—, stimulant of nerve endings 52
acetylcholinesterase 52—56
—, after axotomy 67
—, — denervation 63
—, concentration of 135
—, DFP 52
—, preganglionic 65
adenohypophysis 105—108
adrenalectomy, blood pressure 5, 7, 32, 34, 36, 37, 38
—, compensatory hypertrophy 106
adrenaline see epinephrine
adrenal medulla 32—34
— —, blood pressure 34, 35, 39, 42, 43
— —, bretylium and xylocholine 38
— —, comparison of cat and dog 34
— —, perfused and non-perfused 6, 7, 31, 32
— —, release of catecholamines 30, 50
— —, sensitivity to non-nicotinic agents 35, 36
adrenocorticotrophin 105, 106
AHR-602 see also muscarinic agents
—, adrenal medulla 32

AHR-602, blood pressure 36—38
—, chemical name 20
—, morphine and methadone 29
—, nicotine block 26
—, nicotine- or TMA injection 25
—, range of doses 21
—, skeletal muscle 39
α receptors see receptors, α
amino acid decarboxylase see NSD 1055
amphetamine 19, 60, 96, 120, 139
amygdala 7
anaphylatoxin, actions of 175
—, assay of 179
—, definition 161
—, dose-response curve 175
—, effects on isolated organs and tissues 175
—, formation of 162
—, historical 161
—, in vivo effects 177
—, preparation of 168
—, properties of 169
— formation by plasma enzyme 164
— —, inhibition of 167
— —, role of complement in 173
anaphylatoxin-forming enzymes 163, 165, 167, 168
— — —, activation of 170
— — —, inhibition of 171
— — — of cobra venom 175
anaphylatoxinogen 163, 166, 167, 168
anaphylatoxin system, components of 165
angiotensin, adrenal medulla 7, 9, 32, 33, 34
— analogues 42, 43
—, 217 AO 58
—, atropine 29
—, blood pressure 38, 39
—, cocaine 27
—, denervation of ganglion 23
—, ganglionic stimulation 21
—, histamine 24
—, morphine and methadone 29
—, nicotine 25, 26
—, preganglionic stimulation 22, 27
—, pyrilamine 29
—, range of doses 21

angiotensin, subthreshold doses 24
—, tachyphylaxis 24
anterior commissure 131
2-anthrylguanidine 41
anticholinesterase agents 52—56, 138, 139
— — see also physostigmine, neostigmine, DFP
antidiuretic hormone 92
antihistaminic agents 29
217 AO 53, 58
apomorphine 93
arcuate nucleus 105—108
arecoline 123
ascorbic acid 106
atropine, acetylcholine 43, 45, 46
—, adrenal medulla 34
—, blood pressure 48, 51
—, chlorisondamine 48
—, DFP 53, 55
—, feeding area 112
—, hexamethonium 49
—, 5 HT 8, 31
—, hypnogenic area 129
—, LN-wave 48
—, low amplitude depolarization 23
—, metacholine 46
—, muscarine 6, 40
—, nicotine 13
—, physostigmine 50, 53
—, pilocarpine 5
—, posttetanic potentiation 48
—, superior cervical ganglion 29
—, synaptic potentials 47, 48
Auerbach plexus 60, 71
autoinhibition 13
axotomy 66, 67, 70

Bechterew's nucleus 129
benzylguanidine 41
2-β-aminoethylpyridine 41
3-β-aminoethyl-1,2,4 triazole 41
β receptors see receptors, β
blood brain barrier 90, 97, 98
botulinum toxin 31, 47
bradykinin, adrenal medulla 7, 8, 9, 32, 33, 34
—, atropine 29
—, cocaine 27
—, morphine and methadone 29
—, nicotine 25, 26, 27
—, pyrilamine 29
—, subthresold doses 24
— derivates 43
bretylium 19, 34, 38, 59
brom-LSD 8, 30, 31, 41

bromphenol blue 97
bufotenidine 42
bufotenine 42
—, chemical name 124

calcium 28, 29, 55
Cannon's law of denervation 64
cannulation of the lingual artery 9, 10
carbachol 15, 19, 20
—, anterior commissure 131
—, drinking responses 104
—, feeding area 111—113
—, reticular formation, injection into 121—124
carbonyl reagents 168, 173
cardiac ganglia 31
catecholamines see also epinephrine, norepinephrine, isoproterenol
—, adrenal medulla 34, 35
—, blood pressure 36
—, bretylium 38
—, feeding area 111—113
—, ganglionic transmission 59
—, hyperpolarization 47
—, intravenous injection
—, — —, EEG 94, 95
—, — —, uptake in organs 94
—, intraventricular injection 98
—, nicotine 30, 50
—, nictitating membrane 63
—, reticular formation, injection into 118, 120, 121, 125
—, xylocholine 38
caudate nucleus 122
— —, acetylcholine, release of 136
— —, blood brain barrier 98
— —, estradiol implantation 107
cerebral cortex, release of acetylcholine 136
— —, stimulation of 136
chemitrode cannula 99, 100
chemoreceptors 39
chlorisondamine 37, 48
choline acetylase 62
chromaffin cells 58
cobra enzyme 165, 166, 181
cobra venom, anaphylatoxin formation by 163
— —, anaphylatoxin forming enzyme of 175
cocaine 27, 28
—, adrenal medulla 33
—, blood pressure 5, 35, 36, 38
—, 5-HT 31
—, McN-A-343 31
coeliac ganglion 58, 67, 71

competitive block 14, 15
complement, role in anaphylatoxin forma-
    tion 173
compound 48/80 41
cortical photic responses 117, 118
cortisol 106
cross circulation 133
crystalline substances, implantation of,
    advantages 101—104

DCI 57
decamethonium 64
DFP 52—56
dibenamine 38, 47, 56, 58, 59
dibozane 38
diethylamide-LSD 30
dihydro-beta-erythroidine 97
dihydroergotamine 8, 56, 57
dimethylaminoethanol, feeding area 112
dimethylhistamine 6, 41
diphenhydramine 29
direct application of drugs into the brain
    tissue, methods 99, 100
DMPP 19
—, adrenal medulla 9, 38
—, adrenergic nerve terminals 19
—, block by nicotine 15, 16, 26, 33
—, cocaine 28
—, dose response curve 12, 13
—, hexamethonium 31
—, 5-HT 10
—, McN-A-343 25
—, morphine and methadone 29
—, preganglionic stimulation 22
—, range of doses 21
—, reserpine 19
—, 5-hydroxy-3-indol-acetamidine 30
—, non-nicotinic agents 30
DOPA-decarboxylase 42, 90
dopamine 19, 60, 112
D-receptors see receptors
d-tubocurarine, asynchronic discharge
    54
—, blood brain barrier 97, 98
—, nicotinic receptors 40
—, physostigmine 53
—, preganglionic denervation 65
—, synaptic potentials 47, 48

early firing 45, 46, 65
EDTA 109, 167, 172, 246, 247
EEG, acetylcholine injected to carotid
    artery 96, 97
—, amphetamine effect on 89
—, blood pressure effect on 94, 95

EEG, catecholamines injected to carotid
    artery 94—97
—, caudate nucleus, electrical stimulation
    of 122
—, electrical stimulation at different
    frequencies 103, 104, 132, 133
—, 5-HT effect on 96
—, 5-hydroxytryptophan effect on 96
—, hypertonic solutions injected to carotid
    artery 92
—, LSD effect on 89
—, medial forebrain bundle, stimulation of
    129
—, physostigmine effect on 88
—, preoptic region, stimulation by 5-HT 124
—, reticular formation, stimulation by
    acetylcholine 118—126
—, — —, — — carbachol 121, 124
—, — —, — — catecholamines 118, 120,
    121
—, — —, — — electrical 114—116
—, — —, — — 5-HT 124, 125
—, — —, — — oxotremorine 123
—, — —, — — physostigmine 121, 124
—, — —, — — procaine 118
—, — —, transsections of 114—116
electrical stimulation at different frequen-
    cies 114, 115, 132—140
— — compared to chemical stimulation
    102, 103
electrolytic ablation method compared to
    chemical stimulation 103
elodoisin 32, 34
enzyme, anaphylatoxin forming 163, 165
epinephrine 56—60
— see also catecholamines
—, adrenal medulla, stimulation 32
—, after axotomy 66
—, blood pressure 35, 37
—, hexamethonium 33
—, 5-HT, antagonism 98
—, splanchnic stimulation 50
EPSP (slow) see LN-wave
ergotamine 4, 5, 7, 38, 56
ergotoxin 38
ergotropic zone 127, 132
estradiol 106—109
ethoxybutamoxane 112

feeding area 110—113
food intake 109—113
formation of anaphylatoxin 162
forebrain descending inhibitory system
    127—132
— hypnogenic structures 127, 128

forebrain limbic — midbrain limbic pathway of Nauta 128—132
fornix-hippocampus 107
fourth ventricle, floor of osmopotentials 92

gamma-hydroxybutyric acid 134
ganglion, block 14, 15
—, —, compound 48/80 41
—, —, non-depolarizing 53
—, —, physostigmine 54
—, blocking agents 36, 37
—, cells, recording of transmembrane potentials 10
—, isolation 9
—, sensitivity to acetylcholine 64
—, stimulation, delay of onset of response 23
—, surface potential 46, 47
—, sympathetic see sympathetic ganglion
ganglionic transmission 47—52
— —, adrenergic mechanisms 71, 124, 134
— —, axotomy 66
— —, block of cholinesterases 52—55
— —, bretylium 59
— —, guanethidine 59
— —, preganglionic denervation 61, 62
— —, reinnervation 67
— —, sympathomimetic agents 56—60
gonadotrophin 106—108
gonadal steroids 106—109
guanethidine 19, 38, 59, 136
guanidine derivates 41, 42
Gudden's nuclei 128, 129

habenular nucleus, lateral 131
habenulointerpeduncular tract 131
heart-lung preparation 50, 53
hemicholinium 20
hexamethonium, acetylcholine 43, 44, 45
—, adrenal medulla 33, 34, 50
—, anticholinesterases 53—55
—, blood pressure 9, 37, 48
— DMPP 19, 31
—, early firing 64, 65
—, guanidines 42
—, histamine 7
—, 5-HT 8, 31, 39
—, muscarine 5
—, nicotine 19, 25, 39, 46
—, — block 26, 27
—, nicotinic receptors 40
—, non-nicotinic agents 25
—, pacemaker 49
—, pilocarpine 5
—, preganglionic stimulation 22, 23
—, superior cervical ganglion 47

hexamethonium, synaptic potentials 47, 48
—, vagal stimulation 50, 53
—, viscero-visceral reflexes 70
hindbrain, osmoreceptors 92
hippocampus 91
—, blood brain barrier 98
histamine 6, 7
—, adrenal medulla 6, 7, 32, 33, 34
—, atropine 29
—, blood pressure 36, 37, 38
—, CNS 39
—, cocaine and procaine 27, 28
—, compound 48/80 41
—, diphenhydramine 29
—, dose-response curve 21, 22, 24, 25
—, epinephrine 30
—, isolated ileum 31
—, morphine and methadone 29
—, nicotine 25, 26, 27
—, perfused ganglion 23, 30, 31
—, postganglionic stimulation 24
—, preganglionic stimulation 22, 23, 24
—, pyrilamine 29
—, subthreshold doses 24
—, tachyphylaxis 24
hordenine 60
5-hydroxy-3-indoleacetamidine 30, 31, 42
3-hydroxytryptamine 136
4-hydroxytryptamine 42
5-HT, adrenal medulla 7, 32
—, antagonists 30
—, atropine 8, 29
—, chemoreceptors 7, 8, 39
—, cholinesterase inhibitors 8
—, cocaine 27, 28
—, EEG, effect on 96
—, epinephrine, antagonism 98
—, forebrain limbic system 132
—, ganglionic action 8, 10, 30, 31
—, guanidines 41, 42
—, morphine and methadone 29, 30
—, neuromuscular junction 40
—, nicotine 25, 26
—, postganglionic stimulation 24
—, preganglionic stimulation 8, 22
—, preoptic region, injection to 124
—, range of doses 21
—, receptors 8, 30, 32
—, reticular formation, injection into 124, 125, 131
—, smooth muscle 10
—, subthreshold doses 24
—, tachyphylaxis 24
6-hydroxytryptamine 42
5-hydroxytryptophan 42, 95, 131, 132

hypogastric nerve 136, 137
hypothalamic-hypophyseal tract 135
hypothalamus 104—113
—, anterior 109, 127
—, — medial 106, 109
—, dorsocaudal 127
—, lateral 128
—, —, control of adenohypophysis 106—109
—, —, feeding area 109—113
—, medial 110
—, osmoreceptors 92
—, posterior 106, 108, 128
—, receptor areas 93, 94
—, uptake of catecholamines 94
—, ventromedial 107—111

ileum, isolated 31, 43
indol alkylamines and amidines 42
inferior mesenteric ganglion, acetylcholin-esterase 67
— — —, adrenergic synapses 71
— — —, afferent and efferent fibers 70
— — —, axotomy 66
— — —, muscarinic agents 31, 40, 42
inhibitors of monaminoxidase 42, 60
interpeduncular nucleus 128, 129
intestinal parasympathetic ganglia 31
— plexus 59
intracranial chemoreceptors 93, 94
— pressoreceptors 93
intraventricular injections 97, 98
isoproterenol 56—60

kallidin 32, 34

late firing 45, 65
limbic hypnogenic pathway 128—132
LN-wave 47, 48
lobeline 39
lordosis reflex 109
LSD 8, 30

magnesium, acetylcholine effect on 253
—, acid base balance 199
—, active and passive transport 238, 239, 240
—, adrenocortical hormones effect on 249, 250
—, age and sex 191, 197
—, alcohol effect on 252
—, alcoholism 235, 255
—, anemia 257
—, atherosclerosis 255, 256
—, calcitonin effect on 249

magnesium, cardiac glycosides, effect on 252, 253
—, complexed 200, 201, 238
—, contents of bone 204
—, — — cerebrospinal fluid 204
—, — — digestive secretions 204
—, — — erythrocytes 201, 202
—, — — hard tissues 204
—, — — heart 202
—, — — kidney 202, 246
—, — — liver 202
—, — — milk 205
—, — — muscle 202
—, — — plasma 191—201
—, — — salivary glands 245, 246
—, — — soft tissues 202—204
—, — — sweat 205
—, — — tooth 204
—, — — transcellular fluids 204, 205
—, — — whole animals 190, 191
—, deficiency 217—238
—, —, acetylcholine action 219
—, —, anorexia 219
—, —, body 220, 231, 232
—, —, bone 220, 221, 232
—, —, calcium and phosphorus metabolism 222—224, 227, 228, 234—236
—, —, cardiovascular system 225
—, —, causes 229—231
—, —, diabetic acidosis 230
—, —, definition 218, 228, 229
—, —, diarrhea 230
—, —, growth, during 218—228
—, —, gut 226
—, —, histamine 219
—, —, hormonal factors 228
—, —, hypoparathyroidism 231
—, —, kidney 225
—, —, liver 226
—, —, mast cells 219
—, —, muscular changes 218, 219, 220, 225
—, —, neurological changes 218, 219, 225
—, —, plasma 219, 220, 231
—, —, potassium metabolism 224, 234
—, —, soft tissues 221, 222, 232, 233
—, —, thyroid 225
—, desoxypyridoxine effect on 253
—, distribution in organism 190—205
—, diuretics, effect on 252, 253
—, eating, effect on 198
—, effect on atherosclerosis 258, 259
—, — — ATP concentration 216
—, — — blood pressure 215, 258
—, — — calcium concentration 216, 260
—, — — calcium excretion 217

magnesium, effect on caries 259
—, — — cholesterol concentration 217
—, — — enzymatic reactions 211, 212
—, — — glomerulonephritis 258
—, — — heart 214, 215, 259
—, — — hypertension 258
—, — — kidney 214, 215
—, — — nephrolithiasis 259
—, — — neural tissues 212, 213, 216
—, — — parathyroid function 216
—, — — phosphate concentration 216
—, — — potassium concentration 216
—, — — skeletal muscle 213, 214
—, — — smooth muscle 213
—, — — strontium excretion 217
—, — — subcellular structures 203, 212
—, — — uterus 216
—, excretion, dermal losses 209
—, —, fecal 207—209
—, —, urine 209
—, extracellular space 202, 203
—, hibernation 198
—, hormonal effects on 247—252
—, hypomagnesemia 235—238
—, hypothermia 198
—, hypertension 256, 257
—, insulin effect on 252
—, intake 205, 206
—, isotopic exchangeability 203, 204
—, lithium effect on 253
—, mental defects 257
—, muscular dystrophy 257
—, parathyroid hormone effect on 247—
      249
—, pregnancy and lactation 191, 199
—, protein-binding 199, 200
—, pump 241, 242
—, renal failure 253—255
—, salicylate effect on 253
—, subcellular structures 203, 212
—, temperature, effect on 198, 199
—, therapeutic uses 258, 259
—, thyroid effect on 250, 251
—, transport, anions, influence of 243
—, —, calcium, influence of 241—243
—, —, discussion 260, 261
—, —, diuresis 240, 241
—, —, erythrocytes 246
—, —, gall bladder 246
—, —, intestinal 243—245
—, —, kidney cortex 246
—, —, muscle 246, 247
—, —, renal 238—243, 246
—, —, salivary glands 245, 246
—, —, sodium, influence of 241—243

magnesium, transport, stomach 246
—, —, subcellular organelles 247
—, vitamin D effect on 249
mamillary region 7, 8
4-(m-chlorphenylcarbamoyloxy)-2-butynyl-
      trimethyl-ammonium chloride see
      McN-A-343
McN-A-343 see also muscarinic agents
—, adrenal medulla 32, 33
—, blood pressure 36, 37, 38, 40
—, cocaine and procaine 28
—, denervation of ganglion 23
—, diphenhydramine 29
—, dose-response curve 24
—, ganglionic effect 31
—, intracellular recording 32
—, late potentiation 25
—, morphine 29
—, muscarine 29
—, nicotine and TMA 25
—, preganglionic stimulation 22
—, range of doses 21
—, skeletal muscle 39
—, subthreshold doses 24
—, tachyphylaxis 24
—, TEA 25
mecamylamine 25, 37
medial forebrain bundle 127, 128, 129
median eminence 105—109
melatonin, chemical name 124
mescaline 60
methacholine 17, 21, 29, 39, 40, 46
methadone 29, 33, 36, 38
— see also morphine
methamphetamine 60
methodological remarks, microinjection 99,
      100
microinjection, compared to implantation
      of crystalline substances 101—104
—, methods 99, 100
—, side effects 100, 101
midline thalamic nuclei 131
miniature synaptic potentials 61, 62, 63
monoaminoxydase 90
morphine 29, 36, 38, 49
—, adrenal medulla 33
—, acetylcholine 45
—, 5-HT 8, 10, 31
—, McN-A-343 31
—, M-receptors 30
muscarine 4—6
— see also muscarinic agents
—, adrenal medulla 32
—, blood pressure 35
—, cocaine 27

muscarine, ganglionic effects 31
—, morphine and methadone 29
—, nicotine and TMA 25, 26
—, range of doses 21
muscarinic agents 37, 39, 40
— — see also muscarine, pilocarpine, McN-A-343, AHR-602
— —, anticholinesterases 53
— —, atropine effect on 29
— —, epinephrine 30, 51
— —, feeding area 111
— transmission 49
myenteric plexus of intestine 70

2-naphthylguanidine 41
Nauta, pathway of 128—132
N-benzyl-3-pyrrolidyl acetate methobromide see AHR-602
neostigmine 52—56
— see also anticholinesterases
—, acetylcholine 45
—, axotomy 66
nerve terminals 52, 54, 61, 62, 63, 66, 68, 70
—, adrenergic 34, 58, 71, 125, 135
neuromuscular junction 40, 60
nicotine 11—17
—, adrenal medulla 9, 32, 33, 38
—, atropine 13, 29, 30
—, depolarizing block 36, 45
—, — —, late phase of 25, 26, 27, 33, 36, 37, 45, 46, 49
—, — —, nicotine effect on 16
—, — —, ouabain-sensitive 18
—, — —, TMA 18
—, blood pressure 5, 36, 37, 39
—, catecholamines 30, 50
—, cocaine 5, 28
—, depolarization of ganglion cells 12, 14, 15
—, dose-response curve 12, 23, 24, 25
—, ganglion stimulation 21, 22, 31
—, 5-HT 10
—, muscarine 11
—, pilocarpine 5
nicotinic agents 11—17
nictitating membrane 63, 64, 69
— —, acetylcholine 45
— —, block by nicotine 49
— —, chromaffin cells 137
— —, denervated 32
— —, pilocarpine 4
— —, receptors 30
— —, superior cervical ganglion 9, 10, 21
Nissl substance 66
nitroglycerin 58

N-methyl-5-hydroxytryptamine 42
N,N-dimethyltryptamine 42
N,N,N-trimethyltryptamine 42
nodose ganglion 67
noradrenaline see norepinephrine
norepinephrine 56—60
— see also catecholamines
—, adrenal medulla, stimulation 9, 32
—, blood pressure 35, 37
—, bretylium and xylocholine 38
—, eating response 104
—, electrical stimulation, release of 136, 137
—, hexamethonium 33
—, midline thalamic nuclei 131
—, nerve terminals 34
—, preoptic region 130, 132
—, splanchnic stimulation 50
NSD 1055 41
nucleus centralis medialis 124, 130, 131
— reticularis pontis 131
N-wave 47

osmoreceptors 92, 110
ouabain 17, 18
oxotremorine 40, 58
—, chemical name 123
oxytocin 43

pacemaker 48, 50, 53, 54
papaverine 58
paradoxical sleep 120, 121, 123, 127, 131
paraventricular nuclei, osmopotentials 92
partial denervation 68, 69
pelvic nerve bladder preparation 41
pendiomid 64, 70
pentamethonium 64
peptone 8
perceptual integration 116, 117
perifornical area 131
periventricular system 128
phenoxybenzamine 8, 30, 56, 112
phentolamine 95
phenyldiguanide 39, 41, 42
phenylethylamine 60
phenylpropanolamine 60
physostigmine 52—55
— see also anticholinesterases
—, after axotomy 66
—, caudate nucleus, injection to 122
—, EEG, effect on 88
—, hexamethonium 50
—, 5-hydroxytryptophan 132
—, intracarotid injection 125
—, intravenous injection 122

physostigmine, late firing 45, 65
—, reinnervation 67, 68
—, reticular formation, injection to 121, 124
pilocarpine 4—6
— see also muscarinic agents
—, adrenal medulla 32, 33, 34
—, atropine 29
—, block by nicotine 25, 26, 27
—, blood pressure 5, 7, 36, 37, 38
—, chemoreceptors 39
—, cocaine 27, 28
—, dose-response curve 22, 24, 25
—, epinephrine 51
—, inferior mesenteric ganglion 31
—, methacholine 17
—, morphine and methadone 29
—, perfused ganglion 30, 31
—, postganglionic stimulation 24
—, reganglionic stimulation 22, 24
—, pyrilamine 29
—, range of doses 21
—, smooth muscle 31
—, subthreshold doses 24
—, tachyphylaxis 24
pituitary 105—109
—, uptake of catecholamines 94
pituitary-gonadal axis 105—108
plasma enzyme, anaphylatoxin formation by 164
— —, inactivation by carbonyl reagents 167
— —, preparations from rat plasma 166
polypeptides 28, 42, 43
postganglionic discharge 53—55, 57
postoptic region 106
posttetanic potentiation 48
potassium, adrenal medulla 7
—, atropine 29
—, cocaine 27, 28
—, content of ganglion cells 18
—, ganglionic block 14, 15, 25, 26
—, histamine 39
—, morphine and methadone 29
—, perfused ganglion 31, 64, 65
—, preganglionic stimulation 22
—, TEA 5
preganglionic nerve terminals see nerve terminals
— denervation 60—66
— —, adrenergic nerve terminals 58
— —, sprouting 68
— —, viscero-visceral reflex 70
— stimulation, acetylcholine 46
— —, facilitation of preganglionic effects 22

preganglionic stimulation, high and low frequencies 57
— —, 5-HT 42
— —, miniature synaptic potentials 62
— —, morphine and methadone 29
— —, physostigmine 53
— —, postganglionic discharge 48
— —, stellate ganglion 49
— —, substance P 43
— —, superior cervical ganglion 31, 58, 59
preoptic region, estradiol implantation 107—109
— —, 5-HT, implantation of 124
— —, hypnogenic area 127—132
— —, osmopotentials 92
prevertebral ganglia 58, 59, 70
procaine 28, 118
prolactin 107
pronethalol 38, 57
propanolol 34
P-wave 47
pyrilamine 29, 34

quaternary ammonium compounds, blood brain barrier 97

radiatio grisea tegmenti 130
rat uterus, isolated 41, 42
receptor, theory 15, 16
receptors, acetylcholine 46
—, — see also receptors, nicotinic, and receptors, muscarinic
—, α blocking agents 35, 36, 38, 47, 56, 58
—, —, sympathomimetic amines 57, 60, 71
—, angiotensin 42
—, β 34, 38, 56, 57, 58, 60, 71
—, body temperature 93
—, carotid sinuses 92
—, chemoreceptors 93, 94
—, D- 8, 21, 41
—, definition 3
—, food intake 93
—, histamine 47
—, 5-HT 29, 30, 42, 47
—, innervated 45
—, M- 8, 30, 41, 42
—, muscarinic, acetylcholine 43
—, —, block by nonspecific neuronal stabilizing agents 29
—, —, ganglionic transmission 50
—, —, in different organs 40
—, —, preganglionic denervation 65
—, —, reticular formation 123, 124
—, —, superior cervical ganglion 45
—, —, TEA 48

receptors, negative feedback 105, 107, 108
—, nicotinic, block by nicotine 18, 49, 50
—, —, cocaine 28
—, —, epinephrine-containing cells 30
—, —, ganglionic transmission 47
—, —, neostigmine 54
—, —, neuromuscular junction 40
—, —, preganglionic denervation 65
—, —, reticular formation 123, 124
—, —, superior cervical ganglion 45
—, —, TEA 48
—, non-nicotinic 28, 30
—, pressoreceptors 92, 93
—, reticular formation 123, 124
—, sexual behaviour 93
—, tryptamine 8, 124
—, water intake 93
red nucleus 131
reinnervation 61, 65, 67, 68, 69
renin 43
Renshaw cells 90, 91, 126
reserpine 19, 38, 47, 59, 135, 136
reticular activating system, acetylcholine
    effect on 96, 97
— formation, acetylcholine injection
    118—126
— —, ascending systems 96, 97, 113—126,
    129
— —, — —, electrical stimulation
    114—116
— —, — —, sensory transmission 116,
    117
— —, — —, surgical ablation 114—116
— —, — tonic inhibition 117, 119
— —, blood pressure effect on single
    units of 93
— —, carbachol injection 121—124
— —, catecholamines injection 118, 120,
    121
— —, cortical photic responses 117, 118
— —, descending inhibitory influences
    116, 117
— —, emetic center 93
— —, 5-HT injection 124, 125
— —, oxotremorine injection 123
— —, physostigmine injection 121, 124
— —, procaine injection 118
reticulo-cortical arousal 116, 117
reticulopetal stimuli 91
rhinencephalon, osmopotentials 92

salivary gland 62
satiety center 110
Schwann cells 61
self-amplifying system 134

self-re-excitation 52
septal fibers 128
skeletal muscle 39, 63, 66
sleep hormone 133, 134
smooth muscle 35, 38, 40, 43, 66
splanchnic stimulation 30, 32, 48, 50, 51
sprouting 68, 69
stellate ganglion 31, 49, 66, 71
steroid hormones, central effects of
    104—109
stria medullaris 131
substance P 43
superior cervical ganglion see also
    sympathetic ganglia
— — —, acetylcholinesterase effect on
    52, 61, 63, 67
— — —, adrenergic synapses 58, 71
— — —, amphetamine effect on 60
— — —, anticholinesterase effects on 54,
    55
— — —, atropine effect on 43
— — —, compared to other ganglia
    35—37
— — —, histamine content of 41
— — —, methacholine 12
— — —, muscarinic agents, effect of 4,
    45, 48, 50
— — —, non-nicotinic agents, sensitivity
    to 35, 36, 37
— — —, perfused 5, 6, 29, 30, 31
— — —, pilocarpine effect on 4
— — —, preganglionic denervation
    63—66
— — —, receptors of 45
— — —, reinnervation of 67—69
— — —, substance P, effect on 43
— — —, sympathomimetic agents 56, 57
— — —, synaptic potentials 47, 48
supraoptic region, osmopotentials 92
sweat gland 61, 62
sympathetic ganglia, adrenal medulla,
    comparison 35
— —, blood pressure 34, 36, 39
— —, denervated 61, 63
— —, 5-HT, content of 42
— —, non-nicotinic agents 37
— —, physostigmine 55
— —, superior cervical ganglion, compari-
    sion 38
sympathomimetic agents 56—60
— — see also epinephrine, norepi-
    nephrine, isoproterenol
synaptic barrier 90—92
synthetic polypeptides 8
syrosingopine 59

tachyphylaxis 5, 42
—, anaphylatoxin action 176, 177, 178
TEA, axotomy 67
—, blood pressure 5, 37, 48
—, chemical name 5
—, DFP 53
—, histamine effect, potentiation 7
—, 5-HT, antagonism 31
—, McN-A-343 25
—, muscarinic agents 40
—, non-nicotinic agents 25
—, preganglionic denervation 65
—, viscero-visceral reflex 70
TEPP 52
tertiary amines, blood brain barrier 97
testosterone 104, 106—109
thalamic nuclei, unspecific 91, 121, 130
— projection system, unspecific 128
thalamocortical responses 116
TMA 17, 18
—, block by nicotine 16
—, block by 25, 26
—, blood pressure 37, 54

TMA, cocaine effect on 28
—, depolarization of ganglion cells 14, 15
—, dose-response curve 18
—, 5-HT, antagonism 10
—, nicotine, antagonism to 15
—, non-nicotinic agents 30
trimethylhistamine 6, 41
trophotropic zone 127, 132
tryptamine 42
tyramine 20, 60

unmasking of muscarinic mechanisms 50
urinary bladder preparation 31

vagus 50, 53, 67, 68
valethamate 48
vasopressin 9, 43, 58
viscero-visceral reflex 70
vomiting center 93

water intake 109—113

xylocholine 34, 36, 38

ERGEBNISSE DER PHYSIOLOGIE
BIOLOGISCHEN CHEMIE UND
EXPERIMENTELLEN PHARMAKOLOGIE

REVIEWS OF PHYSIOLOGY
BIOCHEMISTRY AND
EXPERIMENTAL PHARMACOLOGY

HERAUSGEGEBEN VON

R. JUNG   K. KRAMER   O. KRAYER   E. LEHNARTZ
FREIBURG/BR   MÜNCHEN   BOSTON   MÜNSTER/WESTF.

F. LYNEN   A. v. MURALT
MÜNCHEN   BERN

U. TRENDELENBURG   H. H. WEBER   O. WESTPHAL
BOSTON   HEIDELBERG   FREIBURG/BR

SONDERDRUCK AUS BAND 59

U. TRENDELENBURG

SOME ASPECTS OF THE PHARMACOLOGY
OF AUTONOMIC GANGLION CELLS

WITH 9 FIGURES

NICHT IM HANDEL

SPRINGER-VERLAG
BERLIN HEIDELBERG GMBH 1967

Inhaltsverzeichnis

59. Band

Some Aspects of the Pharmacology of Autonomic Ganglion Cells. By
U. TRENDELENBURG, Boston, Mass. With 9 Figures

Topical Application of Drugs to Subcortical Brain Structures and Selected
Aspects of Electrical Stimulation. By T. J MARCZYNSKI, Chicago, Ill.

The Anaphylatoxin-Forming System. By W. VOGT, Göttingen. With
1 Figure

Magnesium Metabolism. By M. WALSER, Baltimore, Md. With 6 Figures

Namenverzeichnis

Sachverzeichnis

ERGEBNISSE DER PHYSIOLOGIE
BIOLOGISCHEN CHEMIE UND
EXPERIMENTELLEN PHARMAKOLOGIE

REVIEWS OF PHYSIOLOGY
BIOCHEMISTRY AND
EXPERIMENTAL PHARMACOLOGY

HERAUSGEGEBEN VON

R. JUNG   K. KRAMER   O. KRAYER   E. LEHNARTZ
FREIBURG/BR   MÜNCHEN   BOSTON   MÜNSTER/WESTF

F. LYNEN   A. v. MURALT
MÜNCHEN   BERN

U. TRENDELENBURG   H. H. WEBER   O. WESTPHAL
BOSTON   HEIDELBERG   FREIBURG/BR.

SONDERDRUCK AUS BAND 59

T. J. MARCZYNSKI

TOPICAL APPLICATION
OF DRUGS TO SUBCORTICAL BRAIN STRUCTURES
AND SELECTED ASPECTS
OF ELECTRICAL STIMULATION

SPRINGER-VERLAG
BERLIN HEIDELBERG GMBH 1967

Inhaltsverzeichnis

59. Band

Some Aspects of the Pharmacology of Autonomic Ganglion Cells. By
U. TRENDELENBURG, Boston, Mass. With 9 Figures

Topical Application of Drugs to Subcortical Brain Structures and Selected
Aspects of Electrical Stimulation. By T. J MARCZYNSKI, Chicago, Ill.

The Anaphylatoxin-Forming System. By W. VOGT, Göttingen. With
1 Figure

Magnesium Metabolism. By M. WALSER, Baltimore, Md. With 6 Figures

Namenverzeichnis

Sachverzeichnis

ERGEBNISSE DER PHYSIOLOGIE
BIOLOGISCHEN CHEMIE UND
EXPERIMENTELLEN PHARMAKOLOGIE

REVIEWS OF PHYSIOLOGY
BIOCHEMISTRY AND
EXPERIMENTAL PHARMACOLOGY

HERAUSGEGEBEN VON

R. JUNG   K. KRAMER   O. KRAYER   E. LEHNARTZ
FREIBURG/BR   MÜNCHEN   BOSTON   MÜNSTER/WESTF

F. LYNEN   A. v. MURALT
MÜNCHEN   BERN

U. TRENDELENBURG   H. H. WEBER   O. WESTPHAL
BOSTON   HEIDELBERG   FREIBURG/BR

SONDERDRUCK AUS BAND 59

W. VOGT

THE ANAPHYLATOXIN-FORMING SYSTEM

WITH 1 FIGURE

SPRINGER-VERLAG
BERLIN HEIDELBERG GMBH 1967

Inhaltsverzeichnis

59. Band

Some Aspects of the Pharmacology of Autonomic Ganglion Cells. By U. TRENDELENBURG, Boston, Mass. With 9 Figures

Topical Application of Drugs to Subcortical Brain Structures and Selected Aspects of Electrical Stimulation. By T. J MARCZYNSKI, Chicago, Ill.

The Anaphylatoxin-Forming System. By W. VOGT, Göttingen. With 1 Figure

Magnesium Metabolism. By M. WALSER, Baltimore, Md. With 6 Figures

Namenverzeichnis

Sachverzeichnis

ERGEBNISSE DER PHYSIOLOGIE
BIOLOGISCHEN CHEMIE UND
EXPERIMENTELLEN PHARMAKOLOGIE

REVIEWS OF PHYSIOLOGY
BIOCHEMISTRY AND
EXPERIMENTAL PHARMACOLOGY

HERAUSGEGEBEN VON

R. JUNG   K. KRAMER   O. KRAYER   E. LEHNARTZ
FREIBURG/BR   MÜNCHEN   BOSTON   MÜNSTER/WESTF

F. LYNEN   A. v. MURALT
MÜNCHEN   BERN

U. TRENDELENBURG   H. H. WEBER   O. WESTPHAL
BOSTON   HEIDELBERG   FREIBURG/BR

SONDERDRUCK AUS BAND 59

M. WALSER

MAGNESIUM METABOLISM

WITH 6 FIGURES

SPRINGER-VERLAG
BERLIN HEIDELBERG GMBH 1967

Inhaltsverzeichnis

59. Band

Some Aspects of the Pharmacology of Autonomic Ganglion Cells. By
U. TRENDELENBURG, Boston, Mass. With 9 Figures

Topical Application of Drugs to Subcortical Brain Structures and Selected
Aspects of Electrical Stimulation. By T. J MARCZYNSKI, Chicago, Ill.

The Anaphylatoxin-Forming System. By W. VOGT, Göttingen. With
1 Figure

Magnesium Metabolism. By M. WALSER, Baltimore, Md. With 6 Figures

Namenverzeichnis

Sachverzeichnis

SPRINGER-VERLAG
BERLIN·HEIDELBERG NEW YORK

Brain and Conscious Experience

Study Week September 28 to October 4, 1964 of the Pontificia Academia Scientiarum

Edited by Professor Sir **John C. Eccles**, Institute for Biomedical Research, Education and Research Foundation, American Medical Association, Chicago, Illinois; formerly Professor of Physiology, Australian National University, Canberra, Australia

With 147 figures
XXII, 591 pages 8vo 1966
Cloth DM 67,20; US $ 16 80

Though there have been in recent years many symposia concerned with the so-called higher functions of the brain, for example with perception, learning and conditioning, and with the processing of information in the brain, there has been no symposium specifically treating with brain functions and consciousness since the memorable laurentian conference of 1953, which was later published in 1954 as the book "Brain Mechanism and Consciousness"

The subject matter of this Study Week raises the most important questions that man can ask about himself and his relation to the material world. Besides the published lectures, the discussions have a special merit because in the five days' meeting under excellent conditions the participants were prepared to talk with an unusual ease and freedom This book therefore serves as a record of an intimate disputation among scholars of quite different approaches to the central theme

Experimental Brain Research
Experimentelle Hirnforschung
Expérimentation Cérébrale

Editorial Board: P. Dell, Paris, J. C Eccles, Chicago; D M. MacKay, Keele; D. Ploog, Munchen; J. Szentágothai Budapest; D. B Tower, Bethesda

In 1967 there will be 2 volumes (approx 4 issues per volume) Price per volume DM 72,—; US $ 18.00 plus postage

Brain and Conscious Experience

Study Week September 28 to October 4, 1964 of the Pontificia Academia Scientiarum

Edited by Professor Sir John C. Eccles, Institute for Biomedical Research, Education and Research Foundation, American Medical Association, Chicago, Illinois; formerly Professor of Physiology, Australian National University, Canberra, Australia

With 147 figures
XXII, 591 pages 8vo. 1966
Cloth DM 67,20; US $ 16 80

Though there have been in recent years many symposia concerned with the so-called higher functions of the brain, for example with perception, learning and conditioning, and with the processing of information in the brain, there has been no symposium specifically treating with brain functions and consciousness since the memorable laurentian conference of 1953, which was later published in 1954 as the book "Brain Mechanisms and Consciousness".
The subject matter of this Study Week raises the most important questions that man can ask about himself and his relation to the material world Besides the published lectures, the discussions have a special merit because in the five days' meeting under excellent conditions the participants were prepared to talk with an unusual ease and freedom. This book therefore serves as a record of an intimate disputation among scholars of quite different approaches to the central theme.

Experimental Brain Research

Experimentelle Hirnforschung

Expérimentation Cérébrale

Editorial Board: P. Dell, Paris, J. C. Eccles, Chicago, D. M. MacKay, Keele, D. Ploog, München, J. Szentágothai, Budapest, D. B. Tower, Bethesda

In 1967 there will be 2 volumes (approx. 4 issues per volume)
Price per volume DM 72,—; US $ 18.00 plus postage

SPRINGER-VERLAG
BERLIN HEIDELBERG NEW YORK